PRACTICAL MATHEMATICS

SEVENTH EDITION

CLAUDE IRWIN PALMER

LEONARD A. MRACHEK
Hennepin Technical Centers
Minneapolis, Minnesota

GLENCOE

Macmillan/McGraw-Hill

New York, New York
Columbus, Ohio
Mission Hills, California
Peoria, Illinois

Sponsoring Editor: Peggy Adams
Editing Supervisor: Ira C. Roberts
Design and Art Supervisor: Meri Shardin
Production Supervisor: Laurence Charnow

Text Design: Piñeiro & Tsao Design Associates, Inc.
Production Services: York Production Services, Inc.
Cover Designer: Edward Smith Design, Inc.

Library of Congress Cataloging in Publication Data
  Main entry under title:

  Practical mathematics.

    Includes index.
    1. Mathematics—1961–   . I. Palmer, Claude Irwin,
(date)
QA39.2.P685   1986   513′.14   84-29731
ISBN 0-07-048254-3

Practical Mathematics, Seventh Edition

Imprint 1993

Send all inquiries to:
Glencoe Division
Macmillan/McGraw-Hill
936 Eastwind Drive
Westerville, Ohio 43081

  6 7 8 9 10 11 12 13 14 15   RRD-C   00 99 98 97 96 95 94 93

ISBN 0-07-048254-3

DEDICATION

The seventh edition of *Practical Mathematics* is dedicated to the late Dr. John Jarvis, University of Wisconsin—Stout, Menomonie, Wisconsin, for his contributions to education and mankind.

Leonard A. Mrachek

C O N T E N T S

6 MEASUREMENT 74

7 POWERS AND ROOTS 91

PART 2   ALGEBRA   101

8 INTRODUCTION TO ALGEBRA 102

19 QUADRATIC EQUATIONS 255

20 VARIATION 269

PART 3 GEOMETRY 281

21 INTRODUCTION TO GEOMETRY 282

22 POLYGONS 295

23 TRIANGLES 304

24 THE CIRCLE 322

28 TRIGONOMETRIC TABLES 398

29 RIGHT TRIANGLES 414

30 GRAPHICAL REPRESENTATION OF TRIGONOMETRIC FUNCTIONS 433

31 OBLIQUE TRIANGLES 438

Appendix   474

Answers to Odd-Numbered Exercises   511

Index   537

PREFACE

Practical Mathematics has been a classic in the field of technical-vocational mathematics since its first edition. In the seventh edition, I have maintained the traditional, practical approach to teaching technical-vocational mathematics while using up-to-date terminology and examples as well as encouraging the use of the hand-held calculator.

There are no formal proofs in the seventh edition. Each concept is developed by a reasonable explanation; a sketch or drawing; worked-out applied examples; a summary, rule, or procedure; and drill and applied exercises. This is a time-tested method for teaching mathematics.

Several of the chapters have been streamlined and updated. The use of the hand-held calculator is encouraged to allow more time to investigate additional problems. Rather than including a chapter on the hand-held calculator, I recommend that the calculator be used at the appropriate time—once the basic arithmetic skills are mastered. However, remember the calculator is only a computational aid; the user must have the mathematical understanding to know which buttons to push.

In the arithmetic section (Part 1), the chapter on percentage has been updated and rearranged, and the measurement chapter has been clarified, specifically the metrics portion. The revised algebra section (Part 2) includes a more complete explanation of simple equations; more and better examples of the properties and application of levers; changes in the exponent and logarithms chapter, including applications; and more applied problems concerning logarithms and the hand-held calculator. The geometry section (Part 3) has been improved and made more efficient and a polygon classification figure has been added. Although reading of trigonometric tables is covered in Part 4 ("Trigonometry"), the student is encouraged to use the hand-held calculator. The applied trigonometry problems included cover a wide range of difficulty.

Solving practical applied problems after studying and learning the fundamental concept is a sound way to learn mathematics. The seventh edition includes more than 2000 applied problems as well as hundreds of illustrated practical examples.

The seventh edition of *Practical Mathematics* is intended for the post-secondary student in a community college or technical-vocational institute, for the adult education student, or for the 4-year-college student who wants to learn practical mathematics for application in the world of work and technology. The broad coverage of the material is intended to hold the interest of the student. Every effort has been made to maintain the relation between essential occupational competencies and the content of the previous edition.

The author acknowledges and thanks the many people who have contributed to this revision. A special thanks to Jacqueline Mrachek and John Mrachek who helped prepare the manuscript and made many helpful suggestions. The greatest thanks goes to the students who reported what they need to know and who continue to help develop the concept. The author has worked with numbers for many years and realizes that, unfortunately, there will be errors. For these errors, the author assumes all responsibility and hereby solicits comments and suggestions for the improvement of future editions.

Leonard A. Mrachek

PART 1

ARITHMETIC
ARITHMETIC
ARITHMETIC
ARITHMETIC
ARITHMETIC
ARITHMETIC

INTRODUCTION TO ARITHMETIC

The language of mathematics is a universal language; all mathematical symbols have the same meaning to people who speak many different languages throughout the civilized world. Unfortunately, many students do not master the mathematical subjects they study because they fail to learn and understand the language of mathematics, including certain essential definitions and technical terms. To succeed in mathematics, you must realize that there are no short cuts to learning this important subject; you must work hard. As technology plays an increasing role in our society, mathematics becomes even more important for those who want to succeed.

Use of the hand-held calculator is becoming more and more widespread. We recommend the use of a calculator for lengthy calculations. However, everyone should know how to perform the fundamental operations of arithmetic, which we have included in the early chapters. The hand-held calculator is a tool of technology and should not be expected to substitute for a thorough understanding of arithmetic.

We have found the following study suggestions to be effective:

1. Study each assignment until you understand it thoroughly before proceeding to the next lesson.

2. Read each explanation carefully. Then rework each completed example in the text so that you will understand completely the general principle explained.

3. Study each rule or procedure until you can rewrite it in your own words.

4. Verbal problems are the real test of your understanding of a mathematical principle. In working a verbal problem, make sure you first understand the problem before attempting to solve it. Then, if possible, make a sketch or drawing. In advanced mathematics, solving many problems is difficult or impossible without the aid of a sketch.

5. Check the solution you obtain. First, decide if your answer is reason-

able. Then substitute the solution into the equation and check it against the statement of the problem.

1-1   SYMBOLS USED IN WORKING WITH WHOLE NUMBERS

The Arabic numerals are 0, 1, 2, 3, 4, 5, 6, 7, 8, and 9. They are the *digits* of arithmetic.

The fundamental operations of mathematics are:

Addition +

Subtraction −

Multiplication ×

Division ÷

Some signs of grouping are:

Parentheses ( )

Brackets [ ]

Braces { }

Vinculum _____

Some signs that indicate order are:

Is less than <, as 4 < 7 (The sign points to the smaller quantity.)

Is greater than >, as 9 > 5 (The sign points to the smaller quantity.)

Is equal to =, as 5 = (10 ÷ 2)

1-2   NUMBERS

Arithmetic uses combinations of numbers and symbols to solve practical problems. Every day you use numbers to count. Numbers *represent things counted*. The teeth on a gear and the power consumed by an electric light

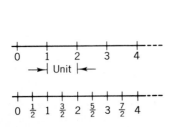

Fig. 1-1. Spaces on a line.

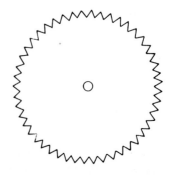

Fig. 1-2. Circular saw blade with 48 teeth.

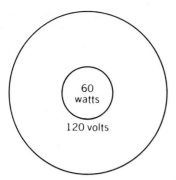

Fig. 1-3. 60-watt electric light bulb.

bulb are expressed by numbers. Numbers may also represent distances from a fixed point on a straight line, as shown in Fig. 1-1. In this case we are counting the spaces on the line.

Pictured in Fig. 1-2 is a circular saw blade with 48 teeth. Figure 1-3 shows a light bulb with 60 watts (W) of power.

1-3 DEFINITIONS OF TERMS

The greatest single cause of failure to understand and appreciate mathematics is not knowing and understanding the definitions of the terms. In mathematics, more than in any other subject, each word used has a definite and fixed meaning. The following are definitions of some basic terms you should know.

An *integer,* or an *integral number,* is a whole number. Thus 1, 2, 3, 4, 5, 6, 7, 8, 9, 10, 11 are the first eleven positive integers.

A *factor,* or a *divisor,* of a whole number is any other whole number that divides evenly into the number. Thus, 3 and 4 are factors of 12. Name other factors of 12.

A *prime number* in arithmetic is a number that has no factors except itself and 1. Examples of prime numbers are: 2, 3, 5, 7, and 11.

A *composite number* is a number that has factors other than itself and 1. Examples of composite numbers are: 4, 6, 8, 9, and 12. Composite numbers may be expressed as the product of two or more prime numbers.

A *common factor,* or *common divisor,* of two or more numbers is a factor that will divide evenly into each of them. If this factor is the largest factor possible, it is the *greatest common divisor* (gcd). Thus, 4 is a common divisor of 16 and 24, but 8 is the gcd of 16 and 24.

A *multiple* of a given number is a number that is evenly divisible by the given number. If a number is evenly divisible by two or more other numbers, it is a common multiple of them. The least (smallest) such number is the *lowest common multiple* (lcm). Thus 36 and 72 are common multiples of 12, 9, and 4; however, 36 is the lcm.

An *even number* is a number exactly divisible by 2. Thus, 2, 4, 6, 8, 10, and 12 are even integers.

An *odd number* is an integer that is not evenly divisible by 2. Thus, 1, 3, 5, 7, 9, and 11 are odd integers.

A *product* is the result of multiplying two or more numbers together. Thus, 21 is the product of 3 × 7. Also, 3 and 7 are factors of 21.

A *quotient* is the result of dividing one number by another. For example, 7 is the quotient of 21 divided by 3.

A *dividend* is a number to be divided; a *divisor* is a number that divides. For example, in 100 ÷ 25 = 4, 100 is the dividend, 25 is the divisor, and 4 is the quotient.

▉ EXERCISES

1-1 Is the day of the month on which you were born an even integer or an odd integer? Is it a prime integer or a composite integer? Answer these questions for the year of your birth.

1-2 Write all the prime numbers under 25. 1-100

1-3 Give all the prime factors of the following numbers (except 1 and the number itself): 1188, 24, 120, 720, 5040, 10, 70, 770.

Suggestion: *You can find the prime factors most efficiently by selecting the smallest prime factor (2), dividing by that prime factor as many times as can be done evenly, and then repeating the process on the quotient using the next-larger prime factor. For example:*

$$
\begin{array}{r}
2)\overline{1188} \\
2)\overline{594} \\
3)\overline{297} \\
3)\overline{99} \\
3)\overline{33} \\
11
\end{array}
$$

Thus, 1188 = 2 × 2 × 3 × 3 × 3 × 11. Therefore, the prime factors of 1188 are 2, 2, 3, 3, 3, and 11.

1-4 Are 39, 915, and 800 divisible by 2? By 3? By 4? By 5? By 6? By 7? By 8? By 9? By 10? By 11?

Find all the common divisors and the gcd of each of the groups of numbers in Exercises 1-5 to 1-11.

1-5 Illustration: Find the gcd of 72, 108, and 180.

$$72 = 2 \times 2 \times 2 \times 3 \times 3$$
$$108 = 2 \times 2 \times 3 \times 3 \times 3$$
$$180 = 2 \times 2 \times 3 \times 3 \times 5$$

The common factors are 2, 2, 3, 3; their product, 36, is the gcd.

1-6 12, 16, 24

1-7 16, 24, 32

1-8 9, 27, 144

1-9   110, 220, 330

1-10   24, 720, 5040

1-11   65, 780, 1190

1-12   A coffee dealer wants to bag 345 pounds (lb) of one kind of coffee, 483 lb of another kind, and 609 lb of a third kind. He wants to use bags of equal size and as large as possible and to fill them exactly with each kind (no coffee left over and all bags full). How many pounds must each bag hold? How many bags will he need?

1-13   What is the largest number that will divide evenly into 120 and 150?

1-4   SEQUENCE OF OPERATIONS

In a series of additions the terms may be placed in any order and grouped in any way. Thus, $4 + 5 = 9$ and $5 + 4 = 9$; $(4 + 5) + (7 + 3) = 19$, $(7 + 5) + (4 + 3) = 19$, and $[7 + (5 + 4)] + 3 = 19$.

In a series of subtractions, changing the order or the grouping of the terms may change the result. Thus, $100 - 20 = 80$, but $20 - 100 = -80$; $(100 - 20) - 10 = 70$, but $100 - (20 - 10) = 90$. When no grouping is given, subtractions are performed in the order written—from left to right. Thus, $100 - 20 - 10 - 3 = 67$ (by steps, $100 - 20 = 80$, $80 - 10 = 70$, $70 - 3 = 67$).

In a series of multiplications the factors may be placed in any order and in any grouping. Thus, $[(2 \times 3) \times 5] \times 6 = 180$ and $5 \times [2 \times (6 \times 3)] = 180$.

In a series of divisions, changing the order or the grouping may change the result. Thus, $100 \div 10 = 10$, but $10 \div 100 = 0.1$; $(100 \div 10) \div 2 = 5$, but $100 \div (10 \div 2) = 20$. Again, if no grouping is indicated, the divisions are performed in the order written—from left to right. Thus, $100 \div 10 \div 2$ is understood to mean $(100 \div 10) \div 2$.

When there is no grouping in a series of mixed mathematical operations, multiplications and divisions are to be performed in the order written, then additions and subtractions in the order written.

EXAMPLE 1-1 $12 + 3 - 2 - 9 + 7 - 3 = 8$, by performing operations in the order in which they are given.

EXAMPLE 1-2 $(12 \div 3) + (8 \times 2) - (6 \div 2) + (7 \times 2 \times 3) - 9 = $?
$$4 \quad + \quad 16 \quad - \quad 3 \quad + \quad 42 \quad - 9 = 50$$

First perform the multiplications and divisions then the additions and subtractions from left to right.

EXAMPLE 1-3 $120 \div 3 \times 5 \times 2 \div 2 = 200$. Perform the multiplications and divisions in the order in which they occur.

EXAMPLE 1-4   In a series of different operations, parentheses ( ) and brackets [ ] can be used to group the operations in the desired order. Thus, $120 \div 3 \times 5 \times 2 \div 2 = [120 \div (3 \times 5 \times 2)] \div 2 = [120 \div 30] \div 2 = 4 \div 2 = 2$.

■ EXERCISES

Note: In Exercises 1-14 to 1-22, remember that the operations indicated by the inmost groups should be performed first.

1-14 $(18 + 19 - 8) + (12 - 8 - 4) = ?$

1-15 $[21 \div (7 \times 3)] - (4 \times 5) = ?$
       $[21 \div 21] \quad ⧡ \quad (20) \quad = ?$
          $1 \qquad ⧡ \quad 20 \quad = 21 \; ⧡ \; 19$

1-16 $18 - [(3 - 6 - 9) \div (9 - 6)] - 12 = ?$

1-17 $[(25 - 4 - 6) \div (3 \times 5)] - 4 \times 3 = ?$

1-18 $(27 - 7) - [(2 \times 3) \div (3 \times 2)] - 3 \times 6 = ?$

1-19 $[(55 - 10) \times (3 \times 6 \times 9)] \div [(4 \times 10) + 5] = ?$

1-20 $[256 \div (16 \times 16)] - [225 \div (15 \times 1)] = ?$

1-21 $(720 \div 360) + (180 \times 2) - 1 = ?$

1-22   Do the following divisions. Check your work by finding the product of the divisor and the quotient and then adding the remainder. The result should equal the dividend.

(a) $93,462 \div 79$

$$
\begin{array}{r}
1\ 183 \\
79\overline{)\ 93,462} \\
79 \\
\hline
14\ 4 \\
7\ 9 \\
\hline
6\ 56 \\
6\ 32 \\
\hline
242 \\
237 \\
\hline
5
\end{array}
$$

The answer is $1183\frac{5}{79}$; check it as follows:

$$
\begin{array}{r}
1183 \\
\times\ 79 \\
\hline
10647 \\
8281 \\
\hline
93457 \\
+\ 5 \\
\hline
93,462
\end{array}
$$

The result of multiplying 79×1183 and adding 5 is the dividend, 93,462; therefore, the division is correct.

(b) $637,842 \div 2327$

(c) $647,847 \div 353$

(d) $9,963,486 \div 573$

1-23 The sum of two numbers is 76. One of the numbers is 12; what is the other?

1-24 The product of two numbers is 863. One of the numbers is 3; what is the other?

1-25 A circular saw has 75 teeth 1 inch (in.) apart. To cut satisfactorily, the saw teeth must travel about 9225 feet per minute (ft/min). How many revolutions per minute (rpm) must the saw make?

Solution: This problem is worked to show the value of a sketch in understanding a problem (Fig. 1-2).

If there are 75 teeth 1 in. apart, when the saw blade goes around once it equals 75 in. To find the number of inches which the sawteeth must travel per minute to do satisfactory work, multiply 9225 ft by 12 in./ft; $9225 \times 12 = 110,700$ in. To find the number of revolutions, divide 110,700 by 75. $110,700 \div 75 = 1476$ rpm.

1-26 The circumference of a drive wheel of a locomotive is 7 meters (m). How many revolutions will the wheel make in 210 kilometers (km) (1000 meters = 1 kilometer)?

1-27 In the United States, 21 accidents occur per minute. How many accidents will there be in the United States in 365 days?

1-5   APPLYING RULES AND PROCEDURES

Solving problems frequently requires using rules and procedures. You may learn the rule from a fellow worker or may find it in a handbook or textbook. Sometimes the logic of a rule is apparent, but often it is not. Rules whose logic is not obvious may be the result of experience, experiment, or merely "rules of thumb."

Read carefully each rule given, then work the problems that follow. You should be able to apply the principles illustrated to solve other problems.

Rule: *To find the number of revolutions made by a driven gear in a given time, multiply the number of teeth of the driving gear by the number of revolutions the driving gear makes in the given time, and divide by the number of teeth of the driven gear.*

Teeth $A \times$ rpm $A =$ teeth $B \times$ rpm B

EXAMPLE 1-5 A gear with 16 teeth, making 30 rpm, is driving a gear with 32 teeth. Find the rpm of the driven gear B.

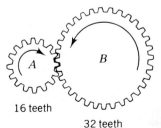

16 teeth

32 teeth   **Fig. 1-4.** Sketch of two gears.

Solution Eliminate mistakes in a problem of this type by drawing a sketch. Will gear B revolve more rapidly or more slowly than gear A? It *must* revolve more slowly. Why?

Apply the rule: teeth A × rpm A = teeth B × rpm B

$$16 \times 30 = 32 \times \text{rpm } B$$

$$\frac{16 \times 30}{32} = \text{rpm } B$$

$$\text{rpm } B = 15$$

▨ EXERCISES

1-28 A gear with 24 teeth, turning 100 rpm, is driving a gear with 96 teeth. Find the rpm of the driven gear.

1-29 A gear with 42 teeth is being driven by a gear with 84 teeth. If the driving gear is turning 850 rpm, how fast is the driven gear revolving?

1-30 A motor with a 23-tooth gear splined to its shaft will run at 1700 rpm. The motor is driving a machine which must rotate at 425 rpm. How many teeth will be on the gear splined to the shaft of the machine?

Hint: Draw a sketch. Will the gear to be placed on the machine have more than or fewer than 23 teeth?

1-31 A part in a peanut-packing machine should revolve at 360 rpm. A motor to run this machine is available and is designed to run at 1800 rpm. If the gear on the machine has 120 teeth, how many teeth will the gear selected for the motor have?

1-32 A gear train is shown in Fig. 1-5.
  (a) If gear A is revolving at 1200 rpm, how fast is gear C revolving?
  (b) If gear C is revolving at 600 rpm, how fast is gear A revolving?
  (c) If gear B is revolving at 1800 rpm, how fast are gears A and C revolving?

This rule may be revised so that you can apply it to pulley problems.

> *Rule:* To find the number of revolutions made by a driven pulley in a given time, multiply the diameter of the driving pulley by the number of revolutions the driving pulley makes in the given time, and divide by the diameter of the driven pulley.

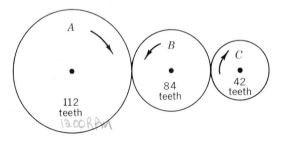

Fig. 1-5. Gear train. Exercise 1-32.

HW →

1-33 A pulley 45 in. in diameter, making 68 rpm, is driving a pulley 51 in. in diameter. Find the rpm of the driven pulley.

1-34 Find the rpm of a pulley 9 in. in diameter driven by a 24-in.-diameter pulley making 48 rpm.

1-35 Find the rpm of a pulley 52 centimeters (cm) in diameter driven by a 36-cm-diameter pulley making 78 rpm.

Rule: According to Ohm's law, the voltage drop in a circuit is equal to the current in amperes (*A*) multiplied by the resistance in ohms (Ω). If this law is rewritten in a kind of mathematical shorthand, it is more easily understood:

$$\text{Voltage drop} = \text{amperes} \times \text{ohms}$$

Solve Exercises 1-36 through 1-38 using Ohm's law.

1-36 Calculate the voltage drop in a circuit when the current flowing is 6 amperes and the resistance is 2.5 ohms.

1-37 Find the current flowing in a circuit if the voltage drop is 12.5 volts (V) and the resistance is 5 ohms.

1-38 Find the resistance of a circuit if the current flowing is 225 amperes and the voltage drop is 25 volts.

1-39 Write a rule for Exercises 1-37 and 1-38. Write a "shorthand version" of these rules in terms of voltage drop, amperes, and ohms.

1-40 Write a rule for gear problems in terms of the following: driven gear, rpm, teeth, and driving gear.

Solution:

$$\text{rpm driven gear} = \frac{\text{teeth driving gear} \times \text{rpm driving gear}}{\text{teeth driven gear}}$$

HW →

1-41 Write a rule for pulley problems in terms of the following: driven pulley, rpm, diameter, driving pulley.

COMMON FRACTIONS

2·1 FRACTIONS AND CALCULATORS

Calculations with fractions are essential to the solution of many problems, and thus it is important to understand and to be able to perform calculations with fractions. Although electronic calculators are used for many types of calculations, they cannot replace a thorough understanding of fractions.

2·2 COMMON FRACTIONS—DEFINITIONS

The number 6 divided by 3 gives an exact quotient of 2. This may be written $\frac{6}{3} = 2$. However, if you attempt to divide 6 by 7 you are unable to calculate an exact quotient. This division may be written $\frac{6}{7}$ (read "six-sevenths"). This is a *fraction*. The fraction $\frac{6}{7}$ represents a number, but it is not a whole number.

"Sevenths" indicates that a unit is divided into seven equal parts. The fraction $\frac{6}{7}$ indicates that six of the seven equal parts are included in the number. This is true if the number below the line and the number above the line are both integers.

A fraction such as $\dfrac{A}{B}$, where A and B are integers, indicates division:

$$A \div B$$

The divisor, or the number below the line in the fraction, is the *denominator* of the fraction. The dividend, or the number above the line in the fraction, is the *numerator* of the fraction. The denominator tells into how many parts the unit is divided. The numerator tells how many of these parts are included.

The fraction $\frac{3}{4}$ is read "three-fourths." Its denominator tells you that a unit has been divided into four equal parts, and its numerator tells you that three of the four parts are included.

On the ruler shown in Fig. 2-1, each inch is divided into four equal

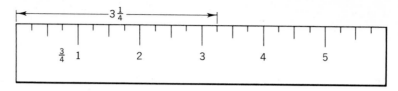

Fig. 2-1. Ruler.

parts. Three-fourths of an inch ($\frac{3}{4}$ in.) would, therefore, be three of these equal fourths; this length is marked by the fraction $\frac{3}{4}$ in.

2-3　MIXED NUMBERS

Just as there are whole numbers and fractions, there are numbers made up of both whole numbers and fractions. The mixed number $3\frac{1}{4}$, which is read "three and one-fourth," means $3 + \frac{1}{4}$ ($3\frac{1}{4}$ is also marked on the ruler in Fig. 2-1). A *mixed number* is a whole number and a fraction.

2-4　PROPER AND IMPROPER FRACTIONS

If the numerator of a fraction is less than the denominator, then the value of the fraction is less than 1. If the numerator and denominator are equal, the value of the fraction must be 1. If the numerator of a fraction is greater than the denominator, the value of the fraction must be greater than 1. For example, the fraction $\frac{3}{4}$ has a value less than 1, the fraction $\frac{4}{4}$ has a value of 1, and the fraction $\frac{7}{4}$ has a value greater than 1.

A *proper* fraction's numerator is less than the denominator. For example, $\frac{3}{4}$ is a proper fraction.

An *improper* fraction's numerator equals or is greater than the denominator. Thus, $\frac{4}{4}$ and $\frac{7}{4}$ are improper fractions.

2-5　REDUCING A COMMON FRACTION TO ITS LOWEST TERMS

A fraction is in its lowest terms when its numerator and denominator have no common factor. The fraction $\frac{4}{8}$ is a proper fraction, but both numerator and denominator may be reduced by dividing each by 4.

EXAMPLE 2-1　Reduce $\frac{4}{8}$.

Solution　Divide both numerator and denominator by the common factor or divisor 4.

$$\frac{4}{8} = \frac{4 \div 4}{8 \div 4} = \frac{1}{2}$$

EXAMPLE 2-2　Reduce $\frac{16}{48}$.

Solution　Divide both numerator and denominator by the common factor 16.

$$\frac{16}{48} = \frac{16 \div 16}{48 \div 16} = \frac{1}{3}$$

EXAMPLE 2-3 Reduce $\frac{15}{48}$.

Solution Divide both numerator and denominator by the common factor 3.

$$\frac{15}{48} = \frac{15 \div 3}{48 \div 3} = \frac{5}{16}$$

2-6   REDUCING AN IMPROPER FRACTION

Improper fractions have numerators that equal or are greater than their denominators. An improper fraction is converted to a mixed number by division. The fractional part of the resulting mixed number should be reduced to its lowest terms.

EXAMPLE 2-4 Convert $\frac{47}{9}$.

Solution Perform the division:

$$
\begin{array}{r}
5 = 5\frac{2}{9} \\
9\overline{)\,47} \\
45 \\
\hline
2
\end{array}
$$

The resulting mixed number is $5\frac{2}{9}$.

EXAMPLE 2-5 Convert $\frac{22}{3}$.

Solution Perform the division:

$$
\begin{array}{r}
7 = 7\frac{1}{3} \\
3\overline{)\,22} \\
21 \\
\hline
1
\end{array}
$$

▮ EXERCISES

Reduce the fractions in Exercises 2-1 to 2-4 to lowest terms and convert the improper fractions in Exercises 2-5 to 2-12 to mixed numbers.

2-1 $\frac{5}{15},\frac{4}{12},\frac{15}{25},\frac{8}{24}$	**2-4** $\frac{120}{1680}$	**2-7** $\frac{36}{7}$	**2-10** $\frac{14}{3}$
2-2 $\frac{8}{40},\frac{35}{45},\frac{28}{35},\frac{14}{64}$	**2-5** $\frac{17}{8}$	**2-8** $\frac{125}{20}$	**2-11** $\frac{19}{6}$
2-3 $\frac{270}{360}$	**2-6** $\frac{19}{12}$	**2-9** $\frac{88}{9}$	**2-12** $\frac{27}{2}$

2-7   CHANGING A WHOLE OR MIXED NUMBER TO AN IMPROPER FRACTION

EXAMPLE 2-6 Change 5 to an improper fraction with a denominator of 4.

Solution Since each one of the five units equals $\frac{4}{4}$, then 5 converted to a fraction with 4 as a denominator is

$$\frac{5}{1} \times \frac{4}{4} = \frac{20}{4}$$

Check this answer by referring to the 6-in. ruler in Fig. 2-1. Each unit (here an inch) is made up of 4 divisions of $\frac{1}{4}$ (one-fourth) of an inch; so 5 in. will include 20 of these fourths, or $\frac{20}{4}$.

EXAMPLE 2-7   Change $3\frac{1}{4}$ to an improper fraction with a denominator of 4.

Solution   Multiply 4 by 3, add 1, and place the result over 4.

$$3\frac{1}{4} = (4 \times 3) + 1 = \frac{13}{4}$$

Check this answer by referring to Fig. 2-1; $3\frac{1}{4}$ in. includes 13 fourths of an inch or $\frac{13}{4}$ in.

2-8   CHANGING TWO OR MORE FRACTIONS TO FRACTIONS WITH THE SAME DENOMINATOR

When fractions are added or subtracted, they must all have the same denominator. You can always find a common denominator by multiplying all the original denominators, as illustrated in the example below.

EXAMPLE 2-8   Change $\frac{1}{4}$ and $\frac{1}{2}$ to fractions having the same denominator.

Solution   Multiply the denominators: $2 \times 4 = 8$. The common denominator is 8. Change the $\frac{1}{4}$ to a fraction with 8 as a denominator $\left(\frac{?}{8}\right)$. Since the denominator has been multiplied by 2, the numerator must be multiplied by 2.

$$\frac{1}{4} \times \frac{2}{2} = \frac{2}{8}$$

Repeat the process for the $\frac{1}{2}$:

$$\frac{1}{2} = \frac{?}{8} = \frac{1}{2} \times \frac{4}{4} = \frac{4}{8}$$

The two fractions with denominators of 8 are $\frac{2}{8}$ and $\frac{4}{8}$. Notice that the values of the two fractions have not changed, because the numerator and denominator are multiplied by 1 written as $\frac{2}{2}$ and $\frac{4}{4}$.

2-9   FINDING THE LOWEST COMMON DENOMINATOR

When adding or subtracting fractions, we change all the denominators into what is called the *lowest common denominator* (lcd). This means we must find the smallest (lowest) number by which all the denominators can be divided evenly. The lcd of a group of fractions is the lcm (lowest common multiple) of all the denominators.

EXAMPLE 2-9   In Example 2-8 we changed $\frac{1}{4}$ and $\frac{1}{2}$ to $\frac{2}{8}$ and $\frac{4}{8}$, respectively, so that the addition could be completed.

$$\frac{1}{4} + \frac{1}{2} = \frac{2}{8} + \frac{4}{8} = \frac{6}{8} = \frac{3}{4}$$

We could have used the lcd, 4, and obtained the same answer.

$$\frac{1}{4} + \frac{1}{2} = \frac{1}{4} + \frac{2}{4} = \frac{3}{4}$$

EXAMPLE 2-10 Convert $\frac{4}{9}$, $\frac{7}{12}$, and $\frac{13}{24}$ to fractions with the lowest common denominator.

Solution The lcm of 9, 12, and 24 is 72. If you divide 72 by each of the denominators, you will obtain the numbers to be used as multipliers.

$$\frac{4}{9} = \frac{32}{72}$$

$$\frac{7}{12} = \frac{42}{72}$$

$$\frac{13}{24} = \frac{39}{72}$$

Note: *Frequently the lcd of the given fractions can be seen by inspection. If you cannot find an lcd by inspection, use the following method.*

  Factor each denominator into its prime factors. Then take each prime factor the greatest number of times it appears in any one of the denominators. The product of these "greatest numbers" will be the lcd of the fractions.

EXAMPLE 2-11 Find the lcd of $\frac{11}{30}$, $\frac{7}{45}$, $\frac{14}{135}$, and $\frac{13}{25}$.

Solution

$$30 = 2 \times 3 \times 5$$
$$45 = 3 \times 3 \times 5$$
$$135 = 3 \times 3 \times 3 \times 5$$
$$25 = 5 \times 5$$

To determine the lcd use 2 once, 3 three times, and 5 twice, and calculate the product. The lcd is $2 \times 3 \times 3 \times 3 \times 5 \times 5 = 1350$.

2-10 ADDITION OF FRACTIONS

Rule: *Fractions must have common denominators to be added.*

If fractions do not have a common denominator, their lcd must be determined. The following examples will illustrate.

EXAMPLE 2-12 Add $\frac{7}{12}$, $\frac{5}{12}$, and $\frac{11}{12}$.

Solution Just as 7 apples + 5 apples + 11 apples = 23 apples, so 7 twelfths + 5 twelfths + 11 twelfths = 23 twelfths.

The work may be arranged as follows:

$$\frac{7}{12} + \frac{5}{12} + \frac{11}{12} = \frac{7 + 5 + 11}{12} = \frac{23}{12} = 1\frac{11}{12}$$

EXAMPLE 2-13 Find the sum of $\frac{7}{12}$, $\frac{8}{15}$, and $\frac{17}{30}$.

Solution Here the fractions must first be converted to fractions with an lcd. The lcm of 12, 15, and 30 is 60. Then

$$\frac{7}{12} + \frac{8}{15} + \frac{17}{30} = \frac{35}{60} + \frac{32}{60} + \frac{34}{60} = \frac{35 + 32 + 34}{60} = \frac{101}{60} = 1\frac{41}{60}$$

EXAMPLE 2-14 Find the sum of $3\frac{3}{4}$, $5\frac{4}{7}$, $2\frac{9}{14}$, and $7\frac{1}{2}$.

Solution Add the whole numbers and the fractions *separately,* and then combine these sums. Write the work as shown:

$$3\frac{3}{4} + 5\frac{4}{7} + 2\frac{9}{14} + 7\frac{1}{2} = 3\frac{21}{28} + 5\frac{16}{28} + 2\frac{18}{28} + 7\frac{14}{28}$$

$$= 3 + 5 + 2 + 7 + \frac{21 + 16 + 18 + 14}{28}$$

$$= 17 + \frac{69}{28} = 17 + 2\frac{13}{28} = 19\frac{13}{28}$$

A more convenient method of arranging numbers for adding mixed numbers is to place them beneath each other and then add. This technique is similar to adding whole numbers.

$$3\frac{3}{4} = 3\frac{21}{28}$$
$$5\frac{4}{7} = 5\frac{16}{28}$$
$$2\frac{9}{14} = 2\frac{18}{28}$$
$$+ 7\frac{1}{2} = 7\frac{14}{28}$$
$$17\frac{69}{28} = 19\frac{13}{28}$$

Note that $\frac{69}{28} = 2\frac{13}{28}$.

▓ EXERCISES

2-13 Add the following and express the sums in the simplest form:
    (a) $\frac{3}{7} + \frac{8}{7} + \frac{11}{7} + \frac{4}{7}$ (g) $14\frac{3}{4} + 30\frac{1}{2} + 4$
    (b) $\frac{2}{9} + \frac{5}{9} + \frac{19}{9} + \frac{23}{9}$ (h) $7\frac{2}{3} + 9\frac{3}{4} + 11\frac{1}{2}$
    (c) $\frac{5}{14} + \frac{9}{14} + \frac{27}{14} + \frac{11}{14}$ (i) $\frac{9}{10} + \frac{7}{12} + \frac{5}{4} + \frac{2}{3}$
    (d) $\frac{4}{5} + \frac{5}{6} + \frac{7}{16}$ (j) $\frac{7}{2} + \frac{2}{7} + \frac{9}{14} + \frac{11}{28}$
hw ⇀   (e) $8 + \frac{7}{4} + \frac{8}{3} + \frac{7}{2}$ (k) $\frac{222}{8} + \frac{333}{2} + \frac{100}{6}$
    (f) $9\frac{1}{4} + 8\frac{5}{8} + 7$ (l) $\frac{13}{7} + \frac{23}{4} + \frac{111}{14}$

2-14 A brass rod was cut into lengths of $4\frac{1}{4}$, $3\frac{3}{8}$, $6\frac{1}{2}$, $7\frac{9}{16}$, and $2\frac{3}{4}$ in. How long was the original rod if $\frac{1}{16}$ in. was wasted in each cut?

Hint: Draw a sketch before attempting to solve this problem.

2-15 Three lamps are connected in series. Their resistances are $1\frac{1}{3}$ ohms, $\frac{2}{5}$ ohm, and $\frac{1}{6}$ ohm. If the total resistance is the sum of the individual resistances, what is the total resistance in ohms of the three lamps?

2-16 On successive holes a golfer drives a golf ball $205\frac{1}{3}$, $197\frac{1}{2}$, $182\frac{3}{4}$, and $220\frac{1}{6}$

yards (yd). Find the total number of yards that she drove on these four holes.

2-17 Five poured castings weigh $31\frac{1}{2}$, $102\frac{1}{5}$, $97\frac{1}{3}$, $88\frac{1}{10}$, and $203\frac{1}{5}$ lb. Find the total weight in pounds of the castings.

2-11 SUBTRACTION OF FRACTIONS

Fractions may be subtracted as shown in the following examples.

EXAMPLE 2-15 Subtract $\frac{4}{11}$ from $\frac{9}{11}$.

Solution Only fractions with identical denominators can be subtracted, and thus it is possible to subtract 4 elevenths from 9 elevenths; the remainder is 5 elevenths. This may be written

$$\frac{9}{11} - \frac{4}{11} = \frac{9-4}{11} = \frac{5}{11}$$

EXAMPLE 2-16 Subtract $\frac{7}{11}$ from $\frac{2}{3}$.

Solution First change both fractions to the same denominator. This may be written

$$\frac{2}{3} - \frac{7}{11} = \frac{2}{3}\left(\frac{11}{11}\right) - \frac{7}{11}\left(\frac{3}{3}\right) = \frac{22}{33} - \frac{21}{33} = \frac{22-21}{33} = \frac{1}{33}$$

EXAMPLE 2-17 $7\frac{2}{3} - 3\frac{3}{5} = ?$

Solution Remember that each fraction must have a common denominator.

$$7\frac{2}{3} = 7 + \frac{2}{3} = 7 + \frac{10}{15}$$

$$3\frac{3}{5} = 3 + \frac{3}{5} = 3 + \frac{9}{15}$$

Then the solution is

$$7\frac{2}{3} = 7\frac{10}{15}$$
$$-3\frac{3}{5} = 3\frac{9}{15}$$
$$\overline{\phantom{-3\frac{3}{5} = }4\frac{1}{15}}$$

Note: $\frac{2}{3}(\frac{5}{5}) = \frac{10}{15}$ *and* $\frac{3}{5}(\frac{3}{3}) = \frac{9}{15}$.

EXAMPLE 2-18 $7\frac{1}{2} - 3\frac{2}{3} = ?$

Solution

$$7\frac{1}{2} = 7\frac{3}{6} = 6\frac{9}{6}$$
$$-3\frac{2}{3} = 3\frac{4}{6} = 3\frac{4}{6}$$
$$\overline{\phantom{-3\frac{2}{3} = 3\frac{4}{6} = }3\frac{5}{6}}$$

In this case the numerator (4) of the fraction to be subtracted is greater than the numerator (3) of the fraction from which it is subtracted, so we borrow 1 from the 7, change it to sixths ($1 = \frac{6}{6}$), and add it to the $\frac{3}{8}$. This results in $6\frac{9}{8}$ instead of $7\frac{3}{8}$. The subtraction is then made as before.

EXAMPLE 2-19 $8 - 2\frac{3}{16} = ?$

Solution Borrow 1 from 8 and write 8 as $7\frac{16}{16}$; then subtract as shown.

$$8 \quad = 7\tfrac{16}{16}$$
$$\underline{-2\tfrac{3}{16} = 2\tfrac{3}{16}}$$
$$5\tfrac{13}{16}$$

■ EXERCISES

2-18 Subtract the following and reduce the results to simplest form:

(a) $\frac{2}{3} - \frac{1}{3}$ (g) $7\frac{3}{5} - 3\frac{9}{10}$

(b) $\frac{2}{3} - \frac{1}{6}$ (h) $13 - 12\frac{11}{12}$

(c) $\frac{7}{8} - \frac{3}{4}$ (i) $16 - 14\frac{23}{32}$

(d) $\frac{9}{10} - \frac{3}{5}$ (j) $\frac{47}{18} - \frac{74}{90}$

(e) $\frac{11}{16} - \frac{1}{2}$ (k) $\frac{19}{131} - \frac{2}{19}$

(f) $3\frac{1}{3} - 2\frac{2}{3}$ (l) $\frac{111}{36} - \frac{31}{72}$

2-19 Perform the additions and subtractions and reduce the results to their simplest forms:

(a) $21\frac{3}{4} + 19\frac{2}{7} - 10\frac{13}{28}$ (d) $5\frac{12}{13} - \frac{5}{26} - \frac{11}{13} + 19$

(b) $6\frac{6}{7} + 5\frac{13}{20} - 10$ (e) $13\frac{3}{5} + \frac{7}{8} - 7\frac{3}{4} - \frac{1}{8} + 2$

(c) $7\frac{3}{4} - \frac{4}{5} + 2\frac{17}{20}$ (f) $13 - (6\frac{7}{8} - \frac{3}{8}) + \frac{1}{2} - \frac{5}{8}$

2-20 If 6 is added to the numerator and to the denominator of $\frac{2}{3}$, is the value of the fraction increased or decreased? By how much?

2-21 Find the sum and difference of $3\frac{5}{9}$ and $1\frac{7}{24}$.

2-22 A municipal bond is to be paid as follows: $\frac{1}{3}$ the first year, $\frac{2}{5}$ the second year, and the remainder the third year. How much is paid the third year?

2-23 The distance from outside to outside of two holes in a steel plate is $6\frac{2}{3}$ in. If one hole is $1\frac{1}{6}$ in. in diameter and the other $2\frac{1}{24}$ in. in diameter, what is the length of metal between the holes?

2-24 The total length of a piece of square bar stock is $18\frac{7}{8}$ in. If $9\frac{11}{16}$ in. of this bar stock is turned to a cylinder, what length will remain square?

2-25 A series electric circuit is made up of three resistances of $2\frac{1}{8}$, $3\frac{1}{4}$, and $15\frac{1}{2}$ ohms. What is the resistance of the circuit if the total resistance is the sum of the individual resistances?

2-26 Three castings weigh $4\frac{1}{6}$, $9\frac{1}{3}$, and $18\frac{1}{4}$ lb. These castings are machined, and a total of $15\frac{1}{16}$ lb of metal is removed from the three castings. What is the total weight of the three machined castings?

2-12   MULTIPLICATION OF FRACTIONS

To multiply two fractions, multiply the numerators to obtain the numerator of the product and multiply the denominators to obtain the denominator of the product.

EXAMPLE 2-20   Multiply $\frac{2}{3} \times \frac{5}{7}$.

Solution   Multiply the numerators and the denominators.

$$\frac{2 \times 5}{3 \times 7} = \frac{10}{21}$$

EXAMPLE 2-21   Multiply $4 \times \frac{3}{5}$.

Solution   Write 4 over 1; multiply the numerators and the denominators, then reduce to a mixed number.

$$\frac{4 \times 3}{1 \times 5} = \frac{12}{5} = 2\frac{2}{5}$$

EXAMPLE 2-22   Multiply $8\frac{1}{3} \times 3\frac{2}{5}$.

Solution   Change both mixed numbers into improper fractions and proceed as before.

$$8\frac{1}{3} = \frac{(3 \times 8) + 1}{3} = \frac{25}{3} \text{ and } 3\frac{2}{5} = \frac{(5 \times 3) + 2}{5} = \frac{17}{5}$$

$$\frac{25}{3} \times \frac{17}{5} = \frac{25 \times 17}{3 \times 5} = \frac{425}{15} = 28\frac{5}{15} = 28\frac{1}{3}$$

EXAMPLE 2-23   Multiply $5\frac{1}{4} \times 3 \times 3\frac{1}{3}$.

Solution   Change to improper fractions and multiply.

$$5\frac{1}{4} \times 3 \times 3\frac{1}{3} = \frac{21}{4} \times \frac{3}{1} \times \frac{10}{3} = \frac{21 \times 3 \times 10}{4 \times 1 \times 3} = \frac{630}{12} = 52\frac{6}{12} = 52\frac{1}{2}$$

EXAMPLE 2-24   Multiply $7\frac{3}{5}$ by 6.

Solution   Change to improper fractions and multiply.

$$7\frac{3}{5} \times 6 = \frac{38}{5} \times 6 = \frac{38 \times 6}{5 \times 1} = \frac{228}{5} = 45\frac{3}{5}$$

EXAMPLE 2-25   Multiply 47 by $16\frac{1}{5}$.

Solution   Here is another method for multiplying a whole number by a mixed number.

$$\begin{array}{r} 47 \\ \times 16\frac{4}{5} \\ \hline 5\overline{)188} \\ 37\frac{3}{5} \\ 282 \\ 47 \\ \hline 789\frac{3}{5} \end{array}$$

Procedure:

1. Multiply 47 by 4 and divide by 5. This is the same as multiplying 47 by $\frac{4}{5}$.
2. Multiply 47 by 16, using the ordinary method for multiplying whole numbers.
3. Add the three products to obtain $789\frac{3}{5}$.

You can use this same procedure to multiply two mixed numbers.

EXAMPLE 2-26 Multiply $25\frac{2}{5}$ by $6\frac{1}{3}$.

Solution Here is another way to multiply a mixed number by a mixed number.

$$\begin{array}{r} 25\frac{2}{5} \\ \times 6\frac{1}{3} \\ \hline \frac{2}{15} \\ 8\frac{1}{3} \\ 2\frac{2}{5} \\ 150 \\ \hline 160\frac{13}{15} \end{array}$$

Procedure:

1. Multiply the fractions $\frac{1}{3} \times \frac{2}{5}$ (this = $\frac{2}{15}$).
2. Multiply $25 \times \frac{1}{3}$ (this = $8\frac{1}{3}$).
3. Multiply $\frac{2}{5} \times 6$ (this = $2\frac{2}{5}$).
4. Multiply 25×6 (this = 150).
5. Add the partial products to obtain $160\frac{13}{15}$.

■ EXERCISES

Find the product of each of the following:

2-27 $\frac{3}{4} \times 5$	**2-32** $\frac{3}{20} \times \frac{1}{6}$	**2-37** $5\frac{1}{3} \times 6$
2-28 $\frac{2}{5} \times 10$	**2-33** $\frac{11}{5} \times \frac{2}{3}$	**2-38** $8\frac{1}{4} \times 5\frac{2}{6}$
2-29 $\frac{7}{8} \times 3$	**2-34** $\frac{7}{2} \times \frac{3}{4}$	**2-39** $5 \times 1\frac{1}{6} \times \frac{7}{3}$
2-30 $\frac{5}{9} \times 9$	**2-35** $1\frac{1}{3} \times 12$	**2-40** $1\frac{3}{4} \times 2\frac{2}{3} \times 5\frac{1}{2}$
2-31 $\frac{7}{21} \times 5$	**2-36** $1\frac{7}{10} \times 1\frac{1}{3}$	**2-41** $2\frac{1}{8} \times 7 \times 7\frac{3}{4}$.

2-13 CANCELLATION

In solving problems, a fractional form often results.

EXAMPLE 2-27 Multiply $\frac{64}{48} \times \frac{25}{15}$.

Solution If you *first* do the multiplications of the numerators and the denominators and *then* reduce to simplest form, you will obtain the correct result. However, whenever possible reduce by dividing equal quantities into the numerator and the denominator. The value of a fraction is not changed when its numerator and denominator are divided by the same number.

The following problem illustrates this principle:

$$\frac{\overset{4}{\cancel{64}} \times \overset{5}{\cancel{25}}}{\underset{3}{\cancel{48}} \times \underset{3}{\cancel{15}}} = \frac{20}{9} = 2\frac{2}{9}$$

Procedure:

1. Divide 48 and 64 by 16.
2. Divide 25 and 15 by 5.
3. Multiply 4 × 5 and 3 × 3.
4. Convert the answer to a mixed number by dividing 20 by 9.

EXAMPLE 2-28 Multiply $\frac{8}{32} \times \frac{12}{17} \times \frac{17}{24}$.

Solution Reduce and multiply.

$$\frac{\overset{1}{\cancel{8}} \times \overset{1}{\cancel{12}} \times \overset{1}{\cancel{17}}}{\underset{4}{\cancel{32}} \times \underset{1}{\cancel{17}} \times \underset{2}{\cancel{24}}} = \frac{1}{8}$$

Procedure:

1. Divide the 17 by 17.
2. Divide 12 and 24 by 12.
3. Divide 8 and 32 by 8.
4. Multiply 1 × 1 × 1 = 1 and 4 × 1 × 2 = 8.

▨ EXERCISES

Find the product of each of the following, canceling wherever possible:

2-42 $18 \times \frac{2}{9}$	**2-45** $5\frac{1}{3} \times 6$	**2-48** $9\frac{5}{8} \times 1\frac{1}{11}$
2-43 $7 \times \frac{1}{3}$	**2-46** $8\frac{1}{3} \times 5\frac{1}{5}$	**2-49** $2\frac{1}{6} \times 1\frac{1}{13}$
2-44 $4\frac{1}{2} \times \frac{1}{9}$	**2-47** $10\frac{1}{2} \times \frac{3}{7}$	**2-50** $18\frac{1}{2} \times \frac{3}{74}$

2-51 1 lb is equal to about $\frac{9}{20}$ kg. How many kilograms are in 20 lb? 40 lb? 15 lb? 50 lb? A 2000-lb car? How much do you weigh in kilograms?

2-52 1 ft is equal to about $\frac{3}{10}$ m. How many meters are in 2 ft? 1 yd? A 300-ft football field? A 14-ft boat? A 30-ft-tall house? How tall are you in meters?

2-53 The circumference of a circle is about $3\frac{1}{7}$ times its diameter. Find the circumference of a circle if the diameter is 14 cm; if diameter is 28 cm; if diameter is $\frac{1}{22}$ cm.

2-54 The diagonal of a square is very nearly $1\frac{5}{12}$ the length of one side. Find the diagonal when one side is 24 in.; when one side is 18 in.; when one side is 840 ft.

2-55 An alloy used for bearings in machinery is $\frac{24}{29}$ copper, $\frac{4}{29}$ tin, and $\frac{1}{29}$ zinc. How many kilograms of each make up 174 kg of alloy?

2-56 A metal alloy called antifriction is $\frac{37}{1000}$ copper, $\frac{111}{125}$ tin, and $\frac{3}{40}$ antimony. Find the weight of each metal in a mass of the alloy weighing 1250 lb.

2-57 If a motor makes 2100 rpm, how many revolutions does it make in $\frac{3}{4}$ hour (h)?

2-58 A color television sells for $450; $\frac{2}{3}$ of that selling price represents cost of labor and materials and $\frac{1}{50}$ represents profit. If the remainder is spent on advertising, how much of the selling price is spent for each—labor and materials, profit, and advertising?

2-59 A tank holds 300 gallons (gal). If a pipe empties $\frac{1}{4}$ of the tank in an hour, how many gallons will be left in the tank at the end of 2 h?

2-60 If a city block is $\frac{1}{5}$ mile (mi), how far has a man walked when he has gone 5 blocks east and $11\frac{1}{2}$ blocks north?

2-61 A company employs 165 workers at $8\frac{2}{5}$ per hour, 228 workers at $6\frac{1}{2}$ per hour, and 560 workers at $4\frac{1}{2}$ per hour. What is the total payroll for a workday of 8 h?

2-62 What is $\frac{1}{4}$ of $12? Will the answer be greater or smaller than $12?

2-63 A tank is $\frac{5}{8}$ full of gasoline. If $\frac{1}{8}$ of that amount is drawn off, what part of the whole tank is drawn off? Calculate how much gasoline remains in the tank.

2-14 DIVISION OF FRACTIONS

Rule: *To divide by a fraction, invert the fraction following the division sign and proceed as in multiplication.*

EXAMPLE 2-29 Divide $\frac{3}{4}$ by $\frac{1}{2}$.

Solution Invert the fraction following the division sign ($\frac{1}{2}$) and proceed as if multiplying.

$$\frac{3}{4} \div \frac{1}{2} = \frac{3}{4} \times \frac{2}{1}$$

$$\frac{3 \times \overset{1}{\cancel{2}}}{\underset{2}{\cancel{4}} \times 1} = \frac{3}{2} = 1\frac{1}{2}$$

EXAMPLE 2-30 Divide $\frac{3}{7}$ by 4.

Solution Write 4 over 1, invert $\frac{4}{1}$, and proceed as if multiplying.

$$\frac{3}{7} \div 4 = \frac{3}{7} \div \frac{4}{1} = \frac{3}{7} \times \frac{1}{4} = \frac{3}{28}$$

EXAMPLE 2-31 Divide 6 by $\frac{2}{3}$.

Solution Write 6 over 1, invert $\frac{2}{3}$, and proceed as in multiplication.

$$\frac{6}{1} \div \frac{2}{3} = \frac{6}{1} \times \frac{3}{2} = \frac{\overset{3}{\cancel{6}} \times 3}{1 \times \cancel{2}} = \frac{9}{1} = 9$$

EXAMPLE 2-32 Divide $7\frac{2}{3}$ by $9\frac{1}{5}$.

Solution Convert both mixed numbers to improper fractions, invert $\frac{46}{5}$ (the fraction following the division sign), and proceed as in multiplication.

$$7\frac{2}{3} \div 9\frac{1}{5} = \frac{23}{3} \div \frac{46}{5} = \frac{23}{3} \times \frac{5}{46} = \frac{\overset{1}{\cancel{23}} \times 5}{3 \times \underset{2}{\cancel{46}}} = \frac{5}{6}$$

Procedure:

1. If either of the numbers to be divided is a whole number, place that number over 1.
2. If either number is a mixed number, change the mixed number to an improper fraction.
3. Invert the fraction following the division sign and proceed as in multiplication of fractions.

■ EXERCISES

2-64 Divide the following, using your pencil only when necessary.

(a) $\frac{5}{8} \div 10$ (h) $\frac{16}{3} \div 32$ (n) $\frac{27}{32} \div \frac{3}{4}$ (t) $52\frac{1}{2} \div 4$

(b) $\frac{15}{14} \div 5$ (i) $13 \div \frac{1}{3}$ (o) $\frac{18}{5} \div \frac{16}{5}$ (u) $84\frac{2}{3} \div 7$

(c) $\frac{33}{13} \div 3$ (j) $14 \div \frac{7}{8}$ (p) $16\frac{1}{2} \div 4$ (v) $321\frac{3}{5} \div 3$

(d) $\frac{8}{13} \div 6$ (k) $\frac{1}{3} \div \frac{4}{3}$ (q) $362\frac{4}{5} \div 2$ (w) $72\frac{5}{8} \div 5$

(e) $\frac{24}{17} \div 8$ (l) $\frac{5}{8} \div \frac{15}{16}$ (r) $674\frac{2}{3} \div \frac{1}{3}$ (x) $46\frac{3}{8} \div 3$

(f) $\frac{225}{11} \div 25$ (m) $\frac{3}{5} \div \frac{4}{15}$ (s) $27\frac{3}{5} \div 9$ (y) $159\frac{7}{8} \div 4$

(g) $\frac{169}{15} \div 13$

2-65 If the denominator of a fraction is multiplied by 4, how is the value of the fraction changed? How is the value changed if the denominator is multiplied by 8? By 7?

2-66 If $\frac{1}{10}$ in. on a map represents 50 miles (mi), how many miles are represented by 3 in. on the map?

2-67 In the blueprint of a house, $\frac{1}{4}$ in. in the print represents 1 ft in the actual house. Find the dimensions of rooms that measure as follows on the blueprint: $2\frac{1}{2}$ in. by $2\frac{1}{2}$ in.; $4\frac{1}{8}$ in. by $4\frac{5}{8}$ in.; $5\frac{3}{8}$ in. by 6 in.; $3\frac{1}{16}$ in. by $4\frac{5}{16}$ in.

2-68 What is the difference between the sum of $\frac{2}{21}$ and $\frac{3}{35}$, and the product of $\frac{3}{7}$ and $\frac{28}{183}$?

2-15   COMPLEX FRACTIONS

By definition a *complex fraction* is a fraction that may contain one or more fractions in the numerator and the denominator. Many practical problems require work with complex fractions. When solving complex fractions, do all the indicated operations in the numerator, then all the indicated operations in the denominator, then simplify. Study the examples; they will teach you to simplify complex fractions.

EXAMPLE 2-33   Find the total resistance in ohms of a parallel circuit made up of three resistances of 5, 10, and 15 ohms. Use the following formula:

$$\text{Total resistance} = \frac{1}{1/R_1 + 1/R_2 + 1/R_3}$$

Solution   Substitute 5, 10, and 15 for R_1, R_2, and R_3.

$$R = \frac{1}{\dfrac{1}{5} + \dfrac{1}{10} + \dfrac{1}{15}}$$

Determine the lowest common denominator (30), convert to equivalent fractions, add the fractions in the denominator, and simplify.

$$R = \frac{1}{\dfrac{6}{30} + \dfrac{3}{30} + \dfrac{2}{30}} = \frac{1}{\dfrac{11}{30}}$$

$$= 1 \div \frac{11}{30} = \frac{1}{1} \times \frac{30}{11} = \frac{30}{11} = 2\frac{8}{11} \text{ ohms}$$

EXAMPLE 2-34   Evaluate $(4\frac{1}{3} + 3\frac{1}{6})/(2 + 3\frac{1}{6})$.

Solution   Perform the additions in the numerator and the denominator and then simplify.

$$\frac{4\frac{1}{3} + 3\frac{1}{6}}{2 + 3\frac{1}{6}} = \frac{\frac{13}{3} + \frac{19}{6}}{\frac{12}{6} + \frac{19}{6}} = \frac{\frac{26}{6} + \frac{19}{6}}{\frac{12}{6} + \frac{19}{6}} = \frac{\overset{45}{\cancel{\frac{45}{6}}}}{\underset{1}{\frac{31}{6}}} = \frac{45}{\cancel{6}} \times \frac{\cancel{6}}{31} = \frac{45}{31} = 1\frac{14}{31}$$

EXAMPLE 2-35   Evaluate the following:

$$\frac{(4\frac{1}{3} - 3\frac{1}{6}) \div 2\frac{3}{4}}{\left(3\frac{1}{2} + \dfrac{1}{3}\right) \div \left(23/\dfrac{1}{6}\right)}$$

Solution   Perform the indicated operations in the numerator and the denominator and simplify.

$$= \frac{\left(\dfrac{13}{3} - \dfrac{19}{6}\right) \div \dfrac{11}{4}}{\left(\dfrac{7}{2} + \dfrac{1}{3}\right) \div \left(23/\dfrac{1}{6}\right)} = \frac{\left(\dfrac{26}{6} - \dfrac{19}{6}\right) \div \dfrac{11}{4}}{\left(\dfrac{21}{6} + \dfrac{2}{6}\right) \div \left(23/\dfrac{1}{6}\right)} = \frac{\dfrac{7}{6} \div \dfrac{11}{4}}{\dfrac{23}{6} \div \left(23/\dfrac{1}{6}\right)}$$

$$= \frac{\dfrac{7}{6} \div \dfrac{11}{4}}{\dfrac{23}{6} \div \left(23 \times \dfrac{6}{1}\right)} = \frac{\dfrac{7}{6} \div \dfrac{11}{4}}{\dfrac{23}{6} \div 138} = \frac{\dfrac{7}{6} \times \dfrac{4}{11}}{\dfrac{23}{6} \times \dfrac{1}{138}} = \frac{\dfrac{28}{66}}{\dfrac{23}{828}} = \frac{28}{66} \times \frac{828}{23}$$

$$= \frac{14}{33} \times \frac{\overset{36}{\cancel{828}}}{\underset{1}{\cancel{23}}} = \frac{14}{\underset{11}{\cancel{33}}} \times \frac{\overset{12}{\cancel{36}}}{1} = \frac{168}{11} = 15\tfrac{3}{11}$$

Note that the first steps in the solution to this problem are the adding and subtracting of the fractions included in the numerator and denominator. It is possible to arrive at the same answer by varying the preceding steps. You should read and study this illustration and work it out for yourself.

■ EXERCISES

Perform the operations indicated in the following problems and express the resulting fractions in their simplest terms.

2-69 $\dfrac{\frac{17}{50}}{\frac{34}{100}}$

2-70 $\dfrac{5\frac{1}{2}}{\frac{11}{10}}$

2-71 $\dfrac{5\frac{1}{4}}{7\frac{7}{8}}$

2-72 $\dfrac{\frac{2}{3}}{2\frac{2}{3}}$

2-73 $\dfrac{1}{2\frac{3}{4} + 5\frac{6}{7} + 1}$

2-74 $1\frac{1}{2} + 3\frac{1}{4} - 1\frac{1}{8}$

2-75 $(4\frac{1}{3} - 3\frac{1}{6}) \div 2\frac{3}{4}$

2-76 $(3\frac{1}{2} + \frac{1}{3}) \div \dfrac{23}{\frac{1}{6}}$

2-77 $2 - \left(\dfrac{\frac{1}{2}}{4} \div 5\right)$

2-78 $\left(2\frac{5}{8} \div \dfrac{4}{\frac{5}{10}}\right) \times 2 = {}^{21}\!/_{32}$

2-79 $\frac{4}{17} \times 2\frac{7}{12} \div 20\frac{2}{3} = \frac{1}{34}$

2-80 $\frac{7}{4} \div (\frac{2}{3} + \frac{5}{6}) - \frac{1}{4} = \frac{11}{12}$

2-81 $\dfrac{2\frac{1}{3}}{\frac{5}{2}} = \frac{14}{15}$

2-82 $\dfrac{1}{\frac{1}{7} + \frac{1}{8} + \frac{3}{28}} = 2\frac{2}{3}$

2-83 $\dfrac{(\frac{21}{8} \div 8) \times 2}{2 - (\frac{1}{8} \div 5)} = \frac{105}{316}$

2-84 $\dfrac{\frac{2}{3} + \frac{2}{9}}{4\frac{1}{3} - 3\frac{2}{3}} = 1\frac{1}{3}$

2-85 $\dfrac{1}{2\frac{3}{4} + 5\frac{6}{7} - 1} = \frac{28}{213}$

2-86 $\dfrac{8\frac{1}{3} \times 3}{12\frac{1}{2} + 4\frac{1}{9}} = 1\frac{151}{299}$

2-87 $\dfrac{(2\frac{1}{2} \times 12\frac{1}{3}) - 3}{8\frac{1}{9} + 2\frac{1}{2}}$

2-88   The cost of a clock radio includes materials costing $12, labor costing $45, advertising costing $4, and transportation costing $3. What fraction of the total is each separate cost?

2-89   A piece of forged metalwork weighed $214\frac{1}{2}$ lb.; after machining it weighed $156\frac{3}{4}$ lb. The forging cost is $16\frac{1}{2}$ cents per pound and the material machined off is sold as scrap at $3\frac{1}{4}$ cents per pound. Find the net cost of the metal in the finished piece.

2-90   A $\frac{7}{16}$ twist drill has a speed of 130 rpm when cutting steel. How long will it take to drill through a $\frac{5}{8}$-in. steel plate if 120 revolutions are required to drill a hole 1 in. deep?

34.6 sec

2-91   In drilling through mild steel 3 cm thick, a drill press operator can drill a hole 1 cm in diameter in $1\frac{3}{4}$ min. Find the distance drilled per minute.

2-92   If the resistance in ohms of a series-parallel electric circuit is expressed by these fractions

$$5\frac{1}{8} + 2\frac{1}{3} + \cfrac{1}{\cfrac{1}{6} + \cfrac{1}{2} + \cfrac{2}{3}}$$

what is the resistance of this circuit?

2-93   In Fig. 2-2, the distance f and F across the flats in a bolt head or nut, either square or hexagonal, is equal to $1\frac{1}{2}$ times the diameter of the bolt plus $\frac{1}{8}$ in.
 If the diameter of the bolt is $\frac{3}{8}$ in., what is f?
 If the diameter of the bolt is $1\frac{3}{4}$ in., what is F?

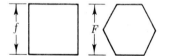

Fig. 2-2. Bolt flats.

2-94   The inside height of a bookcase is 6 ft $3\frac{1}{4}$ in. Six shelves are to be inserted, each one $\frac{7}{8}$ in. thick, equally spaced. Find the clearance between the shelves. (Note: Six shelves will give you seven spaces for books.)

2-95   A carpenter is to cut 9 shelves, each one 2 ft 8 in. long. He has boards 10, 12, 14, and 16 ft in length from which to cut the shelves. Which boards should he choose to cut to have the least possible waste?

2-96   A skilled painter can paint a garage in 8 h; her daughter can paint the same garage in 12 h. What part of the garage can each paint in 1 h? What part can both, working together, paint in 1 h? If mother and daughter work together, how long will it take them to paint the entire garage?

2-97   Convert the following temperature readings.

Note: *To convert a temperature reading from Fahrenheit to Celsius or vice versa, substitute the known value in this equation:* $C = \frac{5}{9}(F - 32°)$.

 For example, the boiling point of water is 212°F. What is the boiling point in Celsius?

$$C = \frac{5}{9}(212° - 32°) = \frac{5}{\overset{}{\underset{1}{9}}}(\overset{20°}{\cancel{180°}}) = 100°$$

To check, solve $100° = \frac{5}{9}(F - 32°)$ for F.

$$\frac{9}{\underset{1}{\cancel{5}}} \times \overset{20°}{\cancel{100}} = \frac{\overset{1}{\cancel{9}}}{\underset{1}{\cancel{5}}} \times \frac{\overset{1}{\cancel{5}}}{\underset{1}{\cancel{9}}}(F - 32°)$$

$$180° = F - 32°$$

$$F = 180° + 32° = 212°.$$

(a) 32°F (water's freezing point)
(b) 98⅗°F (normal body temperature)
(c) 70°F (normal house temperature)
(d) 0°F
(e) 105°F
(f) −10°C
(g) 35°C
(h) 65°C

DECIMAL FRACTIONS

3-1 INTRODUCTION TO DECIMAL FRACTIONS

Calculations with fractions can be cumbersome and time consuming. Thus decimal fractions or "decimals" have been developed. Decimals are a great modern invention for shortening calculations. They streamline calculations, and they are essential both in the metric system and in the use of the electronic calculator. Decimals are a part of our everyday life, and a sound knowledge of them is important for everyone.

Although the electronic hand-held calculator can be used for many calculations involving decimals, seeing how the results are obtained will provide a better understanding of the concepts involved.

3-2 DEFINITIONS OF DECIMAL FRACTIONS

A *decimal fraction* is a fraction written with a denominator of 10 or a multiple or power of 10. Thus $\frac{7}{10}$, $\frac{53}{100}$, $\frac{76}{1000}$, $\frac{4326}{1000}$, and $\frac{3756}{10000}$ are examples of decimal fractions.

In writing a decimal fraction, omit the denominator and indicate what the denominator is by placing a period (.), called a *decimal point,* in the numerator so that there are as many digits to the right of this point as there are zeros in the denominator. Thus $\frac{7}{10}$ is written 0.7, $\frac{53}{100}$ is written 0.53, $\frac{76}{1000}$ is 0.076, $\frac{4326}{1000}$ is 4.326 and $\frac{3756}{10000}$ is 0.3756.

In writing decimals, a zero is placed to the left of the decimal point when there is no whole-number figure. Thus, .53 may be written 0.53. (This notation is used in this book.)

When there are fewer figures in the numerator than there are zeros in the denominator, zeros are added to the left of the figures to make up the required number. Thus, $\frac{76}{1000} = 0.076$, and $\frac{3}{10000} = 0.0003$.

Changing the position of a decimal point changes the value of a fraction. For each place the decimal point is moved to the right, the value of the decimal fraction is multiplied by 10; for each place it is moved to the left, the value is divided by 10. Thus, 2.75 becomes 27.5 when the point is moved one place to the right and 0.275 when the point is moved one place to the left. In the first case, 2.75 is multiplied by 10; in the second case, it is divided by 10.

Remember that when you have a number in which the same figure is used throughout, such as 3333, the values expressed by the figures (here, 3s) vary greatly. For every place a 3 is moved to the left, its value is increased 10 times; however, as it is moved from the left to the right, its value is divided by 10 with each step. Thus, each 3 has 10 times the value of the 3 to the right of it, and $\frac{1}{10}$ the value of the 3 to the left of it.

These relations also hold when you pass to the right of a decimal point, that is, when you pass from the figures representing units to those representing decimal fractions. We have the following values of the positions:

Thousands	Hundreds	Tens	Units	Decimal point	Tenths	Hundredths	Thousandths	Ten-thousandths	Hundred-thousandths	Millionths	Ten-millionths	Hundred-millionths
0	0	0	0	.	0	0	0	0	0	0	0	0

3-3 READING NUMBERS

The whole number 23,676 is read "twenty-three thousand six hundred seventy-six." Note the word "and" is never used in reading a whole number.

A decimal is read like a whole number except that the value of the righthand places is added. For example, the number 0.7657 is read "seven thousand six hundred fifty-seven ten-thousandths."

When a whole number and a decimal fraction are combined, the word "and" is used between the two parts. Thus 73.2658 is read "seventy-three and two thousand six hundred fifty-eight ten-thousandths." You may find it more convenient to read 73.2658 as "seventy-three point twenty-six fifty-eight" or "seven three point two six five eight."

3-4 CHANGING A COMMON FRACTION TO A DECIMAL FRACTION

A common fraction is an indicated division. To change a common fraction to a decimal fraction, divide the denominator into the numerator.

EXAMPLE 3-1 Change the common fraction $\frac{2}{3}$ to a decimal fraction.

Solution   Divide the denominator (5) into the numerator (2).

$$
\begin{array}{r}
.4 \\
5\overline{)\,2.0} \\
\underline{2\,0}
\end{array}
$$

This decimal fraction is read "four-tenths." The common fraction $\frac{4}{10}$ is also read "four-tenths"; it reduces to the fraction $\frac{2}{5}$.

Many common fractions can be changed to an exact decimal fraction. Examples 3-2 through 3-6 illustrate the procedure for changing fractions to decimals.

EXAMPLE 3-2   Change the common fraction $\frac{7}{8}$ to a decimal fraction.

Solution   Divide the denominator (8) into the numerator (7).

$$
\begin{array}{r}
.875 \\
8\overline{)\,7.000} \\
\underline{6\,4} \\
60 \\
\underline{56} \\
40 \\
\underline{40}
\end{array}
$$

EXAMPLE 3-3   Change $\frac{3}{4}$ to a decimal fraction.

Solution   Divide the denominator (4) into the numerator (3).

$$
\begin{array}{r}
.75 \\
4\overline{)\,3.00} \\
\underline{2\,8} \\
20 \\
\underline{20}
\end{array}
$$

EXAMPLE 3-4   Change $\frac{1}{11}$ to a decimal fraction.

Solution   Divide the denominator (11) into the numerator (1).

$$
\begin{array}{r}
.09\overline{09}09 \\
11\overline{)\,1.000000} \\
\underline{99} \\
100 \\
\underline{99} \\
100 \\
\underline{99} \\
1
\end{array}
$$

Note: The 09 continues to repeat, and therefore the answer can be written $0.\overline{09}$, since the bar above the 0.09 indicates that it is a repeating decimal.

Procedure to change a common fraction to a decimal fraction:

1. Place the numerator of the fraction under a division sign.

2. Place a decimal point after that number.

3. Add zeros to the right of the decimal point.

4. Place a decimal point above the division sign directly over the decimal point under the division sign.

5. Carry the division until the remainder is zero, or as far as conditions of the problem require.

6. All common fractions end with a remainder of zero, as in Examples 3-1 through 3-3, or as a repeating decimal, as in Example 3-4.

Improper fractions and mixed numbers can also be changed to decimal fractions by division. Examples 3-5 and 3-6 illustrate the procedure.

EXAMPLE 3-5   Change $\frac{11}{8}$ to a decimal fraction.

Solution   Divide the denominator (8) into the numerator (11).

$$
\begin{array}{r}
1.375 \\
8\overline{)11.000} \\
\underline{8} \\
3\,0 \\
\underline{2\,4} \\
60 \\
\underline{56} \\
40 \\
\underline{40}
\end{array}
$$

EXAMPLE 3-6   Change $1\frac{4}{5}$ to a decimal fraction.

Solution   (a) Change to an improper fraction and divide the denominator (5) into the numerator (9):

$$
\begin{array}{r}
1.8 \\
5\overline{)9.0} \\
\underline{5} \\
4\,0 \\
\underline{4\,0}
\end{array}
$$

or   (b) $1 + \frac{4}{5}$

$$
\begin{array}{r}
.8 \\
5\overline{)4.0} \\
\underline{4\,0}
\end{array}
$$

Thus $1 + 0.8 = 1.8$.

3-5   CHANGING A DECIMAL FRACTION TO A COMMON FRACTION

Changing a decimal fraction to a common fraction is not always as easy as changing a common fraction to a decimal fraction. To change a decimal fraction, first write that fraction as a common fraction with 10, 100,

1000, 10,000, etc., as the denominator. Reading the decimal fraction will then indicate the proper denominator; then reduce the common fraction to lowest terms. For example, the decimal fraction 0.625 is read "six hundred twenty-five thousandths." The denominator, then, will be 1000.

Study how the following decimal fractions are changed to common fractions:

EXAMPLE 3-7 Write 0.625 as a fraction.

Solution Write 0.625 as $\frac{625}{1000}$ and reduce.

$$\frac{625}{1000} = \frac{625 \div 25}{1000 \div 25} = \frac{25 \div 5}{40 \div 5} = \frac{5}{8}$$

EXAMPLE 3-8 Write 0.25 as a fraction.

Solution Write as a common fraction and reduce.

$$0.25 = \frac{25}{100} = \frac{1}{4}$$

EXAMPLE 3-9 Write 0.4375 as a fraction.

Solution Write as a common fraction and reduce.

$$0.4375 = \frac{4375}{10,000} = \frac{175}{400} = \frac{7}{16}$$

A common fraction can be changed to a decimal and that decimal fraction changed back to the original fraction only if, in the indicated division, the remainder is 0; that is, if the denominator is contained in the numerator an exact number of times. For example:

$$\frac{3}{8} = 8) \overline{\begin{array}{l} .375 \\ 3.000 \end{array}} \qquad \text{and} \qquad 0.375 = \frac{375}{1000} = \frac{15}{40} = \frac{3}{8}$$
$$\begin{array}{r} 2\ 4 \\ \hline 60 \\ 56 \\ \hline 40 \\ 40 \\ \hline \end{array}$$

$$\frac{2}{3} = 3) \overline{\begin{array}{l} .666 \\ 2.000 \end{array}} \qquad \text{and} \qquad 0.666 = \frac{666}{1000} = \frac{333}{500}$$
$$\begin{array}{r} 1\ 8 \\ \hline 20 \\ 18 \\ \hline 20 \\ 18 \\ \hline 2 \quad \text{(remainder)} \end{array}$$

The fraction $\frac{333}{500}$ is not in agreement with the original fraction $\frac{2}{3}$, because the *remainder 2 was dropped*.

However, it is possible to obtain perfect agreement by carrying the remainder. This process is illustrated for the same problem:

$$\frac{2}{3} = 3\overline{)\begin{array}{l} .666 \\ 2.000 \end{array}} = 0.666\frac{2}{3}$$

$$\begin{array}{r} 1\ 8 \\ \hline 20 \\ 18 \\ \hline 20 \\ 18 \\ \hline 2 \end{array}$$

$$0.666\frac{2}{3} = \frac{666\dfrac{2}{3}}{1000} = \frac{\dfrac{2000}{3}}{\dfrac{1000}{1}} = \frac{2000}{3} \div \frac{1000}{1} = \frac{2000}{3000}$$

$$= \frac{2}{3} = \text{the original fraction}$$

Remember $666\frac{2}{3}$ changed to an improper fraction is $\dfrac{666 \times 3 + 2}{3}$ or $\dfrac{2000}{3}$.

3-6 DECIMAL EQUIVALENTS

In changing common fractions to decimal fractions, using a table of decimal equivalents is more convenient than carrying out the division process explained in Sec. 3-4. A table of decimal equivalents is a very useful tool for drafting rooms, machine shops, etc.

TABLE 3-1 DECIMAL EQUIVALENTS

$\frac{1}{64} = 0.0156$	$\frac{17}{64} = 0.2656$	$\frac{33}{64} = 0.5156$	$\frac{49}{64} = 0.7656$
$\frac{1}{32} = 0.0312$	$\frac{9}{32} = 0.2812$	$\frac{17}{32} = 0.5312$	$\frac{25}{32} = 0.7812$
$\frac{3}{64} = 0.0468$	$\frac{19}{64} = 0.2968$	$\frac{35}{64} = 0.5468$	$\frac{51}{64} = 0.7968$
$\frac{1}{16} = 0.0625$	$\frac{5}{16} = 0.3125$	$\frac{9}{16} = 0.5625$	$\frac{13}{16} = 0.8125$
$\frac{5}{64} = 0.0781$	$\frac{21}{64} = 0.3281$	$\frac{37}{64} = 0.5781$	$\frac{53}{64} = 0.8281$
$\frac{3}{32} = 0.0937$	$\frac{11}{32} = 0.3437$	$\frac{19}{32} = 0.5937$	$\frac{27}{32} = 0.8437$
$\frac{7}{64} = 0.1093$	$\frac{23}{64} = 0.3593$	$\frac{39}{64} = 0.6093$	$\frac{55}{64} = 0.8593$
$\frac{1}{8} = 0.1250$	$\frac{3}{8} = 0.3750$	$\frac{5}{8} = 0.6250$	$\frac{7}{8} = 0.8750$
$\frac{9}{64} = 0.1406$	$\frac{25}{64} = 0.3906$	$\frac{41}{64} = 0.6406$	$\frac{57}{64} = 0.8906$
$\frac{5}{32} = 0.1562$	$\frac{13}{32} = 0.4062$	$\frac{21}{32} = 0.6562$	$\frac{29}{32} = 0.9062$
$\frac{11}{64} = 0.1718$	$\frac{27}{64} = 0.4218$	$\frac{43}{64} = 0.6718$	$\frac{59}{64} = 0.9218$
$\frac{3}{16} = 0.1875$	$\frac{7}{16} = 0.4375$	$\frac{11}{16} = 0.6875$	$\frac{15}{16} = 0.9375$
$\frac{13}{64} = 0.2031$	$\frac{29}{64} = 0.4531$	$\frac{45}{64} = 0.7031$	$\frac{61}{64} = 0.9531$
$\frac{7}{32} = 0.2187$	$\frac{15}{32} = 0.4687$	$\frac{23}{32} = 0.7187$	$\frac{31}{32} = 0.9687$
$\frac{15}{64} = 0.2343$	$\frac{31}{64} = 0.4843$	$\frac{47}{64} = 0.7343$	$\frac{63}{64} = 0.9843$
$\frac{1}{4} = 0.25$	$\frac{1}{2} = 0.5$	$\frac{3}{4} = 0.75$	$1 = 1.0$

You can use Table 3-1 to find the value of a decimal fraction if the common fraction is known; conversely, the opposite procedure is possible. Some of the decimal fractions are exact equivalents. For instance, 0.3750 is exactly equivalent to $\frac{3}{8}$. However, some of the decimal equivalents are only approximate values; for example, 0.6718 is approximately equal to $\frac{43}{64}$. The table is accurate to the nearest 0.0001 for all fractions listed.

Because the results of most computations appear in decimal form, workers using decimal scales or micrometers can readily lay out or measure the necessary dimension. However, anyone using a fractional rule (a carpenter, for example) must convert the decimal into fractional form, and for this reason it is often necessary to express decimals of an inch to the nearest eighth or sixteenth.

This table is useful for making such conversions. Thus, 2.375 is changed to $2\frac{3}{8}$ by locating 0.375 in the table, finding $\frac{3}{8}$, and then adding that fraction to the whole number 2.

The decimal 1.7885 is reduced to the nearest sixteenth by locating the fraction to the nearest sixteenth to 0.7885. By observation we find that $\frac{3}{4}$, or $\frac{12}{16}$, is 0.7500 and $\frac{13}{16}$ is 0.8125, and thus $\frac{13}{16}$, or 0.8125, is closer to 0.7885 than $\frac{12}{16}$, or 0.7500; thus the decimal fraction 1.7885 to the nearest sixteenth is $1\frac{13}{16}$.

▨ EXERCISES

3-1 Write the following in decimal form:
  (a) Twenty-three thousandths; three hundredths; ninety-seven hundred-thousandths
  (b) Six hundred and forty-one ten-millionths
  (c) Ten and nineteen ten-thousandths
  (d) Fifty and eighty-five billionths
  (e) Eight hundred million and ninety-six ten-millionths

3-2 Read the following and then write in figures:
  (a) Three tenths
  (b) Five hundredths
  (c) Four and two-tenths
  (d) Three and three hundredths
  (e) Two hundred fifty-six ten-thousandths
  (f) Three hundred fifty-six ten-thousandths
  (g) Two hundred fifty-six and twenty-three thousandths
  (h) One hundred fifty-five and twenty-three thousandths
  (i) Four hundred fifty-six thousandths
  (j) Three hundred twenty-five and twenty-five ten-thousandths

3-3 Read the following and then write in words:
  (a) 0.8   (c) 0.407   (e) 45.0013   (g) 5.213 57
  (b) 0.90   (d) 0.005 009   (f) 21.202 002   (h) 12,000.000 12

Hw ➙ **3-4** Change the following fractions to decimal fractions:
  (a) $\frac{3}{5}$ (c) $\frac{3}{8}$ (e) $\frac{3}{16}$ (g) $\frac{7}{125}$
  (b) $\frac{1}{4}$ (d) $\frac{3}{20}$ (f) $\frac{6}{11}$ (h) $1\frac{8}{9}$

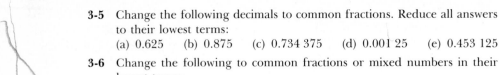

3-5   Change the following decimals to common fractions. Reduce all answers to their lowest terms:

(a) 0.625   (b) 0.875   (c) 0.734 375   (d) 0.001 25   (e) 0.453 125

3-6   Change the following to common fractions or mixed numbers in their lowest terms:

(a) 0.125   (c) 0.37½   (e) 12.095⅝

(b) 0.18¾   (d) 0.49⅛   (f) 22.71⅔

3-7   The following are U.S. standard values for the gage and thickness of sheet metal: No. 00, 0.343 75 in.; No. 2, 0.265 625 in.; No. 4, 0.234 375 in.; No. 7, 0.1875 in.; No. 13, 0.093 75 in.; No. 28, 0.015 625 in. Using the table of decimal equivalents, find the approximate thickness of each in a common fraction to the nearest thirty-second of an inch.

3-8   Using the table of decimal equivalents, change the following decimals of an inch to the nearest sixty-fourth of an inch: 0.394; 0.709; 1.416; 1.89.

3-7   ADDING DECIMALS

The following examples illustrate the addition of decimal fractions.

EXAMPLE 3-10   Add 36.036, 7.004, 0.002 36, and 723.002 6.

Solution   Arrange the numbers to be added so that the decimals are in a vertical line.

$$
\begin{array}{r}
36.036 \\
7.004 \\
0.002\ 36 \\
+723.002\ 6 \\
\hline
766.044\ 96
\end{array}
$$

Procedure for adding decimals:

1. Place the numbers in a column so that the decimal points are beneath one another.

2. Add as in whole numbers.

3. Place the decimal point in the sum beneath the other decimal points.

3-8   SUBTRACTING DECIMALS

Example 3-11 solves a subtraction problem involving decimals.

EXAMPLE 3-11   Subtract 46.8324 from 437.421.

Solution   Arrange the numbers so that the decimals are under each other.

$$
\begin{array}{r}
437.421 \\
-46.8324 \\
\hline
390.5886
\end{array}
$$

Procedure for subtracting decimals:

1. Write the numbers so that the decimal points are beneath one another.
2. Subtract as in whole numbers.
3. Place the decimal point in the answer beneath the other decimal points.

3-9   MULTIPLYING DECIMALS

Multiplying decimal fractions and multiplying whole numbers are the same operations except for the placement of the decimal point in the answer. The placement of that decimal point is illustrated in Examples 3-12 and 3-13.

EXAMPLE 3-12   Multiply 7.32 by 0.032.

Solution   Multiply as whole numbers and point off as many decimal places in the product as there are places in the factors.

$$
\begin{array}{r}
7.32 \text{ (factor)} \\
\times 0.0\,32 \text{ (factor)} \\
\hline
1\,4\,64 \\
21\,9\,6 \\
\hline
0.23\,4\,24 \text{ (product)}
\end{array}
$$

EXAMPLE 3-13   Multiply 0.002 64 by 0.000 314.

Solution   Multiply as whole numbers and point off as many decimal places in the product as there are places in the factors.

$$
\begin{array}{r}
0.00264 \text{ (factor)} \\
\times 0.000314 \text{ (factor)} \\
\hline
1056 \\
264 \\
792 \\
\hline
0.00000082896 \text{ (product)}
\end{array}
$$

Procedure for multiplying decimals:

1. Multiply the factors as whole numbers.
2. Point off as many decimal places in the product as there are places in the two factors.

   Note that in Example 3-12 there are five decimal places in the two factors, so the decimal point is placed to the left of the fifth number in the product. In Example 3-13 there are 11 decimal places in the two factors, and the decimal point must be placed to the left of the eleventh number. After multiplying, there are five numbers (82896) in the product, and therefore six 0s must be added before placing the decimal point.

   Always count from right to left when determining decimal places in a product.

3-10   DIVIDING DECIMALS

Dividing decimals, as with multiplying decimals, involves the same procedure as for whole numbers, except for the placement of the decimal point.

Two examples illustrate this type of problem.

EXAMPLE 3-14   Divide 0.4375 by 0.125.

Solution   Place the dividend 0.4375 inside the division symbol and the divisor 0.125 outside. Move the decimal in the divisor 0.125 three places to the right so it is a whole number. Then move the decimal point in the dividend .4375 three places to the right and divide.

$$
\begin{array}{r}
3.5 \leftarrow \text{quotient} \\
\text{Divisor} \rightarrow 0{\times}125.)\overline{0{\times}437.5} \leftarrow \text{dividend} \\
\underline{375} \\
62\ 5 \\
\underline{62\ 5}
\end{array}
$$

EXAMPLE 3-15   Divide 4365 by 0.005.

Solution   Place the dividend 4365 inside the division symbol and the divisor 0.005 outside. Move the decimal in the divisor 0.005 three places to the right, so it is a whole number. Then move the decimal in the dividend 4365 three places to the right by adding three zeros and divide.

$$
\begin{array}{r}
873\ 000. \leftarrow \text{quotient} \\
\text{Divisor} \rightarrow 0{\times}005.)\overline{4365{\times}000.} \leftarrow \text{dividend} \\
\underline{40} \\
36 \\
\underline{35} \\
15 \\
\underline{15}
\end{array}
$$

Procedure for dividing decimals:

1. Move the decimal point in the divisor to the right so that the divisor is a whole number.
2. Move the decimal point to the right in the dividend the same number of places as it was moved in the divisor.
3. Place the decimal point in the answer above the decimal point in the dividend.
4. Divide as in whole numbers.

Note that in Example 3-14 the decimal point is moved three places in the divisor and in the dividend. In Example 3-15 the decimal point is moved three places in the divisor and therefore must be moved three places in the dividend. This is accomplished by adding three zeros, then placing the decimal point after the third zero. The (x) over the decimal points is used to show that the decimal points have been moved.

3-11   ROUNDING OFF NUMBERS

You will often be asked to give an answer or a result which is correct to a certain number of decimal places. If you have an answer of 47.264 735 and want it rounded to three places, your answer is 47.265; rounded to two places, it is 47.26; rounded to one place, 47.3.

Procedure for rounding off a number to a given number of decimal places:

1. Look one digit beyond the number of places the result is to be rounded.
2. If the first number dropped is 5 or more, add 1 to the last remaining digit.
3. If the first number dropped is 4 or less, do not change the last remaining digit.

EXERCISES

3-9   Add and check by adding up.

(a) 4.1	(b) 146.9	(c) 47.25
67.5	0.004 12	5.006 95
42.001	31.416	193.5
13.18	125.001	5.875
0.0004	231.8	9.000 010 5

3-10   Change to decimals as necessary and add:
(a) $10\frac{1}{4} + 8.9\frac{1}{2} + 3.02 + 135.24 + 185.64$
(b) $1.35 + 16\frac{1}{5} + 2.37\frac{1}{2} + 56\frac{1}{4} + 2000$
(c) $13\frac{1}{4} + 14\frac{1}{3} + 77\frac{1}{2} + 12.5 + 28.675 + 15\frac{2}{3} + \frac{1}{2}$

Note: *$15\frac{2}{3}$ should be rounded to 15.667, since the other addends are given in thousandths.*

(d) $11\frac{1}{9} + 66\frac{2}{3} + 1\frac{2}{9} + 125.125 + 375.375 + 10\frac{1}{4}$
(e) $78.808 + 202.202 + 62\frac{1}{2} + 98\frac{3}{20} + 10\frac{3}{4} + 111.1$

3-11   Subtract the following and check by adding the difference to the number that was subtracted. Granted, this type of problem is easily solved with a calculator; however, working them out gives you a better understanding.
(a) $1 - 0.668\ 97$

Solution:

$$1.000\ 00$$
$$-0.668\ 97 \text{ (number subtracted)}$$
$$0.331\ 03 \text{ (difference)}$$

Add to check; the sum should be 1.00000.
(b) $1 - 0.698\ 97$ (e) $3.1416 - 1.4142$
(c) $2 - 1.301\ 03$ (f) $1.732\ 05 - 1.442\ 25$
(d) $4.641\ 588 - 4.626$ (g) $75.7575 - 55.1\frac{1}{5}$

3-12   From one hundred take seven thousandths.

3-13   From one hundred and one tenth take one and ten thousandths.

3-14   1 quart (qt) liquid measure has 57.75 cubic inches (in.³), and 1 quart dry measure has 67.200 625 in.³. How many cubic inches larger is the dry quart than the liquid quart?

3-15 Multiply the following:
    (a) 3.62×0.0037 (d) 2.236×799
    (b) 7.789×4.924 (e) $0.000\ 76 \times 0.0015$
    (c) $2.53 \times 0.006\ 35$ (f) 2.967×2.967

3-16 $8.943 \times 1\frac{2}{3} = ?$

Solution: Change $1\frac{2}{3}$ to $\frac{5}{3}$; reduce and multiply.

$$\frac{\overset{2.981}{\cancel{8.943}}}{1} \times \frac{5}{\cancel{3}} = 14.905$$

Note: *Would the correct answer be obtained if we rounded $1\frac{2}{3}$ to 1.667? Why?*

3-17 Change the fractions to decimals as needed and multiply.
    (a) $2.55 \times 4\frac{2}{5} = ?$
    (b) $1\frac{2}{3} \times 1.4142 \times 61 \times 6.5 = ?$
    (c) $0.0506 \times 10\frac{1}{2} = ?$
    (d) $8.4 \times 0.0105 \times 1.0055 = ?$
    (e) $9876.5 \times 0.0011 \times 0.091 = ?$
    (f) $0.6\frac{6}{8} \times 8\frac{5}{8} \times 6.6705 = ?$

3-18 Multiply $5\frac{3}{4}$ thousandths by $5\frac{5}{4}$ hundredths.

3-19 Complete the following:
    (a) $33(6.25) \div 8.25 = ?$
    (b) $90.58 \times 2.2046 = ?$

3-20 Divide 43.769 by 4.76 correct to four decimal places.

Solution: Divide to five places, adding zeros as needed, and round to four places.

$$
\begin{array}{r}
9.19516 \\
4 \times 76\,)\,\overline{43 \times 76.90000} \\
42\ 84 \\
\hline
92\ 9 \\
47\ 6 \\
\hline
45\ 30 \\
42\ 84 \\
\hline
2\ 460 \\
2\ 380 \\
\hline
800 \\
476 \\
\hline
3240 \\
2856 \\
\hline
\end{array}
$$

Explanation: Since the fifth decimal figure (6) in the quotient is greater than 5, the answer is 9.1952.

3-21 Solve the following to four decimal places.
    (a) $9.375 \div 4.76 = ?$ (d) $43.45 \div 3.1416 = ?$
    (b) $89.7201 \div 3.276 = ?$ (e) $3.1416 \div 6.67 = ?$
    (c) $34.675 \div 4.375 = ?$ (f) Divide 324.8 by 4000.

3-22 Divide $3.1416 \times 1.25 \times 50$ by $0.8 \times 2.75 \times 3$.

Solution: Arrange as shown; cancel and solve.

$$\frac{3.1416 \times 1.25 \times 50}{0.8 \times 2.75 \times 3} = \frac{\overset{\overset{\overset{0119}{\cancel{1309}}}{\cancel{3927}}}{\cancel{31416}} \times \overset{5}{\cancel{125}} \times 50}{\underset{1}{\cancel{08}} \times \underset{\underset{1}{\cancel{11}}}{\cancel{275}} \times \underset{1}{\cancel{3}}} = 29.750$$

Note that the canceling may be done as with whole numbers, disregarding the decimal point. When the canceling is complete, point off as many places in the result as the difference between the sum of the decimal places in the numerator and the sum of the decimal places in the denominator. In the example, there are six places above and three below the line; therefore the result has three decimal places. If the sum of the places below the line is greater than the sum above the line, then, and only then, add enough zeros above so that the number of places above is the same as those below. The final result in this case will be a whole number. Thus,

$$10 \div 0.005 = \frac{10.000}{0.005} = 2000$$

(This is the same as multiplying both the numerator and the denominator by 1000.) Obviously, this problem could be done very efficiently with a hand-held calculator.

3-23 Find the value of the following:

(a) $\dfrac{37.5 \times 60.6 \times 200}{2.5 \times 303 \times 0.2}$ to the tens place

(b) $\dfrac{3.1416 \times 2.2 \times 25 \times 88}{1.25 \times 0.11 \times 40}$ to three places

(c) $\dfrac{8.2 \times 2.5 \times 10.8 \times 0.96}{41 \times 200 \times 1.2}$ to four places

(d) $\dfrac{0.7854 \times 60 \times 12.5 \times 5280}{231 \times 0.025 \times 300}$ to two places

(e) $\dfrac{5.8 \times 8.25 \times 10.1 \times 1.732}{60.60 \times 0.25}$ to two places

3-24 An oil exploration company hires 12 workers for $9.25 per hour each. The workers are to be paid time and one-half for all hours over 40. What would be the weekly payroll if they work $6\frac{1}{2}$ days for 8 h each?

3-25 What is the inside diameter of a pipe whose outside diameter is 6.84 in. if it is made of iron $\frac{3}{16}$ in. thick?

3-26 A pump delivers 61.41 liters (l) per stroke. What weight of water will it deliver in 120 strokes? (1 l of water weighs 1 kg.)

3-27 In 1 lb of phosphor bronze, 0.925 lb is copper, 0.07 lb is tin, and 0.005 lb is phosphorus. How much of each element is there in $25\frac{1}{8}$ lb of phosphor bronze?

3-28 Add $5\frac{1}{8}/7\frac{1}{5}$ to $4\frac{1}{4}/8\frac{1}{8}$, then divide the sum by $(\frac{8}{65} \div \frac{9}{10})$. (First change the fractions to decimals.)

3-29 A child paid 60 cents for $\frac{1}{4}$ lb of cookies. The cookies were mixed equally from one box marked \$1.28 per pound and another box marked \$1.12 per pound. Explain how the error was made. What was the correct price for the $\frac{1}{4}$ lb of cookies?

3-30 A car is going 62.5 miles per hour (mph). How long will it take this car to go $468\frac{3}{4}$ mi?

3-31 A layer of No. 8 wire, which is 0.162 in. in diameter, is wound on a pipe $24\frac{3}{4}$ in. long. How many turns of wire are wound on the pipe?

3-32 1 kilogram equals about 2.2 lb. How many pounds are in 106.5 kg? A 1203-kg car? A 77-kg person?

3-33 1 km equals 1093.613 yd. How many kilometers are in 1 mi (1760 yd)? $\frac{1}{4}$ mi? A football field (100 yd)?

3-34 1 British thermal unit (Btu) per hour equals 0.000 292 87 kilowatts (kW). How many kilowatts equal 100 Btu/h? 250 Btu/h? An 11,000-Btu/h air conditioner?

3-35 An iron bar is 10.18 in. long, 3.45 in. wide, and 0.87 in. thick. Find its weight if 1 in.^{3} of iron weighs 0.28 lb.

3-36 Nickel steel will stand a pull of about 6230 kg per square centimeter (kg/cm^2) in cross section. What pull will a bar stand if that bar is 2.8125 cm wide and 2.1875 cm thick?

3-37 The composition of white metal as used by the Navy Department is: tin, 7.6 parts; copper, 2.3 parts; zinc, 83.3 parts; antimony, 3.8 parts; and lead, 3.0 parts. Find the number of pounds of each in 1270 lb of white metal.

3-38 A reamer that is 6 in. long is 1.2755 in. in diameter at the small end and 1.4375 in. at the larger end. Find the taper per foot. (The taper per foot means the decrease in diameter per foot of length.)

3-39 A woman lent \$256 to a friend. At different times she was paid $\frac{1}{4}$, $\frac{1}{5}$, $\frac{3}{20}$, $\frac{1}{10}$, and $\frac{1}{20}$ of the total amount. How much has been paid? How much remains to be paid?

3-40 1 cubic foot (ft^3) water weighs 62.5 lb; find the volume of 1 lb water and of 23 lb water.

3-41 1 ft^3 ice weighs 57.5 lbs; find the volume of 1 lb ice and of 49.3 lb ice.

3-42 The commission merchants of Minneapolis receive 150 truck loads of Christmas trees each December. If each truck holds 1800 trees and the average price per tree is \$8.25 delivered to the final buyer, what gross amount will the merchants realize if they pay \$950,000 for the trees?

3-43 Number 8 (B. & S.) gage sheet steel is 0.1285 in. thick and weighs 5.22 lb per square foot (ft²). Find the thickness of a pile of 48 such sheets. Find the nearest whole number of sheets to make a pile 1 ft thick. Find the weight of this number of sheets if each sheet measures 6.25 ft².

3-44 Number 25 (B. & S.) gage sheet copper is 0.0179 in. thick and weighs 0.811 lb/ft². Answer the same questions as in Exercise 3-43.

3-45 How much must be paid for 1600 ft of steel bar weighing 1.87 lb/ft and costing $37.35 per hundred pounds?

3-46 If a steel tape expands 0.000 16 in. for each inch when heated, how much will a tape 100 ft long expand?

3-47 A round piece of work being turned on a lathe is 2.2448 cm in diameter. What is its diameter after a cut 0.06 cm deep is made on the work?

3-48 The area of the cross section of a steel wire is 0.976 square inches (in.²). If a total load of 79,056 lb was applied to this wire, what would be the load per square inch of cross section? If the wire was 8 in. long and was stretched 0.70 in., what would be the elongation (stretch) per inch in length?

3-49 The cost of making an electric motor was reduced from $16.25 to $15.97. How much would be saved in the cost of manufacturing 12,000 of these motors?

3-50 An interior decorator has three rooms to carpet. One room is 10 ft by 12 ft; another is 12 ft by 16 ft; and the third is 6 ft by 8 ft. If the carpet is $21.90 per square yard (yd²), what will be the cost to carpet the three rooms? (9 ft² = 1 yd²)

3-51 A drill makes 280 rpm and drills a hole 0.9 cm deep in $\frac{4}{5}$ min. What is the feed? (*Feed* is the distance the drill advances with each revolution.)

3-52 A drill $1\frac{1}{4}$ in. in diameter makes 120 rpm and has a feed of 0.012 in. How long will it take to drill 12 holes in a cylinder head $1\frac{1}{2}$ in. thick, if 1 min is allowed for setting for each hole?

3-12 MEASURING INSTRUMENTS

The remainder of this chapter discusses the use of several measuring instruments. Some of these measuring tools are found in the home, and others are highly specialized; however, the reading of all of them requires the application of common fractions or decimal fractions.

3-13 THE FRAMING SQUARE

The *framing square* is used by carpenters and other construction workers. A portion of this tool is shown in Fig. 3-1. The outside scale of a framing square can be used to measure to the nearest sixteenth of an inch; the two inside scales measure to the nearest eighth of an inch. A square can measure to the nearest sixteenth or eighth of an inch, because each inch is divided into eight or sixteen equal parts.

The back side of a framing square is shown in Fig. 3-2. The inside scales on the back side measure to the nearest eighth of an inch. The

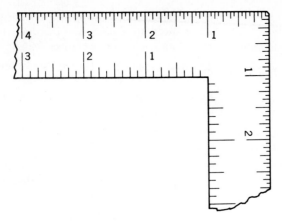

Fig. 3-1. Framing square, face side.

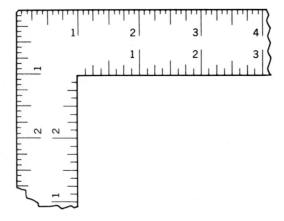

Fig. 3-2. Framing square, back side.

outside scale measures to the nearest twelfth of an inch and is used to lay out rafters. The length of each division, $\frac{1}{12}$ in., can be determined by counting on the outside scale of the square the divisions between an inch.

Note that in both Fig. 3-1 and Fig. 3-2 the marks indicating the half inch are the greatest in length, the marks indicating the quarter inch are shorter, the marks indicating the eighth inch are still shorter, and the marks indicating the sixteenth inch are shortest. These variations are designed to assist you in reading the square; the same system is used on many similar measuring instruments.

3-14 THE CIRCUMFERENCE RULE

Sheet metal workers use a measuring device called a *circumference rule*. The face of the rule has two scales (see Fig. 3-3). The upper scale on the rule is a conventional scale and for the rule pictured can be used to

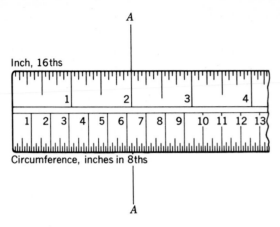

Fig. 3-3. The circumference rule.

measure to the nearest sixteenth of an inch. The lower scale is the circumference of a round pipe whose diameter is found on the upper scale. For example, a pipe 2 in. in diameter will be approximately $6\frac{1}{4}$ in. in circumference (Fig. 3-3, *A-A*). Therefore, to make a round pipe 2 in. in diameter, at least a $6\frac{1}{4}$-in. width of material will then be necessary. Some allowance must be made for the type of seam to be used, and that allowance is added to the $6\frac{1}{4}$ in.

It is possible to reverse the preceding process to locate the diameter of a round pipe if its circumference is known. First locate the circumference on the lower scale and then read the diameter on the upper scale.

3-15 THE MICROMETER CALIPER

Accuracy is very important in certain types of measurement. The physical principle of the screw is used in many mechanical devices. Its concept of measuring small distances where great accuracy is required is shown in a *micrometer caliper* in Fig. 3-4.

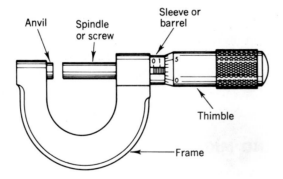

Fig. 3-4. Micrometer caliper.

The object to be measured is placed between the anvil and the spindle. The spindle has a thread cut of 40 threads to the inch on the part inside the sleeve. The thimble is outside the sleeve and turns the spindle. It also protects the thread from dust and wear. One complete turn of the thimble changes the opening of the micrometer by $\frac{1}{40}$ in. = 0.025 in., the same as the pitch of the thread.

The sleeve is graduated on a line along the length of the spindle into divisions of $\frac{1}{40}$ in. each, every fourth of which is numbered. The numbered marks then represent tenths of an inch.

The thimble has a beveled end that is divided into 25 equal divisions. A turn from one of these divisions to the next moves the spindle $\frac{1}{25}$ of $\frac{1}{40}$ in., or 0.001 in. When the end of the thimble is at 0 on the sleeve, the end of the spindle should just touch the anvil.

In Fig. 3-4, the thimble has been turned away from the 0 point. To determine the complete reading, (1) note the reading on the sleeve as uncovered by the thimble, which here shows one numbered division and three small divisions; (2) notice that the third division from 0 on the beveled edge of the thimble is on the centerline of the spindle. From this we have:

1 numbered division	= 0.100 in.
3 small divisions, 0.025 in. each	= 0.075 in.
3 divisions on thimble	= 0.003 in.
Complete reading	= 0.178 in.

3-16 MICROMETER WITH VERNIER

A micrometer caliper that reads to thousandths of an inch may be made to read to ten-thousandths of an inch with a *vernier* on the sleeve, so that 10 divisions on the vernier correspond to 9 divisions on the thimble. Thus 11 parallel lines on the sleeve occupy the same space as 10 lines on the thimble. These lines are numbered 0, 1, 2, 3, 4, 5, 6, 7, 8, 9, 0. The difference between one of the 10 spaces on the sleeve and one of the 9 spaces on the thimble is $\frac{1}{10}$ of a space on the thimble or $\frac{1}{10000}$ in. in the micrometer reading.

To read a micrometer graduated to ten-thousandths, note the thousandths as usual, then observe the number of divisions on the vernier until a line coincides with a line on the thimble. If it is the second line, marked 1, add $\frac{1}{10000}$, if it is the third, marked 2, add $\frac{2}{10000}$; and so on. Thus the reading for Fig. 3-5 is 0.2000 in. + 0.0250 in. + 0.0007 in. = 0.2257 in.

3-17 DIRECT-READING MICROMETERS

A direct-reading micrometer, shown in Fig. 3-6, will measure objects from 0 to 1 in. to the nearest ten-thousandth (0.0001) of an inch.

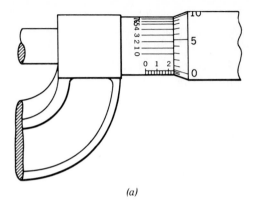

(a)

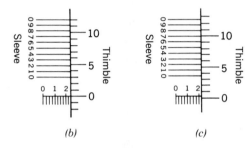

(b)   *(c)*

Fig. 3-5. Reading the micrometer caliper.

The first three places are shown in the window of the micrometer in Fig. 3-6. The fourth place is read from the vernier scale as described in Sec. 3-16. The reading on the instrument shown in Fig. 3-6 is 0.3500 in.

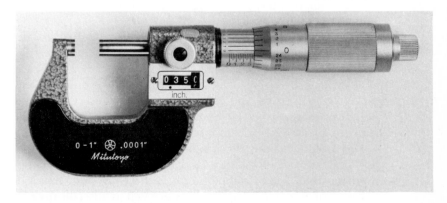

Fig. 3-6. Direct-reading micrometer (*Courtesy Mitutoyo Mfg.*).

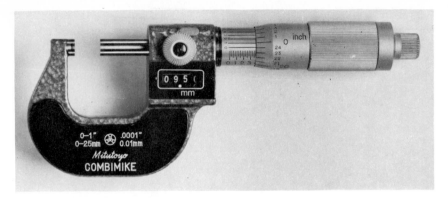

Fig. 3-7. Dual-reading micrometer (*Courtesy Mitutoyo Mfg.*).

3-18   DUAL-READING MICROMETERS

A micrometer calibrated to provide measurements in both English units and metric system units is shown in Fig. 3-7. This instrument has two scales—one scale on the thimble, sleeve, and vernier for English system measurements, and the second scale in the window for direct metric readings. The micrometer shown in Fig. 3-7 will measure lengths from 0 to 1 in. to the nearest 0.0001 in. and from 0 to 25 millimeters (mm) to the nearest 0.01 mm. The readings as shown on the two scales are 0.3740 in. and 9.50 mm.

3-19   THE VOLTMETER

The measuring instruments discussed thus far all measure length. In contrast, the voltmeter measures electrical pressure. The voltmeter in Fig. 3-8 measures direct current and may be used to determine any voltage up to 20 volts to the nearest 0.5 volt. The electrical pressure is indicated by the pointer. The voltage is then read on the scale directly under the pointer.

In using a measuring instrument of this type, first inspect the scale and determine the value of each major division. For the voltmeter in Fig. 3-8, the major divisions are labeled to indicate 0, 5, 10, 15, and 20 volts. The distance between each two of these five divisions is, in turn, divided into five equal parts and each of these parts into two halves which are indicated on the scale. The longer lines equal 1 volt and the shorter lines equal 0.5 volt. To read this instrument, first note the number to the left of the pointer, count the major divisions to the right of this number, include the smaller division, and then add the result. In Fig. 3-9, the number to the left of the pointer is 5; there are three major divisions to the left of the pointer. The pointer itself is directly over the shorter division mark. The voltage, then, is 5 + 3 + 0.5 = 8.5. If the pointer is

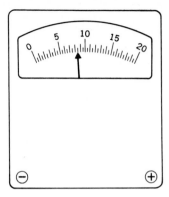

Fig. 3-8. A voltmeter.

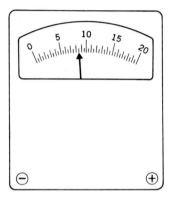 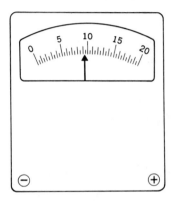

Fig. 3-9. Reading a voltmeter.   **Fig. 3-10.** Reading a voltmeter.

not directly over a mark, you may read the voltage to the nearest half volt or estimate the reading. In Fig. 3-10 the voltage measured is 9.2 by estimation or, to the nearest half volt, 9.0.

3-20   SUMMARY OF MEASURING INSTRUMENTS

Use of the measuring instruments described in this chapter illustrates several everyday applications of fractions and decimal fractions and also emphasizes the necessity of knowing the relationship between these two kinds of numbers.

The following summarizes the steps to follow in reading these various measuring instruments. Apply these steps to almost all measuring instruments:

Procedure for reading measuring instruments:

1. Inspect the entire scale of the instrument you are using.
2. Note the printed numerals on its scale.

3. Count the divisions between each printed numeral.

4. Determine the value of each division.

5. Continue determining the value of all divisions marked on the scale until you know the value of the shortest division.

6. Read the instrument by performing the necessary addition.

▇ EXERCISES

3-53 to 3-62   A barrel and thimble of a micrometer are shown in Figs. 3-11 to 3-20. What is the length of the object being measured in each of these problems?

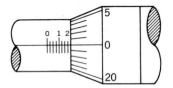

Fig. 3-11. Exercise 3-53.

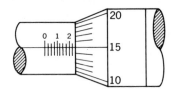

Fig. 3-12. Exercise 3-54.

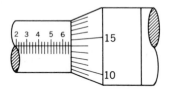

Fig. 3-13. Exercise 3-55.

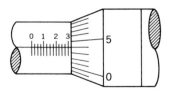

Fig. 3-14. Exercise 3-56.

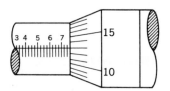

Fig. 3-15. Exercise 3-57.

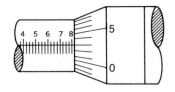

Fig. 3-16. Exercise 3-58.

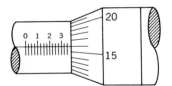

Fig. 3-17. Exercise 3-59.

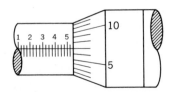

Fig. 3-18. Exercise 3-60.

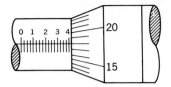

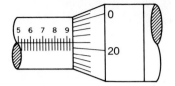

Fig. 3-19. Exercise 3-61.   **Fig. 3-20.** Exercise 3-62.

3-63 to 3-66   Read the direct-current voltages in Figs. 3-21 to 3-24.

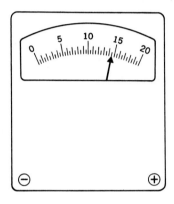

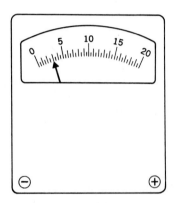

Fig. 3-21. Exercise 3-63.   **Fig. 3-22.** Exercise 3-64.

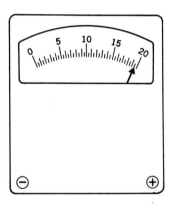

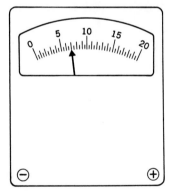

Fig. 3-23. Exercise 3-65.   **Fig. 3-24.** Exercise 3-66.

C H A P T E R 4

PERCENTAGES

4-1 INTRODUCTION TO PERCENTAGES

If you understand both fractions and decimal fractions, percentage problems will be easy to learn. Percent is one of the best ways to compare one quantity to another quantity.

The word "percent" means "by the hundred." Percentage is often designated by the symbol %. Thus 10% means 10 percent, $\frac{10}{100}$, 0.10, or "ten out of one hundred." These equivalents may be written in the reverse order: $0.10 = \frac{10}{100} = 10\%$.

4-2 PERCENTAGES AS FRACTIONS

To change a fraction to an equivalent percent, do the indicated division, move the decimal point two places to the right, and add the percent symbol. Thus,

$$\frac{2}{5} = 5) \overline{\begin{array}{r} .40 \\ 2.00 \\ \underline{2\ 0} \end{array}} = 0.40 = 40\%$$

The decimal fraction is equivalent to 40 percent (40%) or "40 out of one hundred." A second example:

$$\frac{7}{8} = 8) \overline{\begin{array}{r} .875 \\ 7.000 \\ \underline{6\ 4} \\ 60 \\ \underline{56} \\ 40 \\ \underline{40} \end{array}} = 0.875 = 87.5\%$$

Decimal fractions may be changed to percentages by moving the decimal point two places to the right and adding a percent symbol; percents may be changed to decimal fractions by moving the decimal points two places to the left and dropping the percent symbol. For example, $0.05 = 5\%$; $0.0005 = 0.05\%$; $1.07 = 107\%$; $4.33\frac{1}{3} = 433\frac{1}{3}\%$.

Percentages less than 1 need special attention. For example, $\frac{2}{5}$ percent is $\frac{2}{5}$ of 1 percent and is not equivalent to the fraction $\frac{2}{5}$. (The percent equivalent of $\frac{2}{5}$ is 40 percent.) The $\frac{2}{5}$ percent is equal to 0.4 percent or to the decimal fraction 0.004.

Percentages are used in many ways. An advertisement may proclaim: "All items in the store have been reduced 20% for quick clearance!" Or you may read in the report of a financial institution: "We pay 10.5% interest on all deposits." Decimal fractions and percentages are useful for determining and comparing batting averages and pitching records in baseball, gains or profits, and other computations. Two examples will illustrate.

EXAMPLE 4-1 During the 1983 baseball season, John Castino of the Minnesota Twins had 156 hits in 563 times at bat. Tom Brunansky of the same team had 123 hits in 542 times at bat. Since the number of "at bats" is not the same for both players, no exact comparison can be made on the basis of the number of hits. An exact comparison is possible if the batting percentages for both players are computed. John Castino's batting average is $\frac{156}{563}$ or 0.277 or 27.7 percent. Tom Brunansky's average is $\frac{123}{542}$ or 0.227 or 22.7 percent. It is now obvious that Castino has the superior batting average.*

EXAMPLE 4-2 A chair whose original selling price was $125 is on sale at a 20% discount at one store. It is possible to buy the same chair at another store for $90. Would you buy the chair at the first store or the second?

Solution To determine the cost of the chair at the first store, subtract the discount of 20% from the original selling price of $125.

$$20\% = 0.20, \ 0.20 \times 125 = \$25$$
$$\$125 - \$25 = \$100$$

Thus you will save $10 if you purchase the chair from the second store.

The most common percentage problems occur in one of the three following typical forms.

FORM 1 What is 20% of $7200? A dealer paid $7200 for a car and sold it at a profit of 20%. Find this profit. The $7200 is the *base;* the 20% is the *rate.*

$$20\% \text{ of } \$7200 = 0.20 \times \$7200 = \$1440, \text{ the profit}$$

In this problem the base and rate were given, and it was then necessary to determine the profit.

FORM 2 240 is what percent of 1200? A student paid $1200 for a car and later sold it at a price which netted a $240 profit. What was the percentage of profit? $1200 is the base; $240 is the profit.

$$\text{Rate} = \frac{240}{1200} = 0.20 \text{ or } 20\%$$

*Tom Brunansky led the American League in game winning "runs batted in" in 1983.

FORM 3 85 is $62\frac{1}{2}\%$ of what number? A store sold a radio at a profit of $85. If the profit was $62\frac{1}{2}\%$, what did the radio originally cost? $85 is the profit; $62\frac{1}{2}\%$ is the rate.

$$\text{Cost} = \frac{85}{62\frac{1}{2}\%} = \frac{85}{0.625} = \$136$$

The selling price of the radio was $221; $136 (the cost) + $85 (the profit) = $221.

In all three of the preceding problems, two values were known, and those values were used to determine a third or unknown value. All three problems make use of one important formula:

$$\text{Base} \times \text{rate} = \text{percentage (profit, discount, etc.)}$$

The number of which the percentage is taken is the *base*. The percentage taken is the *rate*. The part of the base determined by the rate is the percentage.

PROBLEM IN FORM 1 What is $37\frac{1}{2}\%$ of $720? Solve this problem by changing the percentage to a decimal fraction and then multiplying ("of" means multiply). Find $37\frac{1}{2}\%$ of $720.

$$0.375 \times \$720 = \$270$$

A rule-of-thumb for checking this type of problem is: If the rate is less than 100 percent, the answer must be less than the base (in this case, less than the $720). If the rate is greater than 100 percent, then the answer will be greater than the base.

PROBLEM IN FORM 2 45 is what percent of 450? You are given the base, 450, and the percentage, 45; you are to determine the rate. To determine rate, divide the percentage, 45, by the base, 450: 45/450 = 1/10 = 0.10 or 10%. Because 45 is less than 450, the answer must be less than 100 percent. If the percentage were larger than the base, the answer would then be greater than 100 percent.

PROBLEM IN FORM 3 85 is $62\frac{1}{2}\%$ of what number?

$$v = b \times r$$
$$85 = b \times 62\frac{1}{2}\%$$
$$85 = 0.625 \times b$$

$$\frac{85}{0.625} = b$$

$$
\begin{array}{r}
1\ 36 \\
0.62\ 5\overline{)85.0\ 00} \\
62\ 5 \\
\hline
22\ 50 \\
18\ 75 \\
\hline
3\ 750 \\
3\ 750 \\
\hline
\end{array}
$$

$$b = 136$$

EXERCISES

Note: *Solve Exercises 4-1 to 4-13 in your head.*

4-1 What is 25% of 20? of 48? of 88? of 140?

4-2 What is 33⅓% of 45? of 90? of 120? of 360?

4-3 What is 8% of 8? of 20? of 88? of 800?

4-4 What is 10% of 8? of 20? of 88? of 800?

4-5 4 is what percent of 8? of 16? of 20?

4-6 9 is what percent of 9? of 18? of 72?

4-7 40 is what percent of 60? of 480? of 800?

4-8 What percent of 4 is 72? 7½ of 24?

4-9 What percent is ⅔ of 6½? 27½ of 150?

4-10 10% of what number is 4? 8? 15? 90?

4-11 33⅓% of what number is 6? 16? 25? 96?

4-12 62½% of what number is 5? 20? 60? 500?

4-13 66⅔ of what number is 2? 20? 80? 6000?

4-14 20% from what number leaves 48?

Suggestion: *In 4-14, if 20% has been subtracted from some number, then 48 must be 100% − 20% or 80% of that number.*

4-15 10% from what number leaves 9?

4-16 20% from what number leaves 16?

4-17 33⅓% from what number leaves 20?

4-18 40 is 20% less than what number?

4-19 20 is 40% less than what number?

4-20 40 is 20% more than what number?

4-21 20 is 40% more than what number?

4-22 37½ is 87½% less than what number?

4-23 An automobile salesperson allowed $8000 for a car on a trade-in and later sold that car for $8760. What was her profit in percent?

4-24 A 1976 automobile sold new for $7600; a 1985 model of the same type of car sold for $13,300. Compute the percentage increase.

4-25 A family spent 24% of their salary for board and room. If they spent $150 per week for board and room, what was their yearly salary? (52 weeks in a year.)

4-26 Gasoline is purchased at $1.10 per gallon and sold for $1.265 per gallon. What fractional part of the cost is profit? What percent?

4-27 Some lumberyards will offer a 5% discount to individuals who build their own home and act as their own contractors. If the cost of supplies in a house is $88,900, how much does the builder save?

4-28 An 8% tax is levied on all airplane tickets. If the price of a ticket without tax is $230, what is the total cost of the ticket?

4-29 If automobile production in March is 576,000 cars, a figure 20% above that for February production, how many cars were built in February?

4-30 A quantity of metal was purchased for $3600; three-fourths of that quantity was then sold for the cost of the whole. What percentage of gain would have resulted had the entire amount been sold at the same rate as the three-fourths?

4-3   RELATIONS SHOWN BY PERCENTAGE

As previously explained, the percentage of gain or loss is often more significant than the actual gain or loss. In fact, comparisons of increase or decrease are usually possible only after the percentage of increase or decrease has been computed. Three examples illustrate this point.

EXAMPLE 4-3 The number of motor-vehicle traffic deaths in Illinois in 1971 was 2400; it decreased to 2254 in 1972. The number of deaths due to traffic accidents in Maine decreased from 271 to 258 during these same years. Which decrease is greater? Note that a direct comparison is not possible until the percentage decrease for each state is computed.

Solution Determine the amount of decrease, place the amount of decrease over the original, and find the percentage.

Illinois: $2400 - 2254 = 146$ (actual decrease)

$$\text{Percent decrease} = \frac{146}{2400} = 0.061 \text{ or } 6.1\%$$

Maine: $271 - 258 = 13$ (actual decrease)

$$\text{Percent decrease} = \frac{13}{271} = 0.048 \text{ or } 4.8\%$$

The actual decrease in deaths was 146 for Illinois and 13 for Maine. Illinois also showed a greater percentage decrease: 6.1% compared with Maine's decrease of 4.8%

EXAMPLE 4-4 The population of Atlanta increased from 1,596,000 in 1970 to 2,010,000 in 1980, a gain of 414,000. In the same period, the population of Denver increased from 1,240,000 to 1,615,000, a gain of 375,000. Atlanta experienced the larger gain in actual number of residents, but these figures give no indication of the rate of increase in the population of the two areas.

Computing the percent increase for these two areas gives the following:

Denver: $$\frac{375,000}{1,240,000} = 30.2\%$$

Atlanta: $$\frac{414,000}{1,596,000} = 25.9\%$$

It is now obvious that the percentage of Denver's population rise was greater than that of Atlanta, although Atlanta's increase was greater in numbers.

Note: The percentage of increase or decrease is based on the original numbers.

$$\text{Rate (percent of increase or decrease)} = \frac{\text{amount (actual increase or decrease)}}{\text{base (original number)}}$$

EXAMPLE 4-5 Two automatic machines are scheduled to be replaced, one during the current year and the second a year later. Listed are the production records for these machines:

	Machine *A*	Machine *B*
Pieces produced	1280	2520
Defective pieces	8	20

From the data, determine which machine should be replaced first (comparison by percentage of defective pieces produced by each machine is a good indicator of which machine to replace first).

$$\text{Machine } A\text{: Percent defective} = \frac{8}{1280} = 0.62\%$$

$$\text{Machine } B\text{: Percent defective} = \frac{20}{2520} = 0.79\%$$

On the basis of the percent comparison, machine *B* should be replaced first.

4-4 AVERAGES AND PERCENTAGE OF ERROR

Data for practical calculations are obtained by measuring quantities or the result of experimental observations; in each case, the data are liable to error. To obtain a reliable result, several measurements or observations are averaged.

An average (mean) result is obtained by adding measured results and dividing the sum by the number of measurements. This average may be accepted as an approximation. The error in any particular observation is determined by finding the difference between the given observation and the average. The amount of error can be conveniently expressed as a percentage and is then spoken of as the "percentage of error."

EXAMPLE 4-6 By measuring the diameter of a steel rod with a micrometer, a technician obtains these measurements: 0.8905 cm, 0.8907 cm, 0.8905 cm, 0.8904 cm, and 0.8906 cm. Find the average measurement and the percentage of error in the largest and smallest measurements.

Solution

0.8905 cm + 0.8907 cm + 0.8905 cm + 0.8904 cm + 0.8906 cm = 4.4527 cm
4.5527 cm ÷ 5 = 0.89054 cm average
0.8907 cm − 0.89054 cm = 0.00016 = error in largest measurement

Using the formula $r = p \div b$ (from base × rate = percentage) gives

$r = 0.00016$ cm $\div$ 0.89054 cm = 0.00018 = 0.018%
= percent of error in largest measurement
0.89054 cm $-$ 0.8904 cm = 0.00014 cm = error in smallest measurement
$r = 0.00014$ cm $\div$ 0.89054 cm = 0.00016 = 0.016%
= percent of error in smallest measurement

Remember that the percent of error in any measurement is always found by using the average measurement as the base and the amount of error as the percentage.

4-5   LIST PRICES AND DISCOUNTS

Prices of equipment and materials in catalogs and price lists are usually subject to discounts. In budgeting and cost forecasting, it is necessary to know the amount of the discount.

In business and industry, discounts are occasionally given as "60% and 10% off," or simply "60 and 10," or perhaps "sixty and ten." This does not mean a discount of 70%, but rather a discount of 60% and then a discount of 10% on the remainder, which amounts to a total discount of 64% of the original cost.

For example, if the list price is $3.50 with 60% and 10% off, take 60% of $3.50, which is $2.10. Then deduct that $2.10 from $3.50, leaving $1.40. Now take 10% of $1.40, which is $0.14, and deduct it from $1.40, which leaves $1.26, the actual cost.

Similarly, we may have discounts of 40%, 10%, and 2%, or "40, 10, and 2 off." These are calculated in the same manner as were the two discounts.

▨ EXERCISES

4-31 48% of 2000 = ?

4-32 What is $\frac{5}{8}$% of $24.40?

4-33 What is $62\frac{1}{2}$% of $24.40?

4-34 30 is $2\frac{1}{2}$% of what number?

4-35 What percent of $70 is $12.50?

4-36 10% of 30 is 3% of what number?

4-37 An estimate indicates that an 18% reduction in traffic deaths in one year would save 7200 lives. What were traffic fatalities for the previous year if the estimate is accurate?

4-38 An electric range regularly priced at $350 is on sale at a 10% discount, and for a cash purchase an additional discount of 6% is given after the 10% has been computed; what is the selling price of the range?

4-39 A rough casing weighs 20.475 kg; after machining in a lathe, it weighs 19.575 kg. The loss in finishing is what percentage of its weight in the rough?

4-40 A steel I beam expands 0.015% of its length when exposed to the sun.

Find the increase in the length of such a beam 7.58 m long after expansion.

4-41 A firm increases the wages of its employees $12\frac{1}{2}\%$. Find the new wages of a person who was earning $96.25 per day.

4-42 An airline company carried 31,223 passengers in 1 month. A year later, during the same month, this company carried 47,722 passengers. What was the percentage increase?

4-43 The usual allowance made for shrinkage on the casting of iron pipe is $\frac{1}{8}$ in./ft. What percent is this?

4-44 Water, in freezing, expands its volume by 9%. How many cubic meters (m^3) of water are therefore necessary to make 7.2 m^3 of ice?

4-45 The profits of a business during the past year totaled $14,656, 28% more than those of the previous year. What were the profits the previous year?

4-46 According to tests of career plans and self-appraisal of high school students, 105 out of 1000 want to become physicians. Of this number, 5 actually do become physicians. What percentage want to become physicians? What percentage actually do?

4-47 An iron meteorite found in 1908 weighed 3275 lb. If 91.63% of its weight was iron and 7.33% nickel, what was the weight of the iron? Of the nickel? Of other materials?

4-48 A used car is priced at $8240, and this price includes a federal tax of 3.5%. What was the price of the car? How much was the tax?

4-49 The 6% sales tax on a car is $328.84. What is the price of the car, including the sales tax?

4-50 A dealer sold two used cars at $5500 each. On one he lost 25%, and on the other he gained 25%. How much did he lose or gain?

4-51 The following measurements of the diameter of a small steel shaft were made with a micrometer: 0.5677, 0.5674, 0.5671, 0.5678, and 0.5673. Find the percentage of error in the greatest and smallest measurement.

4-52 The rate of delivery of a letter mailed in Chicago and delivered in Washington, D.C., is: first class, 22 h 45 min; special delivery, 14 h 30 min; and priority mail, 12 h 6 min. What percentage of time is saved by using special delivery over straight first class? Over priority mail?

4-53 The actual cost of removing 1 cubic yard (yd^3) of rock in excavating a certain canal is $4.73. What price should be included in the estimate if 12% is to be allowed for supervising and 10% on the cost, including supervising, is to be allowed for profit?

4-54 A manufacturer sold a suit to a retailer at a profit of 20% over the cost of production. The retailer then sold the suit to a customer for $144 and made a profit of $33\frac{1}{3}\%$. How much did the suit cost the retailer, and what was the cost of production?

4-55 The composition of white metal is, by weight, 4 parts of copper, 9 of antimony, and 97 of tin. Express these parts as percents. Find the weight of each material required to make 2.376 kg of the alloy.

4-56 A ton of coal from the Rock Island field contained 11.57% moisture and 6.27% dry coal ash. How many pounds of ash are in a ton of that coal?

4-57 A merchant buys rubber doormats at $48 per dozen, less discounts of 40%, 15%, and 5%. At what price per mat should she sell them to make 35%?

4-58 If the author receives 8.5 percent of the selling price of a book, how many books selling at $29.95 each must be sold to pay the author $8500?

4-59 In a compound, the weight of two substances, A and B, are in the ratio of 1.3498 to 1 (1.3498 parts of A to 1 part of B). What is the percent of each in the compound?

4-60 Find the selling price of a boat listed at $5500 and subject to discounts of 40 percent, 10 percent, and $7\frac{1}{2}$ percent.

4-61 The recorded measurement of a city block is 528 ft. However, careful measurement reveals the length to be 527.75 ft. Find the percentage of error in the recorded length.

4-62 A sample of nickel steel contains 24.51 percent nickel and 0.16 percent carbon. How much nickel and carbon are in 2240 kg of nickel steel?

4-63 In an analysis of the best-quality crucible cast steel, the following elements were found: carbon, 1.2%; silicon, 0.112%; phosphorus, 0.018%; manganese, 0.36%; sulfur, 0.02%; iron, 98.29%. Compute the number of pounds of each substance if the total weight is 76.5 lb.

4-64 The grade of a railroad track is given in percent. Thus a grade of 1% is a rise of 1 ft in 100 ft. If a railroad has a constant grade of $1\frac{1}{4}$%, what is the rise in $3\frac{1}{2}$ mi?

4-65 The total rise in a $1\frac{3}{4}$% grade is 43.6 ft. Determine the length of track having this grade.

4-66 A railroad rises 112.7 ft in $3\frac{1}{2}$ mi. Find the percent grade.

4-6 SIMPLE INTEREST

Interest is money paid for the use of money, that is, "rent" paid for the use of someone else's capital. It is usually calculated at a certain percent per year.

The base on which interest is calculated is the *principal*. The total amount of interest paid when money is borrowed depends on three factors: the sum borrowed, the rate at which that sum was borrowed, and the length of time that sum is used.

The interest to be paid or received is equal to the *principal* (the sum borrowed) multiplied by the *rate of interest* and by the *amount of time in years* during which the money is used:

$$\text{Simple interest} = \text{principal} \times \text{rate} \times \text{time in years}$$

EXAMPLE 4-7 Calculate the interest to be paid on $500 borrowed for 1 year at 12% interest.

Solution

$$\begin{aligned}
\text{Interest} &= \text{principal} \times \text{rate} \times \text{time in years} \\
&= 500 \times 0.12 \times 1 \\
&= \$60
\end{aligned}$$

The total which the borrower must repay at the end of 1 year is, then, the principal ($500) plus the interest ($60), or $560. This sum of the principal plus interest is called the *amount*.

EXAMPLE 4-8 $200 is deposited for 2 years at an interest rate of 6.5% in a savings bank. How much interest will the bank pay at the end of the 2 years?

Solution

$$\text{Interest} = \text{principal} \times \text{rate} \times \text{time in years}$$
$$= \$200 \times 0.065 \times 2$$
$$= \$26$$

EXAMPLE 4-9 Find the interest to be paid if $750 is borrowed for 2 years 7 months at 12% interest.

Solution Here the time, expressed in years, is $\frac{31}{12}$ years. (2 years 7 months = 31 months).

$$\text{Interest} = \text{principal} \times \text{rate} \times \text{time in years}$$

$$= 750 \times 0.12 \times \frac{31}{12}$$

$$= \overset{15}{\cancel{750}} \times \frac{\overset{1}{\cancel{12}}}{\underset{2}{\cancel{100}}} \times \frac{31}{\underset{1}{\cancel{12}}} = \frac{465}{2}$$

$$= \$232.50$$

Note: *In computing interest, it is common practice to regard 360 days as a year, 12 months as a year, and 30 days as a month.*

EXAMPLE 4-10 Compute the interest, at 6 percent, on $375 for 2 years 5 months 15 days.

Solution

$$2 \text{ years } 5 \text{ months } 15 \text{ days} = 885 \text{ days}.$$
$$2 \times 360 + 5 \times 30 + 15 = 885$$

$$\text{Interest} = 375 \times 0.06 \times \frac{885}{360} = \frac{375}{1} \times \frac{6}{100} \times \frac{885}{360}$$

$$= \frac{\overset{15}{\cancel{375}}}{1} \times \frac{\overset{1}{\cancel{6}}}{\underset{4}{\cancel{100}}} \times \frac{885}{\underset{\underset{4}{\cancel{60}}}{\cancel{360}}}$$

$$= \frac{885}{16} = \$55.31$$

■ EXERCISES

4-67 Find the interest for the following:
   — (a) $375.15 for 4 years at 15%
   —(b) $496.84 for 5 years at 18%
   (c) $250.50 for 4 years at 16%
   (d) $695.49 for 10 years at 12.5%
   (e) $453.20 for 2 years 8 months at 14.5%
   (f) $425.60 for 5 years 7 months at 13.5%
   (g) $317.42 for 4 years 3 months at 16.5%
   (h) $3,180 for 2 years 10 months 16 days at 18.5%

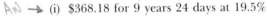 (i) $368.18 for 9 years 24 days at 19.5%
   (j) $125.80 for 4 years 8 months 20 days at 17.5%

4-68 Find the interest and amount for each of the following:
   (a) $700 for 3 years at 14.5%
   (b) $14.30 for 2 years 9 months at 18%
   (c) $245.60 for 2 years 7 months 21 days at 12%
   (d) $436.75 for 1 year 2 months 15 days at 15%
   (e) $325.25 for 2 years 9 months 12 days at 17.5%
   (f) $87.50 for 3 years 3 months at 5.25%
   (g) $480 for 6 years 3 months at 15%
   (h) $14.20 for 9 years 9 months at 14.75%

4-69 A note for $225 at 12% runs for 9 months. What is the amount of the note when due?

4-70 A note for $390 at 18% runs for 3 years 6 months. What is the amount due?

4-71 In a real estate transaction, *A* bought a house from *B* on a contract for deed for $10,000; the purchaser paid $1500 on April 1. The terms of the contract require the payment of $235 on the first of each month to meet the 16 percent interest on the balance of the principal for the previous months; the remainder is applied as a payment on the principal. Make a table showing the interest paid, the payment on the principal, and the balance for the first of each month from May 1 to September 1, inclusive.

RATIO AND PROPORTION

5-1   RATIO

There are many ways to indicate the relationship of one number to another: one method is the calculation of ratio. A *ratio* is the comparison of two numbers by division or an indicated division. The ratio of one number to another is determined when the one number is divided by the other.

A ratio always includes two numbers. For example, if a box has a length of 6 in. and width of 2 in., the ratio of the *length* to the *width* is expressed as $\frac{6}{2}$ or $6:2$. Both expressions have the same meaning.

All ratios are reduced to the lowest possible terms. This is similar to reducing a fraction to the lowest possible terms. The ratio of $\frac{6}{2}$ or $6:2$ should be reduced to its lowest possible terms by dividing the 6 and the 2 by 2. The resulting ratio is $3:1$ or $\frac{3}{1}$. The ratio of the *length* of the box to its *width* is $3:1$, since the box is 3 times as long as it is wide.

This ratio can also be stated as the relationship of the *width* to the *length*. The box is 2 in. wide and 6 in. long. The ratio of the *width* to the *length* is $2:6$ or $\frac{2}{6}$. This ratio when reduced becomes $1:3$ or $\frac{1}{3}$. The width of the box is $\frac{1}{3}$ its length.

Note that both dimensions given for the box were expressed in the same unit of length—inches. All numbers in a ratio must be expressed in the same units.

If \$20 is divided between two people in the ratio of $2:3$, one gets $\frac{2}{5}$ of \$20, or \$8, and the other gets $\frac{3}{5}$ of \$20, or \$12. Therefore the ratio of \$8 to \$12, or \$8/\$12, reduces to the ratio of $\frac{2}{3}$ or $2:3$.

Here is another example: A drawing, such as a blueprint, is usually made to some scale. For instance, $\frac{1}{4}$ in. on the drawing represents 1 ft on the actual object. The ratio here would be $\frac{1}{4}:12$ or $1:48$.

Note that 1 ft was changed to inches in this example; again, remember that all numbers in a ratio must be expressed in the same units. The scale "$\frac{1}{4}$ in. equals 1 ft" and the statement "1 in. equals 48 in." are equivalent.

EXERCISES

5-1 Express the ratio of 6 to 4; of 7 to 21; of 66⅔ to 100.

5-2 Reduce the following ratios to a simple fraction.
(a) 144:24
(b) 16:288
(c) 0.05:100
(d) 240:0.8
(e) 2.5:¾
(f) ⅚:10
(g) 0.7:⅞
(h) 9 h:30 min
(i) $5:20 cents
(j) 3 mi:1320 ft
(k) 6 quarts:3 pints (pt)
(l) 2 pints:16 gallons
(m) 30 mph:88 ft/min

5-3 With an electron microscope a magnification of 200,000 to 1 can be obtained. To get an idea of such magnitude, picture this: a man magnified 200,000 times could lie with his head in Washington, D.C., and his feet in New York City. If the man were 6 ft tall, what is the distance between these two cities?

5-4 During World War II, 325,000 U.S. soldiers died in battle; in the same period, 2 million Americans died of heart disease. What is the ratio of these deaths?

5-5 A mixture is 8 parts alcohol and 6 parts water. How many gallons of each are in 140 gallons of the mixture?

5-6 A room is 18 ft long by 12 ft wide. What is the ratio of its length to its width?

5-7 Two gears have 90 teeth and 20 teeth, respectively. What is their ratio of teeth?

5-8 One city has a population of 8000; another has a population of 20,000. What is the ratio of their populations? What part is the first of the second?

5-9 Write the inverse ratios of the following (to write an inverse ratio means to reverse the quantities): 7:5; 8:2.5; $12:820.

5-10 Divide 80 apples between two people in the ratio of 5:3.

5-11 If people can fly from Chicago to New York in 1 h 55 min and can go by train in 16 h, what is the ratio of flying time to train time?

5-12 If 51 students entered a class and 33 of them finished the course, what percent finished? What is the ratio of the number that finished to the whole number?

5-13 Four residents owning adjoining lots on the same street with frontages of 25, 35, 50, and 75 ft, respectively, are assessed a total of $15,000 for street improvement. Find the amount each should pay.

5-14 The efficiency of a machine is the ratio of its output to its input. That ratio is usually expressed as a percent:

$$\text{Efficiency} = \frac{\text{output}}{\text{input}}$$

The input in a motor is 6000 watts (W) and the output 5300 W. What is its efficiency? (*Note:* When efficiency is expressed as a percentage, the form is a ratio, with 100 as the denominator.)

5-15 If the efficiency of a machine is 91 percent and its input 98 horsepower (hp), what is its output?

5-16 If the efficiency of a motor is 98 percent and its output 730 W, what is its input?

5-2 PROPORTION

A *proportion* is a statement of equality between ratios. Thus $2:3 = 4:6$ is a proportion. It is evident that the two ratios are equal when the proportion is written in fractional form: $\frac{2}{3} = \frac{4}{6}$. Either form may be read as follows: "Two is to three as four is to six."

The general statement of a proportion can be written:

$$\frac{a}{b} = \frac{c}{d} \qquad \text{(fractional form)}$$

$$a:b = c:d \qquad \text{(proportional form)}$$

where a, b, c, and d represent numbers.

A proportion may be written with a double colon (::) in place of the equal sign:

$$a:b::c:d$$

The first and last terms of a proportion are the *extremes;* the second and third terms are the *means.* Thus in the preceding examples, the a and d and the 2 and 6 are the extremes; the b and c and the 3 and 4 are the means.

5-3 PRINCIPLES OF RATIO AND PROPORTION

By studying proportions, or by thinking of proportions as fractions, it becomes evident that if any three of the four numbers are known, then the fourth can be determined.

Given the proportion $a:b = c:d$, written as a fraction, $a/b = c/d$, then $a \times d = c \times b$. If you substitute numbers for the letters and for the proportion, then $\frac{2}{3} = \frac{4}{6}$, $2 \times 6 = 12$, and $3 \times 4 = 12$. Obviously, the product of the means (3×4) equals the product of the extremes (2×6).

$$\overset{\text{extremes}}{\underset{\text{means}}{a:b = c:d}} \qquad \overset{\text{extremes}}{\underset{\text{means}}{2:3 = 4:6}}$$

EXAMPLE 5-1 Find c in the proportion $2:3 = c:12$?

Solution Write the proportion in fractional form:

$$\frac{2}{3} = \frac{c}{12}$$

Since $12 = 4 \times 3$, c must be 2×4, or 8, since the proportion gives equal fractions. The proportion when completed is:

$$\frac{2}{3} = \frac{8}{12}$$

The solution is an application of equivalent fractions from Chap. 2.

A second method uses the principle that the product of the means equals the product of the extremes:

$$2:3 = c:12$$

The product of the means, $3c$, equals the product of the extremes, 2×12 or 24. Thus $3c = 24$, and $1c$ must equal 8.

EXAMPLE 5-2 What is b in the proportion $3:b = 6:20$?

Solution

$$\frac{3}{b} = \frac{6}{20}$$

$6 = 2 \times 3$; thus 20 must be twice the value of b, and b is $20 \div 2$, or 10. The complete proportion: $3/10 = 6/20$.

The alternative solution is

$$6b = 3 \times 20 \qquad 6b = 60 \qquad b = 10$$

The completed proportion is $3:10 = 6:20$.

EXAMPLE 5-3 If 15 tons of coal cost \$900, what will 27 tons cost at the same rate per ton?

Solution The same relation holds between the cost (prices) and the amount of coal, so the ratio of 15 tons to 27 tons must equal the ratio of \$900 to the cost of 27 tons. Let d stand for the number of dollars that 27 tons cost. Then

$$\text{Ratio of the amount of coal } \frac{15}{27} = \frac{900}{d} \text{ ratio of cost}$$

$$15d = 900 \times 27$$

$$d = \frac{900 \times 27}{15}$$

$$d = \$1620$$

Is this a reasonable answer? 27 is almost twice 15, so the cost of 27 tons should be almost twice the cost of 15 tons (\$900). Note that in each ratio, both quantities are given in the same units: coal in tons and cost in dollars. Thus the above proportion could be written:

$$\frac{15 \text{ tons}}{27 \text{ tons}} = \frac{900 \text{ dollars}}{d \text{ dollars}}$$

EXAMPLE 5-4 If 25 workers can do a piece of work in 30 days, in how many days can 35 workers do the same work?

Solution It is evident that 35 workers can do the work in less time than 25 workers, thus the ratio of the number of days required is equal to the inverse ratio of the number of workers. Let $x =$ number of days required. Then

$$\frac{35}{25} = \frac{30}{x}$$

$$35x = 25 \times 30$$

$$x = \frac{\overset{5}{\cancel{25}} \times 30}{\underset{7}{\cancel{35}}} = \frac{150}{7} = 21\tfrac{3}{7} \text{ days}$$

This answer is reasonable; 35 workers should complete a job faster than 25 workers.

EXAMPLE 5-5 If bell metal is 25 parts copper to 12 parts tin, what is the weight of each in a bell weighing 1850 lb?

Solution The total weight of the bell, in parts, is equal to $12 + 25$, or 37 parts. The ratio of the number of parts of each metal to the total number of parts equals the ratio of the weight of each metal to the total weight of metal. To calculate the pounds of copper, set up the following ratio:

$$\frac{25}{37} = \frac{c}{1850}$$

Then

$$25 \times 1850 = c \times 37$$

$$c = \frac{25 \times 1850}{37}$$

$$c = 1250 \text{ lb}$$

For tin,

$$\frac{12}{37} = \frac{t}{1850}$$

$$12 \times 1850 = t \times 37$$

$$t = \frac{12 \times 1850}{37}$$

$$t = 600 \text{ lb}$$

To check the answer, make sure that the weight of the tin and copper when added equals 1850 lb, the original weight given: $600 + 1250 = 1850$.

▨ **EXERCISES**

5-17 Find the last (or fourth) term of the proportion $8:4 = 26:x$.

5-18 Find the second term of the proportion $8:x = 26:4$.

5-19 Find the first term of the proportion $x:80 = \tfrac{1}{4}:1$.

5-20 If 8 yd of cloth cost \$40, what will 24 yd cost?

5-21 How long will it take \$100 to produce \$100 interest, if it produces \$8.50 in 1 year?

5-22 What will be the cost of 9 lb of candy if 5 lb cost \$4.50?

5-23 If 6 people can dig a ditch in 8 days, in what time can 8 people dig it?

Note: *An inverse ratio ($8:6 = 8:x$) is involved here.*

5-24 If 15 people can do a piece of work in 36 days, in how many days can they perform the same work with the assistance of 9 additional workers?

5-25 Two gears, one with 26 teeth and the other with 20 teeth, run together. In how many revolutions of the large wheel will the smaller wheel gain 12 revolutions?

5-26 If sound travels 6160 ft in $5\frac{1}{2}$ seconds (s), how far does it travel in 1 min?

5-27 If 16 gallons of gas will drive a car 288 mi, how many gallons will be required to drive the same car at the same rate of gas consumption from Chicago to Memphis, a distance of 564 mi?

5-28 A mixture for a casting has 4 parts copper, 3 parts lead, and 2 parts tin. How many kilograms of each metal are in a casting weighing 96 kg?

5-29 If the upkeep on 62 trucks for 1 year is $31,000, find the upkeep on 48 such trucks for 1 year at the same rate.

5-30 On a map, $1\frac{1}{2}$ in. represents 50 mi. Find the miles between two cities 5 in. apart on the map.

5-31 An iron casting weighs 142 lb and costs $7.25. At the same rate, what is the cost of a casting weighing 255 lb?

5-32 The number of revolutions of two gears in mesh is inversely proportional to the number of teeth in the gears. If the number of teeth in the gear wheels is 15 and 48 and the rpm of the smaller wheel is 40, find the rpm of the larger wheel.

5-33 The volume of a quantity of gas is inversely proportional to the pressure on it. If the volume of a quantity of gas is 740 m^3 under a pressure of 1.11 kg/m^2, how many cubic meters will be present under a pressure of 2 kg/m^2?

5-34 A 10 percent solution of silver nitrate is formed by dissolving 10 grams (g) of silver nitrate in 100 cm^3 of water. How many grams of silver nitrate should be used in 745 cm^3 of water to make a 10 percent solution?

5-35 A steel rail in a railroad track expands 0.000 006 36 in. for every inch of length for each degree increase in temperature. If a steel rail is 32 ft long, what is the change in length due to a change from a winter temperature of 20° below zero to a summer temperature of 110° above zero?

5-36 What percent is 59.1 of 51.3?

Solution: If x stands for the number of percent required, the proportion is:
$$59.1 : 51.3 = x : 100 \quad \text{or} \quad 51.3\,x = (100)(59.1).$$

$$x = \frac{59.1 \times 100}{51.3} = 115.2$$

59.1 is therefore 115.2 percent of 51.3.

5-37 36 is what percent of 80?

5-38 248 is 21 percent of what number?

5-39 If 4 percent of my money is $2.88, how much money do I have?

5-4 AN APPLICATION: DENSITY

Experience tells us that some bodies are heavier than others; for example, with two bodies of the same size, one may weigh more than the other. Consider a cubic foot of metal and one of wood; suppose the metal weighs 500 lb and the wood 50 lb. The metal is then 10 times as heavy as the wood, or the ratio of their densities is 10 to 1. We can also say that the density of the metal is 500 lb/ft^3.

Water has a density of about 62.5 lb/ft^3. In the metric system, the density of water under standard conditions is 1 g/cm^3.

The density of a body is its mass per unit volume. For our purpose, the mass is the same as the weight. Strictly speaking, however, the weight of a body near the earth is the force with which the earth attracts the mass of that body.

5-5 SPECIFIC GRAVITY

The term *specific gravity* (abbreviated sp gr), or *relative density*, is used for the ratio of the densities of two bodies. Thus in the previous illustration the specific gravity of the metal with reference to the wood is 10; this means that the metal is 10 times as heavy as the wood. Note carefully that the specific gravity of a substance is a pure number, that is, a number without a unit of measurement attached.

5-6 STANDARDS OF SPECIFIC GRAVITY

Water is the standard with which other substances are compared in stating the specific gravities of solids and liquids. The specific gravity of a substance is found by finding the weight of a certain volume of that substance and dividing this weight by the weight of the same volume of water. Thus, to find the specific gravity of a stone, find its weight and the weight of an equal volume of water. The weight of the stone divided by the weight of the water gives the specific gravity of the stone.

To find the specific gravity of a solid that is heavier than water, weigh the solid in air and then in water to find the difference between its weight in air and its weight in water. This difference is the buoyant force of the water on the solid and equals the weight of the water displaced by the solid; it is the weight of that amount of water that has the same volume as the solid. Then divide the weight of the solid in air by the difference, or by the weight of an equal volume of water; the ratio is the specific gravity of the solid.

The specific gravity of any solid heavier than water can be found in the same way. Water is used as the standard because of its abundance. All substances can be compared with it, although gases are usually compared with air or hydrogen gas. This procedure may be stated as follows:

$$\frac{\text{Weight of the body whose specific gravity is to be found}}{\text{Weight of the same volume of the standard}} = \text{specific gravity}$$

If w stands for the weight of the body whose specific gravity is to be found, s for the weight of the same volume of the standard, and sp gr for the specific gravity, the rule may be stated as a formula:

$$\frac{w}{s} = \text{sp gr}$$

5-7   APPLICATIONS OF SPECIFIC GRAVITY

The specific gravities of the more common substances are given in Table 6 in the appendix. To find the weight of a block of iron 2 ft by 3 ft by 1 ft, first find the number of cubic feet in the block, which is length × width × height or 6 ft³. This number times the weight of 1 ft³ of water indicates the weight of an equal volume of water: 62.5 lb × 6 = 375 lb. The weight of the water multiplied by the specific gravity of iron gives the weight of the iron

$$375 \times 7.2 = 2700 \text{ lb}$$

In terms of the letters already used, if $w/s = $ sp gr, then

$$w = s \times \text{sp gr}$$

EXAMPLE 5-6   Find the specific gravity of a rock if 1 ft³ of that rock weighs 182 lb.

Solution   The density of water is 62.5 lb/ft³; therefore, the specific gravity of the rock is calculated

$$\frac{182}{62.5} = 2.912$$

The specific gravity of the rock is 2.912.

EXAMPLE 5-7   What is the weight of 1 ft³ mercury if its specific gravity is 13.6?

Solution   Because 1 ft³ water weighs 62.5 lb, the weight of 1 ft³ mercury will be

$$w = 62.5 \times 13.6$$
$$= 850 \text{ lb}$$

EXAMPLE 5-8   How many cubic inches are there in 1 lb cork if the specific gravity of cork is 0.24?

Solution   1728 in.^{3} = 1 ft^3; therefore 1728 in.^{3} cork will weigh

$$w = 62.5 \times 0.24$$
$$= 15 \text{ lb}$$

Since 15 lb cork is 1728 in.^{3}, 1 lb cork will be $\frac{1}{15} \times$ 1728 in.^{3}, or 115.2 in.^{3}.

■ EXERCISES

Use Table 6 for specific gravity values to solve the following problems:

5-40   Find the weight of 1 in.^{3} of:
  (a) copper, rolled   (d) lead
  (b) gold   (e) nickel
  (c) cast iron   (f) silver

5-41   What is the weight of 1 quart of milk if its sp gr is 1.03? (A gallon of water weighs approximately 8.4 lb.)

5-42   An empty 1000-cm^3 flask weighs 75 g; when one-half is filled with water and the other with glycerin, it weighs 1205 g. What is the sp gr of the glycerin? (Remember, the specific gravity of water is 1 g/cm^3.)

5-43   What is the sp gr of a substance if 50 in.^{3} of the substance weighs 8 lb?

5-44   If the sp gr of sea water is 1.025 and that of ice 0.92, what fraction of an iceberg floating in the sea is underwater?

5-45   Find the number of liters in a vat 2 m by 75 cm by 50 cm. Find the weight in kilograms of the sulfuric acid (sp gr 1.84) required to fill it.

5-46   Find the value of 17 liters of sulfuric acid (sp gr 1.84) at $1.05 per kilogram.

5-47   A casting of iron when immersed in water displaces 2 quarts; find the weight of the casting.

5-48   A steel forging was found to displace 6.75 quarts of water; find the weight of the forging. (Use sp gr of steel = 7.85.)

5-49   The sp gr of petroleum is about 0.8. How many gallons of petroleum can be carried in a tank car whose capacity is 45,000 lb?

5-8   CONVERSION OF UNITS

Proportions can be used to convert a measurement in units of one system to those of another system. For instance, you can use one to translate a measurement in feet and inches to one in meters or centimeters.

EXAMPLE 5-9   Given that 1 ft = 30.58 cm, how long in centimeters is a $4\frac{1}{2}$-ft board?

Solution

$$\frac{x}{4\frac{1}{2}\text{ ft}} = \frac{30.58 \text{ cm}}{1 \text{ ft}}$$

$$x = \frac{4\frac{1}{2}\text{ ft} \times 30.58 \text{ cm}}{1 \text{ ft}} = 137.61 \text{ cm}$$

Note: *In this example, to determine an answer in centimeters, label all numbers, you will find that all units except the unit of the answer cancel out.*

$$\frac{x}{\text{ft}} = \frac{\text{cm}}{\text{ft}}$$

$$x = \frac{\overset{1}{\cancel{\text{ft}}} \times \text{cm}}{\underset{1}{\cancel{\text{ft}}}} = \text{cm}$$

EXAMPLE 5-10 If 1 in. = 2.54 cm and 1 cm = 0.01 m, how long in meters is a yardstick (36 in.)?

Solution

$$1 \text{ cm} = 0.01 \text{ m}$$
$$1 = 0.01 \text{ m/cm}$$
$$\frac{x}{36 \text{ in.}} = \frac{2.54 \text{ cm}}{1 \text{ in.}}$$
$$\frac{x}{36 \text{ in.}} = \frac{2.54 \text{ cm} \times 0.01 \text{ m/cm}}{1 \text{ in.}}$$
$$x = \frac{36 \text{ in.} \times 2.54 \text{ cm} \times 0.01 \text{ m/cm}}{1 \text{ in.}}$$
$$= 0.9144 \text{ m}$$

Note: *Again, all units except the unit of the answer (m) cancel out.*

The next example shows a shortcut. The principle used is the same, but some of the steps are left out.

EXAMPLE 5-11 If 1 in. = 2.54 cm and 1 cm = 0.01 m, how long is a yardstick (36 in.) in meters?

Solution Since 1 in. = 2.54 cm and 1 cm = 0.01 m, 1 in. = 0.0254 m.

$$1 \text{ in.} = 0.0254 \text{ m}$$
$$36 \text{ in.} = 36 \times 1 \text{ in.}$$
$$= 36 \times 0.0254 \text{ m}$$
$$= 0.9144 \text{ m}$$

■ EXERCISES

5-50 If 1 m = 39.37 in., convert the following measurements:
   (a) 1 ft
   (b) 3 ft
   (c) 10 ft
   (d) 5280 ft (1 mi)
   (e) 1000 m (1 km)
   (f) Your height in meters

5-51 If $1 \text{ ft}^2 = 0.0929 \text{ m}^2$, convert the following measurements:
  (a) A bedroom of 120 ft^2
  (b) A living room of 364 ft^2
  (c) A house of 2400 ft^2
  (d) A ball field of $15,000 \text{ ft}^2$
  (e) A desk top of 18 ft^2

5-52 If 1 quart = 0.95 liters, convert the following:
  (a) A gallon of milk (4 quarts = 1 gallon)
  (b) A pint (pt) of cream (2 pints = 1 quart)
  (c) A 4000-gallon pool
  (d) A well's production of 1.5 gallons of oil per minute

5-53 If 1 lb = 0.45 kg, convert the following:
  (a) Your weight in kilograms
  (b) A 3-lb cat
  (c) A 75-lb dog
  (d) A $1\frac{1}{2}$-ton car (1 ton = 2000 lb)
  (e) An 8-lb hammer

MEASUREMENT

6-1   INTRODUCTION TO MEASUREMENT

Simon Stevin, a Dutchman, invented decimal fractions in 1585. He felt it was the duty of governments to establish an international decimal base system of weights and measures. Four centuries later, the United States is the only major industrial country still operating largely on the English system of measurement. We use feet and inches; the rest of the world uses meters and millimeters.

Some of our tools and industrial products cannot be easily used by the rest of the world because of maintenance and repair problems. For instance, a French garage would not normally stock a $\frac{3}{8}$-in. bolt, a standard part in the United States. However, automobile makers throughout the world are standardizing many of the parts.

Due to the vast increase in world trade, individual companies, whole industries, and the federal government are gradually converting to a modernized form of the metric system called the *Systeme International* or SI. However, one should know both systems and be able to convert quantities from one system to the other.

6-2   INTRODUCTION TO THE METRIC SYSTEM

Gabriel Mouton, a Frenchman, proposed the metric system in 1670. The meter is the basic unit of measure in the metric system. The length of the meter was at first determined as one ten-millionth of the distance from the equator to the North Pole. However, the modern metric system of measurement is a decimal system based on the wavelength of the orange light given off by atoms of krypton gas. The meter is now defined as 1,650,763.73 times this wavelength. One meter (1 m) is approximately 39.37 in.; more exactly, 1 yd or 3 ft (or 36 in.) equals 0.9144 m.

The legal system of measurement in the United States is actually the metric system, because the inch, foot, and yard are defined in terms of the meter.

In the metric system the fundamental unit of length is the meter. From the meter are derived the unit of capacity, the liter; the unit of mass, the gram; and the unit of area in measuring land, the are. All other units are decimal subdivisions or multiples of these measures. For practical purposes, 1 cubic decimeter (cu dm) equals 1 liter; 1 liter of water equals 1 kilogram (kg), and 1 are is an area 10 m on a scale side.

6-3 SI PREFIXES

Common SI prefixes are listed in Table 6-1. Table 6-2 shows metric equivalents.

TABLE 6-1 SOME SI PREFIXES

Prefix	Abbreviation	Meaning	
Micro	μ	0.000 001	(10^{-6})
Milli	m	0.001	(10^{-3})
Centi	c	0.01	(10^{-2})
Deci	d	0.1	(10^{-1})
Base unit	—	1	(10^{0})
Deka	da	10	(10^{1})
Hekto	h	100	(10^{2})
Kilo	k	1000	(10^{3})
Mega	M	1 million	(10^{6})

TABLE 6-2 METER TABLE OF EQUIVALENTS

Meter Units		Comparison to Meter		Abbreviation
1 kilometer	=	1000	meters	km
1 hectometer	=	100	meters	hm
1 dekameter	=	10	meters	dam
1 meter	=	1	meters	m
1 decimeter	=	0.1	meters	dm
1 centimeter	=	0.01	meters	cm
1 millimeter	=	0.001	meters	mm

Prefixes added to a basic unit of measurement allow us to convert a measurement to larger or smaller units as needed. For example, the distance of 1000 meters is written as 1 kilometer (1 km). The word "kilometer" is formed from the words "meter" (the basic unit of length) and "kilo" (meaning 1000). Thus, a "kilometer" means "1000 meters." These prefixes are common to any type of measurement—weight, length, time, etc. A kilometer, kilogram, kilosecond, etc., are all 1000 times the basic unit of measurement.

A more complete listing of SI prefixes is found in Table 14 in the appendix.

6-4   BASE UNITS OF MEASUREMENT

There are seven basic types of measurements that we normally make. There is an SI base unit for each of these, as seen in Table 6-3.

TABLE 6-3   SI BASE UNITS

Measurement	Abbreviation	Metric Unit
Length	m	meter
Mass	kg	kilogram
Time	s	second
Electric current	A	ampere
Temperature	K	kelvin
Light intensity	cd	candela
Molecular substance	mol	mole

6-5   THE ENGLISH SYSTEM

The relationships between measures of length in the English system are as follows:

$$12 \text{ inches (in. or '')} = 1 \text{ foot (ft or ')}$$

$$3 \text{ ft} = 1 \text{ yard (yd)}$$

$$5\tfrac{1}{5} \text{ yd} = 1 \text{ rod}$$

$$320 \text{ rods} = 1 \text{ mile}$$

$$5280 \text{ ft} = 1 \text{ mile}$$

$$1760 \text{ yd} = 1 \text{ mile}$$

Lengths of less than 1 in. are expressed as fractions or decimal fractions.

Use these facts to change from one unit of measure to another. For example, what is the relationship between feet and rods? How many feet are in 1 rod?

$$5\tfrac{1}{2} \text{ yd} = 1 \text{ rod}$$

$$3 \text{ ft} = 1 \text{ yd}$$

Therefore

$$3 \times 5\tfrac{1}{2} = \text{ft in 1 rod}$$

$$3 \times 5\tfrac{1}{2} = 16\tfrac{1}{2}\text{ ft} = 1 \text{ rod}$$

6-6   ADDITION AND SUBTRACTION OF ENGLISH SYSTEM MEASURES

If lengths to be added or subtracted are all expressed in one unit of length, then addition or subtraction is carried on in the conventional way. However, if the lengths to be added or subtracted are expressed in different units, it is necessary to treat the addition or subtraction of the partial lengths as separate problems.

EXAMPLE 6-1   Add 3 ft 10 in. and 6 ft 9 in.

Solution   Since there are two units of length involved, place the numbers to be added in two columns and add each column:

$$
\begin{array}{l}
3 \text{ ft } 10 \text{ in.} \\
\underline{6 \text{ ft } \ 9 \text{ in.}} \\
9 \text{ ft } 19 \text{ in.}
\end{array}
$$

Because 19 in. is more than 1 ft, change the 19 in. into 1 ft 7 in., then add the 1 ft to the 9 ft already shown.

$$9 \text{ ft } 19 \text{ in.} = 9 \text{ ft} + 1 \text{ ft } 7 \text{ in.} = 10 \text{ ft } 7 \text{ in.}$$

6-7   ADDITION AND SUBTRACTION OF METRIC MEASURES

Measurements to be added or subtracted must all be in the same units; if they are not, they must be converted to a common unit. Metric measures can be converted by shifting the decimal point.

EXAMPLE 6-2   Subtract 2.50 cm from 18.77 m.

Solution

$$
\begin{array}{rl}
18.77 \text{ m} & = 18.770 \text{ m} \\
- \ 2.50 \text{ cm} & = \underline{-0.025 \text{ m}} \\
& \ 18.745 \text{ m}
\end{array}
$$

The same problem can be worked changing both given numbers to millimeters:

$$
\begin{array}{rl}
18.77 \text{ m} & = 18,770 \text{ mm} \\
2.50 \text{ cm} & = \underline{ 25 \text{ mm}} \\
& 18,745 \text{ mm}
\end{array}
$$

Since 1000 mm = 1 m, 18.745 mm = 18.745 m. Thus, this is the same answer as the first solution. The answers are the same but given in different units.

▪ EXERCISES

In Exercises 6-1 to 6-5, find the sum.

6-1 2.76 cm, 32.97 cm, 0.02 cm, and 4.07 cm.

6-2 1.27 m, 0.097 m, 5.963 m, and 0.003 m.

6-3 2.93 mm, 62.7 cm, 0.038 m.

6-4 11.23 km, 10.838 hm, 1073 dam.

6-5 0.83 dm, 90.2 cm, 197.3 mm.

6-6 Subtract 32.23 km from 46.07 km.

6-7 Subtract 327.3 mm from 675.9 mm.

6-8 Subtract 16.7 cm from 82.3 dm.

6-9 Subtract 78.9 dm from 127.3 m.

6-10 Subtract 50.6 hm from 6.93 km.

6-11 Subtract 8.93 mm from 2.54 cm.

6-12 Subtract 0.075 cm from 8.77 mm.

6-13 Subtract 0.932 dam from 506 dm.

6-14 Subtract 46.2 mm from 29.2 cm.

6-15 Subtract 69.3 dm from 767 cm.

6-16 A door is 194 cm in height. What is its height in meters?

6-8 MULTIPLICATION AND DIVISION OF ENGLISH SYSTEM MEASURES

If the measures to be multiplied or divided by a pure number are expressed in one unit of measurement, you can complete the multiplication or division in the conventional way. When the measures are expressed in more than one unit of measurement, multiply or divide each partial measure separately and convert to the proper unit of measurement.

EXAMPLE 6-3 Multiply 3 yd 2 ft by 6.

Solution Multiply and convert.

$$
\begin{aligned}
6(3 \text{ yd } 2 \text{ ft}) &= 18 \text{ yd } 12 \text{ ft} \\
&= 18 \text{ yd} + 4 \text{ yd} \\
&= 22 \text{ yd}
\end{aligned}
$$

EXAMPLE 6-4 Divide 13 yd 4 ft 4 in. by 4.

Solution Divide and convert.

$$
\begin{aligned}
\frac{13 \text{ yd } 4 \text{ ft } 4 \text{ in.}}{4} &= \frac{13 \text{ yd}}{4} + \frac{4 \text{ ft}}{4} + \frac{4 \text{ in.}}{4} \\
&= 3\tfrac{1}{4} \text{ yd } 1 \text{ ft } 1 \text{ in.} \\
&= 3 \text{ yd} + 9 \text{ in.} + 1 \text{ ft} + 1 \text{ in.} \\
&= 3 \text{ yd } 1 \text{ ft } 10 \text{ in.}
\end{aligned}
$$

Example 6-5 shows a second method for dividing 13 yds 4 ft 4 in. by 4.

EXAMPLE 6-5 Divide 13 yd 4 ft 4 in. by 4.

Solution Divide and convert.

$$
\begin{array}{r}
 3\text{ yd} \quad 1\text{ ft} \quad 10\text{ in.} \\
\hline
4)\ \ 13\text{ yd} \quad 4\text{ ft} \qquad 4\text{ in.} \\
-12\text{ yd} \\
\hline
1\text{ yd} = +3\text{ ft} \\
7\text{ ft} \\
-4\text{ ft} \\
\hline
3\text{ ft} = \ 36\text{ in.} \\
+\ 4\text{ in.} \\
\hline
40\text{ in.} \\
-40\text{ in.} \\
\hline
\end{array}
$$

Problems can be worked by changing the numbers to be multiplied or divided to one unit of measurement.

EXAMPLE 6-6 Divide 13 yd 4 ft 4 in. by 4.

Solution Convert, divide, and convert.

$$
\begin{array}{r}
13\text{ yd} = \ 468\text{ in.} \\
4\text{ ft} = \ \ 48\text{ in.} \\
4\text{ in.} = + \ \ \ 4\text{ in.} \\
\hline
520\text{ in.}
\end{array}
$$

Divide the 520 in. by 4:

$$
\begin{array}{r}
130\text{ in.} \\
\hline
4)\ 520\text{ in.}
\end{array}
$$

Now change 130 in. into feet and yards:

$$
\begin{array}{r}
3\text{ yd} \\
\hline
36\text{ in./yd})\ 130\text{ in.} \\
-108\text{ in.} \\
\hline
22\text{ in.}
\end{array}
$$

$$
\begin{array}{r}
1 \\
\hline
12\text{ in./ft})\ 22\text{ in.} \\
-12\text{ in.} \\
\hline
10\text{ in.}
\end{array}
$$

$$130\text{ in.} = 3\text{ yd } 1\text{ ft } 10\text{ in.}$$

The examples illustrate multiplication and division of measures of length by a pure number, that is, a number with no unit of measure. If two numbers to be multiplied have the same unit of measure (such as inches, feet, or yards), the resulting answer must be in square units (such as square inches, square feet, or square yards).

The area of the rectangle in Fig. 6-1 is found by multiplying 2 ft by 6 ft, and the answer is 12 ft^2. The rectangle includes 12 squares whose sides measure 1 ft each.

Fig. 6-1. Area of a rectangle.

Note: We cannot convert the 6-ft dimension to 2 yd unless we also convert the 2 ft to $\frac{2}{3}$ yd. The area of the rectangle will be 12 ft² or 1$\frac{1}{3}$ yd². When computing an area, do not mix different units of measurement.

EXAMPLE 6-7 What is the area of a rectangle whose sides are 500 ft and 1 mi?

Solution Area equals length times width, but first convert miles to feet.

$$A = L \times W$$

$$= 500 \text{ ft} \times \left(1 \cancel{\text{mi}}^{1} \times 5280 \frac{\text{ft}}{\cancel{\text{mi}}_{1}} \right)$$

$$= 2,640,000 \text{ ft}^2$$

Check:

$$A = L \times W$$

$$= \left(500 \cancel{\text{ft}}^{1} \times \frac{1 \text{ yd}}{3 \cancel{\text{ft}}_{1}} \right) \times \left(1 \cancel{\text{mi}}^{1} \times \frac{1760 \text{ yd}}{1 \cancel{\text{mi}}_{1}} \right)$$

$$= \frac{880,000}{3} \text{ yd}^2$$

$$= 293,333\tfrac{1}{3} \text{ yd}^2$$

There are 9 ft² to 1 yd², and 293,333$\frac{1}{3}$ yd² = 2,640,000 ft².

EXAMPLE 6-8 Find the length of a rectangle with an area of 24 ft² and a width of 48 in.

Solution Solve the area formula for length and solve.

$$\text{Area} = \text{length} \times \text{width}$$

$$\text{Length} = \frac{\text{area}}{\text{width}}$$

$$= \frac{24 \text{ ft}^2}{\left(\frac{48 \cancel{\text{in.}}}{1} \right)\left(\frac{1 \text{ ft}}{12 \cancel{\text{in.}}} \right)}$$

$$= \frac{24 \text{ ft}^2}{4 \text{ ft}}$$

$$= 6 \text{ ft}$$

6-9   MULTIPLICATION AND DIVISION OF METRIC MEASURES

The multiplication and division of measures in the metric system is similar to the English system. If the numbers to be multiplied or divided are expressed in common units of measurement, the multiplication or division can be accomplished in the conventional way. If the numbers to be multiplied or divided are expressed in different units of measurement, change all dimensions to the same unit.

EXAMPLE 6-9   Multiply 62 m by 5.

Solution   Multiply.

$$
\begin{array}{r}
62 \text{ m} \\
\times 5 \\
\hline
310 \text{ m}
\end{array}
$$

EXAMPLE 6-10   Divide 21 cm by 3.

Solution   Divide.

$$21 \text{ cm} \div 3 = 7 \text{ cm}$$

EXAMPLE 6-11   Multiply 17 mm by 3 mm.

Solution   Multiply.

$$
\begin{array}{r}
17 \text{ mm} \\
\times 3 \text{ mm} \\
\hline
51 \text{ mm}^2
\end{array}
$$

EXAMPLE 6-12   Divide 75 cm² by 15 cm.

Solution   Divide.

$$\frac{\overset{5 \text{ cm}}{\cancel{75 \text{ cm}^2}}}{\underset{1}{\cancel{15 \text{ cm}}}} = 5 \text{ cm}$$

EXAMPLE 6-13   Multiply 16 mm by 5 cm.

Solution   These two lengths may not be multiplied, since they are given in different units of length—millimeters and centimeters. The first step is to convert one of the lengths into the units of the other.

$$16 \text{ mm} \times \left(\overset{1}{\cancel{5 \text{ cm}}} \times \frac{10 \text{ mm}}{\underset{1}{\cancel{1 \text{ cm}}}} \right) = 16 \text{ mm} \times 50 \text{ mm}$$
$$= 800 \text{ mm}^2$$

or

$$\left(\overset{1}{\cancel{16 \text{ mm}}} \times \frac{1 \text{ cm}}{\underset{1}{\cancel{10 \text{ mm}}}} \right) \times 5 \text{ cm} = \frac{16 \text{ cm} \times 5 \text{ cm}}{10}$$
$$= 8 \text{ cm}^2$$

The answers can be given in different units of measurement: square centimeters and square millimeters.

▧ EXERCISES

6-17 Multiply 4 yd 1 ft by 7.

6-18 Multiply 6 yd 2 ft by 19.

6-19 Multiply 3 m by 16.

6-20 Multiply 3 cm 9 mm by 21.

6-21 Multiply 6 yd 2 ft by 3 ft.

6-22 Multiply 12 yd 1 ft 6 in. by 2 ft.

6-23 Multiply 2 m 23 cm by 3 m.

6-24 Multiply 5 dam 2 m by 5 cm.

6-25 Multiply 6 dm 5 cm by 13 m 2 dm.

6-26 A rectangle includes 36 in.^{2} One dimension is 9 in. What is the other dimension?

6-27 A rectangle includes 102 cm^2. One dimension is 45 mm. What is the other dimension?

6-28 A rectangle includes 105 m^2. One dimension is 25 dm. What is the other dimension?

6-29 Determine the area in square meters of a plot of land 312 m by 500 m.

6-10 CONVERSION BETWEEN THE ENGLISH AND METRIC SYSTEMS

In measuring a distance, you can use a meter stick, a yardstick, or any other commonly understood length (hands, rods, paces, etc.). The only thing you need is a *standard length* against which to measure this distance.

The method of changing measures of length from one system to another is based on these relationships:

$$1 \text{ in.} = 2.54 \text{ cm (exactly)}$$

$$1 \text{ ft} = 0.3048 \text{ m (exactly)}$$

$$1 \text{ yd} = 0.9144 \text{ m (exactly)}$$

$$1 \text{ m} = 39.37 \text{ in. (approximately)}$$

You can convert between any two standards if you know the ratio between them. These ratios or conversion factors can be used in the following way:

Given: 1 m = 3.28 ft

$$1 \text{ m} = 3.28 \text{ ft}$$

$$1 \text{ ft} = \frac{1}{3.28} \text{ m}$$

$$1 \text{ ft} = 0.305 \text{ m}$$

EXAMPLE 6-14   Convert 7 m to feet.

Solution
$$1 \text{ m} = 3.28 \text{ ft}$$
$$7 \text{ m} = 7 \times 1 \text{ m}$$
$$= 7 \times 3.28 \text{ ft}$$
$$= 22.96 \text{ ft}$$

EXAMPLE 6-15   Convert 12 ft to meters.

Solution
$$1 \text{ ft} = 0.3048 \text{ m}$$
$$12 \text{ ft} = 12 \times 1 \text{ ft}$$
$$= 12 \times 3.6576 \text{ m}$$
$$= 3.6576 \text{ m}$$

EXAMPLE 6-16   The 100-m dash is one of the events included in each Olympic track meet. How many yards are there in the 100-m dash?

Solution
$$1 \text{ yd} = 0.9144 \text{ m}$$
$$100 \text{ m} = 100 \text{ m} \left(\frac{1 \text{ yd}}{0.9144 \text{ m}} \right)$$
$$= \frac{100 \text{ yd}}{0.9144 \text{ m}}$$
$$= 109.36 \text{ yd}$$

Table 6-4 is a list of common conversion factors for metric and English units of measurement. Additional conversion factors are also listed in the appendix in Table 14.

TABLE 6-4   CONVERSION FACTORS

Length	
Feet (ft)	× 0.305 = meters (m)
Inches (in.)	× 25.4 = millimeters (mm)
Kilometers (km)	× 1094 = yards (yd)
Meters (m)	× 3.28 = feet (ft)
	× 39.4 = inches (in.)
	× 1.09 = yards (yd)
Yards (yd)	× 0.91 = meters (m)
Miles (mi)	× 1.61 = kilometers (km)

Area

Acres	$\times$ 4046.9 = square meters (m^2)
	$\times$ 43,560 = square feet (ft^2)
Square feet (ft^2)	$\times$ 0.093 = square meters (m^2)
Square inches $(in.^2)$	$\times$ 6.45 = square centimeters (cm^2)
Square meters (m^2)	$\times$ 10.76 = square feet (ft^2)
Square centimeters (cm^2)	$\times$ 0.155 = square inches $(in.^2)$

Weight

Pounds (lb) $\times$ 0.45 = kilograms (kg)

Kilograms (kg) $\times$ 2.2 = pounds (lb)

Speed/Acceleration

Feet/second (ft/s) $\times$ 1.1 = kilometers/hour (km/h)

Kilometers/hour (km/h) $\times$ 0.621 = miles/hour (mph)

Feet/second^2 (ft/s^2) $\times$ 0.3 = meters/second^2 (m/s^2)

Miles/hour $\times$ 1.61 = kilometers/hour

Volume

Cubic feet (ft^3) $\times$ 0.028 = cubic meters (m^3)

    $\times$ 28 = liters (l)

Cubic inches $(in.^3)$ $\times$ 16,387 = cubic millimeters (mm^3)

    $\times$ 16,387 = milliliters (ml)

Cubic meters (m^3) $\times$ 35.3 = cubic feet (ft^3)

Cubic millimeters (mm^3) $\times$ 0.00006 = cubic inches $(in.^3)$

Cubic yards (yd^3) $\times$ 0.76 = cubic meters (m^3)

Gallons (gal) $\times$ 3.8 = liters (l)

Liters (l) $\times$ 0.26 = gallons (gal)

▨ EXERCISES

6-30 Convert the following measurements to meters:

(a) 1417 mm	(f) 2673 mm
(b) 17 in.	(g) 139 in.
(c) 14 ft	(h) 2 ft
(d) 7 yd	(i) 4 yd
(e) 0.75 km	(j) 0.20 km

6-31 Convert the following measurements to feet:

(a) 18 in.	(e) 13 m
(b) 6.7 m	(f) 1 m
(c) 3.6 yd	(g) 4.5 yd
(d) 7 in.	(h) 12 in.

6-32 Convert the following measurements as indicated:
(a) 3.7 acres to square meters
(b) 1 acre to square meters
(c) 327 ft^2 to square meters
(d) 21 in.^{2} to square centimeters
(e) 1 m^2 to square feet
(f) 140 cm^2 to square inches
(g) 80 ft/s to kilometers per hour
(h) 55 mph to kilometers per hour
(i) 1 in.^{3} to cubic millimeters
(j) 15 gallons to liters
(k) 8 yd^3 to cubic meters
(l) 1728 in.^{3} to cubic millimeters

6-33 Number 24 gage sheet steel is 0.584 mm thick. Find its thickness in inches to the nearest thousandth.

6-34 What is the diameter in inches of the barrel of a 90-mm gun?

6-35 If 1 ft^3 water weighs 62.5 lb, how much will 1 liter of water weigh?

6-36 Calculate the capacity in liters of a rectangular tank 2 m by 9 dm by 8 dm.

6-37 How much time will be required for a pump delivering 2.75 gallons (gal) per stroke and making 84 strokes per minute to pump 500 barrels (bbl) of oil (1 barrel = 31.5 gal)?

6-38 Change a pressure of 14.5 oz/in.^{2} to pounds per square foot.

6-39 Change a speed of 88 ft/s to an equivalent speed in meters per second.

6-40 Find the side of a square rug whose area is 64 ft^2. Calculate the result to the nearest 0.1 cm.

6-41 Find to the nearest cm the side of a square tabletop 16 ft^2 in area.

6-42 A farmer wants to grow 180 lb of seed per acre. The seed has an 85% germination rate and is 90% pure. How much seed should be spread on a 17-acre field?

6-43 The resistance in a piece of copper wire is directly proportional to the length of the wire. If a 1000.0-ft piece of wire has a resistance of 2.525 ohms, how much resistance will there be in a 1-km piece of the same wire.

6-44 An iron casting weighs 142 lb and costs $373. An alloy replacement casting weighs 39 kg and costs $248. Which casting costs more per unit of weight?

6-45 If sound travels 6160 ft in 5½ s, how long will a noise take to travel 10 km?

6-46 A particular casting can be made from iron coating $15.24 per 100 lb or from an alloy costing $27.40 per 100 kg. Which is cheaper to produce?

6-47 The cross-sectional area of a steel wire is 0.976 in.^{2}. If a load of 35,575 kg is applied to the wire, what would be the load per square inch? Per square centimeter? If the wire is 1 ft long and is stretched 0.0175 mm, what would be the stretch per inch? Per centimeter?

6-48 A drill makes 280 rpm and can drill a hole through a 2-in. piece of metal in $\frac{4}{5}$ min. How long will it take to drill through 10 mm of the same metal?

6-49 A drill makes 120 rpm and has a feed of 0.3 mm. (Feed is the distance a drill advances with each revolution.) How long will it take to drill 12 holes in a cylinder head 1½ in. thick if 1 min is allowed for setting each hole?

6-50 Number 25 (B. & S.) gage sheet copper is 0.0179 in. thick and weighs 0.811 lb/ft^2.
(a) How thick would a pile of 1000 of these sheets be to the nearest centimeter?

(b) How many sheets would be needed to make a pile 1 m high?

(c) How many sheets would be needed to make up 1000 kg if each sheet has an area of 6.25 ft²?

6-11   PRECISION

A number used in counting is considered to be an exact figure. We have exactly 5 apples, or $10.35.

When we work with measurements, however, we are not handling exact numbers. For instance, suppose a machinist measures a stainless steel rod as 25 cm long. This rod may be 25 cm long, but more likely it is slightly longer or shorter. The *degree of precision* in such measurements depends on the precision of the measuring instrument and the skill of the person doing the measuring.

EXAMPLE 6-17   How long is line *a* in Fig. 6-2?

Fig. 6-2. Example 6-18.

Solution 1   If we only need an approximate measurement, line *a* could be described as 4 cm long. This measurement is accurate to the nearest 1 cm.

Solution 2   A more exact measurement would indicate that line *a* is 4.4 cm long. This measurement is accurate to the nearest 0.1 cm.

Solution 3   If we take a close look, we can see that line *a* is 4.43 cm long (we are estimating that the interval is $\frac{3}{10}$ of the way from 4.4 to 4.5 cm). This measurement is accurate to the nearest 0.01 cm.

Solution 4   We cannot make a more precise reading with this ruler, which is divided into tenths of a centimeter. However, a precision measuring instrument could tell us that line *a* is 4.4285 cm. This measurement is accurate to the nearest 0.0001 cm.

Note: *The precision of a measurement is indicated by the placement of the last reliable digit in relation to the decimal point.*

6-12   PRECISION AND ACCURACY OF MEASUREMENTS

No matter how precise a measurement is, error will occur. There are two ways of evaluating this error.

Absolute error indicates the magnitude of the error. It is found by subtracting the true value from the measured value of a reading and disregarding the + or − sign of the answers.

Relative error is the ratio of the absolute error to the true value. It is often expressed as a percentage.

EXAMPLE 6-18 The actual length of a brass rod is 4.42857 cm. Find the absolute and relative error of a measurement of 4.43 cm.

Solution

$$\text{Absolute error} = |M_r - A_r|$$
$$= |4.43 \text{ cm} - 4.42857 \text{ cm}|$$
$$= 0.00143 \text{ cm}$$

$$\text{Relative error} = \frac{\text{absolute error}}{A_r}$$
$$= \frac{0.00143 \text{ cm}}{4.42857 \text{ cm}}$$
$$= 0.03\%$$

6-13 SIGNIFICANT FIGURES

Rule: *Significant figures are those digits which are known to be reliable. The position of the decimal point does not determine the number of significant figures.*

EXAMPLE 6-19 How many significant figures are there in a measurement of 1.35 in.?

Solution There are three significant figures: 1, 3, and 5.

EXAMPLE 6-20 How many significant figures are there in a measurement of 0.000135?

Solution There are again three significant figures: 1, 3, and 5. The three zeros are used only to place the decimal point.

EXAMPLE 6-21 How many significant figures are there in a measurement of 103,500?

Solution There are four significant figures: 1, 0, 3, and 5. The remaining two zeros are used to place the decimal point.

EXAMPLE 6-22 How many significant figures are in 27,000.0?

Solution There are six significant figures: 2, 7, 0, 0, 0, 0. In this case, the .0 means that the measurement is precise to $\frac{1}{10}$ unit. The zeros indicate measured values and are not used solely to place the decimal point.

The accuracy of measurement can be inferred from the number of significant figures. This can be seen in Table 6-5.

TABLE 6-5 ACCURACY

Case	Significant Figures	Measurement	Mismeasurement	Absolute Error	Accuracy
1	1	3000.	2000	1000	33%
2	1	3	2	1	33%
3	1	.003	.002	.001	33%
4	2	4500	4400	100	2.2%
5	2	45	44	1	2.2%
6	2	.0045	.0044	.0001	2.2%
7	3	6780	6770	10	0.1%
8	3	67.8	67.7	.1	0.1%
9	3	.00678	.00677	.00001	0.1%

Note: *The mismeasurement in the fourth column is by one significant digit. If the mismeasurement of case 1 had lowered the value from 3000 to 2999, we would have been dealing with a measurement accurate to four significant digits: 3, 0, 0, and 0.*

EXERCISE

6-51 Which measurement is the most accurate? The least accurate? If each measurement was too great by one unit, what would be the absolute error? The relative error?

(a) 283,000 mi   (c) 0.00036 cm   (e) 3.28 ft
(b) 2760 ft   (d) 25.4 cm   (f) 1.609 km

6-14 ROUNDING OFF A NUMBER

Rule: *A number is rounded off by dropping one or more numbers from the right, and adding zeros if necessary to place the decimal point. If the last figure dropped is 5 or more, increase the last retained figure by 1. If the last figure dropped is less than 5, do not increase the last retained figure.*

EXAMPLE 6-23 Round off 10,547 to 4, 3, 2, and 1 significant figures.

Solution

$$10{,}547 = 10{,}550 \text{ to 4 significant figures}$$
$$10{,}547 = 10{,}500 \text{ to 3 significant figures}$$
$$10{,}547 = 11{,}000 \text{ to 2 significant figures}$$
$$10{,}547 = 10{,}000 \text{ to 1 significant figure}$$

6-15 MAINTAINING ACCURACY WHILE MULTIPLYING OR DIVIDING MEASUREMENTS

When multiplying or dividing measurements, there is a tendency to increase the number of decimal places (indicating precision) or the number of significant digits (indicating accuracy) in the answer. This should not be done, because accuracy cannot be created by calculation.

EXAMPLE 6-24 A block of lead measures 1.7 ft × 1.4 ft × 1.1 ft. If lead weighs 712 lb/ft^3, what does this block weigh?

Solution Multiply length × width × height × weight per cubic foot.

$$\begin{aligned} W &= (1.7 \times 1.4 \times 1.1) \times (712) \\ &= (2.618) \times (712) \\ &= 1864.016 \text{ lb} \\ &= 1900 \text{ lb} \end{aligned}$$

Note: *The weight 1864.016 lb indicates this is a measurement precise to 0.001 lb. This is not true; 1900 lb indicates the true accuracy of the data using only two significant figures. The measurements were accurate to only two significant digits; thus the results can be accurate to only two significant digits.*

Rule: *When multiplying or dividing measurements, answers should be rounded off to the same number of significant digits as are in the* ***least accurate*** *factor of the product.*

EXAMPLE 6-25 Multiply 103.56 cm × 0.00013 cm.

Solution Multiply and round.

$$\begin{aligned} 103.56 \text{ cm} \times 0.00013 \text{ cm} &= 0.0134628 \text{ cm}^2 \\ &= 0.013 \text{ cm}^2 \end{aligned}$$

EXAMPLE 6-26 Multiply 103.56 × 13,010,000.

Solution Multiply and round.

$$\begin{aligned} 103.56 \times 13{,}010{,}000 &= 1{,}347{,}315{,}600 \\ &= 1{,}347{,}000{,}000 \end{aligned}$$

EXAMPLE 6-27 Divide 103.56 by 3.791.

Solution Divide and round.

$$\begin{aligned} 103.56 \div 3.791 &= 27.31733 \\ &= 27.32 \end{aligned}$$

EXAMPLE 6-28 What is 10 percent of 123.45 cm?

Note: *10 percent (or 0.10) is an exact, not an approximate, number.*

Solution

$$\begin{aligned} 10 \text{ percent of } 123.45 &= 0.10 \times 123.45 \text{ cm} \\ &= 12.345 \text{ cm} \end{aligned}$$

EXERCISES

6-52 What is the volume of a cube measuring 1.06 cm per edge?

6-53 What is the area of a rectangular tract of land measuring 151 ft × 310 ft?

6-54 What volume of metal is in a sheet of tin measuring 36 in. × 72 in. × 0.04 in.?

6-55 If a 32.0-in. iron rod is cut into three equal lengths, and each cut wastes $\frac{1}{32}$ in., how long will each rod be? If the rod were 32.00 in.?

6-56 What are the precision and accuracy of the following measurements?
  (a) 0.003 cm   (e) 30 cm   (i) 7.5 in.
  (b) 0.03 cm   (f) 300 cm   (j) 75 in.
  (c) 0.3 cm   (g) 0.075 in.   (k) 750 in.
  (d) 3. cm   (h) 0.75 in.

POWERS AND ROOTS

7-1 POWERS

When several numbers are multiplied together, as $3 \times 4 \times 6 = 72$, the numbers 3, 4, and 6 are the *factors;* 72 is the *product.* If all the factors are alike, for example, $3 \times 3 \times 3 \times 3 = 81$, the product is a *power.* Thus, 81 is a power of 3, and 3 is the *base* of the power. A power is a product obtained by using a base a certain number of times as a factor.

If the base number is used twice as a factor, the product is the *second power* or *square;* if the base is used three times as a factor, the product is the *third power* or *cube;* if the base is used four times as a factor, the product is the *fourth power;* and so on, any number of times.

Instead of writing $3 \times 3 \times 3 \times 3$, it is more convenient to use an *exponent* to indicate that the factor 3 is used as a factor four times. The exponent, a small number placed above and to the right of the base number indicates how many times the base is to be used as a factor. Using this system of notation, the multiplication $3 \times 3 \times 3 \times 3$ is written as 3^4. The 4 is the exponent, showing that 3 is to be used as a factor four times.

EXAMPLE 7-1 Rewrite $10 \times 10 \times 10$ and compute the product.

Solution Write using an exponent and evaluate.

$$10 \times 10 \times 10 = 10^3 = 1000$$

EXAMPLE 7-2 Rewrite $2 \times 2 \times 2 \times 2 \times 2$ and compute the product.

Solution Write using an exponent and evaluate.

$$2 \times 2 \times 2 \times 2 \times 2 = 2^5 = 32$$

EXAMPLE 7-3 Rewrite $3 \times 3 \times 4 \times 4 \times 4$ and compute the product.

Solution Write using exponents and evaluate.

$$3 \times 3 \times 4 \times 4 \times 4 = 3^2 \times 4^3$$
$$= 9 \times 64$$
$$= 576$$

EXAMPLE 7-4 Write 12 squared and 12 cubed and evaluate.

Solution Write using exponents and evaluate.

$$12 \text{ squared} = 12^2 = 12 \times 12 = 144$$
$$12 \text{ cubed} = 12^3 = 12 \times 12 \times 12 = 1728$$

■ EXERCISES

7-1 Rewrite the following and compute each product:
(a) 9×9
(b) $7 \times 7 \times 7$
(c) $11 \times 11 \times 11 \times 11$

7-2 Calculate the square and cube of each of the following:
(a) 6
(b) 10
(c) 13

7-3 Calculate the fourth power of 6.

7-4 Calculate the value of each of the following: 792^2; 35^3; 3^4; 2^{16}.

7-2 ROOTS

Roots of numbers are the reverse of powers. In a product of equal factors, the repeated factor is a *root* of the product. The *square root* of a number is one of the two equal factors of that number. Thus 3 is the square root of 9 because $3 \times 3 = 9$. The *cube root* of a number is one of three equal factors of that number. The *fourth root* is one of the four equal factors, and so on for higher roots. For example:

The cube root of 64 is 4 because $4 \times 4 \times 4 = 64$.

The fourth root of 81 is 3 because $3 \times 3 \times 3 \times 3 = 81$.

The fifth root of 32 is 2 because $2 \times 2 \times 2 \times 2 \times 2 = 32$.

7-3 RADICAL SIGN AND INDEX OF ROOT

The sign $\sqrt{}$ indicates a root and is called the *radical sign*. A small number called the *index* of the root is placed in the opening of the radical sign to signify the root to be found. Thus $\sqrt[3]{64}$ indicates that the cube root of 64 is to be calculated; the small 3 is the index of the root. Because the square root is the most frequently calculated root, its index (2) is omitted.

The square root of 625 is written $\sqrt{625}$, not $\sqrt[2]{625}$. Higher roots are written $\sqrt[4]{243}$, $\sqrt[7]{128}$, and so forth.

When you calculate a square root, you are answering the question: "What number multiplied by itself equals the original number?" Thus the square root of 625, $\sqrt{625}$, is 25 because $25 \times 25 = 625$.

To understand the relationship between powers and roots, study these statements:

$$3 \times 3 = 3^2 = 9 \qquad \sqrt{9} = 3$$
$$3 \text{ squared} = 9 \qquad \text{the square root of } 9 = 3$$
$$4 \times 4 \times 4 = 4^3 = 64 \qquad \sqrt[3]{64} = 4$$
$$5 \times 5 \times 5 \times 5 = 5^4 = 625 \qquad \sqrt[4]{625} = 5$$
$$5 \times 5 = 5^2 = 25 \qquad \sqrt{25} = 5$$

The square root of 25 equals the fourth root of 625.

7-4   SQUARE ROOT

The numbers 1, 4, 9, 16, 25, 36, 49, 64, 81 are the squares of the numbers 1, 2, 3, 4, 5, 6, 7, 8, 9, respectively. These are well worth remembering because these are the only whole numbers less than 100 of which we can find integral square roots. Numbers such as these are called *perfect squares*. Among numbers above 100, perfect squares become still more scarce.

For most numbers, integral square roots cannot be found. The square root of 625 is 25, but the square root of 56 can be expressed neither as a whole number, nor as an exact decimal. However, it can be computed to any desired number of decimal places.

7-5   USE OF THE TABLE OF SQUARE ROOTS

The hand-held calculator is the best tool for finding square roots. However, if you do not have a calculator, the following sections show both how to obtain square roots by using a table and how to calculate square root.

Using a table of square roots saves much time and labor. Table 8 in the appendix gives the square root of any number of not more than four significant figures, correct to four figures; that is, the square root can be found for any number from 1 to 9999, with the decimal point in any position. In working with numbers of four figures, the last figure of a root will occasionally be found to be too large or too small.

The first two figures of a number whose root is to be found are in the N column; the third figure is at the top of the page; and the fourth figure, when different from zero, is at the top of the righthand columns. The following examples illustrate the use of the table.

EXAMPLE 7-5 Find $\sqrt{3.876}$.

Solution

1. Locate 3.8 in the N column of Table 8.
2. Locate 7 at top of the table.
3. Read 1.967 to the right of 3.8 and in the column headed with a 7.
4. The $\sqrt{3.87}$ then equals 1.967.
5. To find $\sqrt{3.876}$ we add to the 3.87 the number found in the righthand column headed 6.
6. The number under the 6 and to the right of the $\sqrt{3.87}$ is 2. This 2 is 0.002 and is to be added to the $\sqrt{3.87}$.
7. $1.967 + 0.002 = 1.969$.
8. $\sqrt{3.876} = 1.969$.

EXAMPLE 7-6 Determine $\sqrt{38.76}$.

Solution

1. Locate 38 in the N column in Table 8.
2. Locate the 7 at the top of the table.
3. Read 6.221 to the right of the 38 and under the 7.
4. The $\sqrt{38.7}$ equals 6.221.
5. To find $\sqrt{38.76}$, we add to the 6.221 the number in the righthand column headed 6.
6. This number is 5 and is equal to 0.005, and the 0.005 is added to the value for $\sqrt{38.7}$.
7. $6.221 + 0.005 = 6.226$.
8. $\sqrt{38.76} = 6.226$.

EXAMPLE 7-7 Determine $\sqrt{387.6}$.

Solution

1. Change the number from 387.6 to 3.876 by dividing by 100.
2. The problem after this change reads: Find $\sqrt{3.876} \times \sqrt{100}$.
3. The square root of 100 is 10.
4. The problem now reads: Find $\sqrt{3.876} \times 10$.
5. $\sqrt{3.876} = 1.969$.
6. $\sqrt{387.6} = 1.969 \times 10$.
7. $\sqrt{387.6} = 19.69$.

EXAMPLE 7-8 Determine $\sqrt{3876}$.

Solution

1. Change the number from 3876 to 38.76 by dividing by 100.
2. The problem after this change reads: Find $\sqrt{38.76} \times \sqrt{100}$.

3. $\sqrt{100} = 10$.

4. The problem now reads: Find $\sqrt{38.76} \times 10$.

5. Extract the square root of 38.76: $\sqrt{38.76} = 6.226$.

6. Multiply the result by 10: $6.226 \times 10 = 62.26$.

7. $\sqrt{3876} = 62.26$.

EXAMPLE 7-9 Determine the $\sqrt{0.006\ 93}$.

Solution

1. Change the number 0.006 93 to 69.3 by multiplying by 10,000.

2. The problem after this multiplication reads: Find $\sqrt{69.3} \div \sqrt{10,000}$.

3. $\sqrt{10,000} = 100$.

4. The value of $\sqrt{69.3}$ as determined by using the table is 8.325.

5. $\sqrt{69.3} \div \sqrt{10,000} = 8.325 \div 100 = 0.083\ 25$.

6. $\sqrt{0.006\ 93} = 0.083\ 25$.

The calculation of a square root may be checked by the multiplication of the answer by itself.

You should observe the following from the examples and discussion:

1. The square root for a number larger than 1 is less than the original number.

2. The square root for a number less than 1 is larger than the original number.

3. The numbers 100, 10,000, and 1,000,000 have integral square roots:

$$\sqrt{100} = 10 \qquad \sqrt{10,000} = 100 \qquad \sqrt{1,000,000} = 1000$$

4. The use of a table to find the square root of a number not within the range of the table is made possible by multiplying or dividing the original number by 100, 10,000, 1 million, etc. This is the first step when computing a square root with the assistance of a table—unless the number is already within the range of the table.

▧ EXERCISES

Determine the square root of each of the following using Table 8.

7-5	3025	**7-13**	3056	**7-21**	45.38
7-6	9604	**7-14**	0.5140	**7-22**	76.35
7-7	1322	**7-15**	0.001 225	**7-23**	48.9
7-8	2102	**7-16**	97.81	**7-24**	4567
7-9	5476	**7-17**	900.9	**7-25**	12.12
7-10	1188	**7-18**	1.41	**7-26**	1001
7-11	1464	**7-19**	1.414	**7-27**	1113
7-12	2340	**7-20**	141.4	**7-28**	0.1234

7-29 0.0321	**7-32** 0.1909	**7-35** 60,000
7-30 0.0079	**7-33** 2900	**7-36** 930,800
7-31 0.0202	**7-34** 67,340	

7-6 DETERMINING SQUARE ROOTS BY CALCULATION

Calculation of a square root can best be illustrated by several examples. The first example shows the complete process.

EXAMPLE 7-10 Calculate the square root of 625.

Solution

$$
\begin{array}{r}
2 \\
\times 20 \\
\hline
40 \\
+5 \\
\hline
45 \\
\times 5 \\
\hline
225
\end{array}
\qquad
\begin{array}{r}
2\ 5 \\
\sqrt{6\text{'}25} \\
4 \\
\hline
2\ 25 \\
2\ 25 \\
\hline
\end{array}
$$

Explanation:

1. Beginning at the decimal point, separate the number into "periods" of two figures each. An apostrophe marks these periods:

$$\sqrt{6\text{'}25}$$

2. Then find the largest integer whose square is equal to or less than the first period (6). Place this integer, 2, above the 6. Next, square the 2 and place the 4 (2 squared) below the 6:

$$
\begin{array}{r}
2 \\
\sqrt{6\text{'}25} \\
4
\end{array}
$$

3. Subtract the 4 from the 6 as in long division:

$$
\begin{array}{r}
2 \\
\sqrt{6\text{'}25} \\
4 \\
\hline
2
\end{array}
$$

4. Bring down the next period:

$$
\begin{array}{r}
2 \\
\sqrt{6\text{'}25} \\
4 \\
\hline
2\ 25
\end{array}
$$

5. Multiply 20 by the number above the radical sign:

$$
\begin{array}{r}
2 \\
\times 20 \\
\hline
40
\end{array}
$$

6. Then divide the result of this multiplication, 40, into the 225.

7. Add the result of this estimated division, 5, to the 40:

$$
\begin{array}{rr}
& 2 \\
\times 20 \\
\hline
40 \\
+5 \\
\hline
45
\end{array}
\qquad
\begin{array}{r}
2 \\
\sqrt{6\,'25} \\
\underline{4} \\
2\,25
\end{array}
$$

8. Now divide 225 by 45 and place the result, 5, above the 25; multiply the 45 by 5 and place the product under the 225:

$$
\begin{array}{rr}
& 2 \\
\times 20 \\
\hline
40 \\
+5 \\
\hline
45 \\
\times 5 \\
\hline
225
\end{array}
\qquad
\begin{array}{r}
2\ 5 \\
\sqrt{6\,'25} \\
\underline{4} \\
2\,25 \\
2\,25
\end{array}
$$

9. If the result of the multiplication of 5 × 45 had been larger than 225, it would have been necessary to try the next smaller number.

EXAMPLE 7-11 Calculate $\sqrt{522{,}729}$.

Solution

$$
\begin{array}{rr}
& 7 \\
\times 20 \\
\hline
140 \\
+2 \\
\hline
142 \\
\times 2 \\
\hline
284 \\
\\
72 \\
\times 20 \\
\hline
1440 \\
+3 \\
\hline
1443 \\
\times 3 \\
\hline
4329
\end{array}
\qquad
\begin{array}{r}
7\ \ 2\ \ 3 \\
\sqrt{52\,'27\,'29} \\
\underline{49} \\
3\,27 \\
\underline{2\,84} \\
43\,29 \\
\underline{43\,29}
\end{array}
$$

Explanation:

1. Separate 522,729 into periods of two numbers.

2. Take the largest possible integral square root of the first period, 52. This square root is 7.

3. Place 7 above 52, square that 7, and place 49 below 52.

4. Subtract 49 from 52; bring down the next period, 27.

5. Multiply the number above the radical sign, 7, by 20.

6. Trial-divide this product, 140, into 327.

7. Add the trial divisor, 2, to 140 and multiply this sum: 142 by 2.

8. Place 2 above 27 and subtract the product: 284 from 327.

9. Bring down the next period, 29.

10. Multiply the number above the radical sign 72, by 20.

11. Trial-divide this product, 1440, into 4329.

12. Add 3 to 1440 and place 3 above 29.

13. Multiply 1443 by 3 and place this product under 4329.

The previous explanation also applies to the following problem.

EXAMPLE 7-12   Calculate $\sqrt{6,780.816}$.

Solution

$$
\begin{array}{r}
2 \\
\times 20 \\
\hline
40 \\
+6 \\
\hline
46 \\
\times 6 \\
\hline
276 \\
\\
26 \\
\times 20 \\
\hline
520 \\
\\
260 \\
\times 20 \\
\hline
5200 \\
+4 \\
\hline
5204 \\
\times 4 \\
\hline
20{,}816
\end{array}
\qquad
\begin{array}{r}
2\ 6\ 0\ 4 \\
\sqrt{6'78'08'16} \\
\hline
4 \\
\hline
2\ 78 \\
2\ 76 \\
\hline
2\ 08\ 16 \\
2\ 08\ 16 \\
\hline
\end{array}
$$

7-7   CUBE AND OTHER ROOTS

Cube and other roots are usually calculated by means of logarithms; this process is explained in Chapter 18.

▨ EXERCISES

7-37   Calculate the square root of each of the following:

(a) 3025	(f) 11,881	(k) 0.001 225
(b) 9604	(g) 14,641	(l) 97.8121
(c) 13,225	(h) 23,409	(m) 1,900.96
(d) 21,025	(i) 173,056	
(e) 5476	(j) 0.514 089	

7-38 Calculate to five decimal places the square roots of each of the following:

(a) 18   (d) 12.6⅞   (g) 0.526
(b) 127   (e) 35   (h) 0.004
(c) 7.25   (f) 245

Use the table of square roots to find the square roots of the following; check. Calculate the square roots and check with a calculator.

7-39	1.41	**7-44**	48.9	**7-49**	0.1234	**7-54**	2900
7-40	1.414	**7-45**	4567	**7-50**	0.032 17	**7-55**	13,240
7-41	141.4	**7-46**	12.12	**7-51**	0.0079	**7-56**	67,340
7-42	45.38	**7-47**	1001	**7-52**	0.0202	**7-57**	60,000
7-43	76.35	**7-48**	1113	**7-53**	0.1909	**7-58**	93,080

2

ALGEBRA
ALGEBRA
ALGEBRA
ALGEBRA
ALGEBRA
ALGEBRA

INTRODUCTION TO ALGEBRA

8-1 INTRODUCTION TO ALGEBRA

Algebra is sometimes called "generalized arithmetic" or "arithmetic in disguise." Algebra is the arithmetic of literal expressions. It is an old subject that has been developed over many centuries.

A French lawyer and mathematician named Franciscus Vieta is credited with the first formal use of algebra in the sixteenth century. He started by developing a systematic use of letters to represent numbers. Algebra is a type of symbolic arithmetic that enables us to solve problems by simple operations which would be difficult if not impossible by using arithmetic with just numbers.

8-2 NUMBERS

Frequently numbers are obtained by counting a group of objects or by measuring something. Thus the *measure* of an object, for example, the length of a room, is the *number* of times it is contained in the unit of measure. If we use 1 ft as the unit of measure, the measure of the length of a room 20 ft long is the number 20. When making a measurement, it is customary to give the *measure* and also the *unit of measure* as an answer. For instance, you describe the length of a room as 20 (measure) ft (unit of measure); the weight of a fish is 5 kg (5 is the measure, and kg is the unit of measure).

8-3 DEFINITE NUMBERS

The numerals 0, 1, 2, 3, etc., have *definite* meanings. For example, the symbol "4" represents the idea "four," meaning a quantity of four of something. This four may be 4 yd, $4, 4 lb, or 4 of any units, but, in any case, it is a *definite number*. Therefore, obviously, the numerals 0, 1, 2, 3 are definite numbers.

Some numbers (such as π, pronounced "pie") are definite numbers with a set value. For instance, π is the ratio of a circle's circumference to its diameter and always has a value of 3.1416 (to four decimal places).

8-4   GENERAL NUMBERS

General numbers are represented by letters. Consider the formula for computing the area of a rectangle: $A = bh$. Here A represents the area (in square units), b the length of the base, and h the height of the rectangle. The letters A, b, and h are called general numbers; the formula $A = bh$ is for computing the area of *any* rectangle. The letters b and h may represent *any* length, such as 0.05 in., 2 ft, or 1 mi. Any formula for common use is expressed in letter form so that it may be used in the solution of various problems. In any discussion or explanation, the letter or letters used represent the same value throughout that computation.

8-5   SIGNS OF OPERATION AND GROUPING

The *signs of operation* in algebra are $+$, $-$, $\times$, and $\div$. They have the same meaning as in arithmetic. However, the multiplication sign ($\times$) is not used where a multiplication is indicated between letters. Instead, the center dot ($\cdot$) or parenthesis $(b)(h)$ is used or no sign is written. Thus $b \times h$ may be written $b \cdot h$ or simply bh. Similarly, $2axy$ means 2 times a times x times y.

The signs of grouping are:

The parentheses ( )

The brackets [ ]

The braces { }

The vinculum ——

The first three are placed around the parts grouped, and the vinculum is usually placed over what is grouped. They all indicate the same thing; namely, that the parts enclosed are to be taken as a single quantity.

Thus, $12 - (10 - 4)$ indicates that 4 is to be subtracted from 10 and then the remainder is to be taken from 12. Hence $12 - (10 - 4) = 6$. Exactly the same thing is indicated by

$$12 - [10 - 4] \qquad 12 - \{10 - 4\} \qquad 12 - \overline{10 - 4}$$

The vinculum is most frequently used with the radical sign. Thus, $\sqrt{6425}$.

Note that in the form $\dfrac{7 - 4}{3 + 4}$, the horizontal line serves as a vinculum and as a sign of division. It thus performs three duties: it indicates a division; it binds together the numbers in the numerator; and it binds together the numbers in the denominator.

In performing the operations in a problem containing the signs of grouping, the operations within the grouping signs must be performed first.

8-6 ALGEBRAIC EXPRESSIONS

An *algebraic expression* uses signs and symbols to represent numbers or quantities. A *numerical algebraic expression* is made up entirely of numerals and signs. A *literal algebraic expression* includes both letters and numbers separated by signs.

Thus, $14 + 18 - (4 + 3)$ and $3ab - 4cd$ are algebraic expressions; the first is numerical, and the second is literal.

The *value* of an algebraic expression is the number it represents.

8-7 COEFFICIENTS

In an expression such as $8abx$, the 8, a, b, and x are *factors* of the expression. Any one of these factors or the product of any two or more of them is the *coefficient* of the remaining part. Thus, $8ab$ may be considered the coefficient of x, or $8a$ the coefficient of bx. However, what is usually meant by "the coefficient" is only the numerical portion, called the *numerical coefficient*. If no numerical quantity is expressed, 1 is understood as the numerical coefficient. Thus, $1axy$ is the same as axy.

8-8 POWERS AND EXPONENTS

If all the factors in a product are equal, as in the case: $a \cdot a \cdot a \cdot a$, the product of the factors is called a *power* of one of them. The form $a \cdot a \cdot a \cdot a$ is written a^4. The small number above and to the right indicates how many times a is used as a factor. In this example, a is called the *base* and 4 is the *exponent* or power.

The exponent or power is written to the right of and a little above the base. When it is a positive whole number, it shows how many times the base is to be taken as a factor. Thus, c^2 is read "c squared" or "c to the second power" and indicates that c is used twice as a factor; c^3 is read "c cubed" or "c to the third power" and indicates that c is used three times as a factor; c^4 is read "c to the fourth power" and indicates that c is used four times as a factor; c^n is read "c to the nth power" or "c exponent n" and indicates that c is used n times as a factor. When no exponent is written, the exponent is understood to be 1. Thus, a is the same as a^1.

Exponents came into general use in the seventeenth century. Before that time, a^2 was written aa, and a^3 was aaa.

8-9 TERMS

A *term* in an algebraic expression is a part of the expression not separated by a plus or minus sign. The + sign or the − sign that precedes a term is a part of that term. Thus, in the algebraic expression $4ax + 3c - d$, the terms are $+4ax$, $+3c$, and $-d$. The following names are used for algebraic expressions having various numbers of terms:

A *monomial* is an algebraic expression of one term; $6axy$ is a monomial expression.

A *binomial* is an algebraic expression of two terms; $3x - 4y$ is a binomial expression.

A *trinomial* is an algebraic expression which consists of three terms; $7x + 5y - 2z$ is a trinomial expression.

Any algebraic expression of two or more terms may be called a *polynomial* or a *multinomial*.

Terms that are exactly the same or that differ only in their numerical coefficients are called *like terms* or *similar terms*. Thus, $6a^3x^2$, $-7a^3x^2$, and $16a^3x^2$ are like terms, because in all three terms the letters a and x appear with the same exponents. The only differences are the numerical coefficients 6, -7, and 16.

Terms that differ other than in their coefficients are *unlike* or *dissimilar* terms. The terms $6a^3x^2$ and $7ax^2$ are unlike terms because the exponents of the a's are different. Of course, $7a^3x^2$ and $16ayz$ are also unlike terms because they contain different literal coefficients.

8-10 TRANSLATION INTO ALGEBRAIC EXPRESSIONS

One of the most powerful properties of algebra is that it can be used as a tool to solve real on-the-job problems. To use algebra, one must translate from sentences and phrases to mathematical expressions or equations. Frequently formulas are stated in words and translated to algebraic equations. For example, the formula for the area of a triangle is stated: area equals one-half the base times the height; this is translated to: $A = 0.5b \times h$ where A = the area; 0.5 = one half; b = base and h = height.

Definite numbers, including coefficients and exponents, general numbers, signs of operation, and grouping are all used to translate a sentence into an algebraic expression. The following are translations of several statements into algebraic expressions:

EXAMPLE 8-1 The sum of three numbers is 12.

Solution Let x, y, and z represent the three numbers. Then $x + y + z = 12$.

EXAMPLE 8-2 The area of a square is equal to the length of its side squared.

Solution Let A represent the area of the square and s the length of the side. Then $A = s^2$.

EXAMPLE 8-3 The sum of two numbers doubled equals 20.

Solution Let x and y represent the two numbers. Then $2(x + y) = 20$.

EXAMPLE 8-4 The sum of two numbers multiplied by the difference of these same two numbers is 125.

Solution Let x and y represent the two numbers. Then $(x + y)(x - y) = 125$.

EXAMPLE 8-5 The product of two numbers divided by the sum of the same two numbers is 18.

Solution Let x and y represent the two numbers. Then

$$\frac{xy}{x + y} = 18$$

EXAMPLE 8-6 The cube of the sum of three numbers equals 1025.

Solution Let x, y, and z represent the three numbers. Then $(x + y + z)^3 = 1025$.

■ EXERCISES

8-1 Determine the value of each of the following:
  (a) $2 + 7 + 3 + 5 + 1 + 9$
  (b) $10(9 + 8) - 7(6 - 5) + 11$
  (c) $12 + 21 - 3(705) + 2(60 \times 80)$
  (d) $175 - 152 + 11(8 \times 11) - 5$
  (e) $[144 \div (3 \times 12)] \times 4(2 \div 4)$

8-2 In the following list, name the monomials, the binomials, and the trinomials. Which of them are multinomials?
  (a) $at^2 + vt$ (h) $ax^3 - ay^3$
  (b) $x^2 + bx + c$ (i) $2r^2 + 2rh$
  (c) $ma - nu^2$ (j) $xyz - abc + 4$
  (d) $200a^2bc^3$ (k) $45 + 73x$
  (e) $x^2 - 20 + y^2$ (l) $rt + \frac{2}{3}at^3 - 0.9s$
  (f) $11x^2 - 12y^2$ (m) $at - bt + ct^2$
  (g) $87d^2t^3$ (n) $12c^2 + CD^2 - 32h^2$

8-3 Name the numerical coefficients in Exercises 8-2f and 8-2l.

8-4 Name the exponents in Exercises 8-2d and n.

8-5 How many factors appear in Exercise 8-2d? In 8-2g?

8-6 Translate the following into algebraic expressions:
  (a) 9 times x added to a times y.
  (b) 6 times h subtracted from 34 times k.
  (c) The product of 5 times x^4 times w cube.
  (d) The product of 3 times a to the fifth power times s cube.
  (e) The product of the sum of h and k times the difference found by subtracting h from k.
  (f) The square of the sum of 4, x, and y.
  (g) The quotient of the sum of a and b divided by their difference ($b > a$).

8-7 Express the product of v^3 times the sum of these two fractions: 5 divided by x and 6 divided by y.

8-8 Translate these algebraic expressions into words:
- (a) $10 + ax$
- (c) $(a + bt) \div abt$
- (b) $x^2 - 2pm$
- (d) $7x - 5(4a - b)c$

8-9 Write the following terms in more compact form, using exponents:
- (a) $9aaabbbc$
- (e) $2(2)(2)(a + b)(a + b)$
- (b) $(3)(3) \times yyzzzz$
- (f) $11(11)(ax)(ax)11$
- (c) $4(2)2aac$
- (g) $a(ab)[a(ab)]ab$
- (d) $5abcabc + 10abbc - 15abbc$
- (h) $(ab - c)(ab - c)(ab - c)$

8-10 Express each of the following as a single number without an exponent:
- (a) $2(10)^4$
- (d) $9 \times 37 \times 10^7$
- (b) $2 \div 10^3$
- (e) $6 \times 27 \times 10^2$
- (c) $a(10)^{10}$
- (f) $2 \times 8 \div (10)^{11}$

8-11 Express each of the following large numbers as a small number times a power of 10. For example, $17,000 = 1.7 \times 10^4$.
- (a) 500,000
- (d) 10,100,000
- (b) 480,000,000
- (e) 3,468,100,000
- (c) 123,000
- (f) 9,463,780,000,000,000

8-12 Separate the numbers below into prime factors and then express them by using exponents. (Note: $36 = 4 \times 9 = 2^2 \times 3^2$; 2 and 3 are prime numbers.)
- (a) 60
- (c) 6084
- (b) 1080
- (d) 14,400

8-13 Find the value of the following:
- (a) $2^2 3^3$
- (b) $3^2(10)^3$
- (c) $5^2 6^3 7^2$

8-14 If $a = 4$, what is the value of $2a$? Of $4a^2$? Of $10a^4$?

8-15 What is the value of $12a$ if a is 3.5? 1.2? $6\frac{2}{3}$? $8\frac{1}{3}$?

8-16 If I have d dollars and you have 4 times as many, how many dollars do you have?

8-17 If a team wins 11 games one month and 15 games the second month, how many games have they won? If they win n games one week and m the next week, how many games have they won during these two weeks?

8-18 If s is the number of feet in the length of a line, how many feet are there in twice this length? In half the length? In $\frac{3}{4}$ the length?

8-19 A girl can run 10 yd in 1 s. How far can she run in 10 s? In c s? In 1 min? In d min?

8-20 If a train runs a miles in c h, how far does it run in 1 h? In q h?

8-21 A grocer receives c cents for 1 lb tea. How much does he receive for 20 lb? For p lb?

8-22 A boy can walk m mph. How many miles can he walk in c h? In d min? How many feet can he walk in e min?

8-23 What number does $10t + 1u$ represent if $t = 8$ and $u = 3$?

8-24 What number does $100h + 10t + 1u$ represent if $h = 4$, $t = 5$, and $u = 6$? If $h = 7$, $t = 8$, and $u = 0$?

8-25 What number does $1000a + 100b + 10c + 1d$ represent if $a = 5$, $b = 7$, $c = 9$, and $d = 1$?

8-26   If a, b, c, and d have the same values as in Exercise 8-25, what number does *abcd* represent?

8-11   FORMULAS

A *formula* is a rule written in algebraic language to solve particular problems. For example, the rule for calculating the area of a rectangle is as follows: area equals the product of the base multiplied by the height. This rule expressed in formula form is $A = bh$, which is also an equation. You should understand such equations in order to develop the ability to change a formula to other forms that are more convenient for calculations. Manipulation of an equation also requires an understanding of fundamental operations of arithmetic and algebra.

8-12   EVALUATION OF ALGEBRAIC EXPRESSIONS AND FORMULAS

To *evaluate* an algebraic expression is to determine its numerical value. The value can be a definite number only when values have been assigned to all the letters in the expression.

A numerical algebraic expression has a definite value that may be found by performing the indicated operations. Thus,

$$21 - (10 + 3) + 14 - 10 = 21 - 13 + 14 - 10 = 12$$

A literal algebraic expression has a definite value which depends on the values given to the letters. Thus, the expression *abc*, which means $a \times b \times c$, has a definite value if $a = 3, b = 4, c = 10$. Putting these values in place of the letters, we have $3 \times 4 \times 10 = 120$.

If any other set of values is assigned to a, b, c, a definite value for the product will also be obtained.

EXAMPLE 8-7   Find the value of $x^2 + 2x + y^3$ when $x = 3$ and $y = -2$.

Solution   Substitute the values for the letters and evaluate.

$$
\begin{aligned}
x^2 &+ 2x &+ y^3 &= \\
(3)^2 &+ 2(3) &+ (-2)^3 &= \\
9 &+ 6 &- 8 &= \\
&15 &- 8 &= 7
\end{aligned}
$$

EXAMPLE 8-8   Find the value of $\pi r^2 h$ if $\pi = 3.1416$, $r = 6$, and $h = 10$.

Solution   Substitute the value for the letters and evaluate.

$$
\begin{aligned}
\pi r^2 h &= \\
(3.1416)(6)^2(10) &= \\
(3.1416)(36)(10) &= \\
(3.1416)(360) &= 1130.976
\end{aligned}
$$

EXAMPLE 8-9 Find the value of $a^3 + 3a^2b + 3ab^2 + b^3$ when $a = 2$ and $b = 3$.

Solution Substitute the values for the letters and evaluate.

$$
\begin{aligned}
a^3 &+ 3a^2b &+ 3ab^2 &+ b^3 &= \\
(2)^3 &+ 3(2)^2(3) &+ 3(2)(3)^2 &+ (3)^3 &= \\
8 &+ (3)(4)(3) &+ 3(2)(9) &+ 27 &= \\
8 &+ 36 &+ 54 &+ 27 &= 125
\end{aligned}
$$

EXAMPLE 8-10 Find the value of $\sqrt{s(s-a)(s-b)(s-c)}$ if $s = \dfrac{a+b+c}{2}$, $a = 36$, $b = 22$, and $c = 20$.

Solution Substituting,

$$s = \frac{a+b+c}{2}$$

$$s = \frac{36 + 22 + 20}{2} = 39$$

Substituting,

$$
\begin{aligned}
\sqrt{s(s-a)(s-b)(s-c)} &= \sqrt{39(39-36)(39-22)(39-20)} \\
&= \sqrt{37{,}791} = 194.399
\end{aligned}
$$

▨ EXERCISES

8-27 Write the following without exponents and find their values:
  (a) $2^3 3^2$ (d) $4^2 + 5^3(10)$
  (b) $2^3 + 3^2$ (e) $(4^2 5^3)(10)$
  (c) $4^2 5^3 10$ (f) $8^2 + 9^2 + 12^2$

Evaluate the following when $a = 5$, $b = 2$, and $c = 10$:

8-28 $2abc^2$ **8-34** $(ab)^2(c-b)^2$

8-29 $(2abc)^2$ **8-35** $ab(c-b)^2$

8-30 $3a^2 - b^2$ **8-36** $a + b^2(c-b)^2$

8-31 $(3ab^2 - c)^2$ **8-37** $[c + (a+b)][c - (a+b)]$

8-32 $(3a + b - c)^2$ **8-38** $(a+b)^3$

8-33 $3a + (b+c)^2$ **8-39** $a^3 + 3a^2b + 3ab^2 + b^3$

8-40 Find the value of the following when $d = 10$:
  (a) $C = \pi d$ (b) $S = \frac{1}{4}\pi d^2$ (c) $V = \frac{1}{6}\pi d^3$

8-41 Find the value of the following when $r = 2$ and $h = 10$:
  (a) $S = 2\pi rh$ (b) $V = \pi r^2 h$ (c) $T = 2\pi rh + \pi r^2$

Evaluate the following when $a = 1$, $b = 3$, $c = 5$, and $d = 0$:

8-42 $a^2 + 2b^2 + 3c^2 + 4d^2$

8-43 $a^4 + 4a^3 + 6a^2b^2 - 4ab^3 + b^4$

8-44 $\dfrac{12a^3 - b^2}{3a^2} + \dfrac{2c^2}{a + b^2} - \dfrac{a + b^2 + c^3}{5b^3}$

8-45 $(c^2 - b^2 - a^2) - (abc - b^2)$

8-46 $(c - b - a)^2 - (abc - b)^2$

8-47 $\dfrac{10ab + cd}{(d + 2ab - c)^2} + \dfrac{(2c)^2}{a^2 + b} - \dfrac{8a^2}{2b^2}$

Evaluate the following if $a = 1$, $b = 2$, $c = 3$, $d = 4$, and $e = 5$:

8-48 (a) $\sqrt{(a + c)}\sqrt{(d + e)}$

(b) $\sqrt{(c^2 + d^2)} - e$

8-49 $(a - 2b + 3c)^2 - (b - 2c + 3d)^2 + (c - 2d + 3e)^2$

8-50 (a) $\sqrt{4c^2 + 5d^2 + e}$

(b) $\sqrt{e^2 + d^2 + c^2 - a^2}$

8-51 Evaluate $(ac - bd)\sqrt{a^2bc + b^2cd + c^2ad - 2}$ if $a = 1$, $b = 2$, $c = 3$, and $d = 0$.

8-52 Find the value of the expression $\dfrac{a(b - c)}{d(e - b)}$, when $a = 500$, $b = 98$, $c = 8$, $d = 150$, and $e = 100$.

8-53 If A stands for the number of square units in the area of a circle, and r stands for the number of lineal units in its radius, state in words the following formula: $A = \pi r^2$.

8-54 In the formula in 8-53, can r be any number you want to make it? Can A? If you make $r = 5$ cm, can you then make A anything you want? Can π be any number you want to make it? Is π a general number? Is A? Is r?

8-55 Using A as the area, b as the base, and h as the altitude, state the following as a formula: The area of any triangle equals one-half the base times the altitude. Draw a figure. Find the area of a triangle if $b = 4$ and $h = 7$. How many figures can you draw with these dimensions?

8-56 Write the following as a formula: The horsepower H of an electric motor is found by multiplying the number of volts V by the number of amperes I and dividing by 746.

8-57 Given that P stands for principal, I for simple interest, t for time in years, r for rate in percent, and A for amount, translate (a) to (g) into words:

(a) $I = Prt$

(e) $t = \dfrac{I}{Pr}$

(b) $A = P + Prt$

(f) $r = \dfrac{I}{Pt}$

(c) $A = P + I$

(g) $r = \dfrac{A - P}{Pt}$

(d) $P = \dfrac{I}{rt}$

(h) Someone lends $500 for 2 years at 12 percent. Select the proper formula and find the simple interest.

(i) Someone lends some money for $1\frac{1}{2}$ years at 14.5 percent and receives $75 interest. Select the proper formula and find the principal.

(j) Someone lends $400 for 3 years and receives $144 interest. Select the proper formula and find the rate of interest.

(k) Someone lends $700 for a certain length of time at 14 percent and receives $245 interest. Select the proper formula and find the number of years.

8-58 Write the formula which states a times the sum of b, c, and d equals s.

8-59 Let L be the cost of a city lot and H the cost of the house on that lot. Write in algebraic symbols the fact that if you add $5000 to the cost of the lot and multiply this sum by 4, you will determine the total cost of the house and lot.

8-60 Write the formula stating that the area S of the surface of a sphere equals 4 times π times the square of its radius r. Use that formula and find the area of the surface of a sphere 15 in. in radius.

8-61 The volume V of a rectangular solid equals the length l times the width w times the height h. Write the formula for finding the number of cubic feet in a room 40 ft by 30 ft by 12 ft. First draw a sketch of the rectangular solid.

8-62 If a man is 48 years old, what was his age a years ago? What will it be b years from now?

8-63 If x years was the age of a woman a years ago, what is her age now? What will it be c years from now?

8-64 What is the next whole number larger than 10? If n is a whole number, what is the next larger whole number?

8-65 What is the next larger even number after 6? If $2n$ is an even number, what is the next larger even number? The next smaller?

8-66 If x is the first of two consecutive numbers, what is the second? What will represent each of three consecutive numbers, if x stands for the middle one?

8-67 Represent three consecutive even numbers if x is the middle one.

8-68 If the length of a stick is a ft, how many inches is it in length?

8-69 If d is the number of dollars an article costs, what is the number of cents?

8-70 A box a ft long, b ft wide, and c ft deep will hold how many bushels (bu), if 1 bushel equals 2150.43 in.^{3}?

ADDITION AND SUBTRACTION OF SIGNED NUMBERS

9-1 INTRODUCTION TO SIGNED NUMBERS

The degrees of temperature indicated on a thermometer are counted in two opposite directions from the zero point. Temperature is read as a number of degrees above or below zero. In arithmetic, temperature is also stated in that way, but in algebra an abbreviated form is used. The signs + (positive or plus) and − (negative or minus) have been adopted to describe temperatures above and below zero. This use of these two signs differs from the ordinary use, which indicates subtraction or addition. Here the signs indicate the sense or direction in which temperature is measured or counted. A temperature of 25 degrees above zero is written +25; 25 degrees below zero is written −25.

A number preceded by a + sign is a *positive number*. One preceded by a − sign is a *negative number*.

Another example of signed numbers is in the measurement of force; a force acting in one direction and another force in an opposite direction are designated a + force and a − force, respectively. Still another example is money gained (profit) may be written +; money lost may be written −.

9-2 THE NEED FOR NEGATIVE NUMBERS

The necessity for extending the number system to include negative numbers is indicated in the following subtractions. The minuend (top number) remains the same, but the subtrahend increases (left to right) by steps of 1. The difference, then, diminishes also by steps of 1 from left to right. When the difference becomes *less than zero*, it is indicated by the sign − placed before the number.

$$
\begin{array}{cccccccc}
6 & 6 & 6 & 6 & 6 & 6 & 6 & 6 \\
-3 & -4 & -5 & -6 & -7 & -8 & -9 & -10 \\
\hline
3 & 2 & 1 & 0 & -1 & -2 & -3 & -4
\end{array}
$$

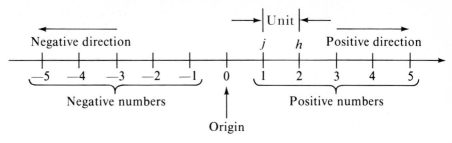

Fig. 9-1. Number line.

9-3 REPRESENTATION OF NEGATIVE AND POSITIVE NUMBERS

For convenience, the positive and negative numbers may be represented on a horizontal line based on a unit of measure *jk*. (See Fig. 9-1.) Traditionally the +, or positive, numbers are placed to the right of a certain point designated zero, and the −, or negative, numbers are placed to the left of this zero point. This setup is very convenient in explaining addition and subtraction.

Toward the right is the *positive* direction and *toward the left* is the *negative* direction, no matter what point one starts from.

The idea of a negative number is opposed to that of a positive number. For example, if a man walks 5 mi east (the positive direction) and then 5 mi west (the negative direction), he ends up at the starting point. The negative distance has canceled out the opposite, or positive, distance.

9-4 DEFINITIONS OF SIGNED NUMBERS

Positive and negative numbers, and zero, form the system of *algebraic numbers*. (In this book the term is used in a restricted sense which is more meaningful to the beginner.)

The *absolute or numerical value* of a number is the value which it has without reference to its sign. Thus +5 and −5 have the same absolute value, 5.

The signs + and −, when used to show direction or sense, are called *signs of quality* to distinguish them from the signs of operation used to indicate an addition or a subtraction. When the + sign is used as a sign of quality, it is usually omitted; but the − sign, when it is a sign of quality, is always expressed.

To show that a notation is a sign of quality, the sign is sometimes enclosed in parentheses with the number: (−3), (+4). Note that (−5) + (+2) indicates that a −5 is to be added to a +2.

The illustrations that follow should help you grasp the idea of negative numbers and to see why they are necessary when a larger number is to be subtracted from a smaller one.

In the study of mathematics, when a new number idea appears, the first thing to do is represent it by a symbol. Next, we find a way of operating with it; that is, we determine how to add, subtract, multiply, and divide with it. One of the first things you learned in arithmetic was methods of operating with positive whole numbers; later, you learned to operate with fractional numbers. Much time spent in studying mathematics is devoted to finding how to add, subtract, multiply, and divide numbers of different kinds: whole, positive, negative, and fractional numbers, and combinations of these.

It is now necessary to develop methods for working with algebraic numbers. Consider carefully each new step which follows to understand it thoroughly.

9-5 ADDITION OF SIGNED NUMBERS

The aggregate value of two or more algebraic numbers is called their *algebraic sum*. The process of finding this sum is addition.

To add 3 to 4, start with 4 and count 3 more, arriving at 7, which is the sum. This technique is fundamental; it seems to be the natural way to add. A child, without being told, will add in this way, using his fingers as counters. If you consider the system of algebraic numbers as arranged on a horizontal line (Fig. 9-2), you will add a positive number by starting with the number to which you want to add, and then count units toward the right—as many units as there are in the number to be added.

For example, using Fig. 9-2, to add 3 to 4, start at 4 and count toward the right to 7. To add $+5$ to -3, start at -3 and count 5 units toward the right, arriving at $+2$. To add $+3$ to -7, start at -7 and count 3 units toward the right, arriving at -4. Because adding -3 to $+7$ is the same as adding $+7$ to -3, the result of adding -3 to $+7$ is 4. In order to start with $+7$ and arrive at $+4$, you must move in the negative direction; that is, toward the left. It is obvious that to add a negative number, you count toward the left. To add -4 to $+9$, then, start at $+9$ and count 4 units toward the left, arriving at $+5$. To add -7 to $+2$, start at $+2$ and go 7 units toward the left, arriving at -5. To add -4 to -5, start with -5 and proceed 4 units toward the left, arriving at -9. These computations are shown below:

$$
\begin{array}{ccccccc}
+4 & -3 & -7 & +7 & +9 & +2 & -5 \\
+(+3) & +(+5) & +(+3) & +(-3) & +(-4) & +(-7) & +(-4) \\
\hline
+7 & +2 & -4 & +4 & +5 & -5 & -9
\end{array}
$$

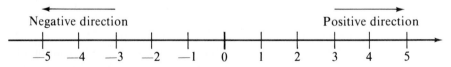

Negative direction — Positive direction

−5 −4 −3 −2 −1 0 1 2 3 4 5

Fig. 9-2. Number line.

Procedure for adding signed numbers:

1. The algebraic sum of two numbers with like signs is the sum of their absolute values, with the common sign prefixed.

2. The algebraic sum of two numbers with unlike signs is the difference between their absolute values, with the sign of the number greater in absolute value prefixed.

To add three or more algebraic numbers with unlike signs, first find the sum of the positive numbers and the sum of the negative numbers by rule 1; then add their sums by rule 2. For example, to find the sum of $+2$, $+10$, -6, -3, -7, $+9$, take $+2 + 10 + 9 = 21$ and $(-6) + (-3) + (-7) = -16$. Then $+21 + (-16) = +5$, the sum.

9-6   SUBTRACTION OF SIGNED NUMBERS

Subtraction is the opposite of addition. If you are given one of two numbers and their sum, subtraction is the process of finding the other number. To illustrate, suppose $7 + b = 10$; then $b = 10 - 7$.

Recall: *10 is called the **minuend** and 7 the **subtrahend***

In arithmetic it is assumed that the minuend is always greater than the subtrahend. In the subtraction of algebraic numbers, it is possible to have a subtrahend larger than the minuend when the numbers are positive, and subtrahend or minuend or both may be negative numbers.

Keep in mind the definition of subtraction as the inverse of addition and the idea of the system of algebraic numbers as arranged along a horizontal line.

1. Subtracting a positive number is the equivalent of adding a numerically equal negative number.

2. Subtracting a negative number is the equivalent of adding a numerically equal positive number.

Thus, according to rule 1,

$$14 - (+8) = 14 + (-8) = 6$$

This is illustrated in Fig. 9-3.

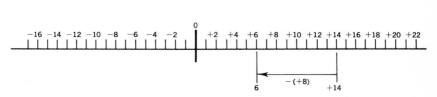

Fig. 9-3. Number line $14 - (+8) = 6$.

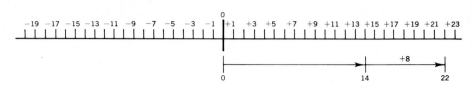

Fig. 9-4. Number line $14 - (-8) = 22$.

Also, according to rule 2,

$$14 - (-8) = 14 + (+8) = 22$$

Try visualizing this example of principle 2 in Fig. 9-4.

Procedure for subtracting signed numbers:

To subtract algebraic numbers, simply change the sign of the subtrahend and proceed as in addition. (Change the bottom sign and add.)

The following six subtraction problems illustrate this rule:

$$\begin{array}{cccccc}
+4 & +7 & +4 & -6 & -3 & -8 \\
\underline{-(+3)} & \underline{-(+3)} & \underline{-(+6)} & \underline{-(-2)} & \underline{-(-7)} & \underline{-(+3)} \\
1 & 4 & -2 & -4 & +4 & -11
\end{array}$$

In these problems, the plus and minus signs within parentheses indicate the quality of each number and do not indicate either addition or subtraction. You should note carefully that $+4$ minus a $+3$ is the same as $+4$ plus a -3.

▓ EXERCISES

9-1 *Add* each of the following and check by using the number line:

$$\begin{array}{ccccccccc}
+6 & -7 & -5 & 14 & +10 & 13 & 19 & 45 & -97 \\
\underline{+8} & \underline{-3} & \underline{+5} & \underline{19} & \underline{-15} & \underline{-18} & \underline{-8} & \underline{-9} & \underline{91}
\end{array}$$

$$\begin{array}{cccccc}
-5 & -2 & +5 & +7 & -5 & +9 \\
+3 & -4 & -6 & -9 & +5 & -4 \\
-2 & +3 & +7 & -2 & -3 & -6 \\
\underline{+4} & \underline{+2} & \underline{-8} & \underline{+4} & \underline{+6} & \underline{+1}
\end{array}$$

9-2 *Subtract* each of the following and check by adding the remainder and the subtrahend. Remember: Change the bottom sign and add.

$$\begin{array}{ccccccccc}
6 & 8 & -11 & 17 & 80 & -95 & -87 & -240 & -0.1 \\
\underline{+3} & \underline{-5} & \underline{+10} & \underline{+13} & \underline{-70} & \underline{90} & \underline{+88} & \underline{+260} & \underline{-1.1}
\end{array}$$

9-3 What number added to 10 equals 5? -1? -10? 9?

9-4 What number subtracted from 15 equals -15? 15? 0? 30? 1? 60?

9-5 Find the *sum* of each of the following:

575	-8008	7070	-1940	\$1.07	95%
-500	1008	$-$ 70	$-$ 60	$-$\$0.77	-15%

9-6 Find the *remainder* in each of the following:

102	-707	-333	5280 ft	\$4.04	$33\frac{1}{3}$%
201	93	-333	$-$ 5200 ft	$-$\$4.96	30 %

Find the *sum* in Exercises 9-7 to 9-13:

9-7 3, -4, -15, 10, 15.

9-8 37, -67, 95, 105, -3.

9-9 1.9, 9.1, -20, 15, -6.

9-10 $\frac{1}{2}$, $\frac{3}{4}$, $\frac{9}{12}$, $-\frac{18}{24}$, $-\frac{4}{8}$.

9-11 $-5\frac{2}{3}$, $-6\frac{3}{4}$, $11\frac{7}{12}$, 10.

9-12 -4.25, -6.74, -10.1, $+3$.

9-13 \$2.90, $-$\$1.90, \$9.75.

9-14 From 12.5% take 8%.

9-15 From -200 take -150.

9-16 From -65 take 100.

9-17 How many degrees of temperature change are there between $+37°$ and $-16°$?

9-7 ADDITION AND SUBTRACTION OF LITERAL ALGEBRAIC EXPRESSIONS

In arithmetic, you learned to add and subtract like numbers using the following rule: When numbers are designated by the same label for example, the same unit of measure, they can be added and subtracted. For example, it is possible to add 5 in., 8 in., and 10 in. and obtain the answer 23 in. However, unlike numbers cannot be subtracted or added. You cannot add 5 in., 8 ft, and 10 yd without converting them to a common unit of measure.

Similarly, in algebra,

$$6d + 4d + 7d = 17d$$
$$4xy + 7xy + 8xy = 19xy$$
$$16x^2y + 23x^2y + 3x^2y = 42x^2y$$

Subtraction:

$$17a - 5a = 12a$$
$$46x^3y^2 - 6x^3y^2 = 40x^3y^2$$

From these examples you can develop the following procedure.

Procedure for adding and subtracting literal algebraic expressions:

Monomials that are alike can be added or subtracted by adding or subtracting their numerical coefficients. But, if the monomials are unlike, the operations can only be indicated.

Several addition problems will illustrate this principle further:

$$
\begin{array}{lll}
+3abc & -16xy^3 & \\
-6abc & +3xy^3 & +3xy^2 \\
+10abc & -4xy^3 & +3xy \\
-16abc & -7xy^3 & -4c^2 \\
-3abc & +28xy^3 & +3a^2 \\
\hline
-12abc & +4xy^3 & 3xy^2 + 3xy - 4c^2 + 3a^2
\end{array}
$$

Note that the expression "like," when applied to algebraic terms, means the literal factors are *exactly alike*. Thus, $3xy^3$ and $3xy$ are unlike because the exponents for the y's are different. Three examples of subtraction follow:

Recall to Subtract Algebraic Expressions: *Change the bottom sign and add.*

$$
\begin{array}{lll}
4ax^2 & -21x^2y & 14ab \\
- \ \oplus 6ax^2 & + \ \ominus 3x^2y & + \ \ominus 6c \\
\hline
10ax^2 & -24x^2y & 14ab + 6c
\end{array}
$$

9-8 ADDING AND SUBTRACTING POLYNOMIALS

Addition and subtraction of polynomials are similar to addition and subtraction of monomials. Arrange the problem in columns so that like terms are in the same column, and combine the terms in each column as with monomials.

EXAMPLE 9-1 Addition.

Solution Arrange the like quantities in the same columns and add.

$$
\begin{array}{l}
+3ax^2 + 14y^2 - \ \ 3z \\
-7ax^2 - 16y^2 + \ \ 7z \\
+10ax^2 - \ \ 4y^2 + \ \ 9z \\
-7ax^2 + 10y^2 - 11z \\
\hline
-ax^2 + \ \ 4y^2 + \ \ 2z
\end{array}
$$

EXAMPLE 9-2 Subtraction.

Solution Arrange the like quantities in the same columns; change the bottom sign and add.

$$
\begin{array}{l}
17xy^2 - 14c^2 + \ \ 4a \\
-10xy^2 + \ \ 5c^2 + \ \ 8a \\
\hline
7xy^2 - \ \ 9c^2 + 12a
\end{array}
$$

9-9 TEST OR PROOF OF RESULTS

It is important to test results. Problems involving addition or subtraction of literal algebraic expressions may be tested by substituting some definite values for the general numbers. If $a = 2$, $b = 3$, and $x = 4$ in the following example, the test is as shown below:

EXAMPLE 9-3

Solution Substitute the given values and evaluate.

$$
\begin{array}{rcl}
-7ab + 4x^2 - 3bx &=& -42 + 64 - 36 = -14 \\
-8ab - 10x^2 - 4bx &=& -48 - 160 - 48 = -256 \\
-9ab + 11x^2 + 6bx &=& -54 + 176 + 72 = 194 \\
\hline
-24ab + 5x^2 - bx &=& -144 + 80 - 12 = -76 \\
& & - 76 = -76
\end{array}
$$

▥ EXERCISES

9-18 Write the sum of each of the following:
  (a) $3a$, $5a$, $7a$
  (b) $10ax^2$, $-3ax^2$, $-5ax^2$, $11ax^2$
  (c) $17m^2n^2$, $13m^2n^2$, $-11m^2n^2$
  (d) $3x + y$, $7x - 2y$, $5x + 4y$
  (e) $2a + 3b - 4c$, $5a - 3b + 8c$
  (f) $3d + 3m - 7z$, $8d - 2m + 2z$
  (g) $10s - 8t + 7u$, $-19s - 12t - 7u$
  (h) $a + b - 3c$, $3a + 4b - 10c$, $-3a - 4b + 17c$

9-19 Subtract $9mn + pq$ from $7mn - 11pq$.

9-20 Subtract $a^2 - 2ab + b^2$ from $a^2 + 2ab + b^2$.

9-21 Add like terms:

$$12a + 12 - a - 4 + 10a - 5 + a - 20a + 14$$

9-22 Add like terms:

$$6x^4 - 7x^3 + 8x^2 - 9x + 5 - 5x^4 + 8x^3 - 7x^2 + 9x + 4$$

9-23 Simplify the following by combining like terms:

$$5x - 7t + 30 + 10t - x + 4t - 20 - x + t - x + 11 - 12t$$

9-24 Add and check by substituting $a = 1$, $b = 3$, $c = 2$:

$$3a - 2b + c, \; a + 20b - 8c, \; -9a + 8b - 7c$$

9-25 Add and check by substituting $s = 2$, $t = 1$:

$$4s - 5t + 6st, \; 7s + 8t - 9st$$

9-26 Subtract $(-13x^3 + 14y^3 - 15z^3)$ from $(8x^3 - 10y^3 + 11z^3)$

9-27 Add $x^3 - 3x^2y + 3xy^2 - y^3$, $x^3 + y^3 - 5x^2y$.

9-28 Add $4abc - abd + 6bcd$, $6abc - 7bcd + 9abd$.

9-29 Add $a^2 - b^2$; $a^2 - 2ab + b^2$; $2a^2 + 3b^2$; $-4a^2 - 6ab - 7b^2$.

9-30 From $8x^3 - 10y^3 + 11z^3 - 12xyz$ take $-13x^3 + 14y^3 - 15z^3$.

9-31 Add $x^3 - 3x^2y + 3xy^2 - y^3$; $x^3 + y^3 - 5x^2y$; $-2x^3 - 4xy^2 + 6x^2y$.

9-32 Add $4abc - 5abd + 6bcd - 7abcd$; $6abc - 7bcd + 8abcd + 9abd$.

9-33 Add $8xy^4 - 4y^5$; $-6x^4y + 8xy^4$; $-3x^3y^2 + 7y^5$; $9xy^4 + 4x^4y - 2x^3y^2$.

9-34 Add $a^2 - b^2$; $a^2 - 2ab + b^2$; $2a^2 + 3b^2$; $-4a^2 - 6ab - 7b^2$; $10a^2$.

9-35 Add $450xyz - 36x^2y + 200xy^2$; $10x^3 - 650y^3 + 140xyz$; $100x^2y + 75xy^2 + 10y^3$; $560x^3 - 100xyz$.

9-36 Add and test by letting $x = 2$ and $y = 3$:

$$3x^2 + xy - 2y^2;\ 4x^2 + 5xy - 3y^2;\ x^2 - 2xy + y^2;\ 2xy - 5y^2$$

9-37 Add and test by letting $m = 2$, $n = 1$, $p = 3$, $q = -1$:

$$9m + 2n - 3p;\ 7p - 2q;\ -5m + 2n - q;\ m + 2q;\ 4n - 3p;\ -m + q$$

9-38 Add $7aby - 4xy + 3ax - naby + 4ax - bz - 3ax + aby + 2bz$.

9-39 Subtract $(2abc - 3ab + 4bc^2 - 10)$ from $(5ab - 4abc - 3bc^2 + 8)$.

9-40 Subtract $(2mn^2 - 3m^2n^2 + 7m^2n - 8mn - 6)$ from $(4m^2n^2 + 8mn^2 - 10)$.

9-41 From $3abx - 2abxy - 5aby^2$ subtract $4abxy + 5aby^2 - 2abx^2$.

9-42 Take $3a^2b - 3ab^2 + b^3$ from the sum of $a^3 - 2b^3 + 3a^2b$ and $4ab^2 - 5a^3 + 2a^2b$.

9-10 SIGNS OF GROUPING

A sign of grouping preceded by a + or − sign indicates that the expression enclosed by the sign of grouping is to be added to or subtracted from that which precedes.

When a *plus* sign precedes a sign of grouping, you should remove the sign of grouping without making any change in signs:

$$a + (b - c) = a + b - c$$

When a sign of grouping is preceded by a *minus* sign, remove the sign of grouping by changing each of the signs within it:

$$a - (b - c + d) = a - b + c - d$$

This change is necessary because the sign of grouping—in this case the parentheses—indicates that the term within the sign of grouping $(b - c + d)$ is to be subtracted from the a. The check is performed below:

$$a - (b - c + d) = a - b + c - d$$

Let $a = 1$, $b = 2$, $c = 3$, $d = 4$; then

$$1 - (2 - 3 + 4) \overset{?}{=} 1 - 2 + 3 - 4$$
$$1 - (3) \overset{?}{=} -2$$
$$-2 = -2$$

Because $-2 = -2$, the change of sign for all terms within the parentheses is correct. When there are several signs of grouping, one within another, you must first remove the innermost set, then the next outer set, continuing until all have been removed.

EXAMPLE 9-4 Simplify $4x^2 - 5y^2 + x - [6x^2 - 3x - (y^2 - x)]$.

Solution Begin with the innermost set of grouping and simplify.

$$4x^2 - 5y^2 + x - [6x^2 - 3x - (y^2 - x)] =$$
$$4x^2 - 5y^2 + x - [6x^2 - 3x - y^2 + x] =$$
$$4x^2 - 5y^2 + x - 6x^2 + 3x + y^2 - x = -2x^2 - 4y^2 + 3x$$

EXAMPLE 9-5 Simplify $8 - \{7 - [4 + (2 - x)]\}$.

Solution Begin with the innermost set of grouping and simplify.

$$8 - \{7 - [4 + (2 - x)]\} =$$
$$8 - \{7 - 4 - 2 + x\} =$$
$$8 - 7 + 4 + 2 - x = 7 - x$$

As soon as like terms appear within a sign of grouping, they should be combined, as, for example, in the first step after removing the parentheses in Example 9-5, the $4 + 2$ within the brackets [ ] may be combined as illustrated below:

$$8 - \{7 - [4 + 2 - x]\} =$$
$$8 - \{7 - [6 - x]\} =$$
$$8 - \{7 - 6 + x\} =$$
$$8 - \{1 + x\} =$$
$$8 - 1 - x = 7 - x$$

■ EXERCISES

9-43 Simplify the following by removing signs of grouping and combining like quantities:
(a) $2a + 5b - (a + 4b)$
(b) $7 - 3x + (-4 - 6x)$
(c) $10y - 10 - (-3y + 4)$
(d) $z + 4x^2 - 14 - (3x^2 + 4 - 2z)$
(e) $m - (3m - 2n + p) - m - n + p$
(f) $x - 2y + 3z - (2x - 3y + 4z) - (-3x + z)$
(g) $-9 - 4n + 3v - 2w - (-9 + 4w) + (5v - 7u - 10w)$
(h) $a - [b - (a + 4)]$
(i) $19 - 3 - [4 - (-6 + 10)]$
(j) $17 - 2[2 + (3 - 7)]$
(k) $2x - [3x - (x - y) - y]$
(l) $-5x - [-3x - (-2x + 3y) - 6y]$
(m) $(a + 2b) - 3a - 4b - [-a - (-b - 10)]$

9-44 $x - y - \{-x + y - [x - y - (-x + y)]\} = ?$ Let $x = 2$ and $y = 3$ and check your answer.

9-45 $[-(3a + 2b) - (a + 9b)] - [a + (-b - 2) + 3] = ?$

9-46 $\{-[(2u + 3v) - u - 3v]\} = ?$

9-47 $10a - 5b + 2c - [-(-a - 2b + 3c) - 7a + 9b - 10c] = ?$

9-48 $8a^2 - 9ab + 10c^2 - [ab + 4a^2 - 3ab - (a^2 + ab + c^2)] = ?$

9-11 INSERTION OF SIGNS OF GROUPING

Any terms of a polynomial may be enclosed, without change of signs, in a sign of grouping preceded by a plus sign and retain the same value. The reason for this is given in Sec. 9-10. Terms may be enclosed in a sign of grouping preceded by a minus sign and retain the same value only if the sign of each term within is changed from − to + or from + to −.

EXAMPLE 9-6 Enclose the last three terms of $ax + by + cd - e$ within parentheses (1) preceded by a + sign and (2) preceded by a − sign:

1. $ax + by + cd - e = ax + (by + cd - e)$
2. $ax + by + cd - e = ax - (-by - cd + e)$

▨ EXERCISES

Insert parentheses around all terms that follow the first − sign in each of the following, but do not change the value of the expression:

9-49 $a - 2b + 3c$

9-50 $3x - x^2 - 2xy - y^2$

9-51 $8a + b - 64a^2 - 16a - 1$

9-52 $3s^2 - 4r - 5s + 2t$

9-53 $16z^2 + 10w - 14z^3w + 15w^2 - 5z^2$

9-54 $-2a - 3b - 4c - 5d + 40abc - 30ab - 20cd$

9-55 Write the following polynomials with the last three terms of each enclosed in parentheses (1) preceded by a plus sign and (2) preceded by a minus sign:

(a) $4 - 4x^2 + 4x - 1$
(b) $10 - 3x + y - 3z$
(c) $9u^2 + 9u - 18v + 24$
(d) $16a^2 + 4b^2 - 16ab - 9abc - c^2$
(e) $81x - 9xy + 4y - 1$
(f) $41ab - 4ac + bc - 4abc$

SIMPLE EQUATIONS

10-1 INTRODUCTION TO SIMPLE EQUATIONS

Many problems solved with algebra involve equations in various forms. Thus, solving equations is one of the most important tools of algebra. Some equations are easy to solve, but the solution of more complex equations can be very involved.

Efficient solving of equations and thorough understanding of their applications require a great deal of time and practice. A considerable part of solving equations may, by sufficient drill, become mechanical. However, there are reasons for each and every step. By understanding these reasons, you will be more capable of solving these problems and more efficient. When studying algebra, the efforts of a good student should be devoted to a sound understanding of the equation, its solution, and its application.

10-2 DEFINITION OF SIMPLE EQUATIONS

An *equation* is a statement where two expressions or quantities are equal. The statement $3x + 4 = 10$ is an equation; it is an algebraic shorthand for "The sum of three times a number plus 4 equals 10." The equation $3x + 4 = 10$ is much easier to work with than the equivalent sentence. Indeed, if we did not have "shorthand" equations and had to work entirely with written statements, solving many problems would be very difficult.

You should consider an equation as similar to a balance scale, as in Fig. 10-1. The equal sign tells you that the two quantities are "in balance" (that is, they are equal). Suppose that the numbers in this equation are pounds; if 4 pounds are removed from each pan, the pans will still be in balance. Fig. 10-1 could then become Fig. 10-2. The corresponding equation would then be $3x = 6$, and the relationship could then be expressed: "Three times an unknown number is equal to six." The unknown number x is 2, because $3(2) = 6$.

123

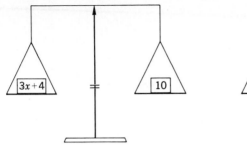 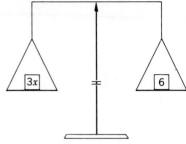

Fig. 10-1. An equation is like a balance. **Fig. 10-2.** $3x = 6$

The above solution could be written in the following three steps:

$$3x + 4 = 10$$
$$3x = 6$$
$$x = 2$$

The first statement expresses the whole equation. In the second, 4 has been subtracted from each side of the equal sign, and, in the last, each side has been divided by 3.

Therefore, to keep an equation in balance the same operation must be done on both sides of the equal sign. An equation is kept in balance by subtracting the same number from both members (sides), adding the same number to both sides, or dividing or multiplying both sides by the same quantity. By understanding this procedure you will find the solution of many equations very easy.

The expression $3x + 4 = 10$ is a conditional equation because it is only true when x has a certain value. Not all equations are conditional; for example, $4x + 3x = 7x$ is true for any value of x. Such an equation is an identical equation or an identity. The value to be found in a conditional equation is called the root of the equation, the unknown number, the unknown quantity, the unknown, or the solution to the equation.

Solving an equation is finding the value of the unknown that makes the equation true.

10-3 AXIOMS FOR SOLVING SIMPLE EQUATIONS

In solving the equation in Figs. 10-1 and 10-2, two basic rules or axioms were used. An axiom is an obvious truth accepted without proof. The following are frequently used in the solution of equations; you should study and understand each one:

1. If equal quantities are added to equal numbers, the sums are equal.
2. If equal quantities are subtracted from equal quantities, the remainders are equal.
3. If equal quantities are multiplied by equal quantities, the products are equal.

4. If equal quantities are divided by equal quantities (except zero), the quotients are equal. (Be sure to note that division by zero is not included.)

5. Quantities that are equal to the same quantities or to equal quantities are equal to each other.

6. Like powers of equal quantities are equal.

7. Like roots of equal quantities are equal.

8. The whole of anything equals the sum of all its parts.

Note: *Axioms 2 and 4 were used to solve the equation $3x + 4 = 10$.*

10-4 SOLUTION OF SIMPLE EQUATIONS

To solve an equation is to find the value of the unknown that will make the equation true. An equation in mathematics is equivalent to a sentence. Solving equations is a step-by-step procedure. Each step is justified by one of the axioms. Study the simple examples to discover the general methods of solving equations.

EXAMPLE 10-1 Solve $x - 5 = 3$ for x. This means, to find the value for x if $x - 5 = 3$. By inspection you can see that $x = 8$, but inspection does not help in solving more complicated equations.

Solution

$$x - 3 = 5$$

Add 3 to each side (axiom 1) $x - 3 + 3 = 5 + 3$
Combine like quantities $x = 8$
Check $8 - 3 = 5$

Note: *To determine that $x = 8$, 3 is added to each member of the equation.*

EXAMPLE 10-2 Solve for x, if $x + 3 = 10$.

Solution

$$x + 3 = 10$$

Subtract 3 from each side (axiom 2) $x + 3 - 3 = 10 - 3$
Combine like quantities $x = 7$
Check $7 + 3 = 10$

EXAMPLE 10-3 Solve for b if $4b = 36$.

Solution

$$4b = 36$$

Divide each member by 4 (axiom 4) $\dfrac{4b}{4} = \dfrac{36}{4}$

$$b = 9$$

Check $4(9) = 36$

EXAMPLE 10-4 Solve for x, if $4x + 5 - 7 = 2x + 6$

Solution

$$4x + 5 - 7 = 2x + 6$$

Collect like terms $(+5)$ and (-7) $4x - 2 = 2x + 6$

Add 2 to both members (axiom 1) $4x = 2x + 8$

Subtract $2x$ from both members (axiom 2) $2x = 8$

Divide both members by 2 (axiom 4) $x = 4$

Check $4(4) + 5 - 7 = 2(4) + 6.$

10-5   CHECKING THE UNKNOWN

An equation always asks the question: What must be the value of the unknown so the two members of the equation are equal? The solution or root to the equation should answer this question.

After you have solved the equation, you should always check it. Do this by substituting the unknown quantity in the given equation. If the two sides are equal, the solution or the root is correct.

EXAMPLE 10-5 Solve and check $47r - 17 = 235 - 37r$

Add 17 to each side $47r - 17 + 17 = 235 - 37r + 17$

                                       $47r = 235 - 37r + 17$

Add $37r$ to each side $47r + 37r = 235 - 37r + 37r + 17$

                                 $47r + 37r = 235 + 17$

Combine like quantities $84r = 252$

Divide both sides by 84 $r = 3$

Check by substituting 3 in the original equation and evaluate

$$47r - 17 = 235 - 37r$$
$$47(3) - 17 = 235 - 37(3)$$
$$141 - 17 = 235 - 111$$
$$124 = 124$$

▨ EXERCISES

Solve and check the following equations:

10-1 $2x + 3 = x + 4$

10-2 $4x - 10 = 2x + 2$

10-3 $9x + 9 + 3x = 15$

10-4 $300x - 250 = 50x + 750$

10-5 $17x - 7x = x + 18$

10-6 $2.5x + 0.5x = 1.5x + 1.5$

10-7 $9y - 19 + y = 11$

10-8 $x + 2x + 3 - 4x = 5x - 9$

10-9 $2y + 3y - 4 = 5y + 6y - 16$

10-10 $75z - 150 = 80z - 300$

10-11 $3.3x + 2.7x - 4.6 = 7.4$

10-12 $2y - 3y + 4y - 5 = 6y - 7y + 15$

10-13 $(4x + 6) - 2x = (x - 6) + 24$

Hint: *First remove the signs of grouping (the parentheses) and then proceed.*

10-14 $15y - [3 - (4y + 4) - 57] = (2 - y)$

10-15 $[2y - (3y - 4) + 5y - 6] + 10y = (12y - 12) + 36$

10-16 $4t - (12t - 24) + 38t - 38 = 0$

10-6 SETTING UP EQUATIONS

The equations discussed previously were expressed in "algebraic language." You should be able to translate a sentence into algebraic language (an equation) and solve the equation. Specific rules and axioms cannot be developed for translation, but the following suggestions and examples should help.

Procedure for setting up literal equations:

1. Read the statement of the problem carefully.
2. Select the unknown quantity and represent it with some letter. If more than one unknown quantity exists in the problem, try to represent those quantities in terms of the same letter.
3. Write the equation using the letter or letters selected and solve.

EXAMPLE 10-6 An unknown number is multiplied by 4; after 8 is subtracted from the product, the result is 0. Find the unknown number.

Solution Let x = the unknown number. Then

$$4x - 8 = 0$$
$$4x = 8$$
$$x = 2$$
$$?$$

Check by substituting 2 for x and evaluate

$$4(2) - 8 = 0$$
$$0 = 0$$

In addition to the check above, test the answer $x = 2$ by comparing it with the conditions described in the original problem. When 8 is subtracted from 4 times the answer (2) is the result zero?

EXAMPLE 10-7 A number multiplied by 8 gives a result of 4. Find the number.

Solution Let y = the number. Then

$$8y = 4$$
$$\frac{8y}{8} = \frac{4}{8}$$
$$y = \frac{1}{2}$$

Check $8\left(\dfrac{1}{2}\right) = 4$

EXAMPLE 10-8 What number divided by 4 equals 8?

Solution Let a = the number. Then

$$\frac{a}{4} = 8$$
$$a = 32$$

Check $\dfrac{32}{4} = 8$

$$8 = 8$$

EXAMPLE 10-9 What number added to twice itself equals 60?

Solution Let x = the number. Then

$$x + 2x = 60$$
$$3x = 60$$
$$x = 20$$
$$?$$

Check $20 + 2(20) = 60$

$$20 + 40 = 60$$
$$60 = 60$$

EXAMPLE 10-10 A room three times as long as it is wide has a perimeter of 32 m. Find its length and width. (First draw a rectangle 3 times as long as it is wide (see Fig. 10-3).

Solution Let x = number of meters in width; then $3x$ = number of meters in length. Therefore $x + x + 3x + 3x$ = perimeter, also 32 m = perimeter.

Thus $x + x + 3x + 3x = 32$ by axiom 5
Combine like terms $8x = 32$
Divide by $8x = 4$, the number of meters in width
$3x = 12$, the number of meters in length

Note: *The statements in the above example are equivalent expressions for the perimeter. These expressions are equal to each other, and the statements provide the equation to be solved.*

$3x$

x [ ] x = width

$3x$ = length   **Fig. 10-3.** Example 10-10.

EXAMPLE 10-11 In a group of 64 persons, the number of children is three times the number of adults. How many children are there, and how many adults?

Solution

$$\text{Let } x = \text{the number of adults}$$
$$3x = \text{the number of children}$$
$$x + 3x = \text{the number of the group (64 = number of group)}$$
$$x + 3x = 64$$
$$4x = 64$$
$$x = 16, \text{ the number of adults}$$
$$3x = 48, \text{ the number of children}$$

EXAMPLE 10-12 The sum of two numbers is 300, and their difference is 200. What are the numbers?

Solution

$$\text{Let } x = \text{the greater number}$$
$$300 - x = \text{the lesser number}$$
$$x - (300 - x) = \text{the difference (also, 200 = the difference)}$$
$$x - (300 - x) = 200$$
$$x - 300 + x = 200$$
$$x + x = 200 + 300$$
$$2x = 500$$
$$x = 250, \text{ the greater number}$$
$$300 - x = 50, \text{ the lesser number}$$

Check: The sum of 250 and 50 is 300, and the difference is 200. Hence, the conditions of the problem are satisfied.

EXERCISES

10-17 A man is s years old. Express in algebraic symbols his age 5 years ago; his age T years ago; and his age $(5 + T)$ years ago. Express his age $(5 + T)$ years from now.

10-18 A woman was s years old 10 years ago. How old will she be 20 years from now? T years from now? When will she be 30 years old?

10-19 A car runs d mi in 8 h. How far does it run in 1 h? In t h? In t h and m min?

10-20 If a car is driven d mi in h h, how long will it take to drive it 100 mi?

10-21 How many cents are there in $20? In a dollars? In 10 dimes? In d dimes? In d dimes and n nickels?

10-22 You have $100 more than I do. If you have x dollars, how much money do I have?

10-23 Express five consecutive odd numbers if a is the middle number.

Write the following as equations and solve for the unknown:

10-24 $4x$ increased by 4 equals 44.

10-25 $10x$ increased by 4 equals $3x$ decreased by $(2 - 3x)$.

10-26 12x decreased by 4 equals 4x increased by 12.

10-27 An automobile runs 29 km on 4 liters gasoline. If it runs 360 km on x liters, find x.

10-28 If 10 is added to twice your money, the result is the same as it would be if your money were subtracted from 43. Let x be your money and find x.

10-29 If 10 is added to 10x and 15 is subtracted from the sum, the result equals 2x plus 3. Find x.

10-30 to 10-33 Form equations for the relations indicated in Figs. 10-4 to 10-7 and solve each equation for the unknown.

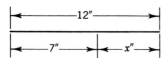

Fig. 10-4. Exercise 10-30.

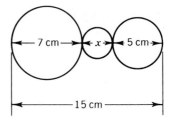

Fig. 10-5. Exercise 10-31.

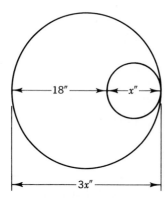

Fig. 10-6. Exercise 10-32.

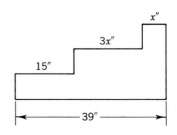

Fig. 10-7. Exercise 10-33.

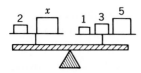

Fig. 10-8. Exercise 10-34.

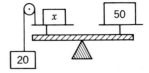

Fig. 10-9. Exercise 10-35.

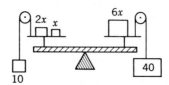

Fig. 10-10. Exercise 10-36.

10-34 to 10-36 Scales will balance when the weights in the pans are equal. Find the values of the unknown weights (in pounds) for the settings shown in Figs. 10-7 to 10-10.

10-37 If x represents a certain number of gallons of oil, and if $140 is the total cost of that many gallons, what is the price per gallon?

10-38 Find the cost per book if eight books at x dollars per book cost $20. State the information as an equation and solve for x.

10-39 A hat and a dress cost, together, $63.00. If the hat costs $11 less than the dress, find the cost of each.

10-40 Noel, Fran, and Toby are working. Noel earns 2 times as much as Fran, and Toby earns 3 times as much as Noel. If the total amount earned by all was $900, how much did each earn?

10-41 Twin steel subway tubes are encased in concrete and sunk under State Street in the Chicago River. The distance from outside to outside of the encasement is 40 ft. The tubes have a diameter of 18 ft. Find the distance between the tubes.

10-42 At a tire sale, the regular price of a tire, plus its sale price, was $77. The sale price of four tires was the same as the regular price of three tires. Find the regular and sale prices of the tires.

10-43 The sum of two numbers is 300; their difference is 200. What are the numbers?

10-44 A number plus its double equals 18. What is the number?

10-45 Find the three consecutive even numbers whose sum is 216.

10-46 Find the four consecutive odd numbers whose sum is 88.

10-47 Find the three consecutive numbers whose sum is 66.

10-48 Divide $210 between A, B, and C so that B will have $35 less than A and $20 more than C.

10-49 Five lamps connected in series have a resistance of 1250 ohms. The first has a resistance of 240 ohms; the second, 260 ohms; and the other three are alike. Find the resistance of each of the like lamps.

10-50 A piece of wire 66 in. long is bent into the form of a rectangle 1 in. longer than it is wide. Find the dimensions of the rectangle.

10-51 A student earned $3600 her first 3 years in college. The second year she earned twice as much as she did the first, and the third year she earned $450 more than the second year. How much did she earn each year?

10-52 In one year three college students earned, together, $4800. The first earned twice as much as the second, and the third earned $200 less than twice as much as the first. How much did each earn?

MULTIPLICATION OF SIGNED NUMBERS

11-1 INTRODUCTION TO MULTIPLICATION OF SIGNED NUMBERS

Multiplication is a shortened process of addition. For example,

$$5 \times 3 = 5 + 5 + 5 = 15$$

can also be stated as follows: *five multiplied by three is equal to adding five three times.* This description of multiplication can also be extended to include fractions. For example,

$$\frac{3}{4} \times 8 = \frac{3}{4} + \frac{3}{4} + \frac{3}{4} + \frac{3}{4} + \frac{3}{4} + \frac{3}{4} + \frac{3}{4} + \frac{3}{4} = \frac{24}{4} \text{ or } 6$$

Note: When fractions are added, only the numerators are added; thus the sum of the addition of eight 3s is 24.

The same sum will, of course, result if this problem is written as a multiplication:

$$\frac{3}{4} \times 8 = \frac{3 \times 8}{4} = \frac{24}{4} \text{ or } 6$$

11-2 MULTIPLICATION OF SIGNED NUMBERS

All the numbers multiplied above are +, or positive, numbers. The resulting answer was positive. Remember that *multiplication of two positive numbers always produces an answer which is positive.* It is often necessary to multiply numbers when one or both are negative, and the rule just stated is the same for two negative numbers. Let us consider the technique for multiplying two negative numbers. For example, multiply $(-3) \times (-5)$. Recall that subtraction is the inverse operation of addition, so this multiplication problem means that -5 is subtracted 3 times or

$$(-3) \times (-5) = -(-5) - (-5) - (-5) = +15$$

This example illustrates the rule that *when two negative numbers are multiplied, the result is a positive number.*

Now consider the multiplication of one positive number and one negative number. For example, $(-5) \times (+3)$:

$$(-5) \times (+3) = (-5) + (-5) + (-5) = -15$$

The result when -5 is multiplied by $+3$ is -15. Since multiplication is a short form of addition, the problem $(-5) \times (+3)$ was rewritten $(-5) + (-5) + (-5)$. The sum, of course, is -15. Therefore, *when two numbers of unlike signs are multiplied, the result, or product, is negative.*

The three situations may be summarized in two statements:

Procedure for multiplying two signed numbers:

When two numbers of like signs are multiplied, the result is positive.

When two numbers of unlike signs are multiplied, the result is negative.

11-3 MULTIPLYING SEVERAL SIGNED NUMBERS

To find the product of multiplying three or more signed numbers, multiply the first two; then multiply the result by the third number. Multiply that result by the fourth number, and so on. Continue until you have multiplied all the numbers. The following examples illustrate this technique:

EXAMPLE 11-1

$$(-3) \times (-4) \times (-5) = (+12) \times (-5) = -60$$

EXAMPLE 11-2

$$(-3) \times (-4) \times (-5) \times (-6) = (+12) \times (-5) \times (-6)$$
$$= (-60) \times (-6)$$
$$= +360$$

Notice in the preceding two examples that the laws of signs are applied: $(-3) \times (-4) = (+12)$ and $(+12) \times (-5) = -60$.

Procedure for multiplying several signed numbers:

The product of an odd number of negative factors is negative.

The product of an even number of negative factors is positive.

The product of any number of positive factors is positive.

▨ EXERCISES

11-1   Find the product of -3, $+4$, and -5.

11-2   Find the product of $+6$, -20, -10, and $+40$.

11-3   Find the cube of -4; of $+5$; of -11.

Find the values of the following:

11-4 $(-3)^3$ **11-6** $(-6)^4$ **11-8** $3(-4)^5(-2)^2$

11-5 $(-1)^{100}$ **11-7** $(-2)^2(-3)^3$ **11-9** $-2(-2)^2(-10)^3$

If $x = -2$, $y = 3$, $z = 4$, find the value of:

11-10 x^2y^2 **11-12** xyz **11-14** $(6)(-5)(-2)y^3$

11-11 y^3z **11-13** $x^2y^2z^2$

Find the values of the following:

11-15 $4(-10)(-3) + 5(+6)(-30)$

11-16 $3(-4)(5) + 5(-3)(-4)$

11-17 $-(3)^2(-1)^2(-4)^3 + 3(-4)^3(-3)(-1)^3$

11-18 $(-12)(-10) - (-10)^2(-2)^2(3)(10)$

11-19 $5(4)(3)(2) - (-2)(-3)(-4)(-5)$

11-20 $4(5)(6)(10) - 1(-4)(-5)(-6)(-10)$

11-4 EXPONENTS

An *exponent* is defined as the number written above and to the right of a base number that indicates how many times the base is taken as a factor. For example, in the term a^5, the 5 is the exponent and indicates that the base number a is to be used as a factor five times: $a^5 = a \times a \times a \times a \times a$. An expression such as a^5 is an *exponential*. Since

$$a^5 = a \times a \times a \times a \times a$$

and
$$a^3 = a \times a \times a$$

then
$$a^5a^3 = a \times a \times a \times a \times a \times a \times a \times a = a^8$$

or
$$a^5a^3 = a^{5+3} = a^8$$

Procedure for multiplying quantities involving exponents:

To multiply like base numbers with exponents simply add the exponents. (Remember: the base number must be the same in each factor.)

EXAMPLE 11-3 $2^3 \times 2^2 = 2^5 = 32$

Check:

$$2^3 = 2 \times 2 \times 2 = 8; \ 2^2 = 2 \times 2 = 4; \text{ and } 8 \times 4 = 32$$

EXAMPLE 11-4 $5^3 \times 5^2 = 5^5$ or 3125

Check:

$$5^3 = 5 \times 5 \times 5 = 125 \text{ and } 5^2 = 5 \times 5 \text{ or } 25$$
$$125 \times 25 = 3125$$
$$3125 = 3125$$

EXAMPLE 11-5 $3^3 \times 3^3 = 3^6 = 729$

Check: $3^3 = 3 \times 3 \times 3 = 27$; $27 \times 27 = 729$

EXAMPLE 11-6 $(+3)^2(+3)^5 = (+3)^{2+5} = 3^7 = 2187$

Check: $(+3)^2 = (+3)(+3) = 9$, $(+3)^5 = (+3)(+3)(+3)(+3)(+3) = 243$,
$9 \times 243 = 2187$

The preceding problems were carried to a final answer (product) to check the solution. Often expressing the final answer in exponent form is more convenient.

Note: *Observe that the base number is the same throughout each example; the base number must not be changed within a problem.*

11-5 MULTIPLYING A MONOMIAL BY A MONOMIAL

The multiplication of a monomial by a monomial is best explained by several examples. First multiply the numerical values, then find the product of the exponentials.

EXAMPLE 11-7 Multiply $+3a$ by $+6b$.

Solution

$$(+3a)(+6b) = 3 \times 6 \times a \times b = +18ab$$

EXAMPLE 11-8 Multiply $+6b^2$ by $+8b^3$.

Solution

$$(+6b^2)(+8b^3) = 6 \times 8 \times b^2 \times b^3 = 48b^5$$

EXAMPLE 11-9 Multiply $-14a^3b^2$ by $3a^4b^3$.

Solution

$$(-14a^3b^2)(3a^4b^3) = (-14)(3)(a^3)(a^4)(b^2)(b^3)$$
$$= -42a^7b^5$$

Here, -14 multiplied by $+3$ produces -42; then a^3 is multiplied by a^4 by adding the exponents ($3 + 4 = 7$); next, b^2 is multiplied by b^3 by adding the exponents ($2 + 3 = 5$). You should understand that the answer, $-42a^7b^5$, can be written $(-42)(a^7)(b^5)$ or (-42) times (a^7) times (b^5). This multiplication cannot be carried further unless a and b are assigned specific values.

11-6 MULTIPLYING A POLYNOMIAL BY A MONOMIAL

In the multiplication of a polynomial by a monomial, multiply each term of the multiplicand (top quantity) by the multiplier (bottom quantity) and obtain the algebraic sum of these partial products.

EXAMPLE 11-10 Multiply $3x^2 - 4b^2$ by $12x^3$.

Solution The first term in the product is obtained by multiplying $12x^3$ by $3x^2$; the second term is obtained by multiplying $12x^3$ by $-4b^2$.

$$\begin{array}{r} 3x^2 - 4b^2 \\ \times 12x^3 \\ \hline 36x^5 - 48b^2x^3 \end{array}$$

EXAMPLE 11-11 Multiply $7ax^3 - 21ab^4 - 3x^2$ by $2a^2b^2x^4$.

Solution Follow the same procedure as in Example 11-10.

$$\begin{array}{r} 7ax^3 - 21ab^4 - 3x^2 \\ \times 2a^2b^2x^4 \\ \hline 14a^3b^2x^7 - 42a^3b^6x^4 - 6a^2b^2x^6 \end{array}$$

▉ EXERCISES

Find the products of the following:

11-21 $5a^2b^3$ and $4ab^2$

11-22 $21m^4n^5$ and $5m(-n)^3$

11-23 $8ab^2c^3$ and $-5b^5c^{10}$

11-24 $-10xyz^4$ and $-6x^4y^6$

11-25 $-(-1)x^2y^4$ and $-(-x^2)(-y)^4$

11-26 $2x^2y$, $-3y^2z$, and $4xy^2z^3$

11-27 $-\frac{4}{5}x^2y^4$, $\frac{5}{8}x^3$, and $-2y$

11-28 $(x^3)(x^3)(x^3)(x^3)$

11-29 $(S^2t^3)^4$

11-30 $2(2a^2b^2)^2$

11-31 $-2(P^3)Q^4$

11-32 $abc(a^2b^2c^2)(abc)^2$

11-33 $3a^3(-2ab^3)(2ab)^3$

11-34 $(3x + 4y)5x$

11-35 $5x^2 + 6x$ times $-6x$

11-36 $4y^2 - 8y + 4$ times $12y$

11-37 $(2y + 3y - 4z)(-4a)$

11-38 $-2(-2xy)(-2x + 3y)$

11-39 $-10xy^2z(-2x^2yz^2 + 5xy^2z - 4xyz)$

11-7 MULTIPLYING A POLYNOMIAL BY A POLYNOMIAL

To multiply a polynomial by a polynomial, multiply each term of the multiplicand by each term of the multiplier; then write the partial products under each other and find the algebraic sum of the like partial products. In short, multiply each term in one polynomial by each term in the other. The following examples illustrate.

EXAMPLE 11-12 Multiply $(y - 2)$ by $(y + 2)$.

Solution Start with the term in the lower left (y) and multiply from left to right the terms in the multiplicand $(y - 2)$; follow the same procedure using $+2$, then add the like terms.

$$
\begin{array}{r}
y - 2 \\
\times\ y + 2 \\
\hline
y^2 - 2y \\
+\ 2y - 4 \\
\hline
y^2 + 0\ - 4 = y^2 - 4
\end{array}
$$

EXAMPLE 11-13 Multiply $(x^2 + 3xy - 2y^2)$ by $2xy - 2y^2$.

Solution

$$
\begin{array}{l}
\ \ x^2 + 3xy\ \ \ - 2y^2 \\
\ \times\ 2xy - 2y^2 \\
\hline
2xy(x^2 + 3xy - 2y^2) = \ \ 2x^3y + 6x^2y^2 - \ \ 4xy^3 \\
-2y^2(x^2 + 3xy - 2y^2) = \ - 2x^2y^2 - \ \ 6xy^3 + 4y^4 \\
\hline
\ \ 2x^3y\ \ + 4x^2y^2 - 10xy^3 + 4y^4
\end{array}
$$

EXAMPLE 11-14 Multiply $(3a^2 + 3b^2 + ab)$ by $(b^3 - 2a^2b + ab^2)$.

Solution

$$
\begin{array}{l}
\ 3a^2 + 3b^2\ \ + ab \\
\ \times\ \ b^3 - 2a^2b + ab^2 \\
\hline
b^3(3a^2 + 3b^2 + ab) = \ \ 3a^2b^3 + 3b^5\ \ + \ ab^4 \\
-2a^2b(3a^2 + 3b^2 + ab) = -6a^2b^3 \ - 6a^4b - 2a^3b^2 \\
ab^2(3a^2 + 3b^2 + ab) = \ \ \ a^2b^3 + 3ab^4 \ + 3a^3b^2 \\
\hline
\ -2a^2b^3 + 3b^5\ \ + 4ab^4 - 6a^4b + \ \ a^3b^2
\end{array}
$$

Note: *You can test or check answers to multiplication problems by substituting numbers for letters. That procedure is illustrated in Example 11-15.*

EXAMPLE 11-15 Multiply $(x^3 - 4)$ by $(x + 8)$.

Solution

$$
\begin{array}{r}
x^3 - 4 \\
x\ + 8 \\
\hline
x^4 - 4x \\
+\ 8x^3 - 32 \\
\hline
x^4 - 4x + 8x^3 - 32
\end{array}
$$

Check: Substituting 3 for x in the above problem,

$$(x^3 - 4) \times (x + 8) = (3^3 - 4)(3 + 8) = (23)(11) = 253$$
$$x^4 - 4x + 8x^3 - 32 = 3^4 - 4(3) + 8(3)^3 - 32$$
$$= 81 - 12 + 216 - 32$$
$$= 253$$

The preceding multiplication is correct because in both cases 253 is the answer.

■ EXERCISES

Multiply each of the following:

Hw ➙ **11-40** $x + 1$ by $x - 1$

11-41 $x - 1$ by $x - 1$

11-42 $x - 2$ by $x + 2$

11-43 $x + 3$ by $x + 4$

11-44 $x + 8$ by $x - 3$

11-45 $x - 11$ by $x + 1$

11-46 $x - 14$ by $x + 5$

11-47 $3x - 3$ by $x - 4$

11-48 $4x + 5$ by $5x - 6$

11-49 $10x - 1$ by $x - 10$

Hw ➙ **11-50** $10x + 1$ by $10x - 1$

11-51 $5x - 4$ by $5x + 4$

11-52 $x^2 - x + 1$ by $x + 1$

11-53 $x^2 + xy + y^2$ by $x - y$

11-54 $a^2 - 4a + 4$ by $a^2 - 2a$

11-55 $b^2 + 5b$ by $b^2 - 5b$

11-56 $3a - 4x$ by $4a - 3x$

11-57 $3x^2 - 4b^2$ by $3x^2 + 4b^2$

11-58 $-2ax + b$ by $2ax + b$

11-59 $-7x^2 + 7x$ by $7x^2 - 7x$

11-60 $-by + ax$ by $ax - by$

11-61 $ax^2 - by^2$ by $ax^2 + by^2$

11-62 $(ax)^2 - by^2$ by $(ax)^2 + by^2$

11-63 $2a + 3x + y$ by $2a + 3x - y$

11-64 $-(-x^2)^2 + (y^2)^2$ by $x^4 + y^4$

11-65 $x^2 - xy + y^2 - 1$ by $x + y$

11-66 $x^2 + xy + y^2 - 1$ by $x - y$

11-67 $2x^2 + y^2 - 2z^2$ by $2x^2 + y^2 + 2z^2$

11-68 $x^2 - 1$ by $x^2 - 4$

11-69 $(2x)^3 - 3(2x)^2 + 3(2x) - 1$ by $(2x + 1)^2$

11-70 $(3a - 4b)^2$ by $(3a + 4b)^2$

11-71 $x^2 - y^2$ by $x^4 + x^2y^2 + y^4$

11-72 $81x^2 - 16y^2 - 25z^2$ by $81x^2 + 16y^2 + 25z^2$

11-73 $(a - x)^3$ by $(a + x)^3$

Note: *The signs of grouping are often used to enclose the factors of a product without any further sign of multiplication. Thus, $(a + b)(a - b)$ means the same as $(a + b)$ times $(a - b)$. To remove the sign of grouping, perform the multiplication.*

Remove the parentheses and simplify the products:

11-74 $(2x + 1)(2x - 1)$

11-75 $(3t - 4u)(3t + 4u)$

11-76 $(ax^2 - by^2)(ax^2 + by^2)$

11-77 $4(x - y)^2 + 4(2xy) - 4(x^2 + y^2)$

11-78 $a^3 + x^3 + 3ax(a + x) - (a + x)(a^2 - ax + x^2)$

11-79 $(x - 2y)(2 + 2y) - 2(-2y + x)$

11-80 Find the area of a rectangle that is s units longer than it is wide.

Hint: *Draw a figure. Let $x = $ the width, then $x(x + s) = $ area.*

11-81 Find the volume of a rectangular solid $2x - 3$ m wide, $7x - 2$ m long, and $x + 4$ m deep. Draw the figure. The formula is $V = lwh$.

11-82 Find the volume of a right circular cylinder where the altitude is h ft and the radius is $2h - 4$ ft. Volume of a cylinder $= \pi r^2 h$, where r is radius and h is altitude. Draw the figure. (Leave answer in terms of π.)

11-83 A hollow square has dimensions as shown in Fig. 11-1. Find its area.

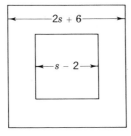
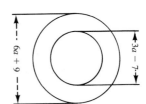

Fig. 11-1. Exercise 11-83.   **Fig. 11-2.** Exercise 11-86.

11-84 A strip of sheet iron w cm wide is bent up at the edges to form an open rectangular gutter d cm deep. Find the area A of the cross section of the gutter.

11-85 A rectangular sheet of tin a cm long and b cm wide is made into an open box by cutting a square c cm on a side from each corner and turning up the sides. Draw the figures. Find the volume of the box.

11-86 A ring has dimensions as shown in Fig. 11-2. Find its area. (Area of circle $= \frac{1}{4}\pi d^2$. Leave answer in terms of π.)

11-8 SOLVING EQUATIONS INVOLVING MULTIPLICATION

Multiplication is often necessary to solve equations. In equations of this type, terms may appear which contain squares or higher powers of the unknown. Thus far you are able to solve such equations only when the terms containing squares or higher powers disappear in the process of transposing and uniting terms.

EXAMPLE 11-16 Solve $(x + 8)(x - 5) = (x - 7)(x + 3) + 9$.

Solution Multiplying $(x + 8)(x - 5)$ and $(x - 7)(x + 3)$ we obtain

$$x^2 + 3x - 40 = x^2 - 4x - 21 + 9$$

Adding or subtracting from both members x^2, $+4x$, $+40$.

$$3x + 4x = 40 - 21 + 9$$

Collecting terms, $\qquad\qquad 7x = 28$

Dividing by 7, $\qquad\qquad x = 4$

Check: Substituting 4 for x in the original equation,

$$(4 + 8)(4 - 5) \overset{?}{=} (4 - 7)(4 + 3) + 9$$
$$(12)(-1) \overset{?}{=} (-3)(7) + 9$$
$$-12 \overset{?}{=} -21 + 9$$
$$-12 = -12$$

EXERCISES

Solve the following equations:

11-87 $(x - 4)^2 = x^2 - 40$

11-88 $(x - 3)^2 = (x + 3)^2 - 24$

11-89 $9(x - 10) = -(x - 10)$

11-90 $20(x + 2) = 2(x + 20)$

11-91 $(x - 3)^2 + 40 = (x + 7)^2 + 200$

11-92 $(2x + 1)^2 = 4(x + 2)^2$

11-93 $(3x - 2)^2 = 3x(3x + 1)$

11-94 $(2x - 3)^2 = 4(x + 2)^2$

11-95 $(6x + 2)(5x - 4) - 30(x - 1)^2 = 34x + 106$

11-96 $6x^2 - 27x + 72 = 3x(2x + 3)$

11-97 $(s + 1)(3s + 1) = 3s^2 + 7s - 13$

11-98 $(h + 1)(h^2 - h + 1) = h^3 - 8h - 31$

11-99 If $x - 14$ is multiplied by $x - 10$, the product is 20 more than x^2. Find x.

11-100 If a number plus 6 is multiplied by the same number minus 13, the product is 27 more than the square of the number. Find the number.

11-101 A rectangle that is 1 m wider and 3 m longer than a square has an area

91 m² greater than the area of that square. Find the dimensions of the square and of the rectangle.

11-102 Prove that the product of the first and the third of three consecutive numbers is always 1 less than the square of the second.

11-103 A person's salary is $12,000 the first year; it increases p dollars each year. Write a formula for the salary s for the nth year.

11-104 n people received the same salary of x dollars per year for t years. For $(m + t)$ years they received together a total sum of P dollars. Write a formula for x; for m; for n.

11-105 Ten students bought a radio. Because four of them were unable to pay their share, each of the others had to pay a total of $8 more. What was the cost of the radio?

DIVISION OF SIGNED NUMBERS AND FACTORING

12-1 INTRODUCTION TO DIVISION WITH SIGNED NUMBERS

Division is repeated subtraction. Division is considered the inverse of multiplication; many mathematical principles and rules have been developed from this fact. As examples, consider these four multiplication problems:

$$(+5) \times (+3) = +15 \qquad (-5) \times (+3) = -15$$
$$(-5) \times (-3) = +15 \qquad (+5) \times (-3) = -15$$

Since division is the inverse operation of multiplication, when the product of two numbers is divided by one of them, the quotient (the answer) is the other. Thus when one of the four answers obtained above is divided by either of the two numbers which were multiplied, the result of that division will be the other number:

$$(+5) \times (+3) = +15 \qquad (+15) \div (+3) = +5$$
$$(-5) \times (-3) = +15 \qquad (-15) \div (-3) = +5$$
$$(+5) \times (-3) = -15 \qquad (-15) \div (+5) = -3$$
$$(-5) \times (+3) = -15 \qquad (-15) \div (+3) = -5$$

From the preceding examples, we can draw the following conclusions.

Procedure for dividing signed numbers:

1. The answer of a division problem is the quotient; it is + when the signs of the dividend and the divisor are alike; the sign of the answer is − when the signs of the dividend and divisor are unlike.

2. The numerical part of the quotient is obtained in the same way as in arithmetic.

12-2 THE LAW OF EXPONENTS

The law applying to exponents in division problems is easily illustrated. Consider the following problem:

$$\frac{a^5}{a^3} = \frac{\overset{1}{\cancel{a}} \times \overset{1}{\cancel{a}} \times \overset{1}{\cancel{a}} \times a \times a}{\underset{1}{\cancel{a}} \times \underset{1}{\cancel{a}} \times \underset{1}{\cancel{a}}} = a \times a = a^2$$

or

$$a^5 \div a^3 = a^{5-3} = a^2$$

The following rule may be derived:

Procedure for dividing quantities with exponents:

To divide powers of the same base, subtract the exponent of the divisor from the exponent of the dividend.

　　This statement is illustrated in each of the following three examples.

EXAMPLE 12-1　$b^4 \div b^4 = b^{4-4} = b^0 = 1.$

Solution　$\dfrac{b^4}{b^4} = \dfrac{\overset{1}{\cancel{b}} \cdot \overset{1}{\cancel{b}} \cdot \overset{1}{\cancel{b}} \cdot \overset{1}{\cancel{b}}}{\underset{1}{\cancel{b}} \cdot \underset{1}{\cancel{b}} \cdot \underset{1}{\cancel{b}} \cdot \underset{1}{\cancel{b}}} = 1$

Note: *Any quantity to the 0 power is 1.*

EXAMPLE 12-2　$c^3 \div c^5 = c^{-2}.$ The answer is c^{-2} or $1/c^2$. This can be easily demonstrated:

Solution　$c^3 \div c^5 = \dfrac{\overset{1}{\cancel{c}} \times \overset{1}{\cancel{c}} \times \overset{1}{\cancel{c}}}{\underset{1}{\cancel{c}} \times \underset{1}{\cancel{c}} \times \underset{1}{\cancel{c}} \times c \times c} = \dfrac{1}{c^2}$

Thus it is possible to change the sign of the exponent by shifting a number from the numerator to the denominator or from the denominator to the numerator:

$$\frac{c^{-2}}{1} = \frac{1}{c^2} \qquad \text{and} \qquad \frac{x^3}{1} = \frac{1}{x^{-3}}$$

EXAMPLE 12-3　$(a + b)^4 \div (a + b)^2$

Solution　$\dfrac{(a + b)^4}{(a + b)^2} = (a + b)^{4-2} = (a + b)^2$

or

$$\frac{\overset{1}{\cancel{(a+b)}}\,\overset{1}{\cancel{(a+b)}}(a + b)(a + b)}{\underset{1}{\cancel{(a+b)}}\,\underset{1}{\cancel{(a+b)}}} = (a + b)^2$$

　　Note from the preceding calculations that any number to the 0 power is equal to 1 and that parentheses may be retained (as in Example 12-3) to shorten what would otherwise be a much longer problem. The term $(a + b)^4$ may be divided by $(a + b)^2$ in the same manner that a^5 is divided by a^3.

Try checking the results of the preceding examples by substituting numbers for the letters.

12-3 DIVIDING ONE MONOMIAL BY ANOTHER

You should learn to perform division in a regular order. The steps in division are as follows:

Procedure for dividing monomials:

1. Determine the sign of the quotient. (Like signs result in $+$, unlike signs in $-$.)

2. Determine the numerical coefficient. (Cancel the factors common to the dividend and the divisor.)

3. Determine literal coefficients. (Subtract the exponents of like letters in the divisor from the exponents of the like letters in the dividend.)

Remember: In division you divide where you multiplied in multiplication, and you subtract exponents where you added exponents in multiplication.

EXAMPLE 12-4 Divide $25a^4x^5$ by $-5a^2x^3$.

Solution Write as shown and make necessary cancellations.

$$\frac{25a^4x^5}{-5a^2x^3} = \frac{\overset{-5}{\cancel{25}}a^{4-2}x^{5-3}}{\underset{1}{\cancel{-5}}} = -5a^2x^2$$

Only the last step need be written in performing the work. The first three steps are mental operations and are shown here for further illustration.

The division of one monomial by another may also be performed as a cancellation. Since an expression like $4a^2b^3$ means $4 \times a \times a \times b \times b \times b$, you can write $16a^3b^5c^3 \div 4a^2b^3$ in the following form:

$$\frac{\overset{1}{\cancel{4}} \times 4 \times \overset{1}{\cancel{a}} \times \overset{1}{\cancel{a}} \times a \times \overset{1}{\cancel{b}} \times \overset{1}{\cancel{b}} \times \overset{1}{\cancel{b}} \times b \times b \times c \times c \times c}{\underset{1}{\cancel{4}} \times \underset{1}{\cancel{a}} \times \underset{1}{\cancel{a}} \times \underset{1}{\cancel{b}} \times \underset{1}{\cancel{b}} \times \underset{1}{\cancel{b}}} = 4ab^2c^3$$

The factors common to the dividend and the divisor are canceled, and the product of the factors remaining in the dividend is the quotient.

This process is too time-consuming for rapid work, but it can clear up points in division which may trouble you.

12-4 CHECKING DIVISION WITH SIGNED NUMBERS AND EXPONENTS

Problems in division can be checked in the same way as multiplication, by substituting convenient values for letters. Division can also be checked by multiplying the divisor by the quotient; the product will be the dividend.

▦ EXERCISES

12-1 $\dfrac{9a}{-3a} =$

12-2 $\dfrac{-12b^2x}{-3b^2x} =$

12-3 $\dfrac{15a^2b^2}{-5a^2b} =$

12-4 $\dfrac{-20u^3v^2}{-5u^2v} =$

12-5 $\dfrac{-24x^4t^3}{-8xt^2} =$

Find the quotients in the following:

12-6 $-25a^2b^2c^3 \div 5abc^2$ **12-10** $-18x^3yz^2 \div 9xy$

12-7 $-20a^5b^6c \div 10abc$ **12-11** $-21vw^2z^2 \div -7vz^2$

12-8 $-30cd^2f \div 15cd^2$ **12-12** $-33r^3sz^2 \div -11rs$

12-9 $-36ax^3y^5 \div 18ay^3$ **12-13** $-35m^2n^3x^4 \div -5m^2x^2$

Divide the following:

12-14 $65ab^5c^8$ by $13ac^6$

12-15 $-21^4ab^4cd^4$ by 21^3abcd^3

12-5 DIVIDING A POLYNOMIAL BY A MONOMIAL

To divide a polynomial by a monomial, divide each term of the dividend by the divisor.

EXAMPLE 12-5   Divide $24a^5y^3 - 96a^5y^6$ by $8a^4y^3$.

Solution

$$8a^4y^3\overline{)\,\begin{array}{l} 3a - 12ay^3 \\ 24a^5y^3 - 96a^5y^6 \\ \underline{24a^5y^3 - 96a^5y^6} \end{array}}$$

The answer, $3a - 12ay^3$, can be checked by multiplication. You could also solve this division problem by writing it in this form:

$$\frac{24a^5y^3}{8a^4y^3} - \frac{96a^5y^6}{8a^4y^3} = 3a - 12ay^3$$

■ EXERCISES

Divide the following and check by multiplication:

12-16 $14ax + 28ay + 84az$ by a; by $14a$

12-17 $12x^4 - 16x^3y + 20x^5$ by $2x^2$; by $-4x^3$

12-18 $-3ab^4 + 6ab^5 - 9ab^6$ by ab^4; by $-3ab^2$

12-19 $5(41)^2 - 4(41) + 7(41)^3$ by $+(41)$

12-20 $3(5^2) - 7(5)^3 + 9(5)(5)$ by 5^2

12-21 $2.5xyz^2 - 0.5x^2yz + 1.5xyz$ by $0.5xz$

12-22 $16a^3b^2c - 20a^2b + 24a^2b^2$ by $-4a^2b$

12-23 $24(xy)^2 - 8(x^2y^2)y - 24(xy)y$ by $-8xy$

12-24 $8a^7 - 7a^6 + 6a^5 - 5a^4$ by $2a^3$

12-25 $y^4 - 1.5y^3 + 0.5y^2$ by $0.5y^2$

12-26 $3x^6 + 2x^4 - x^2$ by $0.5x$

12-27 $0.25a^5b^4 - 2a^4b^3 + a^3b^2$ by $\dfrac{3ab}{2}$

12-28 $4 - 4x - 5x^2 + 7x^3 - 10x^4$ by 0.5

12-29 $a(1 + a)^4 - ab(1 + a)^3 + 7a(1 + a)^2$ by a; by $a(1 + a)$; by $a + 1$

12-30 $24(x + y)^3 - 12(x + y)^2 - 6(x + y)$ by $-6(x + y)$

12-31 $3(m - n)(a + b)^5 - 6(m - n)(a + b)^{10}$ by $-3(m - n)(a + b)^5$; by $(a + b)^2$

12-6 DIVIDING A POLYNOMIAL BY A POLYNOMIAL

The division of a polynomial by a polynomial is similar to the process of long division in arithmetic. Four examples will illustrate.

EXAMPLE 12-6 Divide $x^2 + 7x + 12$ by $x + 3$.

Solution First "decide" that x goes into x^2, x times. Then multiply x by $x + 3$ and subtract the product from $x^2 + 7x$; then "decide" x goes into $4x$, 4 times, multiply 4 by $x + 3$ and subtract the product from $4x + 12$.

$$
\begin{array}{r}
x + 4 \\
x + 3 \overline{)\ x^2 + 7x + 12} \\
\underline{x^2 + 3x } \\
4x + 12 \\
\underline{4x + 12}
\end{array}
$$

Check by multiplying the answer (quotient) by the divisor:

$$
\begin{array}{r}
x + 4 \\
\times\ x + 3 \\
\hline
x^2 + 4x \\
3x + 12 \\
\hline
x^2 + 7x + 12
\end{array}
$$

EXAMPLE 12-7 Divide $x^2 - x^3 + x^4 - 3x + 2$ by $x^2 + x + 2$.

Solution Arrange in order shown and follow the same procedure used in Example 12-6.

$$
\begin{array}{r}
x^2 - 2x + 1 \\
x^2 + x + 2 \overline{) x^4 - x^3 + x^2 - 3x + 2} \\
\underline{x^4 + x^3 + 2x^2} \\
-2x^3 - x^2 - 3x \\
\underline{-2x^3 - 2x^2 - 4x} \\
x^2 + x + 2 \\
\underline{x^2 + x + 2}
\end{array}
$$

Check:

$$
\begin{array}{r}
x^2 - 2x + 1 \\
\times\ x^2 + x + 2 \\
\hline
x^4 - 2x^3 + x^2 \\
x^3 - 2x^2 + x \\
\underline{2x^2 - 4x + 2} \\
x^4 - x^3 + x^2 - 3x + 2
\end{array}
$$

Procedure for dividing polynomials:

1. Arrange the dividend and divisor in descending (or ascending) powers, beginning with the highest (or lowest) power of a common letter.
2. Divide the first term of the dividend by the first term of the divisor.
3. Multiply all terms in the divisor by the result of the preceding division, write the products under like terms, and subtract.
4. Divide the first term of this remainder by the first term of the divisor.
5. Multiply the divisor by the second term of the quotient (the answer obtained in step 4).
6. Continue steps 4 and 5 until the remainder is zero or until the last subtraction results in an answer which cannot be divided.

A third example illustrates this technique of division when the result of the last subtraction is not 0.

EXAMPLE 12-8 Divide $a^3 + 2a^2 + 2a + 5$ by $a + 1$.

Solution Follow the procedure previously stated.

$$
\begin{array}{r}
a^2 + a + 1 \\
a + 1 \overline{) a^3 + 2a^2 + 2a + 5} \\
\underline{a^3 + a^2} \\
a^2 + 2a \\
\underline{a^2 + a} \\
a + 5 \\
\underline{a + 1} \\
4
\end{array}
$$

In this problem, the remainder 4 cannot be divided by the first term of the divisor a. Therefore, 4 is carried in fractional form as in arithmetic. The complete answer for this division problem is, then,

$$a^2 + a + 1 + \frac{4}{a + 1}$$

Check:

$$\begin{array}{rl}
 a^2 + a + 1 & \\
\times \quad a + 1 & \\
\hline
a^3 + a^2 + a \phantom{{}+1} & \\
a^2 + a + 1 & \\
\hline
a^3 + 2a^2 + 2a + 1 & \\
+ 4 & \text{the remainder} \\
\hline
a^3 + 2a^2 + 2a + 5 &
\end{array}$$

EXAMPLE 12-9 Divide $x^3 - 27$ by $x + 3$.

Solution Because the x^2 and x terms are missing in the dividend, it is convenient to make allowance for these missing terms when dividing:

$$\begin{array}{r|l}
 & x^2 - 3x + 9 \\
\hline
x + 3 & x^3 + 0x^2 + 0x - 27 \\
 & x^3 + 3x^2 \\
\hline
 & -3x^2 \\
 & -3x^2 - 9x \\
\hline
 & +9x - 27 \\
 & 9x + 27 \\
\hline
 & -54 \quad \text{remainder}
\end{array}$$

The complete answer is, then,

$$x^2 - 3x + 9 - \frac{54}{x + 3}$$

Check:

$$\begin{array}{r}
x^2 - 3x + 9 \phantom{{}+27} \\
\times \quad x + 3 \phantom{{}+27} \\
\hline
x^3 - 3x^2 + 9x \phantom{{}+27} \\
+ 3x^2 - 9x + 27 \\
\hline
x^3 \phantom{{}- 3x^2 - 9x} + 27 \\
-54 \\
\hline
x^3 - 27
\end{array}$$

EXERCISES

Divide the following and check by multiplication:

12-32 $x^2 + 5x + 4$ by $x + 1$

12-33 $x^2 - 5x + 4$ by $x - 1$

12-34 $14 + 9x + x^2$ by $7 + x$

12-35 $14 - 9x + x^2$ by $7 - x$

12-36 $x^2 + 16x + 63$ by $x + 9$

12-37 $x^2 - 16x + 60$ by $x - 6$

12-38 $x^2 - 2x - 63$ by $x - 9$

12-39 $x^2 + 2x - 63$ by $x + 9$

12-40 $x^3 - 3x^2 + 3x - 1$ by $x - 1$

12-41 $x^3 + 3x^2 + 3x + 1$ by $x^2 + 2x + 1$

12-42 $x^3 + 2x^2 - 5x - 6$ by $x^2 - x - 2$

12-43 $x^3 + x - 4x^2 - 4$ by $x - 4$

12-44 $x^4 - 16$ by $x + 2$

12-45 $27x^6 - 8y^6$ by $3x^2 - 2y^2$

12-46 $2a^3 - 3a^2b + 4ab^2 - 12b^3$ by $a - 2b$

12-47 If the area of a rectangle is expressed as $x^2 + 3x + 2$ and one side is $x + 1$, find the other side. Check by substituting $x + 5$; $x = 0$.

12-48 The area of a rectangle is expressed as $x^2 + x$. In terms of x, find the lengths of the two sides.

12-49 A duck hunter rows x mi up a stream and returns in a total time of t h. She can row u mph, and the stream flows v mph. It is assumed that the two rates and the time are known, and x is to be found. Show by solving the problem that all four fundamental operations (addition, subtraction, multiplication, division) are required. If $u = 1$, $v = .5$, and $t = 6$, find x.

12-7 FACTORING

A *factor* of an algebraic expression is a quantity or term that will divide evenly into the expression. An expression is said to be *prime* when it cannot be divided by any factors other than itself and one. Simply, a prime expression cannot be divided by any quantity except itself or one. An expression has been *factored* when it has been resolved to its prime factors.

Factoring is very closely related to multiplication and division. Actually, *factoring* an algebraic expression means dividing it into its prime factors. Factoring is a useful tool for solving equations and simplifying formulas when necessary or convenient.

12-8 FACTORS OF A MONOMIAL

The factors of a monomial, except for its numeral coefficient, are evident from the meaning of the symbols used. For this reason it is not necessary to discuss the factoring of monomials. Of course, the factoring of the numerical coefficient is simple arithmetic. Thus, the factors of a^4x^2y are a

repeated four times, x repeated twice, and y. The factors of $14x^2y^3$ are $2 \times 7 \times x \times x \times y \times y \times y$.

Because its factors are obvious, a monomial is not usually not separated into prime factors.

12-9 FACTORS OF A POLYNOMIAL WHEN ONE FACTOR IS A MONOMIAL

In factoring $14ax + 28ay + 84az$, we observe that $14a$ can be divided evenly into each term of the polynomial without a remainder. The quotient is $x + 2y + 6z$. The product of $x + 2y + 6z$ and $14a$ is $14ax + 28ay + 84az$. Thus $14a$ and $x + 2y + 6z$ are the factors of $14ax + 28ay + 84az$, or $14ax + 28ay + 84az = 14a(x + 2y + 6z)$.

The factors of a polynomial like the one above are (1) a monomial, containing all that is common to each term of the polynomial and (2) the quotient, found by dividing the polynomial by the monomial.

To factor a polynomial when one factor is a monomial, follow these steps:

Procedure for factoring a monomial from a polynomial:

1. Inspect the terms of the polynomial and determine a monomial that will divide into all these terms. This is one factor.

2. Divide the polynomial by this monomial and find the quotient. This result is the other factor.

EXAMPLE 12-10 Factor $4a^2x - 2ax^2 + 6a^2x^2$.

Solution The monomial factor, determined by inspection, is $2ax$. Divide the polynomial by $2ax$; the quotient is $2a - x + 3ax$, the other factor. Factors are written in the following form:

$$4a^2x - 2ax^2 + 6a^2x^2 = 2ax(2a - x + 3ax)$$

The result may be proved by finding the product of the factors; the product should equal the expression which was factored.

▮ EXERCISES

Factor the following polynomials by inspection, and check by multiplication:

12-50 $3b + 12$

12-51 $7x - 21$

12-52 $15xy + 30z$

12-53 $12xy - 30xz$

12-54 $9x^2y + 21x$

12-55 $4u^2v^3 - 12uv^2$

12-56 $7ab - 14ac + 21ad$

12-57 $12abc^2 - 4a^2bc + 6ab^2c$

12-58 $5axy^4 - 6ax^4y + 7a^4xy$

12-59 $13 - 26hk - 39uv$

12-60 $15ap^3 - 30a^2p^2 + 5p^4$

12-61 $100m^2 - 200mn + 300mn^2$

12-62 $250x^3 - 1000x^6y$

12-63 $17A^2 - 51B^2$

12-64 $15A^2B^2 + 30A^3B^3$

12-65 $(x - 2)a + (x - 2)b$

Suggestion: *Consider $(x - 2)a$ and $(x - 2)b$ as the terms of the polynomial. Then $(x - 2)$ is common to the two terms and is the factor to use as a divisor in finding the other factor.*

12-66 $(a + b)x^2 - (a + b)y^2$

12-67 $(2x - 3y)xy + (2x - 3y)$

12-68 $5(7ax + 5) - 15(7ax + 5)x$

12-69 $10x^2(s + t) - 20xy(s + t)^3$

12-70 $5^2a^2(x + y)^2 - 10a(x + y)$

12-10 THE SQUARE OF A BINOMIAL

The square of a binomial is a special product that occurs very frequently. It is, therefore, convenient to write the square without actually multiplying. By actual multiplication:

$$(a + b)^2 = (a + b)(a + b) = a^2 + ab + ab + b^2$$
$$= a^2 + 2ab + b^2$$

Similarly,

$$(a - b)^2 = a^2 - ab - ab + b^2 = a^2 - 2ab + b^2$$

Thus

$$(a + b)^2 = a^2 + 2ab + b^2$$
$$(a - b)^2 = a^2 - 2ab + b^2$$

▧ EXERCISES

Write the product of each of the following. Then test by actual multiplication.

12-71	$(x + 6)^2$	**12-77**	$(2ax - 3by)^2$
12-72	$(2x - 6)^2$	**12-78**	$(2x^3 + 3xy^2)^2$
12-73	$(2x - 6y)^2$	**12-79**	$(3^2 - 2^2)^2$
12-74	$(A^2 - 2)^2$	**12-80**	$(20 + 1)^2$
12-75	$(2b^2 + 1)^2$	**12-81**	$(50 - 1)^2$
12-76	$(2u^2 - av)^2$	**12-82**	$(100 - 1)^2$

12-11 FACTORS OF A TRINOMIAL SQUARE

A *trinomial square* is a trinomial that is the square of a binomial. Thus, $a^2 + 2ab + b^2$ is a trinomial square because it is the square of the binomial $a + b$. Its factors are $(a + b)(a + b)$. Similarly, the factors of $a^2 - 2ab + b^2$ are $(a - b)(a - b)$. The factors of $4x^2 - 12x + 9$ are $(2x - 3)(2x - 3)$. You should learn to tell whether a trinomial is a perfect square.

Procedure to determine a trinomial square:

1. A perfect trinomial square must have two positive terms, each of which is the square of a monomial.

2. The trinomial also must have one term, either positive or negative, which is twice the product of the square roots of the other two terms. If this term is positive, the factors are sums; if negative, the factors are differences.

Thus, $9a^4 - 24a^2y^2 + 16y^4$ is a trinomial square because $9a^4$ and $16y^4$ are each positive and are squares of the monomials $3a^2$ and $4y^2$.

Also $24a^2y^2$ is twice the product of the square roots of $9a^4$ and $16y^4$. Therefore the factors of $9a^4 - 24a^2y^2 + 16y^4$ are $(3a^2 - 4y^2)(3a^2 - 4y^2)$.

By definition, the square root of a number is one of its two equal factors. The square root of a trinomial square, therefore, is one of its two equal factors.

▦ EXERCISES

Determine which of the following are trinomial squares. Factor and find the positive square root when possible.

12-83 $x^2 + 2x + 1$

12-84 $x^2 + 4x + 4$

12-85 $4y^2 - 4y + 1$

12-86 $16u^2 + 16u + 4$

12-87 $9v^2 - 18v + 9$

12-88 $16a^2b^2 - 8ab^2c^2 + b^2c^4$

12-89 $9 + 6x^4 + x^8$

12-90 $-30x + 225 + x^2$

12-91 $4x^2 + 6xy + 8y^2$

12-92 $16abc + 16a^2b^2 - 4c^2$

12-93 $0.16t^2 + 0.8t + 1$

12-94 $0.25x^2 - 0.25x + \frac{1}{16}$

12-95 $4x^6 + 12x^3y^2 + 9y^4$

12-96 $108U^2V^2 + 36U^4 + 81V^4$

12-97 $-40ST + 16S^2 + 25T^2$

12-98 $-(-112R + 49R^2 + 64)$

12-99 $x^2 + 2x(a + b) + (a + b)^2$

12-100 $(a + b)^2 - 2(a + b)(a - b) + (a - b)^2$

12-12 THE PRODUCT OF THE SUM OF TWO QUANTITIES TIMES THE DIFFERENCE OF THE SAME TWO QUANTITIES

A very common multiplication problem involves finding the product when the sum of any two numbers is multiplied by their difference. By actual multiplication,

$$(a + b)(a - b) = a^2 + ab - ab + b^2 = a^2 - b^2$$

Thus

$$(a + b)(a - b) = a^2 - b^2$$

Here a and b are general numbers; therefore this statement can be used as a formula, so that without actual multiplication you can write the product of the sum and the difference of any two numbers. This formula is stated as follows:

Procedure for finding the product of the sum and difference of two quantities:

The product of the sum and the difference of two numbers equals the difference of their squares.

EXAMPLE 12-11 $(2c + 3b)(2c - 3b) = 4c^2 - 9b^2$

EXAMPLE 12-12 $(16 + 2)(16 - 2) = 16^2 - 2^2 = 256 - 4 = 252$

EXAMPLE 12-13 $102 \times 98 = (100 + 2)(100 - 2) = 10{,}000 - 4 = 9996$

▨ EXERCISES

By inspection, find the product of each of the following. Then check by actual multiplication:

12-101 $(x + 5)(x - 5)$

12-102 $(2x + 5)(2x - 5)$

12-103 $(5xy - 6)(5xy + 6)$

12-104 $(12 + 9RS)(12 - 9RS)$

12-105 $(3xyv - 4ab)(3xyv + 4ab)$

12-106 $(3ab^2c - 4ad^3)(3ab^2c + 4ad^3)$

12-107 $(11axt^2v^3 + w^4)(11axt^2v^3 - w^4)$

12-108 $(2c + d + e)(2c + d - e)$

Suggestion: *This may be written*

$$[(2c + d) + e][(2c + d) - e] = (2c + d)^2 - e^2 = 4c^2 + 4cd + d^2 - e^2$$

12-109 $[(a + 4) - b][(a + 4) + b]$

12-110 $[(x - y) + z][(x - y) - z]$

12-111 $(3 - x + y)(3 + x + y)$

Suggestion: *This may be written*

$$[(3 + y) - x][(3 + y) + x] = (3 + y)^2 - x^2 = ?$$

12-112 $(a + b + 5)(a + b - 5)$

12-113 $(a^2 - b^2 - ab)(a^2 + b^2 + ab) = [a^2 - (b^2 + ab)][a^2 + (b^2 + ab)] = ?$

12-114 $(10 + 2a + 3b)(10 - 2a - 3b)$

12-115 $(a + b + 7)(a - b + 7)$

12-116 $(10ax^2 + 9bc)(9bc - 10ax^2)$

12-117 97×103

12-118 44×36

12-119 702×698

12-13 FACTORS OF THE DIFFERENCE OF TWO SQUARES

Remembering the previous explanations it is easy to see that the difference between two squares can be factored into two binomials, of which one is the sum and the other is the difference of the square roots of these squares.

EXAMPLE 12-14 $a^2 - 4 = (a + 2)(a - 2)$

EXAMPLE 12-15 $16a^4 - 9y^2 = (4a^2 + 3y)(4a^2 - 3y)$

EXAMPLE 12-16 $(a + b)^2 - 2^2 = (a + b - 2)(a + b + 2)$

EXAMPLE 12-17 $a^2 - b^2 + 2bc - c^2 = a^2 - (b^2 - 2bc + c^2)$
$$= a^2 - (b - c)^2 = (a + b - c)(a - b + c)$$

EXERCISES

Factor the following and test by multiplication:

12-120 $16 - x^2$

12-121 $9x^2 - y^2$

12-122 $4U^2 - 4V^2$

12-123 $25a^2 - 64c^2$

12-124 $25a^2 - 9b^2$

12-125 $x^2y^2 - 4y^2z^2$

12-126 $(xy)^2 - 9z^2$

12-127 $4(ab)^2 - (3c)^2$

12-128 $(2abc)^2d^2 - 16$

12-129 $A^2(bc)^2 - 64(10)^2$

12-130 $2^2a^2b^4 - 4^2c^4$

12-131 $2^23^4a^2x^4 - 7^2$

12-132 $(a + 2)^2 - x^2$

12-133 $a^2 + 2ab + b^2 - c^2$

12-134 $a^2 - 4b^2 - 4bc - c^2$

12-135 $4 - (x + 2y)^2$

12-136 $100 - (a - b)^2$

12-137 $4b^2 + 9c^2 - 16x^2 - 12bc$

12-138 $(4 - x)^2 - (x - y)^2$

12-14 GENERAL METHOD FOR FACTORING

Many special methods have been developed for determining products and factoring, but it is always possible to find a product by performing the actual multiplication. You must learn how to factor by actual division.

EXAMPLE 12-18 Factor $a^3 + b^3$.

Solution Note that the numerical coefficients of both terms are 1; therefore, it may be possible to divide $a + b$ into $a^3 + b^3$

$$
\begin{array}{r}
a^2 - ab + b^2 \\
a + b \overline{)\, a^3 + 0a^2b + 0ab^2 + b^3} \\
\underline{a^3 + a^2b} \\
-\ a^2b + 0ab^2 \\
\underline{-\ a^2b - ab^2} \\
ab^2 + b^3 \\
\underline{ab^2 + b^3}
\end{array}
$$

Thus

$$a^3 + b^3 = (a + b)(a^2 - ab + b^2)$$

The two factors of $a^3 + b^3$ are then $(a + b)(a^2 - ab + b^2)$. Because these two factors are prime, the expression $a^3 + b^3$ is now completely factored.

EXAMPLE 12-19 Factor $y^2 - 2y - 63$.

Solution This trinomial is not a trinomial square because -63 is a negative number and the square root of 63 is not a whole number. To understand the factoring of this kind of expression, consider the following multiplication problem:

$$
\begin{array}{r}
x\ + 5 \\
\times\ x\ - 2 \\
\hline
x^2 + 5x \\
-\ 2x - 10 \\
\hline
x^2 + 3x - 10
\end{array}
$$

Observe that the x^2 is obtained by multiplying x by x and that the -10 results from multiplying 5 by -2. The $+3x$ was derived when the 5 was multiplied by x, then -2 was multiplied by x, and the products of these multiplications were added together. The factors of $x^2 + 3x - 10$ are evidently $(x + 5)(x - 2)$, but the actual factoring of this trinomial is illustrated because the principle involved will apply to all problems involving factoring.

To factor $x^2 + 3x - 10$ by reasoning, you must first realize that the factors of x^2 are x times x and the factors of -10 are -2 and 5. You can then guess that $(x - 2)$ and $(x + 5)$ are possible factors of $x^2 + 3x - 10$.

Check your guess by dividing either $(x - 2)$ or $(x + 5)$ into $x^2 + 3x - 10$:

$$
\begin{array}{r}
x + 5 \\
x - 2\overline{)\,x^2 + 3x - 10} \\
\underline{x^2 - 2x} \\
5x - 10 \\
\underline{5x - 10}
\end{array}
$$

This division shows that $(x - 2)$ and $(x + 5)$ are factors of $x^2 + 3x - 10$.

A variation of the actual division is also possible. Again, note that $x \times x = x^2$ and $(-2) \times 5 = -10$. Next, draw two sets of parentheses so that you can place a binomial within each set:

$$(\quad)(\quad)$$

Now, place the observed factors within these parentheses:

$$(x + 5)(x - 2) = x^2 + 3x - 10$$

Multiply each term in one binomial by each term in the other and write the products. When you multiply 5 times x and x times -2, they are like terms; therefore, adding $+5x$ to $-2x = 3x$. Write this sum after x^2. The product, $-2 \times 5 = -10$, is written after $3x$.

Study the following illustrations carefully and note the relations between division, multiplication, and the two methods used to factor the trinomial $y^2 - 2y - 63$.

Factoring by the first method,

1. The factors of the first term y^2 are $y \times y$.

2. The factors of the third term are $(-7) \times (+9)$ or $(+7) \times (-9)$.

$$
\begin{array}{r}
y - 9 \\
y + 7\overline{)\,y^2 - 2y - 63} \\
\underline{y^2 + 7y} \\
- 9y - 63 \\
\underline{- 9y - 63}
\end{array}
$$

The two factors of $y^2 - 2y - 63$ are $(y + 7)$ and $(y - 9)$.

Using the second method, try $(y - 7)(y + 9)$:

$$(y - 7)(y + 9) = y^2 + 2y - 63$$

Because the second term of the trinomial is $+2y$, then $(y - 7)$ and $(y + 9)$ are not factors of $y^2 - 2y - 63$.

Try $(y + 7)(y - 9)$:

$$(y + 7)(y - 9) = y^2 - 2y - 63$$

Therefore $y^2 - 2y - 63 = (y + 7)(y - 9)$.

12-15 FACTORING TRINOMIALS WITH NO COMMON TERM

Factoring of trinomials with no common terms can be done by inspection but this will require very careful study. Study the following:

$$(x + 3)(2x + 1) = 2x^2 + 7x + 3$$
$$(2a + 3)(3a - 5) = 6a^2 - a - 15$$
$$(5y - 4)(2y - 7) = 10y^2 - 43y + 28$$
$$(3x - 2y)(x + 4y) = 3x^2 + 10xy - 8y^2$$

The product of each has a term containing the square of the common letter, a term containing the first power and a term that does not contain the letter. These trinomials always contain the square of the letters involved and therefore are called *quadratic trinomials*. The general form of a quadratic trinomial is: $ax^2 + bx + c$ where a, b, and c are any numbers.

Study carefully the multiplication of $(2a + 3)(3a - 5)$

$$
\begin{array}{r}
2a + 3 \\
\times\ 3a - 5 \\
\hline
6a^2 +\ \ 9a \\
-\ 10a - 15 \\
\hline
6a^2 -\ \ \ a - 15
\end{array}
$$

Performing the multiplication in the usual way and note:

The first term of the product, $6a^2$, is the product of the first two terms of the factors.

The last term of the product, -15, is the product of the second terms of the factors.

The middle term of the product, $-a$, is the sum of $3a$ times 3 and -5 times $2a$ and can be called the "cross product."

By following the inverse of this process you can find the factors of such a quadratic trinomial.

EXAMPLE 12-20   Factor $2x^2 + 7x + 3$

The first term of each factor contains x. The product of the numerical parts of the first terms is 2. Hence, they must be 1 and 2, as these are the only two factors of 2.

The product of the second terms of the factors is 3. Hence, they must be 1 and 3.

The sum of the cross products is $+7x$.

Keeping these facts in mind, one can determine the factors, at least after a few trials. Thus

$$2x^2 + 7x + 3 = (x + 3)(2x + 1)$$

EXAMPLE 12-21 Factor $18x^2 + 3x - 10$.

The first term of each factor contains x. The product of the numerical parts of the first terms is 18. Hence, they may be 1 and 18, 2 and 9, or 3 and 6.

The product of the second terms of the factors is -10. Hence they may be 1 and -10, -1 and 10, 2 and -5, or -2 and 5.

The sum of the cross products is $3x$.

Fitting these numbers together, by trial and error we have:

$$18x^2 + 3x - 10 = (3x - 2)(6x + 5)$$

12-16 THE FACTOR THEOREM

Simple binomials and trinomials are easily factored by the methods which have just been explained. However, the factor theorem provides a method which applies to any polynomial, particularly the more complex polynomials. Although this is a trial-and-error process, it is probably the quickest method for a complex expression.

Simply stated, the factor theorem is: *If a quantity can be substituted into the expression to be factored so that the expression equals zero then the quantity with the sign changed is the last term of one of the factors of the expression. The first term of the binomial factor is the letter for which the substitution was made if the coefficient of the first term is 1.*

The following four examples should help you understand the factor theorem:

EXAMPLE 12-21 Factor $k^2 + 2k - 8$.

Solution By "guess" we decide to substitute $k = 2$ into the expression. If $k = 2$, then this expression will equal 0: $2^2 + 2(2) - 8 = 4 + 4 - 8 = 0$. Hence, $(k - 2)$ is one of the factors of $k^2 + 2k - 8$. The second factor is found by dividing $k - 2$ into $k^2 + 2k - 8$.

$$
\begin{array}{r}
k + 4 \\
k - 2 \overline{) k^2 + 2k - 8} \\
\underline{k^2 - 2k} \\
4k - 8 \\
\underline{4k - 8}
\end{array}
$$

Since $(k - 2)$ and $(k + 4)$ are prime, the expression $k^2 + 2k - 8$ has been completely factored:

$$k^2 + 2k - 8 = (k - 2)(k + 4)$$

EXAMPLE 12-23   Factor $y^2 + 9y + 18$.

Solution   Since 9 and 18 are multiples of 3, our guess is to substitute -3 for y. Substituting -3 for y,

$$(-3)^2 + 9(-3) + 18 = 9 - 27 + 18 = 0$$

Then $(y + 3)$ is one of the factors of $y^2 + 9y + 18$:
Therefore, divide $y + 3$ into $y^2 + 9y + 18$:

$$
\begin{array}{r}
y + 6 \\
y + 3 \overline{)\, y^2 + 9y + 18} \\
\underline{y^2 + 3y } \\
6y + 18 \\
\underline{6y + 18}
\end{array}
$$

The prime factors of $y^2 + 9y + 18$ are $(y + 3)$ and $(y + 6)$. Thus

$$y^2 + 9y + 18 = (y + 3)(y + 6)$$

EXAMPLE 12-24   Factor $x^4 + x^3 - x^2 + 1$.

Solution   Since the last term is $+1$, our guess is -1. By substituting -1 for x, the expression equals 0.

$$
\begin{aligned}
x^4 + x^3 - x^2 + 1 &= (-1)^4 + (-1)^3 - (-1)^2 + 1 \\
&= 1 - 1 - 1 + 1 = 0
\end{aligned}
$$

Thus, $(x + 1)$ is one of the factors of this expression. Divide $x^4 + x^3 - x^2 + 1$ by $x + 1$ to obtain the second factor:

$$
\begin{array}{r}
x^3 - x + 1 \\
x + 1 \overline{)\, x^4 + x^3 - x^2 + 0x + 1} \\
\underline{x^4 + x^3 } \\
- x^2 + 0x \\
\underline{- x^2 + x } \\
+ x + 1 \\
\underline{x + 1}
\end{array}
$$

Then $(x + 1)$ and $(x^3 - x + 1)$ are the factors of $x^4 + x^3 - x^2 + 1$, since $x^2 - x + 1$ cannot be factored.

$$x^4 + x^3 - x^2 + 1 = (x + 1)(x^3 - x + 1)$$

EXAMPLE 12-25   Factor $x^3 + x^2 - 10x + 8$.

Solution   Since the last term is an even number, and $2^3 = 8$ our guess is 2. Substituting $x = 2$ in the above expression,

$$(2)^3 + (2)^2 - 10(2) + 8 = 8 + 4 - 20 + 8 = 0$$

Then $x - 2$ is one of the factors. Dividing,

$$
\begin{array}{r}
x^2 + 3x - 4 \\
x - 2 \overline{) x^3 + x^2 - 10x + 8} \\
\underline{x^3 - 2x^2} \\
3x^2 - 10x \\
\underline{3x^2 - 6x} \\
- 4x + 8 \\
\underline{- 4x + 8}
\end{array}
$$

The trinomial $x^2 + 3x - 4$ can be factored. If we substitute $x = 1$,

$$(1)^2 + 3(1) - 4 = 0$$

Then $(x - 1)$ is one of the factors of $x^2 + 3x - 4$. Dividing,

$$
\begin{array}{r}
x + 4 \\
x - 1 \overline{) x^2 + 3x - 4} \\
\underline{x^2 - x} \\
4x - 4 \\
\underline{4x - 4}
\end{array}
$$

Then $x^3 - x^2 - 10x + 8 = (x - 1)(x + 4)(x - 2)$; the expression is completely factored.

▮ EXERCISES

Factor the following and check your results by multiplication:

12-139	$2x^2 + 11x + 12$	**12-158**	$25x^2 - 10xy + y^2$
12-140	$2x^2 - 7x - 30$	**12-159**	$mn^2 - 6mn + 9m$
12-141	$6x^2 + 22x + 20$	**12-160**	$6y^3 - 48$
12-142	$6x^2 - 16x - 6$	**12-161**	$a^2 - b^2 - 4bc - 4c^2$
12-143	$6x^2 + 17x + 10$	**12-162**	$4x^2 - 100$
12-144	$20x^2 + 41x + 20$	**12-163**	$(x - y)^2 - 25$
12-145	$12x^2 - x - 20$	**12-164**	$(x - y)^3 - 125$
12-146	$15x^2 + 34xy - 77y^2$	**12-165**	$27 + (a - 2b)^3$
12-147	$45x^2 - 78xy - 63y^2$	**12-166**	$10x^2 + 23xy$
12-148	$4x + 8y - 12z$	**12-167**	$2x^2y^4 - 16x^2y$
12-149	$4x^2 + 8xy + 4y^2$	**12-168**	$(x^2 + 4)^2 - 16x^2$
12-150	$8x + x^4$	**12-169**	$a^2x^2 - b^2$
12-151	$10x^2 + 23x + 12$	**12-170**	$22x^2 + 69x + 35$
12-152	$100 - x^4$	**12-171**	$Ax^5 - Ax^2$
12-153	$16 - x^4$	**12-172**	$(2x + y)^2 + 2(2x + y) + 1$
12-154	$(x^2 + 4)^2 - (4x)^2$	**12-173**	$-x^2 - y^2 + 2xy + a^2$
12-155	$6a^2 - a - 2$	**12-174**	$4(x - y)^2 - 4(x - y) + 1$
12-156	$5b^2 - 24b - 5$	**12-175**	$(a - 2b)^3 + (a + 2b)^3$
12-157	$14x^2 + 29x - 15$		

12-176 A circular rug has a radius of 1.5 m. A border 0.25 m wide is added to it. How much is the area of the rug increased?

Note: $A = \pi(R^2 - r^2)$ *factor, substitute, and evaluate.*

12-177 A flagpole 36 ft tall was standing on level ground. It broke and a part fell over, leaving 10 ft standing vertically. How far from the foot of the pole would the broken part reach if one end of it remained attached to the standing part? Draw the figure.

12-178 A ball is thrown upward at an arc of 45° with an initial velocity of 80 ft/s. The height of the ball at any time t is given by the equation: $h = 40t - 16t^2$. Find the time when the ball is 25 ft above the ground.

12-179 The area of the reflective surface plus the area of the frame of a square solar panel is 1.44 times the area of the reflective surface. If the length of the side of the reflective surface is 10 ft, how wide is the frame?

12-180 Find the thickness of the angle iron in Fig. 12-1 if the total cross-sectional area is 5.25.

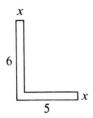

Fig. 12-1. Exercise 12-180.

ALGEBRAIC FRACTIONS

13-1 INTRODUCTION TO ALGEBRAIC FRACTIONS

In the explanations and problems presented thus far, fractions have been avoided except when an indicated division has been used. However, in many problems it is necessary to use algebraic fractions. The same terms are used with fractions in algebra as with fractions in arithmetic, the same principles are applied, and the same operations are performed.

When working with algebraic fractions, you should not calculate without thinking. You should relate the operations of algebra to similar procedures in arithmetic, and then you will have little difficulty with fractions in algebra. In fact, when you are in doubt about an algebraic problem involving fractions, you can perform a similar arithmetical operation, substituting numbers for letters, to determine whether the algebraic methods you are using are correct.

For example,

$$\frac{x}{a} + \frac{y}{b} = \frac{b(x)}{b(a)} + \frac{a(y)}{a(b)} = \frac{b(x) + a(y)}{a(b)} = \frac{bx + ay}{ab}$$

This is the same process as substituting numbers (for example, 2 for x, 3 for y, 3 for a, and 4 for b) and solving arithmetically:

$$\frac{2}{3} + \frac{3}{4} = \frac{4(2)}{4(3)} + \frac{3(3)}{3(4)} = \frac{4(2) + 3(3)}{3(4)} = \frac{8 + 9}{12}$$

13-2 DEFINITIONS OF ALGEBRAIC FRACTIONS

A *fraction* is an indicated division. Thus, $\frac{3}{4}$, $\frac{25}{2}$, $2ab/c$, and $(x^2 + y)/xy$ are fractions. The part of the fraction above the line is the *numerator;* the part below the line is the *denominator.* The numerator and the denominator are called the *terms* of a fraction.

The *degree* of an algebraic expression (such as a fraction) is determined by the largest total number of literal factors in any term of the expression. Thus, a^2 is of the second degree; $x^3y^2 + 1$ is of the fifth degree.

A fraction whose numerator is of a smaller degree than its denominator is a *proper fraction*. If the degree of the numerator is equal to or greater than the degree of the denominator, the fraction is an *improper fraction*. The fractions $a/(a^2 + 1)$, $(x^2 + 2)/(x^3 + 1)$, and $2/(3x^2 - 1)$ are proper fractions; $(x^3 - 1)/(x - 1)$ and $(x + 1)/(x - 3)$ are improper fractions.

The sum of an integral expression and a fraction is a *mixed expression*. Thus, $3x + 2 + x/(x^2 + 1)$ is a mixed expression.

The *value* of a fraction is the number that it represents.

13-3 PRINCIPLES USED IN WORKING WITH FRACTIONS

For reference, here again is the procedure used in working with fractions in arithmetic:

Procedure for simplifying algebraic fractions:

1. Multiplying or dividing both the numerator and the denominator of a fraction by the same quantity (except 0) does not change the value of the fraction.

2. Multiplying the numerator or dividing the denominator of a fraction multiplies the value of the fraction.

3. Dividing the numerator or multiplying the denominator of a fraction divides the value of the fraction.

Thus,

$$\frac{5}{7} = \frac{2(5)}{2(7)} = \frac{10}{14}$$

$$\frac{p}{q} = \frac{m(p)}{m(q)} = \frac{mp}{mq}$$

Note: *A fraction p/q is said to be **changed to higher terms** if p and q are multiplied by a quantity other than 0.*

Five examples illustrate these three principles:

EXAMPLE 13-1 Change $3x/8$ to an equivalent fraction whose denominator is 32.

Solution Multiply the numerator and the denominator by 4.

$$\frac{3x}{8} = \frac{?}{32}$$

To obtain the denominator 32, it is necessary to multiply 8 by 4. If the numerator is also multiplied by 4, then the value of the resulting fraction will equal the value of the original fraction.

$$\frac{3x}{8} = \frac{4(3x)}{4(8)} = \frac{12x}{32}$$

Check: Let $x = 2$. Then

$$\frac{3(2)}{8} \stackrel{?}{=} \frac{12(2)}{32}$$

$$\frac{6}{8} \stackrel{?}{=} \frac{24}{32}$$

$$\frac{3}{4} = \frac{3}{4}$$

EXAMPLE 13-2 Change $5k/(x - y)$ to a fraction whose numerator is $15k$.

Solution To obtain the numerator $15k$, $5k$ is multiplied by 3; therefore the denominator must also be multiplied by 3.

$$\frac{3(5k)}{3(x - y)} = \frac{15k}{3(x - y)} = \frac{15k}{3x - 3y}$$

Notice that the original denominator is $(x - y)$ and that $(x - y)$ must then be multiplied by 3.

EXAMPLE 13-3 Change $(x + y)/(x^2 - y^2)$ to a fraction whose denominator is $(x - y)$.

Solution Remember $x^2 - y^2$ can be factored: $x^2 - y^2 = (x + y)(x - y)$.

$$\frac{\overset{1}{\cancel{(x + y)}}}{(x - y)\cancel{(x + y)}} = \frac{1}{x - y}$$
$$\hspace{4.5em}1$$

Here, both the numerator and the denominator are divided by $(x + y)$.

EXAMPLE 13-4 Reduce $(6x^2y^3)/(12x^4y^4)$ to its lowest terms.

Solution Write the numerator and denominator in expanded form, then cancel and multiply.

$$\frac{6x^2y^3}{12x^4y^4} = \frac{\overset{1}{\cancel{2}} \times \overset{1}{\cancel{3}} \times \overset{1}{\cancel{x}} \times \overset{1}{\cancel{x}} \times \overset{1}{\cancel{y}} \times \overset{1}{\cancel{y}} \times \overset{1}{\cancel{y}}}{\underset{1}{\cancel{2}} \times 2 \times \underset{1}{\cancel{3}} \times \underset{1}{\cancel{x}} \times \underset{1}{\cancel{x}} \times x \times x \times \underset{1}{\cancel{y}} \times \underset{1}{\cancel{y}} \times \underset{1}{\cancel{y}} \times y} = \frac{1}{2x^2y}$$

or

$$\frac{6x^2y^3}{12x^4y^4} = \frac{\overset{1}{\cancel{6}}\overset{1}{\cancel{x^2}}\overset{1}{\cancel{y^3}}}{\underset{2x^2y}{\cancel{12}\cancel{x^4}\cancel{y^4}}} = \frac{1}{2x^2y}$$

EXAMPLE 13-5 Reduce $(x^2 - y^2)/(x^2 + 2xy + y^2)$ to its lowest terms.

Solution Factor and cancel like quantities.

$$\frac{x^2 - y^2}{x^2 + 2xy + y^2} = \frac{(x - y) \overset{1}{\cancel{(x + y)}}}{(x + y) \underset{1}{\cancel{(x + y)}}} = \frac{x - y}{x + y}$$

By canceling, both the numerator and the denominator are divided by the same number, $(x + y)$. It is possible to divide and obtain the answer $(x - y)/(x + y)$, but factoring and then canceling is usually easier.

▨ EXERCISES

13-1 Change $\dfrac{4x}{3}$ to a fraction whose denominator is 21.

13-2 Change $\dfrac{2x + 3u}{4}$ to a fraction whose denominator is 20.

13-3 Change $\dfrac{5t}{6u - 7v}$ to a fraction whose numerator is $15t$.

13-4 Change $\dfrac{3am}{5 + 6n}$ to a fraction whose numerator is $9am^2$.

13-5 Change $\dfrac{2 - 3y}{x + y}$ to a fraction whose denominator is $x^2 - y^2$.

Reduce the following fractions to their lowest terms:

13-6 $\dfrac{36}{64}$

13-7 $\dfrac{175}{275}$

13-8 $\dfrac{28x^2y^3}{35xy^2}$

13-9 $\dfrac{22x^3yz^4}{33x^2yz^2}$

13-10 $\dfrac{360r^3xt^7}{3r^3xt^6}$

13-11 $\dfrac{72a^4b^5c^6}{9a^2b^8c^2}$

13-12 $\dfrac{1028a^8b}{64a^9b^4}$

13-13 $\dfrac{(a - b)(a + b)^2}{a(a + b)}$

13-14 $\dfrac{x^2 - 6x + 9}{x^2 - 5x + 6}$

13-15 $\dfrac{m^2 + 2m - 24}{m^2 - 2m - 48}$

13-16 $\dfrac{n^3(n - m)^2}{n^2 - 2mn + m^2}$

13-17 $\dfrac{2a^2 - 4ab + 2b^2}{10a^2b - 10ab^2}$

13-18 $\dfrac{100 - 25x^2}{50(2 + x)(x - 2)}$

13-19 $\dfrac{x(x^2 + 5x - 14)}{x^3 + 2x^2 - 35x}$

13-20 $\dfrac{36a^2 - 81b^2}{9(2a - 3b)^2}$

13-21 $\dfrac{4 - (a + b)^2}{4(2 + a + b)(2 - a - b)}$

13-22 $\dfrac{(a + b) - (a + b)^2}{(a + b)(1 - a - b)}$

13-23 $\dfrac{(x + y)^2}{x^3 + y^3}$

13-24 $\dfrac{2 - h}{8 - h^3}$

13-25 $\dfrac{24h - 48k}{h^2 - 4hk + 4k^2}$

13-26 $\dfrac{6x^2 - 2xy - 20y^2}{2(x^2 - 4y^2)}$

13-4 ADDITION OF FRACTIONS

In arithmetic, fractions are changed to a common denominator before addition or subtraction. Similarly, literal fractions must be changed into fractions having a common denominator before they can be added or subtracted. To ease computations in arithmetic, the common denominator should be the least (lowest) common denominator. But in algebra the magnitude of a denominator is not always easily determined; therefore, the common denominator should be what is called the *lowest common multiple*. For example, x^3 is of higher *degree* than x^2, but the *value* of these two terms depends upon the value of x. Thus, if $x = 2$, x^3 is the larger; if $x = 1$, the terms are equal; if $x = \frac{1}{2}$, x^3 is the smaller.

13-5 THE LOWEST COMMON MULTIPLE

The lowest common multiple (lcm) (the lowest common denominator in arithmetic) is found by separating the quantities in an expression into prime factors and then finding a quantity that includes all the factors of each quantity. Recall Sec. 2-8 to review finding the lowest common denominator (lcd).

EXAMPLE 13-6 Find the lowest common multiple for 24, 32, and 40.

Solution Write the factors of each quantity, then select the factors for the lcm.

$$24 = 2^3 \times 3$$
$$32 = 2^5$$
$$40 = 2^3 \times 5$$
$$\text{lcm} = 2^5 \times 3 \times 5 = 480$$

EXAMPLE 13-7 Find the lcm for $12x^2y$, $16xy^3$, and $24x^3y$.

Solution Write the factors of each of the quantities, then select the factors for the lcm.

$$12x^2y = 2^2 \times 3 \times x^2 \times y$$
$$16xy^3 = 2^4 \times x \times y^3$$
$$24x^3y = 2^3 \times 3 \times x^3 \times y$$
$$\text{lcm} = 2^4 \times 3 \times x^3 \times y^3 = 48x^3y^3$$

Note that each of the three preceding numbers is divisible into its lcm without a remainder:

$$\frac{48x^3y^3}{12x^2y} = 4xy^2 \qquad \frac{48x^3y^3}{16xy^3} = 3x^2 \qquad \frac{48x^3y^3}{24x^3y} = 2y^2$$

EXAMPLE 13-8 Find the lcm of $x^2 + 2xy + y^2$ and $x^2 - y^2$

Solution Write the factors of each of the quantities, then select the factors of the lcm.

$$x^2 + 2xy + y^2 = (x + y)^2$$
$$x^2 - y^2 = (x + y)(x - y)$$
$$\text{lcm} = (x + y)^2(x - y)$$

13-6 ADDITION AND SUBTRACTION OF ALGEBRAIC FRACTIONS

As stated in Sec. 13-4, literal fractions may be added or subtracted like fractions in arithmetic. The first step in either process is determining the lcd of the denominators. When adding or subtracting fractions, change all fractions to the same common denominator then add or subtract the numerators.

EXAMPLE 13-9 Find the sum of $\dfrac{4}{5} + \dfrac{3}{2x}$

Solution   The lcd of the denominators 5 and $2x$ is $10x$. Then

$$\frac{4}{5} = \frac{4 \times 2x}{5 \times 2x} = \frac{8x}{10x}$$

$$\frac{3}{2x} = \frac{3 \times 5}{2x \times 5} = \frac{15}{10x}$$

and

$$\frac{8x}{10x} + \frac{15}{10x} = \frac{8x + 15}{10x}$$

EXAMPLE 13-10   Find the sum of $x/(a - x)$; $a/(a + x)$; and $(a^2 + x^2)/(a^2 - x^2)$.

Solution   The lcd of the denominators $(a - x)$, $(a + x)$, and $(a^2 - x^2)$ is $(a + x)(a - x)$ or $(a^2 - x^2)$. Then change to fractions with the same denominator.

$$\frac{x}{a - x} = \frac{x(a + x)}{(a - x)(a + x)} = \frac{ax + x^2}{a^2 - x^2}$$

$$\frac{a}{a + x} = \frac{a(a - x)}{(a - x)(a + x)} = \frac{a^2 - ax}{a^2 - x^2}$$

$$\frac{a^2 + x^2}{a^2 - x^2} = = \frac{a^2 + x^2}{a^2 - x^2}$$

Thus

$$\frac{ax + x^2 + a^2 - ax + a^2 + x^2}{a^2 - x^2}$$

Adding the numerators, the sum of the fractions is

$$\frac{2a^2 + 2x^2}{a^2 - x^2}$$

EXAMPLE 13-11   Subtract $(a + 2x)/(a^2 - x^2)$ from $(a + x)/(a^2 - ax)$.

Solution   Find the lcm by factoring: $a^2 - x^2 = (a + x)(a - x)$ and $a^2 - ax = a(a - x)$.

$$\text{lcm} = a(a + x)(a - x) = a^3 - ax^2$$

$$\frac{a + x}{a^2 - ax} = \frac{(a + x)(a + x)}{(a^2 - ax)(a + x)} = \frac{a^2 + 2ax + x^2}{a^3 - ax^2}$$

$$\frac{a + 2x}{a^2 - x^2} = \frac{a(a + 2x)}{a(a^2 - x^2)} = \frac{a^2 + 2ax}{a^3 - ax^2}$$

When the numerator of the second fraction is subtracted from the numerator of the first, the result is

$$\frac{a^2 + 2ax + x^2}{a^3 - ax^2} - \frac{a^2 + 2ax}{a^3ax^2} = \frac{a^2 + 2ax + x^2 - a^2 - 2ax}{a^3 - ax^2} = \frac{x^2}{a^3 - ax^2}$$

▓ EXERCISES

13-27 Add $\dfrac{11}{35}$, $\dfrac{13}{42}$, and $\dfrac{17}{30}$.

13-28 Add $\dfrac{11}{5a}$, $\dfrac{12}{10a}$, and $\dfrac{21}{25a}$.

13-29 Add $\dfrac{3}{a + 2b}$ and $\dfrac{a}{c - 2b}$.

13-30 Add $\dfrac{2}{x}$, $\dfrac{a}{x(2 - y)}$, and $\dfrac{b}{2x^2 - x^2y}$.

13-31 Add $\dfrac{10}{(x + 2)^2}$ and $\dfrac{a}{(x + 2)}$.

13-32 Add $x - 2y$, $x - 2y$, and $\dfrac{x^2 + 4y}{x - 2y}$.

13-33 Add $\dfrac{2x}{3y}$, $\dfrac{3x}{4y}$, $\dfrac{4x}{y}$, and $\dfrac{x}{2y}$.

13-34 Add $\dfrac{2x}{3y}$, $\dfrac{x - y}{6y}$, and $\dfrac{x + y}{9y}$.

13-35 Add $\dfrac{5x - 6}{9}$ and $\dfrac{6x + 5}{18}$.

13-36 Add $\dfrac{4 - h}{4h}$, $\dfrac{h - 8}{8h}$, and $\dfrac{2h + 3}{6h}$.

13-37 Add $\dfrac{x + 3}{10}$, $\dfrac{x - 3}{20}$, and $\dfrac{x}{2}$.

13-38 From $\dfrac{12m - 5n}{8}$ take $\dfrac{3m + 2n}{2}$.

13-39 From $\dfrac{ab - c}{2ab}$ take $\dfrac{ab + c}{4ab}$.

13-40 Combine: $\dfrac{2x - 3y}{7} - \dfrac{4x - 5y}{14} + \dfrac{5x + 6y + 7}{21}$.

13-41 From $6ab + \dfrac{12 - 6ab}{6}$ take $8ab - \dfrac{2ab + 3}{8}$.

13-42 Add $\dfrac{3}{x - 2}$, $\dfrac{4}{x + 2}$, and $\dfrac{5}{x^2 - 4}$.

13-43 Combine: $\dfrac{1}{x^2 - 9x + 18} + \dfrac{1}{x^2 - 8x + 12}$.

13-44 Combine: $\dfrac{1}{x(x - 1)} - \dfrac{2}{x(x + 1)} + \dfrac{3}{x^2 - 1}$.

13-45 Combine: $\dfrac{1}{x^2 - x - 2} + \dfrac{1}{x^2 + x - 6} - \dfrac{1}{6 + 5x + x^2}$.

13-46 Combine: $\dfrac{x + y}{x^2 - y^2} - \dfrac{x - y}{x^2 + 2xy + y^2} + \dfrac{y - x}{(x - y)^2}$.

13-47 Add $\dfrac{1}{f} + \dfrac{1}{f'} - \dfrac{1}{f''} + \dfrac{1}{F}$.

13-48 Combine $\dfrac{6}{x - 7} - \dfrac{8}{x + 9}$ and verify your work by substituting $x = 3$.

13-49 The area of a rectangle is A; one of its sides is b. Compute, in terms of a simple fraction, the perimeter of the rectangle.

13-50 A man can row m mph in still water. How long will it take him to row s mi up a river which flows at a mph? How long will it take him to row s mi down the river? How much longer does he need to row up the river than down?

13-51 A and B are m mi apart on a river that flows at c mph. A woman who can row a mph in still water goes from A to B and back. Compute the time it takes her to make the journey up and down.

13-7 MULTIPLICATION OF FRACTIONS

The product of two or more algebraic fractions is the product of their numerators divided by the product of their denominators, just like fractions in arithmetic.

EXAMPLE 13-12 Multiply $\frac{3}{4}$ by $\frac{5}{9}$.

Solution Reduce the multiplications by canceling all common factors; then, after multiplying the result will be in lowest terms.

$$\frac{3}{4} \times \frac{5}{9} = \frac{\overset{1}{\cancel{3}} \times 5}{4 \times \underset{3}{\cancel{9}}} = \frac{5}{12}$$

EXAMPLE 13-13 Multiply 42 by $\frac{27}{28}$.

Solution

$$\frac{42}{1} \times \frac{27}{28} = \frac{\overset{3}{\cancel{42}} \times 27}{1 \times \underset{2}{\cancel{28}}} = \frac{81}{2} = 40\frac{1}{2}$$

Here, 42 and 28 are divided by 14.

Note: In algebra, the common factors are not always immediately evident, and therefore it is easier to factor most numerators and denominators before attempting to cancel or multiply.

EXAMPLE 13-14 Multiply:

$$\frac{x - y}{x^2 + 2xy + y^2} \times \frac{x \times y}{x^2 - 2xy + y^2}$$

Solution Factor the denominators, cancel, and multiply.

$$\frac{x - y}{x^2 + 2xy + y^2} \times \frac{x + y}{x^2 - 2xy + y^2} = \frac{\overset{1}{\cancel{(x-y)}}\overset{1}{\cancel{(x+y)}}}{\underset{1}{\cancel{(x+y)}}(x + y)\underset{1}{\cancel{(x-y)}}(x - y)}$$

$$= \frac{1}{(x + y)(x - y)}$$

EXAMPLE 13-15 Multiply:

$$\frac{a + 2}{a^2 + 6x + 9} \times \frac{a + 3}{a^2 + 4a + 4} \times \frac{a^2 - 4}{y^3}$$

Solution Factor the denominators, cancel, and multiply.

$$\frac{\overset{1}{\cancel{a+2}}}{\underset{1}{\cancel{(a+3)}}(a + 3)} \times \frac{\overset{1}{\cancel{a+3}}}{\underset{1}{\cancel{(a+2)}}\underset{1}{\cancel{(a+2)}}} \times \frac{\overset{1}{\cancel{(a+2)}}(a - 2)}{y^3} = \frac{a - 2}{y^3(a + 3)}$$

EXAMPLE 13-16 Multiply $\dfrac{1}{a} - \dfrac{1}{b}$ by $a - \dfrac{a^2}{b}$.

Solution Find the lowest common multiple for both quantities.

$$\frac{1}{a} - \frac{1}{b} = \frac{b}{ab} - \frac{a}{ab} = \frac{b - a}{ab}$$

$$a - \frac{a^2}{b} = \frac{ab}{b} - \frac{a^2}{b} = \frac{ab - a^2}{b}$$

Now, factor, cancel, and multiply:

$$\frac{b - a}{ab} \times \frac{ab - a^2}{b} = \frac{b - a}{\underset{1}{\cancel{ab}}} \times \frac{\overset{1}{\cancel{a}}(b - a)}{b} = \frac{(b - a)^2}{b^2}$$

13-8 DIVISION OF FRACTIONS

Like fractions in arithmetic, algebraic fractions may be divided as follows: Invert the divisor, and then proceed as in multiplication. This rule may also be expressed as follows:

Procedure for dividing fractions:

Invert the fraction which follows the division symbol and proceed as in the multiplication of fractions.

The *reciprocal* of any fraction is that fraction inverted. The reciprocal of 4 is, therefore, $\frac{1}{4}$, the reciprocal of $\frac{2}{3}$ is $\frac{3}{2}$; the reciprocal of c/b is b/c.

EXAMPLE 13-17   Divide $\frac{3}{4}$ by $\frac{1}{2}$.

Solution   Invert the fraction which follows the division symbol, cancel, and multiply.

$$\frac{3}{4} \div \frac{1}{2} = \frac{3}{\overset{}{\underset{2}{4}}} \times \frac{\overset{1}{2}}{1} = \frac{3}{2} \text{ or } 1\frac{1}{2}$$

EXAMPLE 13-18   Divide $(x^2 - 11x - 26)/(x^2 - 3x - 18)$ by $(x^2 - 18x + 65)/(x^2 - 9x + 18)$

Solution   Invert the fraction which follows the division symbol, factor, cancel, and multiply.

$$\frac{x^2 - 11x - 26}{x^2 - 3x - 18} \div \frac{x^2 - 18x + 65}{x^2 - 9x + 18} = \frac{x^2 - 11x - 26}{x^2 - 3x - 18} \times \frac{x^2 - 9x + 18}{x^2 - 18x + 65}$$

$$= \frac{\overset{1}{\cancel{(x-13)}}(x + 2)}{\underset{1}{\cancel{(x-6)}}(x + 3)} \times \frac{\overset{1}{\cancel{(x-6)}}(x - 3)}{(x - 5)\underset{1}{\cancel{(x-13)}}}$$

$$= \frac{(x + 2)(x - 3)}{(x + 3)(x - 5)} = \frac{x^2 - x - 6}{x^2 - 2x - 15}$$

▊ EXERCISES

13-52   Multiply $\frac{16}{75}$ by $\frac{125}{64}$.

13-53   Find the product of $13(2 \div 65)(5 \div 6)$.

13-54   Divide $\frac{36}{55}$ by $\frac{12}{77}$.

13-55   Multiply $\dfrac{4x}{5y}$ by $\dfrac{10y}{16}$

Perform the following operations as indicated:

13-56 $\dfrac{3h^3}{49a^2b} \times \dfrac{7ab^3}{9hk}$

13-57 $\dfrac{3xyz^2}{4a^2b} \times \dfrac{7z^3}{2ab^4}$

13-58 $\dfrac{a^2 - ab}{1 + b^2} \times \dfrac{a + ab^2}{a^2 - b^2}$

13-59 $\dfrac{x^2 - 4}{xy + 2y} \times \dfrac{y^2 + y}{x - 2}$

13-60 $\dfrac{x^3}{y^4} \times \dfrac{x^6}{y^5}$

13-61 $\dfrac{5pq^3}{6xy} \div 5p$

13-62 $\dfrac{5a - 3b}{8} \div \dfrac{16ab}{3}$

13-63 $\dfrac{1 - b^2}{3a} \div \dfrac{1 + b}{6a}$

13-64 $\dfrac{3x + 4y}{3x - 4y} \div \dfrac{9x^2 - 16y^2}{3x - 4y}$

13-65 Simplify $\left(\dfrac{3a^2}{5y} \times \dfrac{15ay^3}{7c} \right) \div \dfrac{45a^3}{14c}$

13-66 Simplify $\left(\dfrac{2 - x}{3 + y} \times \dfrac{9 - y^2}{x - 2} \right) \div \dfrac{3 - y}{3 + x}$

13-67 Multiply $\dfrac{x^2 + 2x + 1}{y(x + 1)}$ by $\dfrac{y(x^2 - 1)}{x - 1}$

13-68 Divide $\dfrac{x^2 - 5x + 6}{x^2 - 9x + 20}$ by $\dfrac{x^2 - 3x + 2}{x^2 - 5x + 4}$

13-69 Simplify $\left(\dfrac{x^2 + x - 12}{x^2 + x - 20} \div \dfrac{x^2 - 7x + 12}{x^2 + 9x + 20} \right) \times \dfrac{(x + 4)^2}{(x - 4)^2}$

13-70 Multiply $x^2 + 4x + 4$ by $\dfrac{4}{x^2 - 4}$

13-71 Multiply $\dfrac{a}{b} + \dfrac{b}{a}$ by $a - \dfrac{b^2}{a}$

13-72 Multiply $1 + \dfrac{3a}{1 - a}$ by $1 + \dfrac{a}{1 + a}$

13-73 Multiply $\dfrac{axy}{ax^2 + by^2}$ by $\dfrac{x}{by} + \dfrac{y}{ax}$

13-74 Multiply $\dfrac{x^3y^4z^5}{xy + xz + yz}$ by $\dfrac{1}{x} + \dfrac{y + z}{yz}$

Simplify the following:

13-75 $\left(\dfrac{2 - a}{3 - a} \times \dfrac{5abc}{6xyz} \times \dfrac{9 - a^2}{4 - a^2} \right) \div \dfrac{5ab^2 - 5a^2b}{6x^2y + 6xy^2}$

13-76 $\dfrac{x^2 - 2xy + y^2}{4xy + 4y} \times \dfrac{(x + 1)^2}{2x^2 - 2xy} \times \dfrac{16xy}{x^2 - 1}$

13-77 $\dfrac{x^2 - 5x + 6}{x^2 - 9x + 20} \times \dfrac{x^2 - 16}{x^2 - 9} \times \dfrac{x^2 - 2x - 15}{x^2 + x - 12}$

13-78 $\left[\dfrac{a^2x^2 - x^4}{4a^2 - 5ax + x^2} \div \dfrac{(ax - x^2)^2}{(4a - x)^2} \right] \div \dfrac{a \times x}{4a + x}$

13-79 $\dfrac{x^2 + 8x + 16}{4x + x^2} \div \left(\dfrac{x^2 + 9x + 20}{x^2 - 25} \div \dfrac{x^2 + 5x}{x^2 - 5x} \right)$

13-80 $\left[\dfrac{x^2 - 3x + 2}{x^2 + 3x + 2} \times \dfrac{x^2 + 10x + 16}{(x - 1)^2} \right] \div \dfrac{x^2 + 6x - 16}{x^2 - x - 2}$

13-81 $\dfrac{6x^2 - 7x + 2}{6x^2 + 5x + 1} \times \dfrac{12x^2 - 5x - 3}{12x^2 - 17x + 6} \times \dfrac{2x^2 + x}{4x^2 - 1}$

13-82 $\left(\dfrac{8a}{a^2 - 9} + a \right) \div \left(\dfrac{2a}{a - 3} + a \right)$

13-83 $\left(\dfrac{x}{y} - \dfrac{y}{x} \right) \div \left(\dfrac{y}{x} - 1 \right)$

13-84 $\left(\dfrac{4}{x^2} - 1 \right) \div \left(1 - \dfrac{16}{x^4} \right)$

13-85 $\left(\dfrac{6}{h + 1} - 1 \right) \div \left(\dfrac{8 - 2h}{h^2 - 1} + h \right)$

13-86 $\left(\dfrac{-x - y}{3x} \right) \left(\dfrac{-x^2 + y^2}{-m} \right) \left[\dfrac{-2mx}{(x + y)^2} \right]$

13-87 $\left(\dfrac{a}{a + 1} - \dfrac{1}{1 - a} \right) \div \left(\dfrac{a}{1 - a} + \dfrac{1}{a + 1} \right)$

13-88 $\left(\dfrac{a - b}{a + b} + \dfrac{a + b}{a - b} \right) \div \left(\dfrac{a - b}{a + b} - \dfrac{a + b}{a - b} \right)$

13-89 $\left(\dfrac{4y}{a - y} + 4 \right) \left(\dfrac{a - y}{a + y} - 1 \right)$

13-90 $\left(\dfrac{a^2 + 4}{b + 2} \right) \left(\dfrac{4 - b^2}{a^2 + 2a} \right) \left(\dfrac{a}{1 - a} + 1 \right)$

13-91 $(a^2 + a + 1) \div \left(1 + \dfrac{1}{a} + \dfrac{1}{a^2} \right)$

13-92 $\left(\dfrac{4x^2 - 9x + 2}{3x^2 + 5x - 2} \right) \left(\dfrac{5x^2 + 9x - 2}{6x^2 - 13x + 2} \right) \left(\dfrac{6x^2 - 7x + 1}{20x^2 - 9x + 1} \right)$

13-93 $\dfrac{a^3 - b^3}{a^2 + 5ab + 6b^2} \times \dfrac{a^2 + 3ab}{a^2 - 6ab + 5b^2} \times \dfrac{a^2 - 3ab - 10b^2}{a^2 + ab + b^2}$

13-94 $\dfrac{64a^2b^2 - z^2}{x^2 - 4} \times \dfrac{(x - 2)^2}{8ab + z} \div \dfrac{x^2 - 4}{(x + 2)^2}$

13-95 A master roofer can roof a building in 6 hours, and an apprentice can complete the job in 8 hours. How long would it take them if they worked together?

Solution: Let x = the hours worked together. The master can do $\frac{1}{6}$ of the job each hour and the apprentice $\frac{1}{8}$ of the job per hour. Therefore: masters portion + apprentice portion = 1 job. So,

$$x\left(\frac{1}{6}\right) + x\left(\frac{1}{8}\right) = 1$$

Clear the fractions and solve.

$$\frac{x}{6} + \frac{x}{8} = 1$$

$$24\left(\frac{x}{6}\right) + 24\left(\frac{x}{8}\right) = 1(24)$$

$$4x + 3x = 24$$

$$7x = 24$$

$$x = \frac{24}{7} \text{ or } 3\frac{3}{7} \text{ hours}$$

13-96 Suppose that A can do a piece of work in 10 h and B can do the same amount in 6 h. How many hours will they require, working together, to complete the same amount of work? (Together they can do $\frac{1}{10} + \frac{1}{6}$ of the work per hour.)

13-97 If A can do a piece of work in 10 h and B can do it in m h, how many hours will they need, working together, to complete that work?

13-98 Assume that A can do a piece of work in m h and B can do it in n h. How many hours will they spend, working together, to complete that work?

13-99 A new jet plane could decrease the time needed to cover 4000 miles by one hour by increasing its speed by 200 miles per hour. What is its initial speed?

13-100 Two resistors are connected in parallel so that their total resistance (R) is 3.6 ohms. One resistor's resistance must be 3 ohms greater than the other; find the other's if the formula for total resistance in parallel is: $\frac{1}{R} = \frac{1}{r_1} + \frac{1}{r_2}$.

ADVANCED EQUATIONS

14-1 INTRODUCTION TO ADVANCED EQUATIONS

Equations were first discussed in Chap. 10. Methods for solving more difficult equations will be explained in this chapter. You will apply many of the principles which you studied in previous chapters. For example, factoring must be used to solve some of the equations. You may find it helpful to review previously covered material which you do not thoroughly understand.

14-2 SOLUTION OF ADVANCED EQUATIONS

If an equation has indicated multiplications and signs of grouping, it is usually best to perform the multiplications and remove the signs of grouping before proceeding with the solution of the equation. When solving more difficult equations, do one step at a time and show all necessary work. Study and work through the examples.

EXAMPLE 14-1 Determine the value of c in $4c + 3[2c - 4(c - 2)] = 72 - 6c$.

Solution Remove the parentheses (starting with the innermost set) by multiplying -4 by $(c - 2)$:

$$4c + 3[2c - 4c + 8] = 72 - 6c$$

Remove the brackets by multiplying 3 by $[2c - 4c + 8]$:

$$4c + 6c - 12c + 24 = 72 - 6c$$

Combine like terms:

$$-2c + 24 = 72 - 6c$$

Add $6c$ to both members of the equation:

$$4c + 24 = 72$$

175

Subtract 24 from both members of the equation:

$$4c = 48$$

Divide both members by 4:

$$c = 12$$

Check:

$$4(12) + 3[2(12) - 4(12 - 2)] \stackrel{?}{=} 72 - 6(12)$$
$$48 + 3[24 - 48 + 8] \stackrel{?}{=} 72 - 72$$
$$48 + 3[-16] \stackrel{?}{=} 0$$
$$0 = 0$$

EXAMPLE 14-2 Solve for x:

$$(1 + 3x)^2 = (5 - x)^2 + 4(1 - x)(3 - 2x)$$

Solution Remove parentheses:

$$1 + 6x + 9x^2 = 25 - 10x + x^2 + 4(3 - 5x + 2x^2)$$
$$1 + 6x + 9x^2 = 25 - 10x + x^2 + 12 - 20x + 8x^2$$

Combine like terms:

$$1 + 6x + 9x^2 = 9x^2 - 30x + 37$$

Subtract $9x^2$ from both members:

$$1 + 6x = -30x + 37$$

Subtract 1 from both members:

$$6x = -30x + 36$$

Add $30x$ to both members:

$$36x = 36$$

Divide both members by 36:

$$x = 1$$

Check:

$$(1 + 3)^2 \stackrel{?}{=} (5 - 1)^2 + 4(1 - 1)(3 - 2)$$
$$4^2 \stackrel{?}{=} 4^2 + 0$$
$$16 = 16$$

The steps followed in solving these two equations can be summarized in a procedure.

Procedure for solving advanced equations:

1. Simplify the equation by removing parentheses and other signs of grouping.
2. Collect like terms.
3. By addition or subtraction, transpose all numbers to the right member of the equation and all letters to the left member.
4. Divide (or multiply) to obtain the final answer.

In the preceding examples, you were given equations and then asked to observe how solutions for those equations were obtained. When working with algebra problems, you may be given an equation to solve. But in many problems—particularly in practical applications—it is necessary to set up an equation from a verbal statement or statements. The following two examples illustrate the process of setting up an equation and determining a solution.

EXAMPLE 14-3 Find two numbers whose difference is 20 and whose sum is $\frac{5}{2}$ times that difference.

Solution

Let x = the smaller number.

Then $x + 20$ = the larger number.

$(x + 20) + x$ = the sum of the two numbers.

$(x + 20) - x$ = the difference between the two numbers.

$$(x + 20) + x = \frac{5}{2} [(x + 20) - x]$$

This is the equation we must solve.

$$x + 20 + x = \frac{5}{2} (x + 20 - x)$$
$$2x + 20 = \frac{5}{2} (20)$$
$$2x + 20 = 50$$
$$2x = 30$$
$$x = 15$$
$$x + 20 = 35$$

Check: The sum of the two numbers must be $\frac{5}{2}$ times the difference between the two numbers.

$$35 + 15 \stackrel{?}{=} \frac{5}{2} (35 - 15)$$
$$50 \stackrel{?}{=} \frac{5}{2} (20)$$
$$50 = 50$$

EXAMPLE 14-4 Painter A can paint a floor in 3 h; painter B can paint the same floor in 4 h. If both painters work together, how long will it take them to paint the floor?

Solution (See Fig. 14-1.) Painter A will paint $\frac{1}{3}$ of the floor in 1 h; painter B will paint $\frac{1}{4}$ of the floor in 1 h. Let x = number of hours needed to paint the floor if both work together. Then

$$x\left(\frac{1}{3}\right) = \frac{x(1)}{(3)} = \frac{x}{3} = \text{area of floor painted by } A$$
$$x\left(\frac{1}{4}\right) = \frac{x(1)}{(4)} = \frac{x}{4} = \text{area of floor painted by } B$$

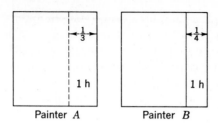

Fig. 14-1. Example 14-4: *A*'s job and *B*'s job.

and

$$\frac{x}{3} + \frac{x}{4} = 1$$

Multiply by 12 each term of both members of the equation:

$$4x + 3x = 12$$
$$7x = 12$$

$$x = \frac{12}{7} \text{ or } 1\frac{5}{7} \text{ h}$$

Is the answer reasonable? It seems reasonable because it is less than 3 h. Any answer equal to or greater than 3 h would be wrong.

The reason the equation was set up as $x/3 + x/4 = 1$ is that we are using the total area of the floor as our unit of area. Hence, the area painted by A in x h plus the area painted by B in x h must equal the total area, or 1.

This example illustrates the value of making a diagram or sketch of a problem before attempting to solve it.

■ EXERCISES

Solve for the unknown and check your answers.

14-1 $2a + 8 = a + 13$

14-2 $16b - 26 = 6b - 6$

14-3 $4u + 15 = 6u + 5$

14-4 $4x + 8 = 2(4x + 2)$

14-5 $14t - 8 + 42 = 10(t + 1)$

14-6 $2(6c + 8) - 1 = 5(3c - 6)$

14-7 $29x - 30 = 6(5x - 4) + 12$

14-8 $2(3h - 15) - 50 = h + 20$

14-9 $4(4s - 6)2 - 10(2s - 6) = -2(4s - 6) - 20$

14-10 $3[-y + 3(2y - 5)] = 4 - 7(5y - 8)$

14-11 $(x - 1)(x + 1) + x(1 - x) = 4x(2x + 1) - 8x(x - 2)$

14-12 $2x(3 - 4x) + (x - 2)(x + 3) = 7x(10 - x) + 12$

14-13 $(x - 4)(x + 5) = 2x(x - 6) - x(x - 10) + 80$

14-14 $14(x - 9) - 4x + 25(1 - x) = 0$

14-15 $(x + 8)(x - 7) - (x - 4)(x + 3) + 5(2x - 10) = 0$

14-16 $(2x - 3)(3x - 2) = (6x - 7)(x - 8) + 9x - 15$

14-17 $10(x - 2) - 10(2 - x) = 4x - 40$

Find the value of a in Exercises 14-18 to 14-23 and check your answers.

14-18 $6a - 3(a + 2)(a - 3) + 3a^2 = (a - 2)(a - 3) - (a + 2)(a + 3)$

14-19 $4.4a + 5.6 = 2.4a - .4$

14-20 $12a - 8 = 10(a - 2)$

14-21 $5 - 5a - 5.2 = 10a + 9.8$

14-22 $9.8a - 9.4 = 6.8a + .6$

14-23 $(a + 4)(5 - 2a) - a(10 - 2a) = 1.5$

Find the value of y in Exercises 14-24 to 14-28:

14-24 $y(y + 4) - (y - 4)(y - 5) + 10 = 8(y + 2)$

14-25 $y^2 - (y - 1)(y + 1) = 4(y + 8)$

14-26 $(y - 1)^2 - (y - 1)(y + 1) + 8 = 7y - 9$

14-27 $4y + 4y(y - 4) + 4(1 - y)(1 + y) = 0$

14-28 $5[y + 24(y - 1) - 20(y + 1)] + 18 = 9$

14-29   Find two numbers whose difference is 10 and whose sum is $4\frac{1}{5}$ times their difference.

14-30   A rectangular field is 5 rods longer than it is wide. If it were 2 rods wider and 3 rods shorter, it would contain 5 square rods less. Find the dimensions of the rectangle.

14-31   The difference between the squares of two consecutive numbers is 15. What are the numbers?

14-32   The difference between the squares of two consecutive even numbers is 44. What are the numbers?

14-3   EQUATIONS SOLVED BY FACTORING

Terms greater than the first degree in the unknown have appeared in some of the equations so far, but these terms have disappeared during the process of collecting like quantities. The equations have always reduced to a form in which the unknown times a certain number equals some number. For example, $6x = 12$ is such a form. Equations in which the unknown appears only in the first degree are *simple* or *linear equations*.

All equations do not reduce to this form. For instance, when an equation has been reduced, the result may be an equation where the square of the unknown equals some number, as in $x^2 = 5$.

An equation where the unknown appears only in the second degree is a *pure quadratic* equation.

In some cases, when an equation is simplified and reduced the simplified equation contains the square and the first power of the unknown, which may equal some number; $x^2 - 5x = 24$ is an example of such an equation. An equation containing the first and second degree of an unknown is an *affected quadratic* equation.

Quadratic equations, and certain other forms, can be solved by factoring.

EXAMPLE 14-5 Solve the affected quadratic equation $x^2 - 5x + 6 = 0$.

Solution and discussion This equation asks the following question: For what values of x does $x^2 - 5x + 6$ equal 0? If the expression in the left member is factored, the equation can be written

$$(x - 2)(x - 3) = 0$$

Now the question is for what values of x does the product $(x - 2)(x - 3)$ equal 0? The product of two factors is 0 if and only if either is 0 or both are 0. Thus the product is 0 if $x - 2 = 0$ or if $x - 3 = 0$. The solution of $x^2 - 5x + 6 = 0$ therefore depends on the solution of two simple equations: $x - 2 = 0$ and $x - 3 = 0$. These equations give the values 2 and 3 for x.

These values of x will prove the solution of the equation when we substitute each value into the original equation:

Substituting $x = 2$ gives $4 - 10 + 6 = 0$ or $0 = 0$.

Substituting $x = 3$ gives $9 - 15 + 6 = 0$ or $0 = 0$.

The values of the unknown that satisfy the equation are the *roots of the equation*. A quadratic equation with one letter has two roots (answers or solutions).

EXAMPLE 14-6 Solve the equation $x^2 - 25 = 0$.

Solution 1 Factoring,

$$(x + 5)(x - 5) = 0$$

Setting each factor equal to 0,

$$x + 5 = 0$$
$$x = -5$$
$$x - 5 = 0$$
$$x = 5$$

Check for the solution $x = -5$:

$$25 - 25 = 0$$

Check for the solution $x = +5$:

$$25 - 25 = 0$$

The two roots (answers) are, then, -5 and $+5$.

Solution 2

$$x^2 - 25 = 0$$
$$x^2 = 25$$
$$\sqrt{x^2} = \pm\sqrt{25}$$
$$x = \pm 5 \text{ or } x = +5, x = -5$$

The answer is read "x equals plus or minus 5" which means that x has two values, $+5$ and -5. The square root of 25 is equal to either $+5$ or -5 because $(+5)^2 = 25$ and $(-5)^2 = 25$. A positive number has two square roots, one positive and one negative.

EXAMPLE 14-7 Solve $(x + 1)(x - 3)(2x - 16) = 0$.

Solution Equating each factor to 0,

$$x + 1 = 0 \qquad x - 3 = 0 \qquad 2x - 16 = 0$$

Solving,

$$x = -1 \qquad x = 3 \qquad x = 8$$

The steps in solving a quadratic equation can be thus:

Procedure for solving quadratic equations:

1. Simplify the equation.
2. Transpose all terms to the left member of the equation.
3. Factor.
4. Equate each factor to 0.
5. Solve each of the equations obtained by step 4.

EXERCISES

The following equations involve only simple factors. Solve by setting each factor equal to 0. Check each solution.

14-33 $(x + 2)(2x - 8) = 0$

14-34 $(x - 4)(4 - 2x) = 0$

14-35 $(2x - 3)(3x + 2)(x - 1) = 0$

14-36 $(x - 4)(4x + 12)(3x - 9) = 0$

Factor the following equations into simple factors and solve:

14-37 $(x^2 + 4x)(2 - 4x) = 0$

14-38 $(x^2 - 9)(x^2 - 36) = 0$

14-39 $(x + 3)(x^2 - 4)(x - 3) = 0$

14-40 $t^2 - 72 = t$

14-41 $s(3s + 2) - 4(3s - 2) + 17 = 2s^2$

Solve the following equations by setting up equations, factoring, and solving for the unknowns.

14-42 Find the number which when added to its square equals 42.

14-43 Twice a number added to its square equals 63. Find the number.

14-44 If 16 is added to the square of a number, the sum equals 10 times the number. Find the number.

14-45 If 44 is subtracted from the square of a number, the difference equals 7 times the number. Find the number.

14-46 A rectangle is 8 in. longer than it is wide. If the area is 560 in.², find the dimensions.

Note: *Only one of the solutions obtained will be reasonable. Often when word problems are solved by quadratic equations, only one of the solutions has any practical significance.*

14-47 A rectangular lot 40 ft by 26 ft is completely surrounded by a path of uniform width. The path has an area of 432 ft². Find its width.

14-48 A rectangular piece of tin is 8 in. longer than it is wide. If a 2-in. square is cut from each corner and the edges turned up, an 18-in.³ box is formed. Determine the lengths of the three edges of the box.

14-49 A rectangle is 4 rods longer than it is wide. If its length is increased by 4 rods and its width by 4 rods, the area is doubled. Calculate the dimensions of the original rectangle.

14-4 FORMULAS

Many practical applied problems are solved by using an appropriate formula. Formulas are nearly always expressed as literal algebraic expressions. The letters in such formulas represent quantities, pounds, dollars, widths, etc.

The formula $I = prt$ is used to compute interest. In this formula, I represents interest; p, principal; r, rate; and t, time. The formula is read: "Interest equals the principal multiplied by the rate multiplied by time." It is easy to solve for any of the letters other than I. Note how p, r, and t are determined in the following illustration.

EXAMPLE 14-8 Solve $I = prt$ for p, r, and t.

Solution Rewrite the formula as shown.

$$prt = I$$

To solve for t, divide both members of the equation by pr:

$$\frac{\cancel{p}\cancel{r}t}{\cancel{p}\cancel{r}} = \frac{I}{pr}$$

Thus

$$t = \frac{I}{pr}$$

The division in the right member of the equation is not performed, it is merely indicated. However, when numbers are substituted for I, p, and r, the solution can be obtained.

To solve for r, divide both members by pt:

$$\frac{prt}{pt} = \frac{I}{pt}$$

$$r = \frac{I}{pt}$$

To solve for p, divide both members by rt:

$$\frac{prt}{rt} = \frac{I}{rt}$$

$$p = \frac{I}{rt}$$

EXAMPLE 14-9 Solve the formula $T = ph + 2A$ for p, h, and A.

Solution Given $T = ph + 2A$; or, $ph + 2A = T$. To solve for p, subtract $2A$ from both members:

$$ph = T - 2A$$

Divide both members by h:

$$p = \frac{T - 2A}{h}$$

To solve for h:

$$ph = T - 2A$$

Divide both members by p:

$$h = \frac{T - 2A}{p}$$

To solve for A:

$$ph + 2A = T$$

Subtract ph from both members:

$$2A = T - ph$$

Divide both members by 2:

$$A = \frac{T - ph}{2}$$

The original formula, $T = ph + 2A$, is used when the quantity T is the answer desired. But if this formula is to be used to calculate p in a series of problems, you can save time by rearranging or solving the equation for p before substituting numbers for the letters. Proper formula arrangement is important for efficient and time-saving calculation, particularly if a calculator or a computer is used.

■ EXERCISES

Solve these formulas for the indicated letters:

14-50 $A = \pi ab$; solve for a and b.
14-51 $S = ph$; solve for p and h.
14-52 $S = 2\pi rh$; solve for r.
14-53 $V = \pi r^2 h$; solve for h.
14-54 $V = \pi R^2 h - \pi r^2 h$; solve for h.

14-55 $A = 4\pi^2 Rr$; solve for R.

14-56 $Z = 2\pi rh$; solve for r.

14-57 $T = 6a^2$; solve for a^2 and a.

14-58 $S = 4\pi r^2$; solve for r.

14-59 $V = \pi r^2 h$; solve for r.

14-60 Given the formula $V = \frac{1}{3}\pi r^2 h$ or $1.0472 r^2 h$ for finding the volume of a circular cone, solve for h and for r.

14-61 The formula $v = \sqrt{2gh}$ expresses the velocity in meters per second that a body attains after falling from a height of h m. Solve the equation for h and derive a formula that expresses the height to which a body will rise if thrown upward with a velocity of v m/s. In this formula, g is a gravitational constant.

Suggestion: First square both members of the equation; then the result is $v^2 = 2gh$.

14-62 The formula $v_t = v_0 + 9.8t$ describes the velocity that a falling body will have where t = time in seconds that the body has fallen; v_0 = velocity in meters per second that the body has at the start (that is, v_0 is the initial velocity); v_t = velocity in meters per second after t s. Solve for v_0 and for t.

14-63 In calculating simple interest,

$$A = Prt + P$$

where A = amount
P = principal
t = time, years
r = rate, percent

Solve for each letter.

14-64 Using the formula in Exercise 14-63, (a) compute t when $P = \$2500$, $r = 10\%$, and $A = \$3125$; (b) compute r when $t = 1.5$ years, $P = \$3280$, and $A = \$3968.80$; (c) compute P when $A = \$87,500$, $t = 6$ years, and $r = 12.5\%$.

14-65 A rectangle 2 cm longer than it is wide has an area of 8 cm². Calculate the length and width. Calculate the length of its edges if its area is (a) 15 cm²; (b) 24 cm²; (c) 35 cm².

EQUATIONS AND APPLICATIONS

15-1 INTRODUCTION TO EQUATIONS AND APPLICATIONS

In preceding chapters, we studied several types of equations which used various principles but did not involve the manipulation of fractions. You will probably remember the steps in the solution of a linear equation; but for your convenience, here is a review of those steps:

1. Simplify both sides of the equation; remove the signs of grouping, and combine like quantities on each side of the equation.
2. Transpose all terms containing the unknown to one side of the equation; transpose all other terms to the other side of the equation.
3. Combine like terms.
4. Divide both members of the equation by the coefficient of the unknown.
5. Test the result by substituting the solution in the original equation.

15-2 CLEARING FRACTIONS IN EQUATIONS

A fraction is an indicated division; in problems, it is often a division that is difficult to perform. Hence, when an equation contains fractions, the fractions must be eliminated by some method other than division.

Procedure for clearing equations with fractions:

An equation can be cleared of fractions by multiplying each term of the equation by the lowest common multiple (lcm) of all fractions in the equation.

EXAMPLE 15-1 Solve:

$$\frac{x}{5} + \frac{x}{8} = 17 - \frac{x}{10}$$

185

Solution Clear each term of the equation by multiplying each term by the lcm 40.

$$40\left(\frac{x}{5}\right) + 40\left(\frac{x}{8}\right) = 40(17) - 40\left(\frac{x}{10}\right)$$

$$8x + 5x = 680 - 4x$$

Add $4x$ to both members: $8x + 5x + 4x = 680$.
Combine the terms: $17x = 680$.
Divide by 17 the coefficient of x: $x = 40$.

Check:

$$\frac{40}{5} + \frac{40}{8} \stackrel{?}{=} 17 - \frac{40}{10}$$

$$13 = 13$$

EXAMPLE 15-2 Given $4abc - 5bc + 16 = 3bc$, solve for a.

Solution Given the equation $4abc - 5bc + 16 = 3bc$, add $5bc$ to, and subtract 16 from, both members:

$$4abc = 5bc + 3bc - 16$$

Combine like terms: $4abc = 8bc - 16$.
Divide by $4bc$:

$$a = \frac{8bc - 16}{4bc} = \frac{8(bc - 2)}{4bc} = \frac{2(bc - 2)}{bc}$$

EXAMPLE 15-3 Given $1/f = 1/p + 1/q$, solve for p.

Solution Given equation:

$$\frac{1}{f} = \frac{1}{p} + \frac{1}{q}$$

Multiply by the lcm fpq:

$$pq = fq + fp$$

Transpose all quantities containing p to the left side of the equal sign:

$$pq - fp = fq$$

Factor p from each term:

$$(q - f)p = fq$$

Divide by $(q - f)$:

$$p = \frac{fq}{q - f}$$

EXAMPLE 15-4 In the equation $S = (E - IR)/0.220$, solve for I.

Solution Given this equation,

$$S = \frac{E - IR}{0.220}$$

First clear the fraction by multiplying both sides by 0.220:

$$0.220S = E - IR$$

Add IR to, and subtract $0.220S$ from, both members:

$$IR = E - 0.220S$$

Divide both sides of the equations by R:

$$I = \frac{E - 0.220S}{R}$$

Checking this type of problem is often as difficult as solving the original problem, but proving your answers is good practice in algebraic operations.
The original equation was

$$S = \frac{E - IR}{0.220}$$

Substitute the answer, $I = (E - 0.220S)/R$, in the original equation:

$$S \overset{?}{=} \frac{E - \dfrac{(E - 0.220S)R}{R}}{0.220}$$

Simplify the numerator by canceling the R's and simplify.

$$S \overset{?}{=} \frac{E - \dfrac{(E - 0.220S)\cancel{R}}{\cancel{R}}}{0.220}$$

$$S \overset{?}{=} \frac{E - (E - 0.220S)}{0.220}$$

$$S \overset{?}{=} \frac{E - E + 0.220S}{0.220}$$

$$S \overset{?}{=} \frac{\dfrac{1}{\cancel{0.220}S}}{\dfrac{\cancel{0.220}}{1}}$$

$$S = S$$

EXERCISES

Solve the following equations for the unknown and check your results:

15-1 $\dfrac{x}{3} + \dfrac{x}{6} = 18 - \dfrac{x}{4}$

15-2 $x + \dfrac{2x}{3} + \dfrac{3x}{4} - 29 = 0$

15-3 $2x - 50 = \dfrac{x}{4} - \dfrac{x}{3}$

15-4 $18 - 3y = \left(\dfrac{-2y}{3} + \dfrac{5y}{6} \right)$

15-5 $4y - 74 = \dfrac{y}{9} - \dfrac{y}{3}$

15-6 $3y - 70 = \left(-\dfrac{y}{6} - \dfrac{y}{12} \right)$

15-7 $26 = \dfrac{u}{8} + \dfrac{u}{4} + \dfrac{u}{6}$

15-8 $1 = -\dfrac{t-2}{5} + \dfrac{2t}{21} + \dfrac{t+9}{12}$

15-9 $29 = \dfrac{5t - 180}{6} + \dfrac{3t}{4}$

15-10 $\dfrac{7v + 2}{10} - 12 = \dfrac{3v + 3}{5} - \dfrac{v}{2}$

15-11 $\dfrac{2h + 4}{3} - 3\tfrac{1}{3} = \dfrac{h - 3}{4} + \dfrac{h + 2}{3}$

15-12 $\dfrac{p - 1}{2} + \dfrac{p - 3}{4} - \dfrac{p - 2}{3} = \dfrac{2}{3}$

15-13 $-(5x + 15) = 5x + 21 - \dfrac{5(2 - x)}{2}$

15-14 $3x^2 + 6x - (x + 1)^2 = -2 + (2x + 3)^2$

15-15 $\dfrac{8(x + 1)^2}{4} - 3(x + 1) = 2(x^2 + 1)$

15-16 $\dfrac{5}{3}(x - 3) = \dfrac{5x + 10}{2}$

15-17 $\dfrac{x}{9}(3x - 2) - \dfrac{2}{3} = \dfrac{x}{6}(2x + 1)$

15-18 $\dfrac{x}{4} + 9 + \dfrac{1}{3}(x + 2) + \dfrac{1}{2}(x + 1) = 0$

15-19 $\dfrac{5}{6}(x) - \dfrac{1}{3}(x) = -0.25x + x - 3$

15-20 $\dfrac{1}{6}(2x - 57) - x + \left(3x - \dfrac{2x - 5}{10} - \dfrac{5}{3} \right) = 0$

15-21 $\dfrac{0.03x - 0.06}{0.9} + 0.5 = \dfrac{0.12}{0.2}$

15-22 Given $3ac - 5c = 17$, solve for c.

Note: $c(3a - 5) = 17$

15-23 Given $(b - x)(b + 2x) = b^2 - 2x^2 - 3b + 4$, solve for b and for x.

15-24 $(a + b)x + (a - b)x = a^2$; solve for x.

15-25 $\frac{1}{2}(a + x) + \frac{1}{3}(2a + x) + \frac{1}{4}(3a + x) = 3a$; solve for x.

15-26 $4(t + b + y) + 3(t + b - y) = y$; solve for t.

15-27 Given $A = \frac{2}{3}hw$, solve for w.

15-28 Given $V = 2\pi^2 Rr^2$, solve for R.

15-29 Given $V = \pi R^2 h - \pi r^2 h$, solve for r^2 and then for r.

15-30 Given $S = \frac{1}{2}ps$, solve for s.

15-31 Given $T = \frac{1}{2}ps + A$, solve for p.

15-32 Given $S = \frac{1}{2}(P + p)s$, solve for p.

15-33 Given $T = \frac{1}{2}(P + p)s + B + b$, solve for s.

15-34 Given $V = \frac{4}{3}\pi r^3$, solve for r.

The following formulas occur in physics and electricity:

15-35 Given $PD = WD_1$, solve for P.

15-36 Given $F = \dfrac{WV}{gr}$, solve for W, g, and r.

15-37 Given $I = \dfrac{En}{R + nr}$, solve for E, R, r, and n.

15-38 Given:

$$I = \frac{E}{R + r} \tag{1}$$

$$I = \frac{nE}{R + nr} \tag{2}$$

$$I = \frac{E}{R + (r/n)} \tag{3}$$

What value of n will make equations (1), (2), and (3) identical? Solve equations (2) and (3) for n.

15-39 If $\dfrac{1}{k} = \dfrac{1}{a} + \dfrac{1}{b} + \dfrac{1}{l}$, solve for k; then calculate the value of k if $a = 19,000$, $b = 90,000$ and $l = 3180$.

15-40 Given $\dfrac{E}{r} = \dfrac{E}{r_1} + \dfrac{E}{r_2}$, solve for r.

15-41 Given $E = RI + \dfrac{rI}{n}$, show $I = \dfrac{E}{R + (r/n)}$.

15-42 Given $nE = RI + \dfrac{nrI}{m}$, show $I = \dfrac{E}{(R/n) + (r/m)}$.

15-43　Given $R_1 = R_0(1 + at)$, solve for R_0, a, and t, in terms of the other quantities.

15-44　Given $T = \dfrac{PRF}{Sp}$, solve for R and S.

15-45　Given $F = \dfrac{4^2 mx}{T^2}$, solve for x.

15-46　Given $E = \dfrac{Ff}{(P - x)p}$, solve for x.

15-47　Given $C = \dfrac{Krr_1}{r + r_1}$, solve for r_1.

15-48　Given $Q = \dfrac{K(t_2 - t_1)at}{d}$, solve for t_1 and a.

15-49　Given $p_t v_t = p_0 v_0 \left[1 + \left(\dfrac{t}{273} \right) \right]$, solve for t.

15-50　Given $H = 1{,}600{,}000 \dfrac{b_1 - b_2}{b_1 + b_2}(1 + 0.004t)$, solve for t.

15-3　SETTING UP EQUATIONS

In solving practical problems, it is frequently necessary either to select an appropriate equation from a handbook or to develop an equation from the stated problem. For solving problems of a purely mechanical type, it is possible to develop and learn definite rules and a set procedure—for example, the steps in the solution of an equation in Sec. 15-2 in this chapter. Unfortunately, it is impossible to outline a definite set of rules for the solution of practical problems. However, the suggestions given below should be of some help to you in setting up equations.

Procedure for setting up equations for word problems:

1. Study the problem carefully until you understand *exactly* the question to be answered.
2. Draw a sketch of the problem (if possible and appropriate).
3. Select the unknown number and represent it by some letter.
4. If there is more than one unknown number, try to express the other unknown numbers in terms of the first one selected.
5. Set up the written statement in the form of an equation.
6. Solve the equation for the unknown.
7. Check the answer or answers obtained with the conditions of the problem. Reject any answers which do not satisfy the conditions of the original equation or which are impossible.

EXAMPLE 15-5　Calculate the length of a side of a square room which, if its length and width are each increased by 3 m, will have its area increased by 81 m².

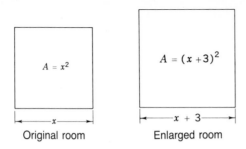

Original room   Enlarged room   **Fig. 15-1.** Example 15-5.

Solution See Fig. 15-1. Let x = original length of a side. Then

$$x + 3 = \text{length of that side after the addition of 3 m}$$
$$x^2 = \text{original area of the room}$$
$$(x + 3)^2 = \text{area of the room after the increase}$$
$$x^2 + 81 = \text{area of the room after the increase}$$

Then

$$(x + 3)^2 = x^2 + 81$$
$$x^2 + 6x + 9 = x^2 + 81$$
$$6x = 72$$
$$x = 12 \text{ m}$$

Check: The area of the original room was 144 m². After the length of a side was increased, the area of the room was 225 m². This total was 81 m² more than the original area of 144 m². The conditions of the problem are satisfied, and thus the answer is correct.

EXAMPLE 15-6 A man invested $30,000 as follows: He paid a certain amount for one city lot and twice as much for a second lot. Then he invested $5000 less in bonds than he paid for the second lot. How much did he invest in each lot and in the bonds? (See Fig. 15-2.)

Solution The total amount of money invested was $30,000 (the sum of the three investments). Let x = amount paid (in dollars) for the first lot, $2x$ = amount paid for the second lot, and $2x - 5000$ = amount for bonds. The following equation results:

$$x + 2x + 2x - 5000 = 30,000$$

Fig. 15-2. Example 15-6.

Solve for x:

$$5x = 35,000$$
$$x = 7000$$
$$2x = 14,000$$
$$2x - 5000 = 9000$$

Thus $7000 was invested in the first lot, $14,000 in the second, and $9000 in bonds. Substituting the answers in the equation, we find they are correct.

▌ EXERCISES

15-51 A man is paid twice as much per day as his daughter is paid. Together, they are paid $120 per day. How much is each paid per day? First solve by inspection; then write an equation and solve.

15-52 A company finds the cost of labor for making a $27 article is three times the cost of the material and the advertising cost is half as much as the cost of the material. Find the cost of labor, material, and advertising.

15-53 The original price of an article was reduced one-third. The sale price was then six-fifths of $320. What was the original price?

15-54 The original price of an article was increased one-fifth; the resultant selling price was 0.75 of $160. Find the original price.

15-55 A number, plus its half, plus its third, is eleven-tenths of 100. What is the number?

15-56 The list price of an electric fan was reduced 20 percent. The fan then sold for $48.00, which was 20 percent more than it cost. Find the original cost and the original list price.

15-57 After reducing the list price of a used car 10 percent, A sold the car to B. Then B increased the price he had paid 10 percent and sold the car to C for $4950. Find the list price.

15-58 A man's weekly salary was increased 50 percent. Then his new salary was increased $33\frac{1}{3}$ percent, so that he made $480 per week. Find his original weekly salary.

15-59 The sum of $11,000 is invested, part at 5 percent and part at 6 percent per year. If total annual income is $590, how much is invested at each rate?

Suggestion: *Let x = number of dollars invested at 5 percent. Then $11,000 - x$ = number of dollars invested at 6 percent. Therefore,*

$$0.05x + 0.06(11,000 - x) = 590$$

15-60 A total of $23,000 is invested, part at 12 percent and part at 12.5 percent per year. The total annual income is $2810. How much is invested at each rate?

15-61 The interest on $1200 for t years at 9.5 percent is $399. Find t.

15-62 The interest on $15,000 for 3.25 years at r percent is $5606.25. Find r.

15-63 If air is a mixture of 4 parts nitrogen to 1 part oxygen, how many cubic feet of each gas can be contained in a 12-ft-high room measuring 40 ft by 30 ft?

Suggestion: *Let x = number of cubic feet of oxygen in the room. Then 4x = number of cubic feet of nitrogen in the room.*

15-64 If a mixture is 3 parts water and 1 part oil, compute the number of liters of oil in 88 liters of the mixture.

15-65 A company employs three people. The first receives 6 parts, the second receives 4 parts, and the third receives 3 parts of a weekly payroll of $1482. Determine the weekly pay of each person.

15-66 A can filled with water weighed 6 lb. When that can was filled with gasoline of specific gravity 0.9, it weighed 5.5 lb. Find the weight of the can.

15-67 If 50 percent of x plus 10 percent of x is $12\frac{1}{2}$ percent of 480, find x.

15-68 For one event in old Soldier Field in Chicago, 110,000 people paid admission. A week later the price of tickets was increased 50 cents. 5000 fewer people were in attendance, but they paid $30,000 more at the box office. Compute the two admission prices.

15-69 A bus averaging 40 mph (including stops) between Omaha and Detroit made the trip in $15\frac{1}{4}$ h less than a truck making the same trip at an average of 25 mph (including stops). Find the distance between the two cities.

15-70 Two cylinders contain water. In one, the water is 26 in. deep and sinks at the rate of $1\frac{1}{2}$ in./min. In the other, the water is 4 in. deep and rises at the rate of $1\frac{1}{4}$ in./min. When is the depth of water in the two cylinders the same; what is that depth?

15-71 A Wheatstone bridge is used to measure electrical resistance when three known resistances are used in connection with the unknown. The unknown may be any one of the resistances r_1, r_2, r_3, and r_4 shown in Fig. 15-3. The three known resistances are satisfied by the proportion:

$$\frac{r_1}{r_2} = \frac{r_3}{r_4}$$

By using this proportion the fourth resistance can be found.
(a) Given $r_1 = 3.6$, $r_2 = 4.7$, $r_3 = 5$; find r_4.
(b) Given $r_1 = 500$, $r_2 = 300$, $r_4 = 125$; find r_3.
(c) Given $r_1 = 19.3$, $r_3 = 27.8$, $r_4 = 17.8$; find r_2.
(d) Given $r_2 = 16.4$, $r_3 = 28.2$, $r_4 = 16$; find r_1.

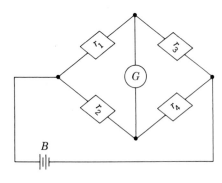

Fig. 15-3. Exercise 15-71.

15-4   EQUATIONS IN ELECTRICITY

The study of electricity involves working with many equations and formulas, some of these formulas will be studied in this section. The following symbols are used in these formulas:

$$R = \text{resistance, ohms } (\Omega)$$
$$I = \text{current, amperes (A)}$$
$$E = \text{voltage or pressure, volts (V)}$$
$$P = \text{power, watts (W)}$$

The relationship between E, I, and R, called *Ohm's law*, is $E = IR$; meaning that voltage equals the product of the current in amperes times the resistance in ohms. To calculate the power in a circuit, use the formula $P = EI$ where P equals the power in watts, (W) and E equals the voltage in volts (V); and I equals the current in amperes (A). It is sometimes convenient to solve Ohm's law for the value to be calculated. For example, $E = IR$ solved for I is $I = E/R$; solved for R, it is $R = E/I$. The formula $P = EI$ may also be solved for any of the values in it. Thus, if the values E and R are known and you want to calculate the power in watts, you can combine two of the formulas derived above. Since $I = E/R$ and $P = IE$, substituting for I its equivalent E/R, the formula $P = IE$ is

$$P = \left(\frac{E}{R}\right)E \text{ or } P = \frac{E^2}{R}$$

A formula can be rearranged or restated to suit the conditions of a particular problem. The following examples illustrate the application of the formulas and the principle discussed above.

EXAMPLE 15-7   What is the power rating of a motor that draws 30 A at 110 V?

Solution   Recall the formula: power equals voltage times current:

$$P = EI$$
$$P = 30 \times 110$$
$$P = 3300 \text{ W}$$

EXAMPLE 15-8   Compute in amperes the current flowing in a circuit if the resistance is 110 Ω and the pressure is 220 V.

Solution   Recall Ohm's Law, voltage equals current times resistance, and solve for current.

$$I = \frac{E}{R}$$

$$= \frac{220}{110}$$

$$= 2 \text{ A}$$

EXAMPLE 15-9 The total resistance of a motor operating on a 110-V circuit is 5.5 Ω. How many watts are expended?

Solution Recall Ohm's Law, voltage equals current times resistance, and power equals voltage times current.

$$P = EI \quad \text{and} \quad I = \frac{E}{R}$$

$$P = E\left(\frac{E}{R}\right) \qquad P = \frac{E^2}{R}$$

$$= \frac{110^2}{5.5}$$

$$= \frac{110 \times 110}{5.5}$$

$$= 2200 \text{ W}$$

An alternative solution is:

$$E = IR$$
$$110 = I(5.5)$$

$$\frac{110}{5.5} = I$$

$$20 \text{ A} = I$$
$$P = EI$$
$$P = 110 \times 20$$
$$P = 2200 \text{ W}$$

The two methods of solution produce the same answer. The first method is more practical when a number of problems involving E and R are to be solved. The second solution is particularly useful when only one problem involving E and R is to be solved.

EXAMPLE 15-10 Compute the horsepower (hp) rating of a motor operating at 440 V and drawing 3 A (746 W = 1 hp).

Solution Recall that power equals voltage times current and 746 watts equals 1 hp. Thus

$$P = EI$$
$$P = 440 \times 3$$
$$P = 1320 \text{ W}$$

Since 746 W = 1 hp,

$$\frac{1320 \text{ W}}{1} \left(\frac{1 \text{ hp}}{746 \text{ W}}\right) = 1.76 \text{ hp}$$

15-5 THERMOMETERS

Two thermometer scales are in common use, Fahrenheit and Celsius. On the Fahrenheit scale, the freezing point is marked at 32° (degrees) and

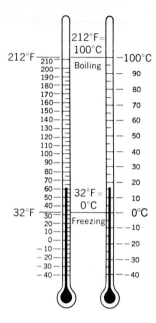

Fig. 15-4. Relationship between Celsius and Fahrenheit scales.

the boiling point at 212° (1°F means 1 degree Fahrenheit). On the Celsius scale, which is used in the metric system and for scientific purposes, the freezing point is designated 0° and the boiling point 100° (1°C means 1 degree Celsius).

Thus, on the Fahrenheit scale there are 180 degrees (212° to 32°) from the freezing point to the boiling point; on the Celsius scale only 100 degrees are indicated in the same range. Hence 180 degrees of the Fahrenheit scale = 100 degrees of the Celsius scale. This relationship is shown in Fig. 15-4.

■ EXERCISES

15-72 If F stands for the number of degrees on the Fahrenheit scale and C for the number of degrees on the Celsius scale, prove (a) and (b) and solve (c).
    (a) $C = \frac{5}{9}(F - 32)$
    (b) $F = \frac{9}{5}C + 32$
    (c) On Washington's Birthday the thermometer fell to $-40°F$ in Alaska. What was the reading on a Celsius scale in Alaska that day?

15-73 The record temperatures for December in St. Paul are: $+63°F$ on December 2, 1982; $-29°F$ on December 19, 1983. Convert the record high temperature to Celsius. Convert the record low temperature to Celsius.

15-74 A reading of 22.5°C is how many degrees Fahrenheit?

15-75 The following metals melt at the temperatures indicated, expressed in Fahrenheit scale: wrought iron, 2822°; steel, 2462°; cast iron, 2210°; silver, 1832°; lead, 620°; and tin, 475°. Find the melting point of each in the Celsius scale.

15-76 60° below 0°F is what temperature on the Celsius scale?

15-77 At what temperature is the reading on the Fahrenheit thermometer the same as the reading on the Celsius thermometer?

Solution: Let x = the reading on each thermometer scale. Then, by the formulas of Exercise 15–72,

$$\frac{5}{9}(x - 32) = \frac{9}{5}x + 32$$

15-78 In the oxyacetylene process of welding, the temperature of the welding flame sometimes reaches 6000°F. What is the corresponding Celsius temperature?

15-6 HORSEPOWER

The term *horsepower* (hp) was first used by James Watt, the inventor of the steam engine. He found that a London draft horse was capable, at least for a short time, of doing work equivalent to lifting 33,000 lb 1 ft in 1 min. Watt chose that value to express the power of his steam engines. It is still used as a unit of measure for power.

The expression *foot pounds* (ft·lb) denotes a unit of work. It is equivalent to a force of 1 lb acting through a distance of 1 ft, $\frac{1}{2}$ lb acting through a distance of 2 ft, $\frac{1}{10}$ lb acting through a distance of 10 ft, etc.

Horsepower is a measure of the rate at which work is performed. Two constants can be used when calculating horsepower: 33,000 ft·lb/min or 550 ft·lb/s. To determine the horsepower rating of a machine, divide by 33,000 the number of foot pounds of work which the machine can do in 1 min, or divide by 550 the number of foot pounds of work which it can do in 1 s. Then

$$\text{Horsepower (hp)} = \frac{\text{ft·lb/min}}{33,000} \quad \text{or} \quad \frac{\text{ft·lb/s}}{550}$$

Note: *Ft·lb/min is read "foot pounds per minute"; ft·lb/s is read "foot pounds per second." The slash mark is often used in place of "per," especially in calculations.*

Electrical power is measured in watts (W) or kilowatts (kW); mechanical power delivered by a motor is measured in horsepower (hp); 1 hp equals 746 W of electrical power.

Note: *1 kW = 1000 W*
        1 hp = 746 W

EXAMPLE 15-11 Convert $2\frac{1}{2}$ kW to watts.

Solution

$$2.5 \text{ kW} \times \frac{1000 \text{ W}}{1 \text{ kW}} = 2500 \text{ W}$$

EXAMPLE 15-12 2625 W is equivalent to how many horsepower?

Solution

$$\frac{2625\ W}{1} \times \frac{1\ hp}{746\ W} = 3.5\ hp$$

EXAMPLE 15-13 An elevator motor must lift 10,000 lb a distance of 50 ft in 4 min. What is the horsepower rating of this motor?

Solution To determine foot pounds per minute,

$$ft \cdot lb/min = \frac{10,000 \times 50}{4} = 125,000$$

$$hp = \frac{125,000}{33,000} = 3.79$$

■ EXERCISES

Convert the following as indicated:

15-79 5.6 kW to watts

15-80 1.5 W to kilowatts

15-81 48 hp to kilowatts

15-82 3 hp to watts

15-83 8000 W to kilowatts

15-84 $2\frac{1}{2}$ hp to watts

15-85 7.875 kW to horsepower

15-86 $\frac{5}{8}$ hp to kilowatts

15-87 How many horsepower are required to lift an elevator weighing 4 tons to the top of a building 72 m/high in $1\frac{1}{2}$ min? (Use 1 ft = 0.3 m.)

15-88 How many horsepower are required to pump 30,000 barrels of water per hour to a height of 45 ft? (Use 1 barrel = 4.21 ft³.)

15-89 The following formula gives the horsepower of an engine:

$$H = \frac{PLAN}{33,000}$$

where H = indicated horsepower
   P = mean effective pressure in pounds per square inch (lb/in.²)
   L = length of stroke, ft
   A = area of piston, in.²
   N = number of strokes of piston (twice the number of revolutions) per minute

Solve the formula for each of the letters (P, L, A, and N) in terms of the others.

Note: *In the formula, H = PLAN/33,000, the numerator contains the number of foot pounds per minute.*

15-90 What horsepower will be developed by an engine having a single cylinder 4 in. in diameter, a stroke of 6 in., a crank making 300 rpm, and a mean effective pressure of 95 lb/in.^{2}? (The area of a circle equals 0.7854 times the square of the diameter.) To solve, use the formula $H = PLAN/33,000$.

$$H = \frac{95 \times \frac{6}{12} \times 0.7854 \times 4 \times 4 \times 300 \times 2}{33,000}$$

15-91 Find the horsepower of a single-cylinder oil well pumping engine with an 8-in. diameter piston with a stroke of 20 in., the crank making 120 rpm, and a mean effective pressure of 120 lb/in.^{2}

15-92 An engine is required to develop 50 hp with an average effective pressure of 138 lb/in.^{2} on a piston 6 in. in diameter and at a crankshaft speed of 500 rpm. Find the length of the stroke.

15-93 Find the mean effective pressure on the piston of a steam engine with a cylinder 12 in. in diameter, a piston stroke of 18 in., 110 rpm, and developing 40 hp.

15-94 In gas engines it is difficult to determine the mean effective pressure, and the formula $H = PLAN/33,000$ cannot be used accurately. However, as a result of research with automotive engines, the formula $H = (D^2N)/2.5$ has been developed. Here D is the diameter of the cylinders in inches; N is the number of cylinders. Find H, if $D = 4\frac{3}{4}$ in. and $N = 6$.

15-95 If D were given in centimeters instead of inches, what would the $H = D^2N/2.5$ formula become?

15-96 The power transmitted by a belt depends on the belt's pull and the rate it travels. Its power may be expressed as a number of foot pounds per minute or as horsepower. Various makers of belts use different values for the working strength of belts. The allowed pull for single belts is from 60 to 100 lb per inch of width. The following formula expresses the horsepower transmitted by a belt:

$$H = \frac{FWS}{33,000}$$

where H = horsepower
    F = pull, lb/in. (pull in pounds per inch of width)
    W = width, in.
    S = speed of belt, ft/min (speed in feet per minute)

Solve for each letter in the formula.

15-97 Compute the horsepower transmitted by a belt 14 in. wide if the pull allowed per inch of width is 90 lb and the speed is 5000 ft/min.

15-98 Compute the horsepower transmitted by a single belt 6 in. wide that runs over a pulley 16 in. in diameter, assuming that the pulley makes 350 rpm and the pull of the belt is 45 lb per inch of width. Allow 2 percent for slipping.

15-99 If a single belt 6 in. wide transmits 7 hp, find its speed in feet per minute, allowing 35 lb pull per inch of width.

15-100 Compute the horsepower transmitted by two 10-in. belts which run over a 36-in. pulley at 420 rpm. Use a pull of 76 lb per inch of width.

15-101 How much work is done in lifting 150 lb from a mine 1100 ft deep? How many horsepower are required to lift this weight that far in $1\frac{1}{2}$ min?

15-7 LEVERS AND MOMENTS

A lever is used to gain advantage in using forces. A lever consists of a rigid bar that may move about a point called the fulcrum. The position of the fulcrum determines the advantage that may be gained by using the lever.

When a bar is balanced on a fulcrum F (Fig. 15-5), at a distance a from one end and b from the other, with weights m and n suspended from either end, the proportion $m: n = b:a$ results. Taking the product of the means and the product of the extremes, we have the equation

$$am = bn$$

The weight at one end of the lever has a tendency to pull that end down; that is, it tends to make the lever turn about the fulcrum in that direction. The amount of this turning effect is known as a *moment* and is the product of the weight by the length of its lever arm. Thus am is the moment of the weight m acting on the lever arm of length a.

Two moments that tend to turn in opposite directions balance each other when they are equal. *Two such moments put equal to each other give the equation used in solving a problem in levers.*

In design problems, the *principle of moments* is valuable for locating the position of forces acting on a beam or for determining the weight of a force. This principle may be stated as follows: *The sum of the moments about a system of forces in balance is 0.* The formula is written $\Sigma M = 0$, where Σ stands for the sum and M is the symbol for moments.

The system of forces explained below and shown in Fig. 15-6 illustrates the principle of moments. In Fig. 15-6, two weights (forces) have been placed at opposite ends of the beam; that beam is then in balance on fulcrum F. The force m is a feet from F, and the force n is b feet from F. The distances a and b are designated "moment arms." Because this system is in balance, $\Sigma M = 0$. The moments for each force are found by multiplying a specific force by its moment arm. The moment for one distance or weight is labeled plus (+), and for the other (which tends to push in the opposite direction) minus (−). Thus, for the system shown, $\Sigma M = 0$; or $+am - bn = 0$; or $am = bn$.

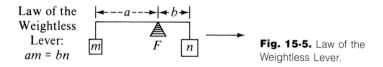

Law of the
Weightless
Lever:
$am = bn$

Fig. 15-5. Law of the Weightless Lever.

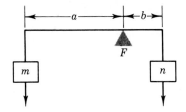

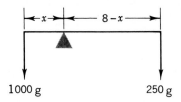

Fig. 15-6. Diagram of forces. **Fig. 15-7.** Diagram of the beam.

EXAMPLE 15-14 A beam 8 cm in length carries a weight of 1000 g at one end and a weight of 250 g at the other end (see Fig. 15-7). Where must the fulcrum be placed so that the beam will balance? (Disregard the weight of the beam.)

Solution See Fig. 15-7.

$$\Sigma M = 0$$
$$x(1000) - 250(8 - x) = 0$$
$$1000x - 2000 + 250x = 0$$
$$1250x = 2000$$
$$x = 1.6 \text{ cm from the 1000 g weight}$$
$$\text{(or 6.4 cm from the 250 g weight)}$$

EXAMPLE 15-15 A lever 10 m in length has a weight of 1000 kg on one end. Where must the support be placed so that a weight of 250 kg at the other end will make it balance? Disregard the weight of the lever.

Solution Let d = the number of m from support to heavier weight. Then $10 - d$ = the number of m from support to lighter weight. The moments are $1000d$ and $250(10 - d)$. Therefore,

$$2500 - 250d = 1000d$$

Transposing,

$$-250d - 1000d = -2500$$

Collecting terms,

$$-1250d = -2500$$

Dividing,

$$d = 2$$

Thus the support in 2 m from the 1000 kg weight to the support.

EXAMPLE 15-16 This example considers the weight of the beam in Fig. 15-8. Place a fulcrum under a beam 12 ft long so that a force of 150 lb at one end will balance a weight of 1750 lb at the other end. (The beam weighs 10 lb/ft.)

Solution When the weight of the beam is taken into consideration, the moment of each arm of the beam is the weight of that arm multiplied by half the length of the arm.

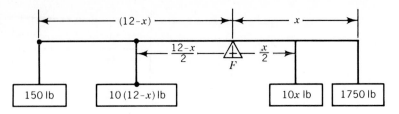

Fig. 15-8. Diagram of the system.

Thus, in this example let x = the length of the short arm in feet.

Then,

$$12 - x = \text{length of the long arm in feet}$$
$$1750x = \text{moment due to weight}$$
$$150(12 - x) = \text{moment due to force}$$

$$\frac{1}{2}x \cdot 10x = \text{moment due to short arm}$$

$$\frac{1}{2}(12 - x) \cdot 10(12 - x) = \text{moment due to long arm}$$

The sum of the moments on one arm is equal to the sum on the other arm if the beam is in balance. Then the equation is

$$1750x + 5x^2 = 150(12 - x) + 5(12 - x)^2$$
$$1750x + 5x^2 = 1800 - 150x + 5(144 - 24x + x^2)$$
$$1750x + 5x^2 = 1800 - 150x + 720 - 120x + 5x^2$$
$$+ 5x^2 - 5x^2 + 1750x + 150x + 120x = 1800 + 720$$
$$2020x = 2520$$
$$x = 1.25$$

▓ EXERCISES

15-102 Two people are carrying a load weighing 350 lb on a pole 10 ft long. Find the weight carried by each if the load is 4 ft from one end of the pole. (Disregard the weight of the pole.)

15-103 If a lever 16 ft long is supported at a point 18 in. from one end, how heavy a weight can be balanced on the long part of the beam if a man weighing 150 lb stands on the short part? (Disregard the weight of the beam.)

15-104 A 200-lb piece of granite is lifted with a 4.5-ft steel rod. What force is required at the end of the rod if a fulcrum is placed 6 in. from the rock?

15-105 A steel bar 3 m long is used to lift a 200-kg weight. If the fulcrum is placed 0.25 m from the weight, what input force is required?

15-106 What is the force of a screwdriver used as a wheel and axle if its blade is 8 mm wide, the handle diameter is 35 mm, and a 20-kg force is applied on the handle?

15-107 A uniform beam 40 ft long weighing 1200 lb rests in a horizontal position upon a support at each end. What is the pressure on each of these supports if the beam has a load of 10,000 lb at a point 10 ft from one end?

15-108 If a 240-lb load in a wheelbarrow is centered at a distance of 20 in. from the wheelbarrow axle, and a worker lifts on the handles at a distance of 4 ft 6 in. from the axle, determine the number of pounds lifted by each hand.

GRAPHICAL METHODS

16-1 INTRODUCTION TO GRAPHS

Graphs are used to illustrate relationships and to solve problems. A good graph can show a pictorial solution to a problem. Graphs are used in magazines, newspapers, and television to indicate facts and relationships which would be difficult to describe by words or to present in a table.

For example, the hourly temperatures in Minneapolis for each of the 24 hours in a day, March 30, 1984, were as shown in Table 16-1.

TABLE 16-1 TEMPERATURES FOR MARCH 30, 1984 IN MINNEAPOLIS

MIDNIGHT	26°	5 A.M. 27°	10 A.M. 40°	3 P.M. 46°	8 P.M.	39°
1 A.M.	25°	6 A.M. 27°	11 A.M. 42°	4 P.M. 46°	9 P.M.	37°
2 A.M.	24°	7 A.M. 30°	NOON 44°	5 P.M. 46°	10 P.M.	34°
3 A.M.	25°	8 A.M. 33°	1 P.M. 44°	6 P.M. 44°	11 P.M.	33°
4 A.M.	26°	9 A.M. 37°	2 P.M. 46°	7 P.M. 40°	MIDNIGHT	32°

By reading all the temperatures listed, you could determine the highest and the lowest temperature for this 24-hour period. You could also determine other facts, such as when temperature declined or rose. But these facts are much easier to see if a graph is constructed. Figure 16-1 is a line graph constructed from the data in Table 16-1. The changes in temperature are shown clearly by this graph. The highest and lowest temperatures are easily determined by inspection. The lowest temperature is 24° at 2 A.M.; the highest is 46° at 2, 3, 4, and 5 P.M.

The dots or points on this graph have been plotted on graph paper from the data that were given in table form. Graph paper is available in many forms and with many scales.

Notice several characteristics of this graph: (1) The graph has a title. (2) Each horizontal space is equal to 1 h. (3) Each vertical space equals 2°. (4) The vertical scale is labeled. (5) The horizontal scale is labeled.

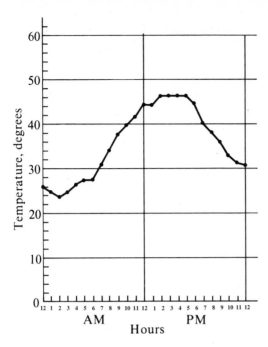

Fig. 16-1. Line graph: Temperature by the hour on March 30, 1984, in Minneapolis.

(6) The vertical scale begins with 0. The labels and the title are meant to assist anyone reading the graph. The value of each horizontal space and vertical space was determined by the person who made the graph. Generally, values are decided on that will show the relationships most clearly. For example, if each vertical space equaled 10°, a difference of 1° would be difficult to plot and read.

Procedure for constructing a line graph:

1. Study the data and determine which of the two sets of facts will make up the horizontal scale.

2. Select the values of the horizontal-scale spaces. (In this graph each horizontal space equals 1 h.)

3. Select the value of a space in the vertical scale. The range of values to be plotted should be considered in determining this scale. (Here the highest temperature to be plotted is 46° and the lowest is 24°, so each vertical space is used to represent an increase or decrease of 2°.)

4. Plot all the data (temperatures).

5. Connect all the plotted points with straight lines.

Fig. 16-2 is a circular graph which shows temperatures over a 24-h period. Here the circular graph paper was revolved under an attachment, connected to a thermometer, which traced the temperature continuously. A circular graph like this eliminates reading an instrument;

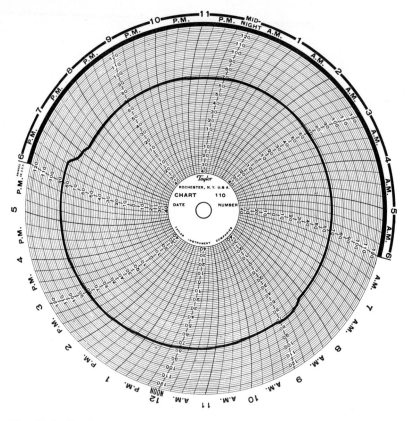

Fig. 16-2. Circular graph: Temperature by the hour *(Taylor Instrument Process Control Division, Sybron Corporation).*

the recording is automatic. Circular graphs have many industrial applications and are widely used.

Inspect the graph in Fig. 16-2 and answer these questions: What is the highest temperature? The lowest? When does each occur?

Note: *A circular graph is* not *the same thing as a circle or pie graph (see Sec. 16-3).*

16-2 BAR GRAPHS

Bar graphs are used to illustrate data which are not mathematically related.

Here is an example: In a 10-year period, a medical component corporation paid a per-share dividend of $1.00 to $2.00. The per-share dividend for each of the 10 years is listed in Table 16-2.

From the data of Table 16-2 a bar graph was constructed; it is shown in Fig. 16-3. This bar graph pictures the growth in dividends per share paid out by a medical component corporation and makes it possible to make several observations very quickly. The same observations could be made from the data as listed in Table 16-2, but only after some study.

TABLE 16-2 YEARLY PER-SHARE DIVIDENDS

Year	Per-share Dividend	Year	Per-share Dividend
1975	$1.00	1980	$1.40
1976	$1.00	1981	$1.50
1977	$1.10	1982	$1.65
1978	$1.20	1983	$1.75
1979	$1.40	1984	$2.00

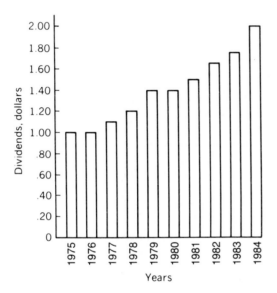

Fig. 16-3. Bar graph: Dividends per share, 1975–1984.

Notice how easily it can be seen that: (1) the dividends paid per share have doubled in 10 years; (2) the greatest increase is for the years 1983–1984; (3) dividends per share did not increase for the years 1975–1976 and 1979–1980; (4) the increase in dividends over the 10-year period has been, except for the years noted, rather steady.

The steps in constructing a bar graph are similar to those followed in making a line graph.

Procedure for constructing a bar graph:

1. Study the data and determine which of the two sets of facts will make up the horizontal scale.
2. Lay out the horizontal scale; that is, the width of the bars.
3. Select the vertical scale and lay out the length of the bars.
4. Construct the bars.

16-3   PIE OR CIRCLE GRAPHS

A pie graph, like a bar graph, shows data which are not mathematically related. It is often used to indicate how a whole or total is divided into parts.

This kind of graph may be used, for example, to illustrate how a tax dollar is spent or what happens to each dollar received by a corporation.

Here is an example: The 1983–1984 general-purpose budget for a midwestern state listed the amounts spent for each service out of each dollar spent. These data are shown in Table 16-3.

TABLE 16-3 HOW THE TAX DOLLAR IS SPENT	
Education	0.49
Health and welfare	0.25
Building program	0.05
Retirement of debt	0.09
General government	0.10
Trade and industry	0.02
Total	$1.00

A pie or circle graph can be used effectively to show the relationships between the sums expended for the services. Figure 16-4 is a pie graph for these data. This graph pictures clearly the proportion of each tax dollar expended for each service.

If the entire circle represents 100 cents or $1.00, what fraction of the circle will represent 25 cents? How much will represent 10 cents?

To construct a circle graph from data, for each item find the fraction of the total amount. Then calculate the central angle by multiplying that

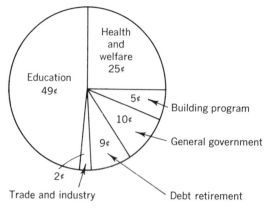

Fig. 16-4. Pie or circle graph: How the tax dollar is spent in a midwestern state.

fraction by 360°. For example:

$$\frac{49}{100} \times 360° = 176.4°$$

16-4 PICTURE GRAPHS

This type of graph uses pictures or drawings of objects to indicate changes—an increase in production, worker-days lost because of accidents, and similar data.

Here is an example: Over a 5-month period a box company constructed and delivered the number of boxes indicated for each month in Table 16-4.

TABLE 16-4 BOX PRODUCTION, JANUARY–MAY	
Month	**Number of Boxes**
January	5000
February	6000
March	8500
April	10,000
May	12,000

Since this was the output of a new factory, the production increase was presented in an employees' magazine. In the magazine article, the data were shown in the form of a picture graph (Fig. 16-5).

The kinds of graphs discussed in these sections are used primarily to picture data or to illustrate a fact or facts; they do not represent mathe-

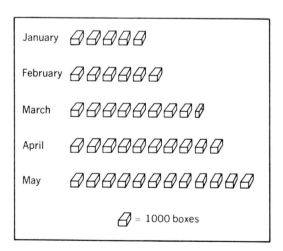

Fig. 16-5. Picture graph: Box production, January–May.

matical relationships. Graphs that depict mathematical relationships and are used to solve problems will be discussed in later sections of this chapter.

EXERCISES

16-1 The hourly temperatures for a 24-h period in Minneapolis are recorded in Table 16-5. Construct a line graph for these data.

TABLE 16-5 EXERCISE 16-1

Hour, A.M.	12	1	2	3	4	5	6	7	8	9	10	11	12
Temperature, °F	45	45	45	45	43	42	41	40	42	51	57	59	62
Hour, P.M.		1	2	3	4	5	6	7	8	9	10	11	12
Temperature, °F		66	70	74	86	86	85	84	83	82	70	69	68

16-2 The population of a Florida city in 1910 was 16,000; in 1920, 16,300; in 1930, 16,850; in 1940, 17,800; in 1950, 19,100; in 1960, 20,700; in 1970, 40,500; and in 1980, 48,500. Construct a line graph illustrating this population growth.

TABLE 16-6 EXERCISE 16-3

Car A	710	Car E	229	Car I	138
Car B	487	Car F	214	Car J	109
Car C	347	Car G	176	Car K	108
Car D	245	Car H	143		

16-3 The number of new automobiles sold in 1 year in a certain county were as shown in Table 16-6. Construct a bar graph for these data.

16-4 Construct (a) a pie or circle graph and (b) a picture graph for the data in Exercise 16-3.

16-5 Construct a bar graph for the data in Exercise 16-2.

16-6 The assets of a company over 80 years were as follows: 1980, 1,450,000,000; 1975, 81,375,000,000; 1970, $1,250,000,000; 1965, $975,000,000; 1955, $554,000,000; 1945, $148,000,000; 1935, $64,000,000; 1925, $25,000,000; 1915, $10,000,000; 1905, $4,000,000. Construct a bar graph for these data.

16-7 Construct a line graph for the data in Exercise 16-6.

16-8 Construct a picture graph for the data in Exercise 16-6.

Note: *The graphs constructed by different students might take quite different forms. One obvious solution is to have a sketch of a labeled bag of money represent a constant number of dollars.*

16-5 MATHEMATICAL GRAPHS

Some of the graphs discussed are used to illustrate data which are not necessarily mathematically related. The graphs described in the next sections are graphs of equations and graphs used to solve problems. These graphs illustrate the mathematical relation between two sets of data.

In business and industry, graphical solutions to problems are used, especially when these two conditions are present: the same problem occurs often and an approximate answer is acceptable.

16-6 GRAPHING DEFINITIONS

Sometimes it is necessary to plot negative as well as positive numbers, but we must first discuss certain definitions and assumptions. Consider Fig. 16-6. If two lines are drawn at right angles to each other, the position of any point, say *P*, may be located by measuring its distance from each of these lines. These lengths in the figure are *OA* and *OB*; they are the *coordinates* of the point *P*. The length *OA* is the *x coordinate,* and *OB* is the *y coordinate.*

The horizontal line is the *x axis* and the vertical line is the *y axis.* Together they are spoken of as the *coordinate system.* The point *O* where the two axes cross is the *origin.*

The *x* axis is also called the *abscissa* and the *y* axis the *ordinate.* The *x* coordinate is called the *abscissa,* and the *y* coordinate the *ordinate,* of a point.

The coordinates are always measured from the origin. An abscissa, *x* value, measured toward the right is *positive,* and an abscissa measured toward the left is *negative.* An ordinate, or *y* value, measured upward is *positive,* and an ordinate measured downward is *negative.*

The axes divide the plane of the graph into four parts called *quad-*

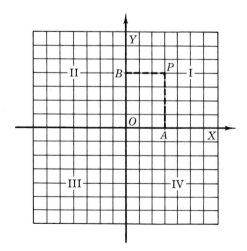

Fig. 16-6. Plotting graphs.

rants: the *first,* *second,* *third,* and *fourth quadrants* as indicated in Fig. 16-6 by the numerals I, II, III, and IV. Both first-quadrant coordinates are positive; in the second quadrant, the abscissa is negative and the ordinate positive; in the third quadrant, both coordinates are negative; in the fourth quadrant, the abscissa is positive and the ordinate negative. These properties are shown in Table 16-7.

TABLE 16-7 QUADRANT SIGNS				
Quadrant	I	II	III	IV
Abscissa	+	−	−	+
Ordinate	+	+	−	−

16-7 PLOTTING POINTS

To *plot* a point is to locate it with reference to the coordinate system. To plot a point whose x coordinate is 3 and whose y coordinate is 4, first draw the axes (see Fig. 16-7), then choose a unit of measure and mark A 3 units to the right of O. Through A draw a line parallel to the y axis. Now mark B on the y axis 4 units above O, and through B draw a line parallel to the x axis. The required point is located where these two lines meet. In the figure it is indicated as $P(3,4)$, which is the form for writing the coordinates of a point. The abscissa is placed first, followed by the ordinate; they are separated by a comma and enclosed in parentheses. This notation is read "the point P whose coordinates are 3 and 4."

The following points, indicated in Fig. 16-7, are located in a similar way: $P(-2,5)$; $P(-4,-3)$; $P(4,-5)$.

Note: *The x value is always written or given first.*

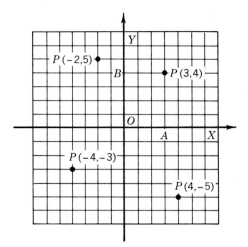

Fig. 16-7. Plotting points.

16-8 GRAPHING AN EQUATION

The equation $3x + 4y = 12$ will be used to demonstrate the graphing of an equation. Remember that for this kind of equation—an equation with two unknowns of the first degree—there are an infinite number of pairs of values for x and y that will satisfy the equation. The first step in graphing this and similar equations is to determine several pairs of values. This is done by assuming several values for one variable and calculating the corresponding values for the second variable.

Construct a table for a number of pairs of values:

x	-6	-4	-2	0	3	5
y	$7\frac{1}{2}$					

Assume the values listed for x, and solve the given equation for the corresponding values of y.

Substitute $x = -6$ in the equation:

$$3x + 4y = 12$$
$$3(-6) + 4y = 12$$
$$-18 + 4y = 12$$
$$4y = 30$$
$$y = 7\frac{1}{2}$$

Complete the table by substituting $x = -4$, -2, 0, 3, and 5 in the equation and calculating the corresponding values for y. The complete table is:

x	-6	-4	-2	0	3	5
y	$7\frac{1}{2}$	6	$4\frac{1}{2}$	3	$\frac{3}{4}$	$-\frac{3}{4}$

Plot the points indicated by the values in the table and connect them with a straight line. This will form a *graph* of the equation. A graph is formed when the plotted points are connected by a straight line (as here) or a curved line (as will be shown in later examples). The line itself is called a graph. The graph for the equation $3x + 4y = 12$ is shown in Fig. 16-8.

When an equation is of the first degree, its graph will be a straight line. Since two points determine a straight line, it is necessary to calculate two pairs of values only. The four additional pairs determined here are only checks on accuracy.

If two points only are to be used to graph an equation, it is customary to take these where the graph crosses the x and y axes. These two points are determined by substituting $x = 0$ and calculating y, and substituting $y = 0$ and calculating x.

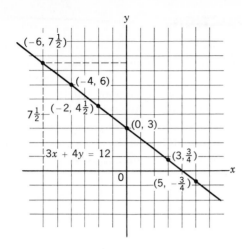

Fig. 16-8. $3x + 4y = 12$.

Procedure for graphing an equation:

1. Construct a table of values.
2. Assume values for either x or y.
3. Calculate the corresponding values for either y or x.
4. Plot the points indicated by the values found.
5. Construct a graph by connecting these points.

16-9   SIMULTANEOUS EQUATIONS

Simultaneous equations are a group of two or more equations containing two unknowns. The word "simultaneous" (at the same time) refers to a set of values—one value for each of the unknowns—which will satisfy the equations "simultaneously."

The graphical solution of two simultaneous equations may be described as follows: Graph both equations; the coordinates of the point of intersection of the two graphs will give the values for the unknowns.

EXAMPLE 16-1   Solve $x - y = 2$ and $x + y = 12$ for x and y.

Solution

1. Graph $x - y = 2$.
2. Graph $x + y = 12$.
3. Read the coordinates of the point of intersection; the coordinates will be the values of x and y.

The table for $x - y = 2$ or $x = 2 + y$:

x	-4	-2	0	2	4
y	-6	-4	-2	0	2

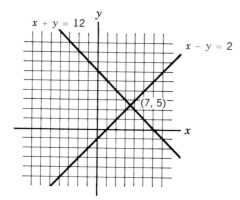

Fig. 16-9. $x - y = 2$ and $x + y = 12$.

The table for $x + y = 12$ or $x = 12 - y$:

x	-5	-3	0	3	5
y	17	15	12	9	7

The values in the table are plotted and the two graphs drawn as shown in Fig. 16-9. The coordinates of the point of intersection are (7,5), and $x = 7$ and $y = 5$.

Checking, $7 - 5 = 2$ and $7 + 5 = 12$.

If the graphs failed to intersect—for example, if they were parallel lines—then the equations would not have a point in common and would be impossible to solve.

The two equations $x + y = 12$ and $3x + 3y = 36$ are equivalent equations; these have an infinite set of solutions. If equivalent equations are drawn, the two graphs will coincide and be the same line or graph.

16-10 THE GRAPH OF AN EQUATION OF ANY DEGREE

The graph of an equation of any degree with two variables (unknowns) can be constructed by the method explained in Sec. 16-8. The graph of an equation of any degree higher than 1 is not a straight line. Therefore more pairs of values must be calculated to determine the graph.

The graph for the equation $x^2 + y^2 = 25$ is constructed as an illustration.

EXAMPLE 16-2 Graph $x^2 + y^2 = 25$.

Solution Solve for y:

$$x^2 + y^2 = 25$$
$$y^2 = 25 - x^2$$
$$y = \pm\sqrt{25 - x^2}$$

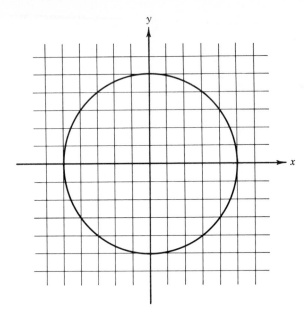

Fig. 16-10. $x^2 + y^2 = 25$.

x	-5	-4	-3	-2	0	2	3	4	5
y	0	± 3	± 4	$\pm\sqrt{21}$	± 5	$\pm\sqrt{21}$	± 4	± 3	0

The values for x and y are plotted and the curve is constructed as shown in Fig. 16-10. The graph in this figure is a circle with a radius of 5.

Note: *When a number greater than 5 is assumed for x, the value under the radical sign in $\pm\sqrt{25 - x^2}$ is negative, and this square root cannot be extracted by any method we have studied.*

EXAMPLE 16-3 Plot the curve of the equation $xy = 1$ or $x = 1/y$.

Solution This will give a curve from which the reciprocal of any number can be read. It is plotted by first finding a number of pairs of values for x and y which satisfy the equation and then plotting the points that have these pairs as coordinates. It should be noted that when x is negative, y is negative, and when x is positive, y is positive. Plotting the pairs of values gives the graph shown in Fig. 16-11.

x	$\pm\frac{1}{16}$	$\pm\frac{1}{8}$	$\pm\frac{1}{4}$	$\pm\frac{1}{2}$	± 1	± 2	± 3	± 6	± 8	± 16
y	± 16	± 8	± 4	± 2	± 1	$\pm\frac{1}{2}$	$\pm\frac{1}{3}$	$\pm\frac{1}{6}$	$\pm\frac{1}{8}$	$\pm\frac{1}{16}$

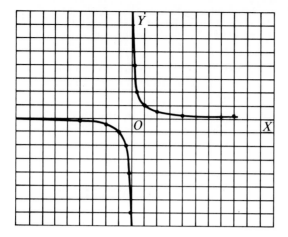

Fig. 16-11. $xy = 1.$

EXERCISES

Plot and solve the following systems of equations:

16-9 $x - y = 1$ and $x + y = 7$

16-10 $x + y = 9$ and $2x - y = 6$

16-11 $2x - 6y = -12$ and $x + 3y = 6$

16-12 $2x + 3y = 0$ and $3x - 2y = 0$

16-13 $x + 2y = 4$ and $3x - 4y = -3$

16-14 $4x + 2y = 6$ and $8x - 6y = -8$

16-15 $2(x - 1) + 2y = 6$ and $2(1 - x) - 8y = -18$

16-16 Plot the curve for $y = x^2$. This gives the curve from which squares and square roots of numbers can be read.

16-17 Plot the curve for $y = 2x^2 - 3$.

16-18 Plot the curve for $y = x^3$. This gives the curve from which cubes and cube roots of numbers can be read. You may wish to use a different scale for x and y.

16-19 Plot the curve for $y = 3x^3 - 1$.

16-20 In calculating simple interest, if P stands for principal, t for time, r for rate, and A for amount, then $A = P(1 + rt)$. If particular numerical values are given to P and r, and if the different values of A are regarded as ordinates and the corresponding values of t as abscissas, then the graph of this equation may be drawn. Draw the graph. What line in the figure represents the principal? What feature in the graph depends upon the rate?

16-21 Using the same formula as in Exercise 16-20, draw a graph in which interest and time are the coordinates of points on the curve.

16-22 Plot the curve of the formula for the area of a circle $A = \pi r^2$. Use radii as abscissas and areas as ordinates. From this can be read the areas of circles of radii from 0 to 6.

16-23 Follow the directions in Exercise 16-22 and plot the curve from which the volumes of spheres can be read. The volume of a sphere is given by the formula $V = \frac{4}{3}\pi r^3$.

16-24 If the temperature of a gas is constant, the pressure times the volume remains constant. Using p for pressure and v for volume, plot a graph showing $pv = 4$ with various values of p and v.

16-11 SLOPE OF A FIRST-DEGREE EQUATION

Figure 16-12 is a graph of the equation $y - x = 0$, or $y = x$. The ratio of the change in y to a change in x is called the *slope of the graph*. In Fig. 16-12 this ratio is $3 \div 3$, or 1. The slope of the equation $y = x$ is, then, 1, since any change in x produces an equal change in y.

The *slope* of a line has been defined as the ratio of the change of y to the change of x. The letter m indicates slope. To a carpenter, slope is the *rise* over the *run*.

$$\text{slope} = \frac{\text{change of } y}{\text{change of } x} = \frac{\text{rise}}{\text{run}}$$

Figure 16-13 is a graph of the equation $y = 2x$. In Fig. 16-13 the slope is $4 \div 2$, or 2; a point moving along the curve is traveling up 2 units for each unit it moves to the right.

Figure 16-14 is a graph of the equation $y = -x$. Compare this graph with Fig. 16-12, the graph of $y = x$. The absolute value of the slope in

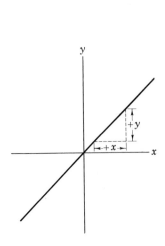

Fig. 16-12. $y - x = 0$, or $y = x$.

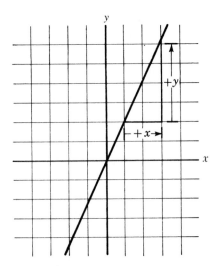

Fig. 16-13. $y = 2x$.

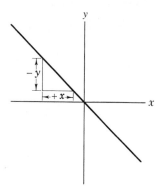

Fig. 16-14. $y = -x$.

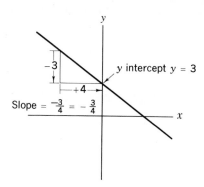

Fig. 16-15. $3x + 4y = 12$.

both graphs is the same: 1. But the slope of the equation $y = -x$ is -1. The negative sign indicates that the curve slopes down from left to right.

The slope, then, can be determined by graphing the equation and computing the ratio of change in y to change in x. It is, however, easier to solve the equation for y and then find the slope by inspection.

EXAMPLE 16-4 Determine the slope of $3x + 4y = 12$.

Solution Solve the equation for y:

$$3x + 4y = 12$$

$$4y = -3x + 12$$

$$y = -\frac{3}{4}x + 3$$

The coefficient of the x term, $-\frac{3}{4}$, is the slope of the equation. $3x + 4y = 12$. The 3 is the y *intercept*—the value of y when the graph intercepts the y axis.

Figure 16-15 is the graph of the equation $3x + 4y = 12$.

Solving an equation for y and expressing it in the form $y = -\frac{3}{4}x + 3$ is known as *placing it in the slope-intercept form.* The general equation for the slope-intercept form is $y = mx + b$, where m is the slope and b is the y intercept—the value of y at the point where the graph intercepts (or "crosses") the y axis. First-degree equations may be plotted quickly if expressed in this form. The slope-intercept form $y = mx + b$ can also be used as a check when an equation is graphed by other means.

16-12 EQUATIONS OF LINES PARALLEL TO AN AXIS

In the equation $x = 2$, x is always equal to 2; the graph of this equation therefore includes only those points whose abscissas are 2. The graph of the equation is, then, a straight line parallel to the y axis whose x intercept is 2. Similarly, $y = 3$ is the equation of a straight line parallel to the x axis with the y intercept 3. In general, $x = a$ is a straight line parallel to the y axis; and $y = b$ is a straight line parallel to the x axis. It follows that $x = 0$ is the equation of the y axis and $y = 0$ is the equation of the x axis.

16-13 THE EQUATION OF ANY STRAIGHT LINE

If the slope and the y intercept of a straight line are known, its equation can be written at once by putting these values for m and b in the equation $y = mx + b$, which is the slope-intercept form of the equation of a straight line.

If two points through which the line passes are known, the equation of the line can be found as follows: Let (x_1, y_1) and (x_2, y_2) be the two points. Then the change in y from y_1 to y_2 is $y_2 - y_1$, and the change in x from x_1 to x_2 is $x_2 - x_1$. Thus the slope of the line through these points is $(y_2 - y_1)/(x_2 - x_1)$, which is the ratio of these changes. This is shown in Fig. 16-16. Now by the definition of a straight line it has the same direction throughout; that is, the slope is always the same.

Then let (x, y) be any other point on the line, and find the slope using one of the given points and the point (x, y). This gives the slope $(y - y_1)/(x - x_1)$. Since these two slopes are equal, the equation of the line is

$$\frac{y - y_1}{x - x_1} = \frac{y_2 - y_1}{x_2 - x_1}$$

This is known as the two-point form of the equation of a straight line. The point (x_2, y_2) could as well be used with (x, y) in finding the slope.

If one point that a line passes through and the line's slope are known, the equation of the line can be found as follows: Let (x_1, y_1) be

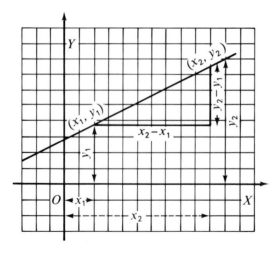

Fig. 16-16. The slope of a straight line.

the point, and let m be the slope. If (x, y) is any point on the line, the slope of the line is $(y - y_1)/(x - x_1)$. But m is also the slope, and therefore the equation of the line is

$$\frac{y - y_1}{x - x_1} = m$$

This is known as the *point-slope form* of the equation of a straight line.

EXAMPLE 16-5 Find the equation of the straight line having a slope of $\frac{2}{3}$ and a y intercept of 2.

Solution Substituting in the equation $y = mx + b$,

$$y = \tfrac{2}{3}x + 2$$

Simplifying,

$$2x - 3y + 6 = 0$$

EXAMPLE 16-6 Find the equation of the straight line passing through the points $(3, 4)$ and $(-1, -2)$.

Solution Here we use the two-point form of the equation of a straight line, where $x_1 = 3$, $y_1 = 4$, $x_2 = -1$, $y_2 = -2$. Substituting these values,

$$\frac{y - 4}{x - 3} = \frac{-2 - 4}{-1 - 3}$$

Simplifying,

$$3x - 2y - 1 = 0$$

EXAMPLE 16-7 Find the equation of a line through $(4, -2)$ with slope $\frac{1}{2}$.

Solution Substituting in the point-slope form of the equation of a straight line,

$$\frac{y + 2}{x - 4} = \frac{1}{2}$$

Simplifying,

$$x - 2y - 8 = 0$$

16-14 GRAPHICAL SOLUTION OF PROBLEMS

Problems which are difficult or impossible to solve by calculation methods can often be solved with a graph.

EXAMPLE 16-8 Design and construct a box of maximum volume from a square sheet of tin 12 in. on a side. (See Fig. 16-17.)

Solution The square box will be made by cutting small squares from each corner and turning up the sides to form the box. See Figs. 16-18 and 16-19. Let

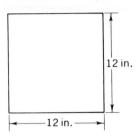

Fig. 16-17. Sheet 12 in. × 12 in.

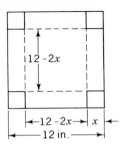

Fig. 16-18. Example 16-8: The sheet of tin.

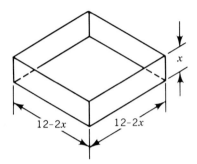

Fig. 16-19. Example 16-8: Dimensions of the box.

y = the volume of the box and x = the height. Then

$$y = x(12 - 2x)(12 - 2x)$$
$$y = x(12 - 2x)^2$$
$$y = x(4x^2 - 48x + 144)$$
$$y = 4x^3 - 48x^2 + 144x$$

x	0	$\frac{1}{2}$	1	$1\frac{1}{2}$	2	$2\frac{1}{2}$	3	$3\frac{1}{2}$	4	5	6
y	0	$60\frac{1}{2}$	100	$121\frac{1}{2}$	128	$122\frac{1}{2}$	108	$87\frac{1}{2}$	64	20	0

From this table, as x increases, y increases up to a value of 2 and then decreases. The value for y when $x = 6$ is 0. Figure 16-20 is the graph of this equation.

The box, as seen from the table and graph, will enclose a maximum volume when the height of the sides is 2 in. ($x = 2$) and the base is 8 in.

*Note: This graph can be used to determine the volume of any box constructed from a square piece 12 in. on a side. For example, a box with a side of $4\frac{1}{2}$ in. will include a volume of approximately 38 in.³ It is also possible to solve this problem for a general case: What will be the height of the side of a box of maximum volume constructed from a square piece **a** inches on the side? The answer will be **a** divided by some number; from the solution presented you should be able to guess this number.*

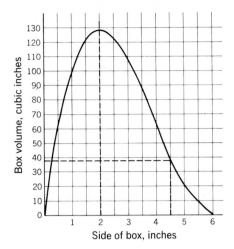

Fig. 16-20. $4x^3 + 48x^2 + 144x$.

If a problem must be frequently solved and an approximate answer is acceptable, it is common practice to use a graphical solution.

EXAMPLE 16-9 Construct a graph to convert Celsius temperature readings to Fahrenheit. The formula is F = 1.4 C + 32°.

Solution F = 1.4 C + 32°.

C	−60	−40	−20	−10	0	10	20	30	100
F	−76	−40	−4	14	32	50	68	86	212

Figure 16-21 is a graph constructed from these values. The graph can be used to find the temperature in Fahrenheit if the Celsius temperature is known or the Celsius temperature if the Fahrenheit temperature is known. The graph is a time-saving device for converting temperatures to either scale.

The scale used in constructing the graph should be as large as possible to permit accurate readings. In the graph in Fig. 16-21, each space equals 20°. If the graph had been constructed using a scale of 1 space = 2°, it would be easier to read and could be more accurately read. Even better would be 1 space = 1°.

EXAMPLE 16-10 Draw the graph of the equation $y = x^3 - 21x + 20$.

x	−6	−5	−4	−3	−2	−1	0	$\frac{1}{2}$	1	$1\frac{1}{2}$	2	3	4	$4\frac{1}{2}$	5
y	−70	0	40	56	54	40	20	$9\frac{5}{8}$	0	$-8\frac{1}{8}$	−14	−16	0	$16\frac{5}{8}$	40

Solution In plotting the x and y points in Fig. 16-22, use one unit per square on the x axis and five units per square on the y axis. Studying the graph shows that a point moving along the curve from the left would at first rise rapidly, rise more

Degrees F

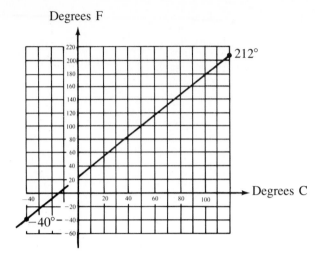

Fig. 16-21. F = 1.8C + 32°.

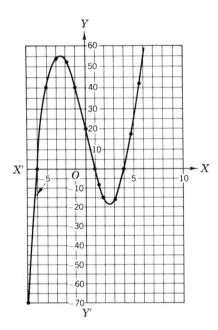

Fig. 16-22. y = x³ − 21x + 20.

slowly, and then stop rising at about $x = -2.6$; then it would fall to where $x = 2.6$, and rise for greater values of x.

A point where the curve ceases to rise and then falls is called a *maximum point of the curve*, and the value of the equation at that point is a *maximum value of the equation*. Here this maximum value of the equation is about 57.

A point where the curve ceases to fall and then rises is called a *minimum point of the curve*, and the value of the equation at that point is a *minimum value of the equation*. The minimum value of the equation is about −17.

▌ EXERCISES

16-25 Plot the following equations, all on the same set of coordinate axes:
(a) $y = \frac{1}{4}x$; (b) $y = \frac{1}{2}x$; (c) $y = 4x$; (d) $y = -2x$.

16-26 Plot the following equations on the same set of axes: (a) $y = 3$; (b) $y = x + 3$; (c) $y = 3x + 3$; (d) $y = -3x + 3$.

16-27 Plot the following equations on the same set of axes:
(a) $y = \frac{1}{2}x + 6$; (b) $y = \frac{1}{2}x + 4$; (c) $y = \frac{1}{2}x - 4$; (d) $y = \frac{1}{2}x$.

16-28 Change the following equations to the slope-intercept form and plot:
(a) $2x - 3y = 6$ (c) $3x - 2y - 6 = 0$
(b) $7x + 6y = 42$ (d) $2x - 6y + 24 = 0$

16-29 Plot the following equations on the same set of axes: (a) $x = 0$; (b) $x = 5$; (c) $x = -6$; (d) $y = 0$; (e) $y = -4$; (f) $y = 2\frac{1}{2}$.

16-30 Find the equations of the following:
(a) through point $(0, 6)$ with slope $\frac{2}{3}$
(b) through $(0, 2)$ with slope $-\frac{2}{3}$
(c) through points $(2, 3)$ and $(4, 6)$
(d) through points $(6, 2)$ and $(-1, 3)$
(e) through points $(0, 2)$ and $(7, 0)$
(f) through points $(-4, -2)$ and $(-1, 0)$
(g) through point $(3, 2)$ with slope -6
(h) through point $(-4, -1)$ with slope $\frac{3}{2}$

16-31 Given two numbers x and $x - 2$, find x so that their product is a minimum. Draw a graph of their product. Find the minimum product.

16-32 Two numbers differ by 4. Find their values so that their product will be a minimum. Draw a graph of their product. Find the minimum product.

16-33 Plot each of the following on different coordinate axes; note the maximum (minimum) point of each. Indicate the points where the graph crosses the x axis; these points are found by setting $y = 0$ and then solving the quadratic equation for its two roots.
(a) $y = x^2 - 2x$ (c) $y = x^2 - 6x + 8$
(b) $y = x^2 + 2x$ (d) $y = x^2 + 6x + 8$

16-34 Find the area of the greatest rectangle that can be bounded by a line 20 cm long.

Suggestion: *Let x = the length of one side. Then $10 - x$ = length of other side. Therefore, area $= x(10 - x)$. Plot, and find the value of x that makes the area a maximum.*

16-35 Find the area of the greatest rectangle that can be bounded by a line 24 cm long. See the suggestions under Exercise 16-34.

16-36 A rectangular sheet of tin 24 cm wide is to be made into a gutter by turning up strips vertically along the sides. Find the width of the strip turned up to get the maximum cross-sectional area. Find the maximum cross-sectional area.

16-37 The sum of two numbers is S. Find the maximum value of their product. Show that the result may be used to prove that when the perimeter of a rectangle is given, the maximum area will be enclosed when the rectangle is a square.

EQUATIONS WITH MORE THAN ONE UNKNOWN

17-1 INTRODUCTION TO EQUATIONS WITH MORE THAN ONE UNKNOWN

The equations presented in previous chapters involved only one un-known quantity or one unknown letter. In this chapter you will study the solution of equations with more than one unknown.

If a problem involves two unknowns, then two equations are re-quired to determine the value of the unknowns. If the problem involves three unknowns, then three equations are required. This chapter will demonstrate the solution of equations with two and three unknowns.

17-2 EQUATIONS WITH TWO UNKNOWNS

Consider the equation $x - y = 2$. This equation has two unknowns, x and y. It is impossible to solve this equation and obtain one and only one value for both x and y, since many values of x and y make it true that $x - y = 2$. For example, if $x = 3$ and $y = 1$, then $x - y = 2$; also, if $x = 10$ and $y = S$, then $x - y = 2$; and if $x = 23$ and $y = 21$, then again $x - y = 2$. It is obvious, then, that many values of x and y satisfy the simple equation $x - y = 2$.

The equation $x - y = 2$ is called an *indeterminate equation* because there is no single solution for it.

EXAMPLE 17-1 Find the width of the rectangle that is 3 ft longer than it is wide. The perimeter is 4.5 times the shorter side.

Solution Let x = length and y = width.

The perimeter of any rectangle is: $P = 2$ length $+ 2$ width, and thus $2x - 2y = 4.5y$, since $2x + 2y$ is the perimeter, and the perimeter is 4.5 times the shorter side (or width). The above equation is an accurate statement of the prob-lem, but from that equation alone it is impossible to arrive at a single value for x and a single value for y. Many values for x and y can satisfy that equation.

It is possible, however, to develop a second equation from the statement "the rectangle is 3 ft longer than it is wide"; this second equation is $x - y = 3$, since the length is 3 ft longer than the width.

The two equations are, then:

$$2x + 2y = 4.5y \qquad (1)$$
$$x - y = 3 \qquad (2)$$

The next step is to solve one of the equations for x:

$$x - y = 3$$
$$x = y + 3$$

Substitute the value $(y + 3)$ for x in the first equation:

$$2x - 2y = 4.5y$$
$$2(y + 3) - 2y = 4.5y \qquad \text{since } x = y + 3$$

This step has eliminated one unknown, and the value for y is then determined by methods already learned.

$$2(y + 3) + 2y = 4.5y$$
$$2y + 6 + 2y = 4.5y$$
$$4y + 6 = 4.5y$$
$$4y = 4.5y - 6$$
$$-0.5y = -6$$
$$y = 12 \text{ ft}$$

The length may be calculated by substituting $y = 12$ into either of the two original equations. The remainder of the solution appears below:

Substituting $y = 12$ in equation (2),

$$x - y = 3$$
$$x - 12 = 3$$
$$x = 15 \text{ ft}$$

Check the values in both original equations:

$$2x + 2y = 4.5y \qquad (1)$$
$$2(15) + 2(12) \overset{?}{=} 4.5(12)$$
$$30 + 24 \overset{?}{=} 54$$
$$54 = 54$$

$$x - y = 3 \qquad (2)$$
$$15 - 12 \overset{?}{=} 3$$
$$3 = 3$$

Do the answers agree with the conditions cited in the problem? Is one side 3 ft longer than the shorter side? Yes. Is the perimeter 4.5 times the shorter side? The perimeter is 54, which is 4.5 times 12 (the shorter side). The answers 12 and 15 ft must therefore be correct.

Note that we could also have solved this problem graphically by graphing the two equations. If we had done this, the graph of $x - y = 3$ would have intercepted the graph of $2x + 2y = 4.5y$ at (15, 12). Thus, $A = 15$, $y = 12$.

In the example, the values $x = 15$ and $y = 12$, when substituted in both equations, produce two identities, $54 = 54$ and $3 = 3$. In this case these two values *satisfy* both equations. A set of two (or more) equations satisfied by one pair (set) of values for the unknowns is called a system of *simultaneous equations*.

Not every set of equations that may appear similar to the above can be solved. A set of equations may not have a pair of values that will satisfy each equation; and a set of equations may be such that *any* pair of values that satisfies one equation will also satisfy the other. Two equations that cannot both be solved by one pair of values are called *inconsistent equations*. Two equations both of which are solved by any pair of values that satisfies one of them are called *equivalent equations* or *dependent equations*.

For example, the equations $2x + 4y = 14$ and $x + 2y = 5$ are inconsistent, since there are no values for x and y that will satisfy both equations. The equations $x + y = 7$ and $3x + 3y = 21$ are equivalent, since any pair of values which satisfies one of these equations will satisfy the other also.

17-3 SOLUTIONS FOR TWO EQUATIONS WITH TWO UNKNOWNS

There are several methods of solving two equations for the values of the two unknowns. The procedure is similar for all methods: eliminate one unknown and then solve for the other unknown. The methods of solution of two unknowns are called: (1) addition, (2) subtraction, (3) substitution, and (4) comparison.

17-4 SOLUTION BY ADDITION

EXAMPLE 17-2 Solve $x - y = 2$ and $x + y = 12$ for x and y.

Solution

$$x - y = 2 \qquad (1)$$
$$x + y = 12 \qquad (2)$$

If you add equations (1) and (2), then the y term is eliminated, since $(+y) + (-y) = 0$. Adding these two equations gives the following:

$$
\begin{aligned}
x - y &= 2 \\
\underline{x + y} &= \underline{12} \\
2x + 0 &= 14 \\
2x &= 14 \\
x &= 7
\end{aligned}
$$

To obtain y, substitute the value found for $x(7)$ in equation (1).

$$
\begin{aligned}
x - y &= 2 \\
7 - y &= 2 \\
-y &= -5 \\
y &= 5
\end{aligned}
$$

Check the values in both original equations:

$$x - y = 2 \tag{1}$$
$$7 - 5 \stackrel{?}{=} 2$$
$$2 = 2$$

$$x + y = 12 \tag{2}$$
$$7 + 5 \stackrel{?}{=} 12$$
$$12 = 12$$

Why is it possible to add these two equations? Study this explanation carefully: In the equation $x - y = 2$, an equal value was added to the left member $(x - y)$ and to the right member (2), since $x + y = 12$. Since we added equal values to both members of the equation $x - y = 2$, the equality is still true. *If equal values are added to or subtracted from both members of the equation, then the equality is still true.*

17-5 SOLUTION BY SUBTRACTION

EXAMPLE 17-3 Solve $x + y = 1$ and $4x + 3y = 0$ for x and y.

Solution

$$x + y = 1 \tag{1}$$
$$4x + 3y = 0 \tag{2}$$

If the coefficient for the y term in equation (1) were 3, then the y term in both equations could be eliminated by subtraction, since $3y - 3y = 0$. Therefore, multiply both members of equation (1) by 3, and arrange the equations, with the change.

$$3x + 3y = 3 \tag{3}$$
$$4x + 3y = 0 \tag{4}$$

Then, subtract equation (4) from equation (3):

$$3x + 3y = 3$$
$$\underline{4x + 3y = 0}$$
$$-x + 0 = 3$$
$$-x = 3$$
$$x = -3$$

Solve for y by substituting the value obtained for $x(-3)$ in equation (1):

$$x + y = 1$$
$$-3 + y = 1$$
$$y = 4$$

Check the values in both original equations:

$$x + y = 1 \tag{1}$$
$$-3 + 4 \stackrel{?}{=} 1$$
$$1 = 1$$

$$4x + 3y = 0 \tag{2}$$
$$4(-3) + 3(4) \overset{?}{=} 0$$
$$-12 + 12 \overset{?}{=} 0$$
$$0 \overset{?}{=} 0$$

17-6   SOLUTION BY SUBSTITUTION

The problem explained in Sec. 17-2 is an example of solving equations by eliminating an unknown by substitution. Here is a problem solved by this method.

EXAMPLE 17-4   Solve $y + 7x = 42$ and $3x - y = 8$ for x and y.

Solution

$$y + 7x = 42 \tag{1}$$
$$3x - y = 8 \tag{2}$$

Solve equation (1) for y:

$$y + 7x = 42$$
$$y = 42 - 7x$$

Substitute the value obtained for y, $y = 42 - 7x$ in equation (2). Note that the minus sign preceding the y in equation (2) is retained since $+y$ (not $-y$) = $42 - 7x$; solve for x.

$$3x - (42 - 7x) = 8$$
$$3x - 42 + 7x = 8$$
$$-42 + 10x = 8$$
$$10x = 50$$
$$x = 5$$

Solve for y by substituting the value obtained for $x(5)$ in equation (1):

$$y + 7x = 42$$
$$y + 7(5) = 42$$
$$y + 35 = 42$$
$$y = 7$$

Check the values in both original equations:

$$y + 7x = 42 \tag{1}$$
$$7 + 7(5) \overset{?}{=} 42$$
$$7 + 35 \overset{?}{=} 42$$
$$42 = 42$$

$$-y + 3x = 8 \tag{2}$$
$$-(7) + 3(5) \overset{?}{=} 8$$
$$-7 + 15 \overset{?}{=} 8$$
$$8 = 8$$

17-7 SOLUTION BY COMPARISON

The fourth method of solving for two unknowns is by *comparison*. In this method both equations are solved for one of the two unknowns. One member of each equation is then set equal to the other. In the resulting equation, only one unknown appears—the other unknown has been eliminated. This method is illustrated below.

EXAMPLE 17-5 Solve $x - 4y = 21$ and $3x - y = 11$ for x and y.

Solution

$$x + 4y = 21 \tag{1}$$
$$3x - y = 11 \tag{2}$$

First, solve both equations (1) and (2) for x:

$$x + 4y = 21 \tag{3}$$
$$x = 21 - 4y$$
$$3x - y = 11$$
$$3x = 11 + y$$
$$x = \frac{11 + y}{3} \tag{4}$$

Since $x = x$, it is possible to equate the right members of equations (3) and (4), thus eliminating the unknown x:

$$\frac{11 + y}{3} = 21 - 4y \tag{5}$$

Solve equation (5) for y, first clearing of fraction by multiplying each side by 3:

$$3\left(\frac{11 + y}{3}\right) = 3(21 - 4y)$$
$$11 + y = 63 - 12y$$
$$13y = 52$$
$$y = 4$$

Substitute this value of y in equation (1) to determine x:

$$x + 4y = 21$$
$$x + 4(4) = 21$$
$$x + 16 = 21$$
$$x = 5$$

Check the values in both of the original equations:

$$x + 4y = 21 \tag{1}$$
$$5 + 4(4) \overset{?}{=} 21$$
$$5 + 16 \overset{?}{=} 21$$
$$21 = 21$$

$$3x - y = 11 \qquad (2)$$
$$3(5) - 4 \stackrel{?}{=} 11$$
$$15 - 4 \stackrel{?}{=} 11$$
$$11 = 11$$

EXAMPLE 17-6 The sum of two numbers is 62, and their difference is 8. What are the numbers?

Solution Let x = the larger number and y = the smaller number. Then the equations are

$$x + y = 62$$
$$x - y = 8$$

To solve, add the equations and solve for x:

$$x + y = 62$$
$$\underline{x - y = 8}$$
$$2x = 70$$
$$\frac{2x}{2} = \frac{70}{2}$$
$$x = 35$$

To obtain y, substitute 35 for x in the equation $x + y = 62$.

$$35 + y = 62$$
$$y = 62 - 35$$
$$y = 27$$

Check the value in both original equations.

EXAMPLE 17-7 Given two grades of zinc ore, the first containing 45 percent zinc and the second 25 percent zinc, find how many pounds of each grade will be used to make a 2000-lb mixture containing 40 percent zinc. (See the table below.)

Grade 1 Zinc Ore (45%)		Grade 2 Zinc Ore (25%)		Mixture Grade 1 + Grade 2
x lb Zn	+	y lb Zn	=	2000 lb
45% Zn (x)	+	25% Zn (y)	=	40% Zn (2000)

Solution Let x = the number of pounds of 45 percent ore and y = number of pounds of 25 percent ore. Then

$$x + y = 2000 \qquad (1)$$

Since the zinc in one grade plus the zinc in the other grade must equal the zinc in the mixture,

$$0.45x + 0.25y = 0.40(2000) \qquad (2)$$

From (1), $y = 2000 - x$. Substitute $y = 2000 - x$ into equation (2) and solve:

$$0.45x + 0.25(2000 - x) = 0.40(2000)$$
$$0.45x + 500 - 0.25x = 800$$
$$0.20x = 300$$
$$x = 1500 \text{ lb of 45 percent ore}$$

From (1), $x + y = 2000$; therefore,

$$1500 + y = 2000$$
$$y = 500 \text{ lb of 25 percent ore}$$

17-8 SOLUTIONS FOR THREE EQUATIONS WITH THREE UNKNOWNS

Solving a set of three equations with three unknowns may be accomplished by any of the four methods shown or by a combination of any of the four. Notice that the solution demonstrated for a set of three equations is longer than the solutions for two equations with two unknowns. The four methods demonstrated may be used for any number of unknowns, but when more than three unknowns are in a problem, other methods, beyond the scope of this textbook, are used.

To solve three equations with three unknowns, you must have three equations. Then, by the methods of elimination illustrated previously, reduce to two equations with two unknowns and then one equation with one unknown. You then substitute this value into the other reduced equations (two unknowns) to obtain the other unknown and then substitute the two unknowns in one of the original equations to obtain the third unknown. Check the values in the three original equations.

EXAMPLE 17-8 Solve $x + y + z = 12$, $-x + y + z = 8$, and $x + y - z = 0$ for x, y, and z.

Solution

$$
\begin{aligned}
x + y + z &= 12 \qquad &(1) \\
-x + y + z &= 8 \qquad &(2) \\
x + y - z &= 0 \qquad &(3)
\end{aligned}
$$

First attempt to reduce from three equations with three unknowns to two equations with two unknowns by using the above methods.

Add equations (1) and (2):

$$
\begin{array}{rl}
x + y + z = 12 & \qquad (1) \\
\underline{-x + y + z = 8} & \qquad (2) \\
2y + 2z = 20 & \qquad (4)
\end{array}
$$

Add equations (2) and (3):

$$
\begin{array}{rl}
-x + y + z = 8 & \qquad (2) \\
\underline{x + y - z = 0} & \qquad (3) \\
2y = 8 & \qquad (5) \\
y = 4 &
\end{array}
$$

Substitute the value for y, 4, in equation (4):

$$2y + 2z = 20 \qquad\qquad (4)$$
$$2(4) + 2z = 20$$
$$8 + 2z = 20$$
$$2z = 12$$
$$z = 6$$

Substitute the values of y, 4, and of z, 6, in equation (1):

$$x + y + z = 12$$
$$x + 4 + 6 = 12$$
$$x = 2$$

Check the values in the three original equations:

$$x + y + z = 12 \qquad\qquad (1)$$
$$2 + 4 + 6 \stackrel{?}{=} 12$$
$$12 = 12$$

$$-x + y + z = 8 \qquad\qquad (2)$$
$$-2 + 4 + 6 \stackrel{?}{=} 8$$
$$8 = 8$$

$$x + y - z = 0 \qquad\qquad (3)$$
$$2 + 4 - 6 \stackrel{?}{=} 0$$
$$0 = 0$$

■ EXERCISES

17-1 Solve each of the following for the unknowns by any method:
(a) $x + y = 4$ and $x - y = 2$
(b) $x + 2y = 7$ and $x - y = 1$
(c) $x + 3y = 1$ and $x - y = 9$
(d) $2x + 3y = 2$ and $2x - 3y = 0$
(e) $2x + 3y = 1$ and $-x + y = 1$
(f) $3x + 5y = 11$ and $15x + 15y = 7$
(g) $2x - y = 3$ and $4x + 2y = 50$
(h) $5x - 11y = 1$ and $5x + 11y = 3$
(i) $x + y = 1$ and $4x - 4y = 6$
(j) $2x + 4y = 10$ and $6x - 2y = 2$
(k) $x + 2y = 5.5$ and $2x - 1.5y = 0$
(l) $5x + 6y = 17$ and $6x + 5y = 16$

17-2 Three times one number plus four times another number is 10, and the second plus four times the first is 9. What are the two numbers?

17-3 What fraction equals $\frac{1}{3}$ when 1 is added to the numerator and equals $\frac{1}{4}$ when 1 is added to the denominator?

Suggestion: *Let x = numerator and y = denominator. Then, x/y is the fraction. The equations are*

$$\frac{x + 1}{y} = \frac{1}{3} \quad \text{and} \quad \frac{x}{y + 1} = \frac{1}{4}$$

Simplify and solve the equations.

17-4 Solve for x and y: $x/2 + y/3 = 3/2$ and $x - 2y = -5$.

17-5 A rectangle is 3 ft longer than it is wide. Its perimeter is 4.5 times its shorter side. Find its width.

17-6 Find two numbers whose sum is 144 and whose difference is 6.

17-7 I bought a suit and a topcoat and had $19 left out of $300. One-sixth of the cost of the suit was $1 more than one-ninth the cost of the coat. Find the price of each.

17-8 Today a father is three times as old as his son. Ten years ago he was seven times as old. Find the ages of both today.

17-9 An interior decorator paid $2400 for an odd lot of rugs. She sold one-third of them at $20 each, one-fourth of them at $30 each, one-sixth of them at $40 each, and one-fourth of them at $50 each. She gained $800 on the sale. How many rugs were sold?

17-10 In a factory where 700 skilled and unskilled are employed, the average daily pay for the skilled is $12.40 and for the unskilled is $9.00. If $8000 is paid daily for the work, find the number of skilled and unskilled workers.

17-11 A small building contractor employed three types of laborers. For one-half of all those employed he cleared $100 each; for one-third of them he cleared $200 each; and for one-sixth of them he cleared $300 each. He cleared a total of $2000. How many laborers were there of each type?

17-12 The sum of two consecutive numbers (x and $x + 1$) exceeds half the smaller number by 25. Find the numbers.

17-13 Given $S = \dfrac{\pi DN}{12}$ and $T = \dfrac{LF}{N}$, eliminate N and find the value of T in terms of the remaining letters.

Suggestion: *Solve each equation for N, and eliminate by comparison.*

Solve for the unknowns in Exercises 17-14 through 17-16.

17-14 $x + 2y - z = -6$
$2x + y - 3z = 9$
$3x - 2y - 4z = -5$

17-16 $8x - 4y + 2z = -12$
$4x + 2y + 6z = -4$
$12x + 8y + 10z = 10$

17-15 $x + y = 0.850$
$x + z = 0.710$
$y + z = 0.765$

17-17 A man has $98 in dollar bills, half dollars, and quarters. Half of the dollars and one-fifth of the half dollars are worth $31; and one-seventh of the half dollars and one-third of the quarters are worth $10. How many pieces has he of each?

17-18 A lever is balanced on a fulcrum with a weight of 20 kg at one end and 25 kg at the other. If 2.5 kg is added to the 20 kg weight, the 25-kg weight will have to be placed 0.3 m farther from the fulcrum to balance the lever. Find the lengths of the two arms of the lever at first.

Suggestion: *Let x = the length of the long arm and let y = the length of the short arm.* *Then*

$$20x = 25y \quad \text{and} \quad 22.5(x - 0.3) = 25(y + 0.3)$$

17-19 A lever is balanced on a fulcrum with a weight w_1 on one arm and a weight w_2 on the other. If p lb is added to w_1, w_2 has to be moved over f ft to balance the lever. Find the lengths of the two arms of the lever.

17-20 A beam that is supported at its ends has a weight of 50 lb placed on it so that it causes an increase in pressure on the support at one end of 20 lb. It is also found that the same pressure is produced at this end by a weight of 60 lb placed 3 ft farther from this end. How long is the beam?

Suggestion: *Consider the support where the increase is 20 lb as the fulcrum. Then there is a pressure of 30 lb at the other end.*

Let y = length in feet of the beam and x = the distance the 50-lb weight is from the fulcrum. Then

$$30y = 50x$$

Similarly, when the 60-lb weight is applied,

$$40y = 60(x + 3)$$

17-21 A glass full of water weighs 18 ounces (oz). When the same glass is full of sulfuric acid of specific gravity 1.75, it weighs 27 oz. Find the weight of the glass when empty.

17-22 How many liters of cream containing 35 percent fat and how many liters of milk containing 4 percent fat should be mixed to give 76 liters of cream containing 25 percent fat?

17-23 Find the number of ounces each of silver 70 percent pure and silver 87 percent pure which will make up 12 oz 82 percent pure.

17-24 The specific gravity (sp gr) of one liquid is 1.75 and that of another is 1.4. How many ounces of each of these two liquids will make up 10 oz of a liquid of sp gr 1.7?

17-25 The sum of three numbers is 12. When the third number is subtracted from the sum of the first two, the result is 2, and when the second number is subtracted from twice the first number and the third number is added, the result is 7. What are the numbers?

EXPONENTS AND LOGARITHMS

18-1 INTRODUCTION TO EXPONENTS AND LOGARITHMS

In previous chapters you studied and worked problems involving positive exponents. In this chapter you will extend your knowledge of exponents by learning about negative, fractional, and zero exponents.

John Napier, from Scotland, is credited with inventing logarithms in 1614. A logarithm is actually an exponent, and the procedures and rules of working with exponents are used when working with logarithms.

By the use of logarithms, the procedures of multiplication, division, raising to a power, and extracting a root of arithmetical numbers are simplified. The process of multiplication is replaced by addition, that of division by subtraction, that of raising to a power to multiplication, and that of extracting a root to division.

Many calculations that are difficult or impossible by regular methods of arithmetic can be readily performed with logarithms. An understanding of logarithms will help you understand the solution of complex problems.

The hand-held calculator has drastically changed the need to learn the rigorous procedures that are necessary for efficient use of logarithms. The hand-held calculator can perform many of the operations that were formerly done by longhand. The calculator should be used whenever possible because it is more efficient and there is less chance for errors. However, understanding the basic operating procedures of logarithms will help you use your calculator to its optimum. Thus you should learn the procedures; however, use the calculator to make your learning easier.

18-2 LAW OF EXPONENTS FOR MULTIPLICATION

The law of exponents for multiplication has been shown and proven for positive integral exponents. As a review:

$$a^2 \times a^3 = (a \times a) \times (a \times a \times a) = a^5$$

or, in a general form,

$$a^n \cdot a^m = a^{n+m}$$

Note: *When multiplying quantities, the bases must be the same in order to add exponents.*

18-3 LAW OF EXPONENTS FOR DIVISION

When quantities with exponents are divided, the exponents are subtracted. (Again, the base numbers *must* be the same.)

$$a^3 \div a^2 = \frac{a \cdot \not{a} \cdot \not{a}}{\not{a} \cdot \not{a}} = a$$

or, in a general form,

$$a^m \div a^n = a^{m-n}$$

When $m = n$, the resulting subtraction gives a zero exponent. The value of any number, other than zero, with a zero exponent is 1. Consider these examples:

$$b^5 \div b^5 = \frac{\overset{1}{\not{b}} \times \overset{1}{\not{b}} \times \overset{1}{\not{b}} \times \overset{1}{\not{b}} \times \overset{1}{\not{b}}}{\underset{1}{\not{b}} \times \underset{1}{\not{b}} \times \underset{1}{\not{b}} \times \underset{1}{\not{b}} \times \underset{1}{\not{b}}} = \frac{1}{1} = 1$$

or

$$b^m \div b^m = b^0 = 1$$

When n is greater than m, the subtraction results in a negative exponent. The meaning of a negative exponent is illustrated by these examples:

$$a^2 \div a^5 = a^{-3} = \frac{\overset{1}{\not{a}} \times \overset{1}{\not{a}}}{\underset{1}{\not{a}} \times \underset{1}{\not{a}} \times a \times a \times a} = \frac{1}{a^3}$$

or

$$a^{-m} = \frac{1}{a^m} \quad \text{and} \quad \frac{1}{b^{-n}} = b^n$$

18-4 LAW OF EXPONENTS FOR THE POWER OF A POWER

To calculate the value of a power of a power, multiply the exponents:

$$(a^3)^2 = (a \times a \times a)(a \times a \times a) = a^6$$

or

$$(a^m)^n = a^{mn}$$

You should recognize this principle as an extension of the rule for multiplying quantities with exponents.

18-5 FRACTIONAL EXPONENTS

A fractional exponent indicates a root. The denominator is the index of the root and the numerator is the exponent of the quantity whose root is indicated (that is, the quantity within the radical sign):

$$a^{m/n} = \sqrt[n]{a^m}$$

EXAMPLE 18-1

$$25^{1/2} = \sqrt{25} = 5$$

EXAMPLE 18-2

$$64^{3/2} = \sqrt{64^3} = \sqrt{262,144} = 512$$

EXAMPLE 18-3

$$64^{3/2} = (\sqrt{64})^3 = (8)^3 = 512$$

Note from Examples 18-2 and 18-3 that the same answer is obtained by either approach. Obviously, extracting the root and then raising to the power is easier, since smaller numbers are manipulated.

Note: *As an index of a root, 2 is not written; the root is understood to be 2 when only the radical sign is used.*

18-6 THE ROOT OF A POWER

Extracting the root of a power is the inverse of finding the power of a power—the exponent of the base number is divided by the root. For example,

$$\sqrt[n]{a^m} = a^{m \div n}$$
$$\sqrt{a^6} = a^{6 \div 2} = a^3$$
$$\sqrt{3^4} = 3^{4/2} = 3^2 = 9 \text{ or}$$
$$\sqrt{3^4} = \sqrt{81} = 9$$

18-7 APPLICATIONS OF EXPONENTS

In the examples illustrating these principles, real numbers are used. Literal algebraic expressions involving powers can also be multiplied, divided, etc., according to these principles.

EXAMPLE 18-4 $a^3b \div a^2 = ab$

Note that the exponent 2 is subtracted from the exponent 3 and that the exponent of b, which is 1, is not affected by this subtraction.

$$a^3b \div a^2 = \frac{a^3b^1}{a^2} = a^{3-2}b^1 = a^1b^1 = ab$$

EXAMPLE 18-5 $(x - y)^3 \div (x - y)^2 = (x - y)$

Note that here the parentheses tie together the expression $x - y$ so that it is manipulated as if it were one quantity. In principle the divisions in examples 18-4 and 18-5 are the same:

$$(x - y)^3 \div (x - y)^2 = (x - y)$$
$$a^3 \div a^2 = a$$

Remember that $(x - y)^3$ means $(x - y)(x - y)(x - y)$. [A common error is interpreting $(x - y)^3$ to mean $x^3 - y^3$.]

EXAMPLE 18-6 Calculate $(3a^2b^3c^2)^3$.

Solution Since

$$(3a^2b^3c^2)^3 = (3a^2b^3c^2)(3a^2b^3c^2)(3a^2b^3c^2)$$

or

$$(3a^2b^3c^2)^3 = 3^{1 \times 3}a^{2 \times 3}b^{3 \times 3}c^{2 \times 3}$$

and

$$(3a^2b^3c^2)^3 = 27a^6b^9c^6$$

18-8 THE POWER OF A PRODUCT

The power of a product is the same as the product of the powers of the factors. That is,

$$(abc)^2 = a^2b^2c^2 \quad \text{since} \quad (abc)^2 = (abc)(abc) = a^2b^2c^2$$
$$(2 \times 4 \times 3)^3 = 2^3 \times 4^3 \times 3^3 = 8 \times 64 \times 27 = 13{,}824$$

Also

$$(2 \times 4 \times 3)^3 = (24)^3 = 13{,}824$$

18-9 POWER OF A SUM OR DIFFERENCE

$$(a + b)^2 = (a + b)(a + b) = a^2 + 2ab + b^2$$
$$(a - b)^2 = (a - b)(a - b) = a^2 - 2ab + b^2$$

and

$$(5 + 4)^2 = (5 + 4)(5 + 4) = 25 + 40 + 16 = 81$$
$$(5 + 4)^2 = (9)^2 = 81$$
$$(8 - 4)^2 = (8 - 4)(8 - 4) = 64 - 32 - 32 + 16$$

and

$$(8 - 4)^2 = (4)^2 = 16$$

18-10 QUANTITIES AFFECTED BY RADICALS

It is impossible to calculate an exact square root of many numbers. Numbers such as 4, 9, 16, and 25 have exact square roots; however, numbers like 2, 3, 5, and 7 do not. It is important to learn to work with numbers whose exact square roots cannot be extracted. For example, 32:

$$\sqrt{32} = \sqrt{16 \times 2} = \sqrt{16} \times \sqrt{2} = 4\sqrt{2}$$

A similar example for a literal expression is:

$$\sqrt{a^4 b^3 c} = \sqrt{a^4 b^2} \times \sqrt{bc} = a^2 b \sqrt{bc}$$

The radical expressions $\sqrt{32}$ and $\sqrt{a^4 b^3 c}$ were *simplified;* the answers were reduced so that the remaining quantity under the radical sign did not have a factor containing an exact root.

18-11 FRACTIONS UNDER THE RADICAL SIGN

When a fraction appears under a radical sign the expression can be changed so that the denominator is a number not affected by the radical sign. The denominator is then said to be *rationalized.*

$$\sqrt{\frac{1}{2}} = \sqrt{\frac{1}{2} \times \frac{2}{2}} = \sqrt{\frac{2}{4}} = \frac{\sqrt{2}}{\sqrt{4}} = \frac{\sqrt{2}}{2}$$

Also

$$\sqrt{\frac{7}{12}} = \sqrt{\frac{7}{12} \times \frac{3}{3}} = \sqrt{\frac{21}{36}} = \frac{\sqrt{21}}{\sqrt{36}} = \frac{\sqrt{21}}{6}$$

In both illustrations the denominators were multiplied by a number which yielded a product with an exact square root. Notice that in each case the numerator and denominator of the fraction were multiplied by the same number, so that the value of the fraction remained unchanged.

18-12 SCIENTIFIC NOTATION

In many computations it is convenient to express very large and very small numbers in scientific notation, which means a number between 1 and 10 times the power of 10. For example, the number 4,250,000,000 is very difficult to work with or compare with another number. Numbers like this may be expressed as one number multiplied by a second number—the second number being 10 with a suitable exponent. The decimal point in scientific notation is placed to the right of the first significant figure, and the remaining number is written as some power of 10. Many quality calculators use scientific notation with large numbers.

Rule: *To change ordinary notation to scientific notation, count the number of places the decimal point is to the right or left of the first significant digit. The number of places is the power of 10; it is positive if the number is greater than 1 and negative if the number is less than 1.*

EXAMPLE 18-7 $4,250,000,000 = 4.25 \times 1,000,000,000 = 4.25 \times 10^9$

EXAMPLE 18-8 $7.26 \times 10^{-6} = 7.26 \times 1/10^6 = 7.26 \times 1/1,000,000$
$$= 0.000\ 007\ 26$$

EXAMPLE 18-9 Multiply $3100 \times 11,000$

Solution Using scientific notation,

$$3100 = 3.1 \times 10^3 \qquad 11,000 = 1.1 \times 10^4$$
$$3100 \times 11,000 = (3.1 \times 10^3)(1.1 \times 10^4)$$
$$3100 \times 11,000 = (3.1 \times 1.1)(10^{3+4})$$
$$3100 \times 11,000 = 34,100,000$$

EXAMPLE 18-10 Simplify $\sqrt[3]{\dfrac{81a^6b^9}{343c^{12}}}$

Solution

$$\sqrt[3]{\frac{81a^6b^9}{343c^{12}}} = \sqrt[3]{\frac{3^3a^6b^9}{7^3c^{12}}} = \frac{3^{3/3}a^{6/3}b^{9/3}}{7^{3/3}c^{12/3}} = \frac{3a^2b^3}{7c^4}$$

■ **EXERCISES**

Perform the indicated operations.

18-1 $(2x^2y^3)^4$

18-2 $(-2ab^2c)^3$

18-3 $(5x^4y^5)^6$

18-4 $(-2a^3b^4c)^2$

18-5 $(4hk^4m)^4$

18-6 $\left(\dfrac{a^4b^3}{c^4}\right)^3$

18-7 $\left(\dfrac{4x}{5y}\right)^6$

18-8 $\left(\dfrac{2^2x^2y}{3abc^3}\right)^4$

18-9 $\sqrt{2^2x^4y^6}$

18-10 $\sqrt[3]{27x^6y^9z^{12}}$

18-11 $\sqrt[4]{\dfrac{2^2x^4y^8}{3^4h^4k^8}}$

18-12 $\sqrt{\dfrac{25c^{10}}{d^8}}$

18-13 $\sqrt[3]{8x^3y^6z^9}$

18-14 $\sqrt[4]{\dfrac{16y^{12}}{x^8}}$

18-15 $\sqrt[3]{\dfrac{64m^6}{n^9}}$

18-16 $\sqrt[5]{3,200,000}$

18-17 $\sqrt{\dfrac{128x^6y^{10}}{2x^2y^4}}$

18-18 $\sqrt[3]{27^2}$

18-13 INTRODUCTION TO LOGARITHMS

The availability of hand-held calculators has greatly reduced the importance of logarithms in calculation. Logarithms are used more today with exponential functions than with calculations. However, you will understand logarithms better by working through arithmetic problems using logarithms.

Learn to read a logarithm table even though most good hand-held calculators will indicate logarithms more efficiently and accurately. Refer to your calculator manual concerning how to obtain logarithms for your particular calculator. In the rest of the chapter we will refer to Table 9 to obtain logarithms although you may wish to use a calculator.

A logarithm is an exponent. Specifically, the *logarithm* of a given number (say, 100) to a given *base* (say, 10) is the exponent necessary to express the number as a power of the base.

$$10^2 = 100 \quad \text{therefore} \quad \log_{10} 100 = 2$$

In general, the logarithm of x to the base b (written as $\log_b x$) is the exponent necessary to express x as a power of b. That is, $\log_b x$ is the number y for which $b^y = x$. Thus,

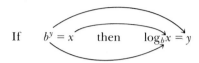

$$\text{If} \quad b^y = x \quad \text{then} \quad \log_b x = y$$

One of the most commonly used bases for logarithms is the base 10. In fact, logarithms to the base 10 are so commonly used that they are called *common logarithms*. When writing the symbol for a common logarithm, the subscript 10 (indicating the base) is generally omitted.

$$\log_{10} x = \log x$$

Throughout most of this chapter we will be dealing with common logarithms.

Up to this point, we have dealt with exponents that were whole numbers or common fractions. However, there will always be an exponent that will raise a given base to any value. You can see this in Fig. 18-1,

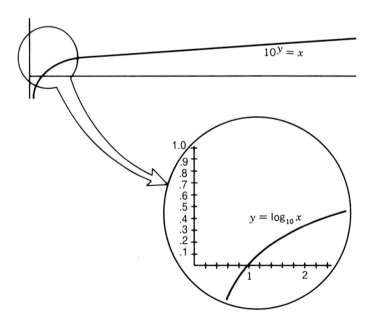

Fig. 18-1. (*a*) the curve $10^y = x$, or $y = \log_{10} x$. (*b*) Enlargement of $10^y = x (y = \log_{10} x)$ from $x = 0$ to $x = 2.4$.

which shows the curve for $y = \log x$. From Fig. 18-1, it appears that $x = 1.2$ when $y = 0.08$.

Table 9 in the appendix is a table for the same equation. From it, we can find that $\log 1.2 = 0.0792$.

There are some important facts to be noted about the logarithms of products, fractions, and powers.

Let M and N be two numbers, and let $m = \log M$ and $n = \log N$. Then $10^m = M$ and $10^n = N$. Also,

$$MN = 10^m 10^n = 10^{m+n}$$

Therefore,

$$\log (MN) = m + n = \log M + \log N$$

Rule: *The logarithm of a product equals the sum of the logarithms of the factors.*

Similarly, since $M \div N = 10^m \div 10^n = 10^{m+n}$,

$$\log (M \div N) = m - n = \log M - \log N$$

Rule: *The logarithm of a fraction equals the logarithm of the numerator minus the logarithm of the denominator.*

Now suppose we wish to find the logarithm of some power of M, say, M^k. We know that

$$M^k = (10^m)^k = 10^{mk}$$

Hence,

$$\log M^k = mk = k \log M$$

Rule: *The logarithm of a power of a number equals the exponent times the logarithm of the number.*

This rule applies whether the exponent is an integer, a fraction, or a decimal; thus the logarithm of the cube root of M equals $\frac{1}{3} \log M$.

The above properties of logarithms give us an easy method of calculating products, quotients, powers, and roots—since it is easier to add than to multiply, to subtract than to divide, and to multiply or divide than to raise to a power or extract a root.

18-14 DETERMINING THE MANTISSA

The decimal portion (.079 18) of the logarithm 2.079 18 is the *mantissa*. The whole-number portion (2 in this case) is the *characteristic*. To determine the mantissa for any number, turn to Table 9 of the appendix. The same basic procedures are followed for any logarithm table. Note that the lefthand column lists a series of numbers; 1.0, 1.1, 1.2 . . . 9.8, 9.9. Also note that the top line of the table lists a series of decimals: 0.00, 0.01, . . . 0.08, 0.09.

To learn how to find a mantissa, read through the following examples carefully, and work each step out for yourself.

EXAMPLE 18-11 Find the mantissa for 3.6.

Solution

1. In Table 9 find 3.6 in the lefthand column.
2. The mantissa for 3.6 is right next to it, in the column headed 0.00.
3. The mantissa of 3.6 is .556 30

EXAMPLE 18-12 Find the mantissa for 3.61.

Solution

1. In Table 9 find 3.6 in the lefthand column.
2. Read across the table until you reach the 0.01 column.
3. The mantissa for 3.61 appears in the 3.6 line and the 0.01 column.
4. The mantissa for 3.61 is .557 51.

EXAMPLE 18-13 Find the mantissa for 3.62.

Work this problem yourself, the answer is .558 71.

18-15 INTERPOLATION OF A MANTISSA

Table 9 can be used to determine the log of a number with four or more significant digits. A special procedure is used to determine the mantissa for a number which cannot be located in the table. The process used to determine the mantissa of numbers of more than three digits is *interpolation*. This process is best explained by an example.

EXAMPLE 18-14 Determine the mantissa of 7.586.

Solution Since 7.586 lies between 7.58 and 7.59 its mantissa must lie between the mantissas of 7.58 and 7.59.

$$
\begin{aligned}
\text{Mantissa of } 7.590 &= .880\ 24 \\
\text{Mantissa of } 7.580 &= \underline{.879\ 67} \\
\text{Difference} &= .000\ 57
\end{aligned}
$$

The tabular difference between these two mantissas is .000 57, and the difference between the two numbers themselves (7.59 and 7.58) is 0.01. The number whose mantissa is to be found is 7.586, and this is 7.58 plus 0.006, which is $\frac{6}{10}$ of the distance between 7.58 and 7.59. Study this process as illustrated in Fig. 18-2.

The mantissa of 7.586 is:

$$.879\ 67 + 0.6 \times .000\ 57 = .879\ 67 + .000\ 34 = .880\ 01$$

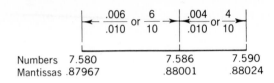

Numbers	7.580	7.586	7.590
Mantissas	.87967	.88001	.88024

Fig. 18-2. Interpolation.

or

$$.880\ 24 - 0.4 \times .000\ 57 = .880\ 24 - .000\ 23 = .880\ 01$$

You may perform interpolation either "up" or "down." In fact, it is a good idea to interpolate both ways to check accuracy. A calculator automatically interpolates for most desired quantities.

18-16 DETERMINING THE CHARACTERISTIC

In Table 9 only numbers between 0.00 and 9.99 are listed. However, we know that:

$$\log 10 = 1\ (10^1 = 10)$$
$$\log 100 = 2\ (10^2 = 100)$$
$$\log 1000 = 3\ (10^3 = 1000)$$

Given a table of logarithms and that $\log_{10} 10^x = x$, we can find the logarithm for any positive number.

EXAMPLE 18-15　Find the log of 36.1.

Solution

$$36.1 = 3.61 \times 10$$

Recall $a^m \times a^n = a^{m+n}$, $\log 36.1 = \log 3.61 + \log 10$.

From the table	$\log 3.61 = 0.557\ 51$
And	$\log 10\ \ \ = \underline{1.000\ 00}$
Therefore	$\log 36.1 = 1.557\ 51$

EXAMPLE 18-16　Find the log of 3,610,000.

Solution

$$3,610,000 = 3.61 \times 10^6$$
$$\log 3.61\ \ \ \ \ \ \ = 0.557\ 51$$
$$\log 10^6\ \ \ \ \ \ \ \ = \underline{6.000\ 00}$$
$$\log 3,610,000 = 6.557\ 51$$

EXAMPLE 18-17　Find the log of 0.000 361.

Solution

$$0.000\ 361 = 3.61 \times 10^{-4}$$
$$\log 3.61 \quad = \quad 0.557\ 51$$
$$\log 10^{-4} \quad = -4.000\ 00$$
$$\log 0.000\ 361 = \quad 6.557\ 51 - 10$$

Note: *The mantissa is positive, while the characteristic is negative. We could subtract 4.000 00 from .557 51 and get −3.442 49.*

When you are doing calculations with logarithms, it is easiest to add 10 to the characteristic; this was done by putting −10 after the logarithm. Observe:

$$\overline{4}.557\ 51$$
$$\underline{10. \qquad - 10}$$
$$6.557\ 51 - 10$$

Another way of writing this is $\overline{4}.557\ 51$, the bar over the 4 indicates this is a negative exponent combined with a positive mantissa. Note that the 4.000 00 has not been subtracted from .557 51; the two have been combined.

18-17 FINDING THE ANTILOG OF A NUMBER

To determine the final answer in any problem involving multiplication, division, raising to powers, or extracting roots with logarithms, it is necessary to reverse the process explained in the last section, that is, to determine a number from the log of the number. The number determined from a logarithm is the *antilogarithm* (antilog).

If the logarithm of 5.196 91 = .715 73; then 5.196 91 is the antilog corresponding to the logarithm .715 73.

Procedure to find the number when its logarithm is given:

1. Determine the characteristic and mantissa of the logarithm.
2. Find the mantissa in the body of the log table.
3. Read the number at the left of the horizontal row the mantissa is found in and the number at the top of the vertical row. Interpolate if necessary.
4. From the characteristic, determine the placing of the decimal point in the answer.

It is rarely possible to find the exact mantissa required in the body of a table; interpolation is therefore almost always necessary. However, a good calculator will automatically determine the antilog of a number.

Several examples will serve to illustrate.

EXAMPLE 18-18 The logarithm of a number is 9.715 73 − 10. What is the number? $\log x = 9.715\ 73 - 10 \qquad x = ?$

Solution The characteristic is −1 (9 − 10) and the mantissa is .715 73.

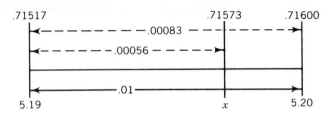

Fig. 18-3. Example 18-19: Interpolation.

Locate .715 73 in the body of the table. The number .715 17 is in the horizontal row that begins with 5.1 at the left and in the vertical column headed by 0.09. The first three numbers in the antilogarithm are, then, 519.

Since the characteristic is −1, the number to be found is between 0 and 1; therefore the decimal point is placed to the left of the 5, and the answer starts as 0.519.

For a more accurate answer, we can interpolate as we did in Sec. 18-15.

$$\text{antilog } .715\ 73 = 5.19 + \left(\frac{.000\ 56}{.000\ 83}\right)\left(\frac{0.01}{1}\right)$$
$$= 5.19 + .006\ 75 = 5.196\ 75$$

Study the following to learn the properties of logarithms:

$\log 4580 = 3.660\ 87$; that is, $4580 = 10^{3.660\ 87}$

$\log 458.0 = 2.660\ 87$; that is, $458.0 = 10^{2.660\ 87}$

$\log 45.80 = 1.660\ 87$; that is, $45.80 = 10^{1.660\ 87}$

$\log 4.580 = 0.660\ 87$; that is, $4.580 = 10^{.660\ 87}$

$\log 0.4580 = \overline{1}.660\ 87$; that is, $0.4580 = 10^{\overline{1}.660\ 87} = 10^{-1+.660\ 87}$

$\log 0.0458 = \overline{2}.660\ 87$; that is, $0.0458 = 10^{\overline{2}.660\ 87} = 10^{-2+.660\ 87}$

$\log 0.004\ 58 = \overline{3}.660\ 87$; that is, $0.004\ 58 = 10^{\overline{3}.660\ 87} = 10^{-3+.660\ 87}$

■ EXERCISES

Determine the logarithms in Exercises 18-19 to 18-38 by using Table 9 in the appendix.

18-19	$\log 2.71 = ?$	**18-29**	$\log 271 = ?$
18-20	$\log 505 = ?$	**18-30**	$\log 12.214 = ?$
18-21	$\log 432.3 = ?$	**18-31**	$\log 1005 = ?$
18-22	$\log 0.101 = ?$	**18-32**	$\log 0.7854 = ?$
18-23	$\log 5924 = ?$	**18-33**	$\log 52.80 = ?$
18-24	$\log 0.01 = ?$	**18-34**	$\log 3.1416 = ?$
18-25	$\log 0.001 = ?$	**18-35**	$\log 1941 = ?$
18-26	$\log 0.0089 = ?$	**18-36**	$\log 0.031 = ?$
18-27	$\log 9278 = ?$	**18-37**	$\log 0.002\ 73 = ?$
18-28	$\log 0.000\ 05 = ?$	**18-38**	$\log 0.0003 = ?$

18-18 MULTIPLICATION WITH LOGARITHMS

Recall that to multiply two numbers with the same base, we add their exponents: $a^m \times a^n = a^{m+n}$. To multiply two or more numbers by logarithms, follow these steps:

1. Determine the logarithm of each number.

2. Add the logarithms (exponents).

3. Determine the answer by finding the antilog, the number corresponding to the sum of the logarithms.

EXAMPLE 18-19 Find the product of 4.53×0.036.

Solution

$$
\begin{array}{ll}
\log 4.53 & = 0.656\ 10 \\
+\log 0.036 & = \overline{2}.556\ 30 \\
\hline
\log \text{ of the product} & = \overline{1}.212\ 40 \\
4.53 \times 0.036 \quad x = & 0.1631
\end{array}
$$

Observe that the addition is performed in the normal way; the 1 carried after adding 6 and 5 is a plus 1, so that when it is added to $\overline{2}$, the sum becomes $\overline{1}$.

18-19 DIVISION WITH LOGARITHMS

Since a logarithm is an exponent, when dividing using logarithms, exponents are subtracted: $a^m \div a^n = a^{m-n}$.

To divide with logarithms,

1. Determine the logarithm of each number.

2. Subtract the logarithms.

3. Determine the answer by finding the antilog, the number corresponding to the logarithm found in step 2.

EXAMPLE 18-20 Divide 38.76 by 7.923.

Solution

$$
\begin{array}{ll}
\log 38.76 & = 1.588\ 38 \\
-\log 7.923 & = 0.898\ 89 \\
\hline
\log \text{ of the quotient} & = 0.689\ 49 \\
38.76 \div 7.923 & = 4.8921
\end{array}
$$

18-20 RAISING TO A POWER WITH LOGARITHMS

Since a logarithm is an exponent, the principle of raising to the power of a power applies: To calculate the power of a power, multiply the exponents: $(a^m)^n = a^{mn}$.

This statement may be revised to apply to calculating a power of a power using logarithms: Raise the logarithm of the base to the power indicated by multiplying the exponents. The result of this multiplication is the logarithm of the required power. The number corresponding to this logarithm (the antilog) is the required power or answer.

EXAMPLE 18-21   Find the value of $(1.756)^7$.

Solution

$$
\begin{array}{ll}
\log 1.756 & = 0.24452 \\
& \underline{\times 7} \\
7 \log 1.756 & = 1.71164 \\
\text{antilog } 1.71164 & = 51.48 \\
(1.756)^7 & = 51.48
\end{array}
$$

18-21 ROOTS WITH LOGARITHMS

To calculate the root of a number, divide the logarithm of the number by the index of the root. The result of this division is the logarithm of the root. The number corresponding to this logarithm (the antilog) is the root. This statement is an application of a principle of exponents: $\sqrt[m]{a^n} = a^{a/m}$.

EXAMPLE 18-22   Find the 5th root of 0.0379.

Solution

$$
\begin{aligned}
\log \sqrt[5]{0.0379} &= \tfrac{1}{5} \log 0.0379 \\
&= \tfrac{1}{5}(\log 3.79 + \log 10^{-2}) \\
&= \tfrac{1}{5}(.578\ 64 - 2) \\
&= \tfrac{1}{5}(8.578\ 64 - 10) \\
&= 1.715\ 73 - 2 \\
&= 0.715\ 73 - 1 \\
\text{antilog } (0.71573 - 1) &= 0.5167 \\
\sqrt[5]{0.0379} &= 0.5167
\end{aligned}
$$

Note:

$$
\begin{array}{ll}
0.578\ 64 - 2 \\
8. \qquad\ \ - 8 \\
\hline
8.578\ 64 - 10
\end{array}
$$

EXAMPLE 18-23   Calculate $\sqrt[6]{0.008\ 673}$.

Solution

$$
\log 0.008\ 673 = \overline{3}.938\ 17
$$

Since the log of 0.008 673 is made up of two parts—the characteristic, which is negative; and the mantissa, which is positive—the division by 6 should not be attempted. The characteristic of the logarithm is changed to a positive number so that the entire logarithm can be divided by 6 and so that the negative portion of the logarithm is exactly divisible by 6.

$$
\log 0.008\ 673 = \overline{3}.938\ 17
$$

Add 60 and subtract 60 from the logarithm as follows:

$$\log 0.008\ 673 = \overline{3}.938\ 17$$
$$\underline{\qquad\qquad\qquad 60. \qquad -60}$$
$$\log 0.008\ 673 = 57.938\ 17 - 60$$

Divide by 6 to find the 6th root.

$$\tfrac{1}{6}\log 0.008\ 673 = \frac{57.938\ 17 - 60}{6}$$
$$\tfrac{1}{6}\log 0.008\ 673 = 9.65636 - 10$$
$$= \overline{1}.65636$$
$$\text{antilog } \overline{1}.656\ 36 = 0.4533$$
$$\sqrt[6]{0.008\ 673} = 0.4533$$

Note: *60 was added and subtracted so that the number would be divisible by 6.*

▓ EXERCISES

18-39 Multiply the following by using logarithms and check by actual multiplication:

(a) (10)(100)(1000)   (g) 4.56 × 11
(b) (80)(60)   (h) 0.841 × 66.6
(c) (25)(25)   (i) 0.0075 × .12
(d) 0.1 × 200   (j) 32.32 × 1002
(e) 0.02 × 50   (k) 845.9 × 9.287
(f) 0.005 × 6   (l) 1034 × 0.024 06

18-40 Divide the following by using logarithms and check by actual division:

(a) 10 ÷ 100   (g) 5280 ÷ 12
(b) 100 ÷ 10   (h) 3.1416 ÷ 40
(c) 200 ÷ 25   (i) 0.15 ÷ 0.30
(d) 225 ÷ 15   (j) 0.1008 ÷ 0.8001
(e) 0.05 ÷ 25   (k) 1.414 ÷ 7000
(f) 0.0064 ÷ 400   (l) 70 ÷ 1.4146

Find these powers and roots:

18-41 $(3.1416)^2$ **18-46** $\sqrt{5.288}$ **18-51** $(4.4)^{3/4}$

18-42 $(0.707)^3$ **18-47** $\sqrt{0.224}$ **18-52** $(0.059)^{2/3}$

18-43 $(0.886)^2$ **18-48** $\sqrt[3]{0.0025}$ **18-53** $(958)^{0.2}$

18-44 $(1.732)^5$ **18-49** $\sqrt{1.014}$ **18-54** $(803.5)^{0.11}$

18-45 $(\tfrac{22}{7})^2$ **18-50** $\sqrt{0.000\ 27}$ **18-55** $(296)^{6.1}$

18-22 NATURAL LOGARITHMS

The same procedures are used for calculations for all logarithms, regardless of the base of the logarithm. In making computations, logarithms to the base 10 are generally used; however, there are many formulas in electricity, physics, chemistry, and other sciences in which another base is used.

This base is a number that cannot be expressed exactly—to seven decimal places, it is 2.718 281 8. It is usually represented by the letter e (just as the ratio of the circumference of a circle to its diameter is represented by the Greek letter π). Logarithms to this base are called *natural logarithms, hyperbolic logarithms,* or *Napierian logarithms.* Tables of natural logarithms are published, but, for purposes of conversion from one base to another, it is necessary to remember only that the natural logarithm of a number is approximately 2.3026 times the common logarithm of the same number (or, that the common logarithm of a number is 0.4343 times the natural logarithm).

The relationships between common logarithms and natural logarithms are stated in the following formulas, where N is any number:

$$\log_e N = 2.3026 \log_{10} N$$
$$\log_{10} N = 0.4343 \log_e N$$

Thus,

$$\log_e 100 = 2.3026 \log_{10} 100$$
$$= 2.3026 \times 2 = 4.6052$$

The following examples use formulas with logarithms to the base e.

EXAMPLE 18-24   The given formula is for finding insulation resistance R by the leakage method.

$$R = 10^6 \times \frac{t}{c} \times \frac{1}{\log_e \left(\dfrac{V_0}{V}\right)}$$

where t = time, sec (t = time in seconds)
$\quad c$ = capacity
$\quad V$ = voltage

Compute the value of R if $t = 120$, $V_0 = 123$, $V = 115.8$, and $c = 0.082$.

Solution   The formula is revised so that a table of common logarithms may be used.

$$R = 10^6 \times \frac{t}{c} \times \frac{1}{2.3026 \log_{10} \left(\dfrac{V_0}{V}\right)}$$

Substituting values,

$$R = 10^6 \times \frac{120}{0.082} \times \frac{1}{2.3026(\log 123 - \log 115.8)}$$

Therefore,

$$R = 10^6 \times \frac{120}{0.082} \times \frac{1}{2.3026(2.08991 - 2.06371)}$$

$$= 10^6 \times \frac{120}{0.082} \times \frac{1}{2.3026(0.02620)}$$

$$= 2.426 \times 10^{10}$$

▮ EXERCISES

18-56 If an open tank kept full of water has a rectangular notch cut on one side, as shown in Fig. 18-4, the number of cubic feet of water that will flow through this notch per second is given by the formula

$$Q = \frac{2}{3} \, cwh^{3/2}\sqrt{2g}$$

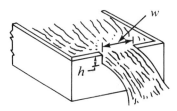

Fig. 18-4. Exercise 18-56.

where Q = amount of flow, ft^3
  c = a constant found by experiment
  w = width of the notch, ft
  h = depth of the notch, ft
  g = 32.2

Find Q when $h = 1\frac{1}{3}$, $w = 2$, and $c = 0.586$.

18-57 Solve $E = LVB \, 10^{-8}$ for E, if

E = electromotive force in volts in moving conductor
L = length of conductor in centimeters
V = velocity in centimeters per second
B = the number of lines of force per square centimeter

Given $V = \dfrac{9 \times 100}{60}$, $B = 8000$, and $L = 0.6 \times 100$, find E.

18-58 The work W done by a volume of gas, expanding at a constant temperature from volume V_0 to a volume V_1, is given by the formula

$$W = p_0 V_0 \log_e \left(\frac{V_1}{V_0}\right)$$

Find the value of W if $p_0 = 87.5$, $V_0 = 246$, and $V_1 = 472$.

18-59 If the mixture in a gas engine expands without gain or loss of heat, it is found that the law of expansion is given by the equation $pv^{1.37} = C$. Given that $p = 188.2$ when $v = 11$, find the value of C.

18-60 The amount of a principal at compound interest for a certain time is given by the following formula:

$$A = P(1 + r)^n$$

where A = amount
  P = principal
  r = rate, %
  n = time, yrs

Find the amount of \$236 at compound interest for 14 years at 8%.

18-61 The amount of money to which a given principal will increase at compound interest for a certain time is determined by the formula:

$$A = P \left(1 + \frac{j}{m}\right)^{mn}$$

where A = amount at the end of a given time
P = principal at the start of a given time
j = nominal rate, %
n = number of yrs
m = number of times compounded per year

Find the amount A if P = $236 is invested at compound interest for 14 years at the nominal rate of 15%:
(a) compounded annually
(b) compounded four times per year

18-62 In biology, when the science of heredity is studied, an important formula is:

$$y = \frac{e^{-x^2}}{\sqrt{\pi}}$$

Solve this formula for x.

18-63 For $63,500 cash, an FHA-approved house can be bought. The buyer agrees to pay $800.00 per month. If the rate of interest is 1% per month, how many years will it take to pay for the house?
Use the formula

$$n = \frac{\log R - \log (R - Ar)}{\log (1 + r)}$$

where A = cash value of the house
r = rate per month
n = number of months
R = amount of each monthly payment

18-64 In an alternating-current (ac) circuit, the current i at any instant is given by the formula

$$i = I(1 - e^{-Rt/L})$$

where I = maximum current
R = resistance
L = coefficient of self-induction
t = time, sec
e = base of the natural system of logarithms

Find i when $I = 120$, $R = 80.5$, $t = 15.8$ and $L = 1140.7$.

18-65 The work W done by a volume of gas, expanding at a constant temperature from a volume V_0 to a volume V_1, is given by the following formula:

$$W = P_0 V_0 \log_e \left(\frac{V_1}{V_0}\right)$$

Find W if $P_0 = 58.94$, $V_0 = 145.6$, and $V_1 = 178.32$.

QUADRATIC EQUATIONS

19-1 INTRODUCTION TO QUADRATIC EQUATIONS

The equations studied in the previous chapters involved unknowns to the first power. Some equations with higher power unknowns were discussed, but they could be solved by factoring. In this chapter, you will study general methods used for solving equations of the second degree.

19-2 DEFINITIONS OF QUADRATIC EQUATIONS

An equation in the form

$$ax^2 + bx + c = 0 \qquad (a \neq 0)$$

is a *quadratic equation*. This equation includes the square and the first power of the unknown x. The coefficients a, b, and c are constants.

As defined earlier, a *pure quadratic equation* contains only the square of the unknown number. An example of this type of equation would be $2x^2 = 4$. An *affected quadratic equation* contains both the square and the first power of the unknown. An example is $x^2 + 3x = 10$.

The equation $2x^2 = 4$ can be solved by methods studied earlier. It is rewritten in an equivalent form below to illustrate another solution.

$$2x^2 - 4 = 0$$
$$ax^2 + bx + c = 0$$

Notice that these two equations have the same form: the coefficient a corresponds to 2; b corresponds to 0; and c corresponds to -4.

The equation $x^2 + 3x = 10$ may be rewritten in a similar way:

$$x^2 + 3x - 10 = 0$$
$$ax^2 + bx + c = 10$$

The coefficient a corresponds to 1; b to 3; and c to -10. Note that the sign of a coefficient is carried with the coefficient.

255

19-3 SOLUTIONS FOR A PURE QUADRATIC EQUATION

The solution of a pure quadratic equation is illustrated in the following examples. They can be checked by substituting the solution into the original equation.

EXAMPLE 19-1 Solve $2x^2 = 4$.

Solution

$$2x^2 = 4$$

Divide both sides by 2 $\qquad x^2 = 2$

Take the square root of each side $\qquad x = \pm\sqrt{2}$

EXAMPLE 19-2 Solve $3x^2 + 8 = 7x^2 - 8$.

Solution

$$3x^2 + 8 = 7x^2 - 8$$

Subtract 8 from both members of the equation. Then

$$3x^2 = 7x^2 - 16$$

Subtract $7x^2$ from both members:

$$-4x^2 = -16$$

Divide both members by -4:

$$x^2 = 4$$

Take the square root of both members:

$$x = \pm 2$$

EXAMPLE 19-3 Find the radius of a circle whose area is 5026.56 in².

Solution The formula for the area of a circle is:

$$A = \pi r^2$$

where A = area of a circle
r = radius
π = 3.1416

Then

$$5026.56 = \pi r^2$$

Divide both sides by 3.1416 $\qquad \dfrac{5026.56}{3.1416} = r^2$

Take the square root of both sides $\qquad \pm\sqrt{\dfrac{5026.56}{3.1416}} = r$

$$\pm 40 = r$$

The radius of this circle is, then, $+40$. (The answer -40 is impossible and is therefore rejected.)

EXAMPLE 19-4 Solve the equation: $\dfrac{x}{x-1} + \dfrac{x-3}{x^2-1} = \dfrac{5}{4}$.

Solution Clear the fractions by multiplying by $4(x-1)(x+1)$, the lowest common denominator.

$$4\overset{1}{\cancel{(x-1)}}(x+1)\left(\frac{x}{\cancel{x-1}}\right) + 4\overset{1}{\cancel{(x-1)}}\overset{1}{\cancel{(x+1)}}\left(\frac{x-3}{\cancel{(x+1)}\cancel{(x-1)}}\right) = \overset{1}{\cancel{4}}(x-1)(x+1)\left(\frac{5}{\cancel{4}}\right)$$

Remove parentheses and combine like quantities:

$$4x(x+1) + 4(x-3) = 5(x+1)(x-1)$$
$$4x^2 + 4x + 4x - 12 = 5(x^2 - 1)$$
$$4x^2 + 8x - 12 = 5x^2 - 5$$

Solve for x:

$$4x^2 - 5x^2 + 8x - 12 + 5 = 0$$
$$x^2 - 8x + 7 = 0$$

Factor and solve:

$$x - 7 = 0 \qquad x - 1 = 0$$
$$x = 7 \qquad\quad x = 1$$

When the roots are checked in the original equation, $x = 1$ results in division by zero, which is undefined; therefore we would accept only the $x = 7$ root.

■ EXERCISES

In Exercises 19-1 to 19-10, solve for x and check the values in the original equations.

19-1 $x^2 = 99 - 10x^2$

19-2 $80 - 2x^2 = 3x^2$

19-3 $-5 + 0.75x^2 - 22 = 0$

19-4 $\dfrac{6}{x+8} - \dfrac{6}{x-8} = 1 - \dfrac{33}{x^2-64}$

19-5 $8 = \dfrac{3}{1+x} + \dfrac{3}{1-x}$

19-6 $(x+3)(x+3) + (x-3)^2 = 4(x^2-9)$

19-7 $\dfrac{15}{8} - \dfrac{x-3}{x-2} = \dfrac{x+3}{x+2}$

19-8 $(x-4)^2 + 4 = -8x + 29$

19-9 $(a-x)^2 - (x-a)(3x+a) = 0$

19-10 $x^2 + a^2x^2 = (a^2-1)^2 - 2ax^2$

19-11 Given $\dfrac{u-2}{u+2} + \dfrac{u+2}{u-2} = \dfrac{40}{u^2-4}$, find the values of u.

19-12 Given $E = 0.5Mv^2$, find the values of v.

19-13 Given $S = 0.5gt^2$, find the values of t if $g = 32.16$ and $S = 100.5$.

19-14 Given $4m^2 = -c^2 + 2(a^2 + b^2)$, solve for m.

19-15 Given $F = \dfrac{mna}{d^2}$, solve for d.

19-16 Given $n^2 = \dfrac{KS^2t}{l^2d}$, solve for S, l, and n.

19-17 Find the radius of a cylinder of altitude 12 cm and volume 1400 cm^3. Use the formula $V = h\pi r^2$, substitute values, and solve for r.

19-18 Find the radius of a right circular cone of altitude 20 in. and volume 145 in.^{3} Use the formula $V = \frac{1}{3}\pi r^2 h$.

19-19 The product of two numbers is 90, and the greater number divided by the smaller is $3\frac{3}{5}$. Find the numbers.

19-20 Solve the equation $P = \dfrac{D^2N}{2.5}$ for D.

19-4 SOLUTION OF AN AFFECTED QUADRATIC EQUATION BY FACTORING

The solution of an affected quadratic equation by factoring has been explained in Chap. 14. This example reviews the procedure.

EXAMPLE 19-5 $x^2 + 23x = -102$.

Solution Add 102 to both members:

$$x^2 + 23x + 102 = 0$$

Factor the left member:

$$(x + 6)(x + 17) = 0$$

Solve by setting each factor equal to 0:

$$x + 6 = 0$$
$$x = -6$$
$$x + 17 = 0$$
$$x = -17$$

Check by substituting both values in the original equations:

$$(-6)^2 + (23)(-6) \overset{?}{=} -102$$
$$36 - 138 \overset{?}{=} -102$$
$$-102 = -102$$

and

$$(-17)^2 + 23(-17) \overset{?}{=} -102$$
$$289 - 391 \overset{?}{=} -102$$
$$-102 = -102$$

▮ EXERCISES

Solve the equations by factoring and then check.

19-21 $x^2 - 3x + 2 = 0$

19-22 $x^2 + 7x + 12 = 0$

19-23 $x^2 = 11x - 18$

19-24 $x = -x^2 + 6$

19-25 $-20 = -x - x^2$

19-26 $x^2 = 19x - 90$

19-27 $18x = 77 + x^2$

19-28 $x^2 + 35x + 250 = 0$

19-29 $6x^2 + 7x + 2 = 0$

19-30 $2x^2 + 7x + 6 = 0$

19-31 $12x^2 + 5 = 19x$

19-32 $x^2 + x = 30$

19-33 $10x^2 = 13x + 30$

19-34 $x^2 - (a + b)x + ab = 0$ (Solve for x.)

19-5 COMPLETING THE SQUARE OF A QUADRATIC EQUATION

It is sometimes impossible to factor an affected quadratic equation; in problems of this type, other methods may be used. One of these methods is *completing the square*.

From Chapter 12, you will recall: $(a + b)^2 = a^2 + 2ab + b^2$. This is a *trinomial square*. The first and last terms of a trinomial square are perfect squares of monomials and the middle term is twice the product of the square roots of the first and last terms. Keeping this in mind, we can find the last term if we know the first two.

Thus, if we know that $a^2 + 2ab$ are the first two terms, we can find the third term by finding the square of the quotient obtained by dividing the second term by two times the square root of the first. Two times the square root of the first term is $2a$. The second term divided by $2a$ gives b, which squared is b^2, the third term. We say then that we have completed the square, which is $a^2 + 2ab + b^2$.

EXAMPLE 19-6 Complete the square of $x^2 + 4x$.

Solution Is it possible to add something to $x^2 + 4x$ that will make a trinomial square like $(a^2 + 2ab + b^2)$? Comparing these two equations, it is obvious that x^2 may be substituted for a^2. This makes the form of the trinomial square become $x^2 + 2bx + b^2$, compared with $x^2 + 4x + ?$.

Look at the coefficient of x in both equations: $2b$ and 4. What must b equal if $2b$ is to equal 4? As b must equal 2, b^2 (the last term in the quadratic square) must equal 4.

Therefore, we can complete the square of $x^2 + 4x$ by adding 4:

$$x^2 + 4x + 4 = (x + 2)(x + 2) = (x + 2)^2$$

EXAMPLE 19-7 Complete the square of $x^2 \ldots + 9$.

Solution The a^2 term is x^2, and its square root is x; the b^2 term is 9, and its square root is 3. Therefore, the $2ab$ term must be $2 \cdot x \cdot 3$, or $6x$.

We can complete the square of $x^2 + 9$ by adding $6x$:

$$x^2 + 6x + 9 = (x + 3)(x + 3) = (x + 3)^2$$

19-6 EQUALIZING A QUADRATIC EQUATION AFTER COMPLETING THE SQUARE

In the two previous examples, two numbers (4 in the first and $6x$ in the second) were added to the original expressions. The original expressions have, then, changed in value (by 4 in one case and by $6x$ in the other). These same numbers *must be subtracted* from the answers so that the values of the answers are equal to the original values. This expression is illustrated in the following example.

EXAMPLE 19-6 Complete the square of $x^2 + 4x$.

Solution

$$x^2 + 4x + 4 - 4$$

Since 4 has been added to the original statement, 4 must also be subtracted so that the value of the original statement is unchanged. That is,

$$x^2 + 4x + 4 \neq x^2 + 4x$$

But

$$x^2 + 4x + 4 - 4 = x^2 + 4x$$

EXAMPLE 19-7 Complete the square of $x^2 \ldots + 9$.

Solution

$$x^2 + 6x + 9 \neq x^2 + 9$$

But

$$x^2 + 6x + 9 - 6x = x^2 + 9$$

The subtraction that completes Examples 19-5 and 19-6 is an important step in solving equations by completing the square.

19-7 SOLUTION OF A QUADRATIC EQUATION BY COMPLETING THE SQUARE

EXAMPLE 19-8 Solve $x^2 + 4x = 12$ for x.

Solution   Divide the numerical coefficient of the b term (4) by 2; square and add to both sides of the equation.

$$x^2 + 4x - 4 = 12 + 4$$

Factor the left member:

$$(x + 2)^2 = 16$$

Extract the square root of both sides of the equation:

$$x + 2 = \pm 4$$

Subtract 2 from both sides to obtain values for x:

$$x = -2 \pm 4$$
$$x = -2 + 4 \quad \text{and} \quad x = -2 - 4$$
$$x = 2 \quad \text{and} \quad x = -6$$

Check by substituting the values obtained into both equations:

$$x^2 + 4x = 12 \qquad\qquad x^2 + 4x = 12$$
$$(2)^2 + 4(2) \stackrel{?}{=} 12 \qquad (-6)^2 + 4(-6) \stackrel{?}{=} 12$$
$$4 + 8 \stackrel{?}{=} 12 \qquad\qquad 36 - 24 \stackrel{?}{=} 12$$
$$12 = 12 \qquad\qquad 12 = 12$$

EXAMPLE 19-9   Solve $x^2 + 12x = -35$.

Solution   Complete the square by dividing the b term (12) by 2; square 6 and add to both sides of the equation.

$$x^2 + 12x + 36 = -35 + 36$$

Factor the left member:

$$(x + 6)^2 = 1$$

Extract the square root of both sides of the equation:

$$x + 6 = +1$$
$$x = -6 + 1 \quad \text{and} \quad x = -6 - 1$$
$$x = -5 \quad \text{and} \quad x = -7$$

Check by substituting the values obtained into both equations:

$$x^2 + 12x = -35 \qquad\qquad x^2 + 12x = -35$$
$$(-7)^2 + 12(-7) \stackrel{?}{=} -35 \qquad (-5)^2 + 12(-5) \stackrel{?}{=} -35$$
$$49 - 84 \stackrel{?}{=} -35 \qquad\qquad 25 - 60 \stackrel{?}{=} -35$$
$$-35 = -35 \qquad\qquad -35 = -35$$

EXAMPLE 19-10   Solve $9x = 4 - 3x^2$.

Solution   Add $3x^2$ to both members:

$$3x^2 + 9x = 4$$

Divide all terms by 3; to complete the square, the coefficient of the x^2 term must be 1.

$$x^2 + 3x = \frac{4}{3}$$

Complete the square by dividing the b term (3) by 2; square and add to both sides of the equation: $(\frac{3}{2})^2 = \frac{9}{4}$.

$$x^2 + 3x + \frac{9}{4} = \frac{4}{3} + \frac{9}{4}$$

$$x^2 + 3x + \frac{9}{4} = \frac{43}{12}$$

Extract the square root of both sides of equation and evaluate:

$$x + \frac{3}{2} = \pm\sqrt{\frac{43}{12}}$$

$$x = -\frac{3}{2} \pm 1.893$$

$$x = -1.5 + 1.893 \qquad \text{and} \qquad x = -1.5 - 1.89$$

$$x = 0.393 \qquad \text{and} \qquad x = -3.393$$

Procedure for completing the square:
The steps in solving a quadratic equation by completing the square are as follows:

1. Transpose all terms containing x to the left member of the equation; transpose all other terms to the right member.
2. Multiply or divide so that the coefficient of x^2 is 1.
3. Complete the square by dividing the b term by 2; square the quantity and add to both sides of the equation.
4. Extract the square root of both sides of the equation.
5. Solve for the unknown values.
6. Prove by substituting the solutions into the original equations.

■ EXERCISES

Supply the missing term in the following quadratic trinomials to form perfect squares:

19-35 $x^2 - \cdots + 4$ **19-39** $36 - \cdots + 9x^2$

19-36 $x^2 - 6x + \cdots$

19-37 $x^2 + 8x + \cdots$ **19-40** $x^2 + \frac{x}{2} + \cdots$

19-38 $9 - \cdots + x^2$ **19-41** $\frac{1}{4} - \cdots + x^2$

                                          19-42 $4ax + a^2 + \cdots$

Solve for x by completing the square:

19-43 $y^2 + 4y = 12$ **19-46** $2x^2 + 3x = 14$

19-44 $x^2 + 6x - 7 = 0$ **19-47** $3x^2 + 4x = 39$

19-45 $x^2 - 30x = 64$ **19-48** $2x^2 + 7x = 39$

19-49	$5x^2 = 50 - 15x$	19-52	$0.5x^2 = x + 0.5$
19-50	$12 + 21x = 6x^2$	19-53	$5x^2 + 2x - 20 = 0$
19-51	$5 = 5x^2 + 10x$	19-54	$11x^2 + 121x - 1 = 0$

19-8 DEVELOPMENT OF THE QUADRATIC FORMULA

The equation $ax^2 + bx + c = 0$ where a, b, and c are any numbers, positive or negative, represents the general form of a quadratic equation. If this general equation is solved, then that solution can be used to determine the unknown value in any quadratic equation. The solution follows.

Solve $ax^2 + bx + c = 0$ for x.

$$ax^2 + bx + c = 0$$

Subtract c from both members:

$$ax^2 + bx = -c$$

Divide each term by a:

$$x^2 + \frac{b}{a}x = -\frac{c}{a}$$

Complete the square by dividing the b term $\left(\dfrac{b}{a}\right)$ by 2; square and add to both sides:

$$x^2 + \frac{b}{a}x + \frac{b^2}{4a^2} = -\frac{c}{a} + \frac{b^2}{4a^2}$$

Factor the left member and place the right member over the lowest common denominator:

$$\left(x + \frac{b}{2a}\right)^2 = \frac{b^2 - 4ac}{4a^2}$$

Extract the square root of both sides of the equation:

$$x + \frac{b}{2a} = \pm\sqrt{\frac{b^2 - 4ac}{4a^2}}$$

$$x + \frac{b}{2a} = \pm\frac{\sqrt{b^2 - 4ac}}{2a}$$

Subtract $b/2a$ from both sides of the equation and simplify:

$$x = -\frac{b}{2a} \pm \frac{\sqrt{b^2 - 4ac}}{2a}$$

Thus the *quadratic formula* is:

$$x = \frac{-b \pm \sqrt{b^2 - 4ac}}{2a}$$

19-9 SOLUTIONS USING THE QUADRATIC FORMULA

Any quadratic equation can be solved by the *quadratic formula*.

$$x = \frac{-b \pm \sqrt{b^2 - 4ac}}{2a}$$

To solve a quadratic equation by this formula, convert the equation to the form $ax^2 + bx + c = 0$, substitute the values a, b, and c in the formula, and evaluate x.

EXAMPLE 19-11   Solve $3x^2 = 2x + 4$.

Solution   Write $3x^2 = 2x + 4$ in the form of $ax^2 + bx + c = 0$:

$$3x^2 - 2x - 4 = 0$$

Then, $a = 3$, $b = -2$, $c = -4$. Substitute these values into the quadratic formula and evaluate:

$$x = \frac{-b \pm \sqrt{b^2 - 4ac}}{2a}$$

$$= \frac{-(-2) \pm \sqrt{(-2)^2 - 4(3)(-4)}}{2(3)}$$

$$= \frac{2 \pm \sqrt{4 + 48}}{6} = \frac{2 \pm \sqrt{52}}{6} = \frac{2 \pm \sqrt{(4)(13)}}{6}$$

$$= \frac{2 \pm 2\sqrt{13}}{6} = \frac{\overset{1}{\cancel{2}}(1 \pm \sqrt{13})}{\underset{3}{\cancel{6}}} = \frac{1 \pm \sqrt{13}}{3}$$

Note: *If a numerical value of x is desired, evaluate $\sqrt{13}$ and complete the arithmetic operations.*

EXAMPLE 19-12   Solve $6x^2 + 7 = -17x$ for x.

Solution

$$6x^2 + 7 = -17x$$

Add $17x$ to both members:

$$6x^2 + 17x + 7 = 0$$

Then $a = 6$, $b = 17$, and $c = 7$; substitute these values into the quadratic formula and evaluate.

$$x = \frac{-17 \pm \sqrt{(17)^2 - 4(6)(7)}}{(2)(6)}$$

$$= \frac{-17 \pm \sqrt{289 - 168}}{12}$$

$$= \frac{-17 \pm \sqrt{121}}{12} = \frac{-17 \pm 11}{12} = -\frac{1}{2} \text{ and } -2\frac{1}{3}$$

EXAMPLE 19-13 Find the value of x in $x^2 + x + 1 = 0$.

Solution

$$a = 1 \qquad b = 1 \qquad c = 1$$

$$x = \frac{-b \pm \sqrt{b^2 - 4ac}}{2a}$$

$$= \frac{-1 \pm \sqrt{1 - 4(1)(1)}}{2(1)}$$

$$= \frac{-1 \pm \sqrt{-3}}{2}$$

The value of the term $\sqrt{b^2 - 4ac}$ in Example 19-13 is negative, and it is impossible to extract the square of a negative number by any of the methods explained in this book. The square root of a number is defined as one of the two *equal* factors into which a number is divided. The square root of 4 is either $+2$ or -2, since $(+2)(+2) = 4$ and $(-2)(-2) = 4$. But can you divide -3 into two equal factors; that is, extract the square root of -3? Since the 3 is negative, one factor must be positive and one must be negative; thus two *equal* factors cannot be found. The square root of a negative number, then, is imaginary. *Consider final solutions only when $b^2 - 4ac$ is positive.* If $b^2 - 4ac$ is negative, leave your answer in the form illustrated in Example 19-13.

EXAMPLE 19-14 Determine the size of a sphere where there are as many square feet in the surface of a certain sphere as there are cubic feet in its volume. Find its radius.

Solution Let r = length of radius, ft. Then, $4\pi r^2$ = area of surface, ft^2, and $\frac{4}{3}\pi r^3$ = volume, ft^3. Therefore,

$$\frac{4}{3}\pi r^3 = 4\pi r^2$$

$$4\pi r^3 - 12\pi r^2 = 0$$

Factoring,

$$4\pi r^2(r - 3) = 0$$

Set each factor equal to 0 and solve:

$$4\pi r^2 = 0 \qquad or \qquad r - 3 = 0$$

The solutions are, then,

$$r = 0 \qquad and \qquad r = 3$$

EXAMPLE 19-15 A mountain climber wants to ascend a distance of 2 mi. This he does at a certain rate. He finds that he could have traveled twice as fast plus 3 mph had he driven his automobile and that he could have traveled the 2 miles in 1 h less time. Find his first rate.

Solution   Let x be the first rate, in miles per hour. Then, since distance in miles divided by this rate gives time in hours,

$$\frac{2}{x} - \frac{2}{2x + 3} = 1$$

Multiply each term by the lowest common denominator and simplify:

$$x(2x + 3)\frac{2}{x} - x(2x + 3)\frac{2}{2x + 3} = x(2x + 3)(1)$$

$$(2x + 3)(2) - x(2) = x(2x + 3)(1)$$
$$4x + 6 - 2x = 2x^2 + 3x$$
$$2x^2 + x - 6 = 0$$

Factor and solve for x:

$$(2x - 3)(x + 2) = 0$$

$$2x - 3 = 0 \qquad\qquad x + 2 = 0$$

$$x = \frac{3}{2} \qquad \text{or} \qquad -2$$

There are two solutions to the equation, but only one $(x = \frac{3}{2})$ makes sense in the context of the problem. Thus his rate would be 1.5 miles per hour.

▮ EXERCISES

19-55   Solve the following equations by the most efficient method:

(a) $x^2 - 5x + 4 = 0$
(b) $2x^2 + 5x + 2 = 0$
(c) $3x^2 - 10x = -3$
(d) $2x^2 - 5x + 3 = 0$
(e) $30x^2 = 17x - 2$
(f) $6x^2 - 13x + 6 = 0$
(g) $u^2 + 2u - 2 = 0$
(h) $h^2 - 3h - 1 = 0$
(i) $2t^2 + 3t - 4 = 0$
(j) $3s^2 = 4s + 5$
(k) $7y^2 = 23y - 6$
(l) $9p^2 - 2p = 1$
(m) $0.25k^2 - 0.2k = 1$
(n) $1.2z^2 + 2.1z - 0.3 = 0$

(o) $\dfrac{U^2}{2} + \dfrac{U}{3} = \dfrac{1}{6}$

(p) $N\left(\dfrac{1}{a} + \dfrac{1}{b}\right) + N^2 + \dfrac{1}{ab} = 0$

19-56   Given $h = c + vt - 16t^2$, solve for a, v, and t.

19-57 Given $T = \dfrac{22R^2}{7} + \dfrac{22RH}{21}$, solve for R.

19-58 If you take twice the square of a certain number and add 6, the sum is seven times the number. What is the number?

19-59 If two consecutive numbers are squared and then added, the sum is 25. What are the numbers?

19-60 Is there a number such that if you add it to its reciprocal the sum is unity? The sum is 2? The sum is any number greater than 2?

19-61 I bought x articles for $1521 and paid x dollars per article for them. How many articles did I buy?

19-62 I worked 4 days, and you worked 3 days; we completed a job that you can do in 2 days less than I can. How long does it take me to do the job?

19-63 I worked 4 days, and you worked 10 days; we completed a job that I can do in 3 days less than you. How long does it take me to do the job?

19-64 A walk containing 784 ft^2 is to be built around a garden 50 ft by 40 ft. How wide must the walk be?

Suggestion: *Let x = width of the walk in feet. Then the number of square feet in both garden and walk is $(50 + 2x)(40 + 2x)$. Therefore, $(50 + 2x)(40 + 2x) = 50 \times 40 + 784$.*

19-65 What is the width of a uniform border around a flower garden 6 ft by 8 ft if its area is 72 ft^2?

19-66 The surface of a cube has as many square centimeters as there are cubic centimeters in its volume. Find the length of an edge.

19-67 There are 728 in.^{3} of wood in a covered cubical box. The boards are 1 in. thick. Find the outside dimensions of the box.

19-68 If a car traveled 4 mph faster, it would take 1 h less to go 360 miles. Find the speed of the car.

19-69 A discount store sold a hat for $11 and gained as many percent as the hat cost. Find the cost of the hat.

19-70 A sporting goods store bought x fishing reels for $120. If the store had bought two fewer for the same money, they would have cost $2 more apiece. Find x.

19-71 I bought a number of radios for $960. If they had cost $16 apiece less, I would have had two more radios for the same money. Find the cost of each radio.

19-72 A rectangular piece of ground is 40 ft by 60 ft. A strip of uniform width is marked off around it which contains one-half the area of the rectangle. Find the width of the strip.

19-73 A window screen has a total area of 10 ft^2, and its length is 4 in. greater than its width. The area inside the frame is 8 ft^2. The frame is of a constant width; find this width.

19-74 An airplane started $2\frac{1}{2}$ h late but finished its flight of 1920 km on time by going 64 km/h faster than usual. What was the usual rate?

19-75 A gasoline tank was filled by two pipes in 3 h and 40 min. The larger pipe alone could fill the tank in $2\frac{1}{2}$ h less time than the smaller pipe alone could. Find the time in which each pipe could fill the tank.

19-76   The circumference of the rear wheel of a road grader is 5 ft more than that of the front wheel. If the hind wheel makes 150 fewer revolutions than the front wheel in going 1 mile, find the circumference of each wheel.

VARIATION

20-1   INTRODUCTION TO VARIATION

All events in this universe are related to other events, and if these relationships can be determined, it is possible to predict what will happen. When relationships can be stated in the language of mathematics, it is easier to predict what will happen. In this chapter you will study some simple relationships and learn how to state them in mathematical language.

20-2   CONSTANTS AND VARIABLES

A number whose value does not change is a *constant*. Certain constants have the same value in mathematics and the sciences. For example, the value of π, the ratio of the circumference to the diameter of a circle, is such a constant. Correct to four places, $\pi = 3.1416$. Other constants remain the same for a given set of circumstances. The value of such a constant does not change within a certain context. For example, in Hooke's law, $f = ks$, the k is a constant whose value may change depending on certain conditions. The f is the force, and s is the strain. These two numbers are *variables*.

A *variable* is a number that may take an unlimited number of values. The f in Hooke's law may be 5 kg, 26 g, 127 oz, 1,000 lb, or any other value. The velocity of a falling body changes from second to second and is therefore also a variable.

In algebra, variables are usually represented by the later letters in the alphabet: x, y, and z. Constants are generally represented by a, b, and c. In physics, geometry, electricity, and mechanics, a variable is often represented by the initial letter of the name of the quantity, as b for base, w for work, v for voltage, and so on.

20-3 DIRECT VARIATION

If two numbers are related in such a way that their ratio is constant—that is, if either increases, the other increases, or if either decreases, the other decreases—the two numbers are said to vary *directly* as each other. If x and y are two variables that vary directly and k is a constant, then $x = ky$.

Ohm's law is an example of direct variation. This law may be written $E = kI$, where E is the voltage, I is the current flowing in a circuit, and k is a constant, the resistance of the circuit. In the same circuit, with k constant, if I increases then E increases.

If either E or I decreases, then the other variable will also decrease. The variation is then *direct*. This electrical law is usually written $E = IR$, where R, the resistance of the circuit, is a constant. This law may be solved for R and written $R = E/I$. The constant R is called the *constant of variation* or the *factor of proportionality*. Each of the following expressions may be used to describe this variation in relation to R: E varies as I; E is directly proportional to I; E is proportional to I. The notation $E \propto I$ may also be used for the statement that E is proportional to I.

EXAMPLE 20-1 A train is leaving a city at a uniform rate (speed). Express the relationship between its distance d from the station and the time t since it left the city.

Solution The distance from the city will vary directly with the time elapsed since the train left the city:

$$d = kt$$

The constant in the expression $d = kt$ is k, the speed of the train. In solving this and similar problems you must be certain that the units substituted in the formula (here, in $d = dt$) are consistent.

If the train is traveling 60 mph (miles per *hour*), how far is it from the city in $2\frac{1}{2}$ h (*hours*)?

$$d = kt \qquad d = 60 \times 2\frac{1}{2} = 150 \text{ mi}$$

The units substituted here are consistent. Note that the speed of the train is in mph, and the time is $2\frac{1}{2}$ h; the answer is, then, expressed in miles.

If the train is traveling 60 mph (miles per *hour*), how far is it from the city in 90 min (*minutes*)?

$$d = kt \qquad d = 60 \times 1\frac{1}{2} = 90 \text{ mi}$$

Note that the 90 min has been changed to $1\frac{1}{2}$ h. Of course, the miles per hour could have been changed to miles per minute instead:

$$d = kt \qquad d = 1 \times 90 = 90 \text{ mi}$$

Notice that the two answers are the same.

EXAMPLE 20-2 Two numbers f and s vary directly when $f = 10$, $s = 4$. Calculate f when $s = 25$.

Solution Since the variation is direct, $f = ks$. Substitute $f = 10$ and $s = 4$ to determine k:

$$f = ks$$
$$10 = k4$$
$$k = 2.5$$

The formula $f = 2.5s$ represents the variation. When $s = 25$.

$$f = 2.5s$$
$$f = 2.5(25) = 62.5$$

EXAMPLE 20-3 The distance traveled by a body falling freely in space varies as the square of the time. If $s = $ the distance in meters, and $t = $ time in seconds, write the relationship. Find the value of k if the distance traveled in $5\,s$ is 122.625 m.

Solution Since the distance varies directly as the square of the time, $s = kt^2$. Substituting the given values for s and t.

$$s = kt^2$$
$$122.625 = k(5)^2$$

$$\frac{122625}{25} = k$$

$$4.905 = k$$

Since the constant k is determined to be 4.905, the formula can be written

$$s = kt^2$$
$$s = 4.905t^2$$

This formula from physics is $s = 4.905t^2$; it is generally written $s = \frac{1}{2}gt^2$. Here g is the acceleration caused by gravity acting on a freely falling object, and its value is 9.81 m/s^2.

20-4 INVERSE VARIATION

Inverse variation may be described as "opposite" variation. As one number in an inverse relationship *increases*, the second number *decreases*.

The resistance to an electric current is less in a thick wire than in a thin wire of the same material; the resistance varies *inversely* as the size of the wire in cross section. "The smaller the wire, the greater the resistance" indicates the same relationship.

The intensity of the light from a lamp decreases as the distance from the lamp increases. Here too the variation is inverse. In fact, the intensity varies inversely as the square of the distance from the lamp.

One number varies inversely with another when their product is a constant. That is, if either increases, the other decreases by a proportional amount; or if either decreases, the other increases.

20-5 THE MATHEMATICAL STATEMENT OF AN INVERSE RELATIONSHIP

If x and y are two variables that vary inversely as each other, and k is a constant, then

$$xy = k \qquad x = \frac{k}{y} \qquad y = \frac{k}{x}$$

The equation $x = k/y$ is read "x varies inversely as y." Similarly, $y = k/x$ is read "y varies inversely as x."

For example, the intensity of illumination from a source of light varies inversely as the square of the distance from the source. If I stands for illumination, d for the distance, and k for the constant, then

$$I = \frac{k}{d^2} \qquad \text{or} \qquad Id^2 = k$$

EXAMPLE 20-4 If x varies inversely as y, state the law and find the value of the constant if $x = 10$ when $y = 0.5$.

Solution $x = k/y$ is the mathematical statement of the variation. Substituting, $10 = k/0.5$. Therefore, $k = 5$. If this value of k is used, the law (or equation) becomes

$$xy = 5$$

EXAMPLE 20-5 If $x = 0.01$, find y from the law of Example 20-4.

Solution Substituting in the law, $0.01y = 5$. Therefore,

$$y = 500$$

EXAMPLE 20-6 The number of vibrations made by a pendulum varies inversely as the square root of its length. A pendulum 39.1 in. long makes 1 vibration per second. How long must a pendulum be to make 4 vibrations per second? To make 1 vibration in 10 s?

Solution The law is $n = k/\sqrt{l}$, where n = number of vibrations per second and l = length of pendulum in inches.

To find k, substitute $n = 1$ and $l = 39.1$.

$$1 = \frac{k}{\sqrt{39.1}} \qquad \text{and} \qquad k = \sqrt{39.1}$$

To find the length of a pendulum which vibrates four times per second, substitute $n = 4$.

$$4 = \frac{\sqrt{39.1}}{\sqrt{l}} = (4)^2 = \left(\frac{\sqrt{39.1}}{\sqrt{l}}\right)^2 \quad \text{and} \quad l = \frac{39.1}{16} = 2.44 + \text{in.}$$

To find the length of a pendulum that vibrates once in 10 s, substitute $n = \frac{1}{10}$. Then,

$$\frac{1}{10} = \frac{\sqrt{39.1}}{\sqrt{l}} \quad \text{and} \quad \left(\frac{1}{10}\right)^2 = \left(\frac{\sqrt{39.1}}{\sqrt{l}}\right)^2$$

and

$$l = 100(39.1) = 3910 \text{ in.} = 325 \text{ ft } 10 \text{ in.}$$

20-6 JOINT VARIATION

One number *varies jointly* as two or more other numbers when it varies directly as the product of the others. Thus, x varies jointly as u and v when $x = kuv$.

A number may vary directly as one number and inversely as another. It then varies as the quotient of the first divided by the second. Thus, if x varies directly as y and inversely as z, the relationship is written $x = k(y/z)$.

■ EXERCISES

20-1 State in the language of variation the relation between the area of a square and its side.

20-2 State the relation between the distance that a man walks at a uniform rate and the time that he is walking.

20-3 State the relation between the volume of a rectangular solid and its three dimensions.

20-4 Does the amount of gasoline used in driving an automobile 100 mi vary directly as the rate per hour?

20-5 Does the weight that a person can lift vary directly as the weight of the person?

20-6 Express in words each of the following relations in the language of variations:

$$S = 4\pi r^2 \quad S = \pi r^2 h \quad V = 4\pi r^3 \quad W \propto \frac{bd^2}{l}$$

$$z \propto \frac{\sqrt{x}}{y^2} \quad v \propto \frac{1}{p} \quad e = \frac{kwl^3}{bd^3} \quad A = \frac{kq}{\sqrt{h}}$$

20-7 Express the following in the algebraic language of variation.
(a) The volume V of a circular cone varies as the altitude h and as the square of the radius r.

(b) The weight of a body above the surface of the earth varies inversely as the square of the distance d of the body from the center of the earth.

(c) The pressure of the wind on an exposed surface varies as the area of the surface and the square of the velocity of the wind.

(d) Newton's law of gravitation states that the force F with which each of two masses m_1 and m_2 attracts the other varies directly as the product of the masses and inversely as the square of the distance d between the masses.

20-8 In $R = xy/(u + v)$, how does R vary with x? Does R become larger or smaller when v becomes smaller?

20-9 In $x/a = b/y$, a and b are constants. How does y vary with x?

20-10 If x varies jointly as y and z, and when $y = 6$ and $z = 3$, $x = 120$, find the constant. Find y when $x = 200$ and $z = 15$.

20-11 The area A of a triangle varies jointly as the base b and the altitude a. Write the law if, when $a = 6$ in. and $b = 4$ in., $A = 12$ in.^{2} What will be the area when the base is 25 in. and the altitude is 40 in.?

20-12 Similar solids vary in volumes as the cubes of their like dimensions. A water pail that is 10 in. across the top holds 12 quarts. Find the volume of a similar pail that is 16 in. across the top.

20-13 The volume of a sphere varies as the cube of its radius, and the volume of a sphere of radius 2 ft is 33.5104 ft^3. Find the volume of a sphere of radius 8 ft.

20-14 If a commercial fishing boat travels at a uniform rate, the amount of fuel consumed varies as the distance and as the square of the rate. A certain boat traveling 150 mi at a rate of 18 mph uses 80 gallons of fuel. Find the amount of coal used by the same ship in going 200 mi at 24 mph.

20-15 A ball starts from rest and rolls down an incline of uniform slope. The distance s that the ball moves varies as the square of the time t. If the ball rolls 10 ft from rest in 6 s, find the distance that it will roll in 12 s. Find the time that it will take to roll 40 ft from rest.

20-16 If the illumination of an object varies inversely as the square of the distance d from the source of light, and the illumination of an object 8 ft from the source of light is 3, find the illumination of an object 40 ft from the source of light. Find the illumination of an object 4 ft from the source of light.

20-17 The number of units of heat H generated by an electric current of I amperes in a circuit varies as the square of the current I, as the resistance R, and as the time t in seconds during which the current passes. Write the law.

20-18 Using the formula in Exercise 20-17, it is found that $H = 388,800$ if $I = 10$, $R = 9$, and $t = 30$ min. Find H when $I = 30$, $R = 50$, and $t = 40$ min.

20-19 A circular sheet of steel 2 ft in diameter increases in diameter by $\frac{1}{200}$ when the temperature is increased by a degree. Find amount of increase in area of the sheet with an increase of 5 degrees.

20-20 A wire cable 1 in. in diameter will lift 10,000 lb. What will a cable $\frac{5}{8}$ in. in diameter lift?

20-21 A styrofoam cone 8 in. high weighs 5 oz. What will be the weight of a cone of the same shape and material but only 6 in. high?

20-22 Two persons of the same "build" are similar in shape. One is $5\frac{1}{2}$ ft tall and weighs 150 lb. Find the weight of the other, who is 6 ft tall.

20-23 The electrical resistance of a substance varies directly as the length L and inversely as the area A of the cross section. If the resistance of a bar of annealed aluminum 1 in. long and 1 in.^{2} in cross section is 0.000 001 144 ohm at 32°F, find the resistance of a wire of the same material 1 ft long and 0.002 in. in diameter.

20-24 Find the resistance of 1 mi of wire $\frac{1}{32}$ in. in diameter, made of the material in Exercise 20-23, at 32°F.

20-25 If the resistance of a coil of wire of the material in Exercise 20-23 at 32°F is 27.3 ohms, and the wire is 0.05 in. in diameter, find the length.

20-26 If the resistance of a wire 9363 ft long is 21.6 ohms, what would be the resistance if its length were reduced to 5732 ft and its cross section made half again as large?

20-27 The quantity Q of electricity that will flow into a radio condenser varies jointly as the capacity C and the voltage E. The constant of variation is 10^{-6} if Q is in coulombs (C), C in microfarads (μF), and E in volts. If $C = 15$ microfarads and $E = 820$ volts, find Q.

20-28 The amount of energy, in joules (J), stored by a condenser varies jointly as the capacity and the voltage squared. The constant of variation is 0.5×10^{-6} if C is in microfarads and E in volts. What is the energy stored in a 2-microfarad condenser with 1000 volts applied?

20-29 What is the capacity of a condenser that is charged to 0.0036 coulomb when 110 volts is applied?

20-30 What energy goes into a condenser in a transmitting station if this condenser has a capacity of 0.0001 microfarad and is charged with a 20,000-volt source?

20-7 A PRACTICAL EXAMPLE OF VARIATION: TRANSVERSE STRENGTH OF BEAMS

Other properties being equal, the strength of a rectangular cross section beam, supported at each end, varies (1) inversely as the length in feet, (2) directly as the width in inches, and (3) directly as the square of the depth in inches.

That the strength varies inversely as the length in feet means that if a beam, supported horizontally as in Fig. 20-1, has its length increased and everything else unchanged, the weight it will support is decreased in the same ratio as the length was increased. Thus, if the length is doubled, it will support one-half as great a weight. If W is the weight and L the length, in the language of variation this fact is stated.

$$W = \frac{k_1}{L}$$

The length L is the distance between the points of support.

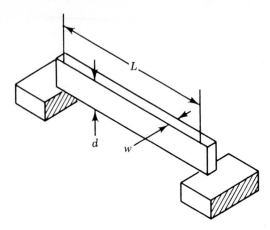

Fig. 20-1. Determining the strength of a beam.

That the strength varies directly as the width in inches means that if the width w is increased in a certain ratio and everything else is unchanged, the weight it will support is increased in the same ratio. Thus if the width is doubled, the weight it will support will be doubled. This law, expressed in symbols, is

$$W = k_2 w$$

The fact that the strength varies directly as the square of the depth in inches means that if the depth d is increased in a certain ratio while other factors remain the same, the weight it will support is increased by the square of the ratio of that increase. If the depth is doubled, the weight that the beam will support is four times as great. In symbols this is expressed

$$W = k_3 d^2$$

Finally, if the length, width, and depth are all involved, we have a combination of all these laws which is expressed

$$W = \frac{Kwd^2}{L}$$

The expression "other things being equal" means that the material must be the same and the beam must be similarly supported and similarly loaded. The nature of the material is an important factor, since material, even of the same kind, varies in strength to a considerable extent. Each beam, therefore, has a *natural constant*, the K of the formula, which must be considered in the calculation of its carrying capacity. This constant is the same for beams when these "other things" are equal; that is, it is the same for beams of exactly the same material, supported and weighted in the same manner.

To find this constant, a bar of similar material is used, for example a bar 1 in. square in cross section and long enough to be placed on supports 1 ft apart. The constant used with beams weighted in the middle is the weight of the central load which is just sufficient to break the bar. The constant may be expressed in pounds, hundredweight (cwt), tons, kilograms etc., and the carrying capacity of the beam to which it is applied will always be expressed in the same unit.

In the practical application of these facts, another important consideration is the ratio of the breaking load to the safe load. This ratio is the *factor of safety*. Its value depends on whether the load is a *live load* (a moving load) or a *dead load* (a stationary load).

The factor of safety for a dead load is usually taken as 5, which means that the safe load on a beam must not be more than one-fifth of the breaking load. The factor of safety for a live load is often taken as 10.

EXAMPLE 20-7 A beam of pine 40 ft long, 1 ft wide, and 1 ft deep will carry a load of 4500 lb. Find the depth of a beam of the same wood similarly loaded which will carry a load of 1200 lb, if its length is 6 ft and its width is 2 in.

Solution Use the formula $W = Kwd^2/L$ and substitute in it $W = 4500$, $L = 40$, $w = 12$, and $d = 12$. This gives

$$4500 = \frac{K \times 12 \times 12 \times 12}{40}$$

Solving for K,

$$K = 104.1\overline{6}$$

Then substituting $W = 1200$, $L = 6$, $w = 2$, and $K = 104.1\overline{6}$ in the formula

$$W = \frac{Kwd^2}{L}$$

$$1200 = \frac{104.1\overline{6} \times 2 \times d^2}{6}$$

and solving for d^2

$$d^2 = \sqrt{34.56}$$
$$d = 5.88 \text{ in.}$$

EXAMPLE 20-8 Given a stick of timber 14 ft long, 8 in. deep, and 3 in. wide, find the depth of a stick of the same material 18 ft long and 4 in. wide that will support five times as much weight as the first stick.

Solution Let W = the weight that the first stick will support. Then

$$W = \frac{Kbd^2}{L} = \frac{K \times 3 \times 8^2}{14} = \frac{96K}{7}$$

In the case of the second stick, d is to be found when the weight is $5W = 480K/7$.

Therefore,

$$\frac{480K}{7} = \frac{K \times 4 \times d^2}{18}$$

Solving, $d = 17.566$.

■ EXERCISES

20-31 Solve the formula $W = \dfrac{Kbd^2}{L}$ for each letter.

20-32 The constant for white pine is 300. Find the weight that a beam of this material, centrally loaded and 10 ft long, 3 in. broad, and 7 in. deep, will support. How much is the safe load if the factor of safety is 5?

20-33 How long may a beam of white pine, centrally loaded, be between supports if it is to have a safe load of 750 lb and is 3 in. by 8 in. set on edge?

20-34 In beams having a load uniformly distributed, the constant is twice as large as when the load is centrally located. Solve Exercise 20-32 if the beam is uniformly loaded.

20-35 Find the width of a beam of oak, 12 in. deep, resting upon supports 18 ft apart, that must carry safely a uniformly distributed load of 5 tons. The constant is 500 if loaded in the center, and the factor of safety is 5.

20-36 The floor of a hall 16 ft wide is supported by joists of pine 3 in. by 12 in. set on edge. Using the constant 300 and a factor of safety of 5, find how far the joists must be placed from center to center to support a load of 140 lb per square foot of floor surface.

20-37 O'Connor gives the following formula for calculating the dead distributed safe load on timber supported at both ends and of rectangular cross section (this includes floor joists):

$$W = \frac{4wd^2K}{2L}$$

where W = load, lb (that is, the load in pounds)
   w = width, in.
   d = depth, in.
   L = length of span, in.
   K = 1900 for oak and 1100 for fir

(Notice how this formula differs from the one previously given.) What safe weight distributed will an oak beam 6 in. by 10 in. set on edge support if the span is 16 ft?

20-38 The joists in a room 14 ft wide and 26 ft long are fir 3 in. by 10 in. How far should their centers be placed apart if the floor is to support a crowd of people? (A crowd of people closely placed averages 140 lb/ft^2).

20-39 The following formula is taken from another source: For white pine beams the formula $W = 2000/3 \times wd^2/L$ gives the safe load when the beam is supported at both ends and loaded in the middle.

W = safe load, lb. less weight of beam
L = length of beam, in.
d = depth of beam, in.
w = width of beam, in.
(a) Given $L = 12$ ft, $w = 3$ in., and $d = 8$ in., find W.
(b) Given $L = 12$ ft, $w = 8$ in., and $d = 3$ in., find W.
(c) Given $L = 16$ ft, $w = 4$ in., and $W = 1900$, find d.
(d) Given $w = 6$ in., $d = 10$ in., and $W = 4100$, find L.

TABLE 20-1 EXERCISE 20-40

Name of Wood	Length, ft	Width, in.	Depth, in.	Breaking Weight, lb	k
White pine	2	2	2	1,430	357.5
Yellow pine	10.75	14	15	68,000	232+
Pitch pine	10.75	14	15	118,500	404.4
Ash	2	2	2	2,052	513
Pitch pine	7	2	2	622	544.25
Ash	7	2	2	772	675.5
Fir	7	2	2	420	367.5

20-40 The data in Table 20-1 are results of experiments on the strength of timbers when loaded in the middle. For each material, find the strength of a stick 1 ft long, 1 in. wide, and 1 in. deep. (This value is called K in the table.)

20-41 Compare the strengths of three pieces of timber of the following dimensions:
(a) 12 ft long, 6 in. deep, and 3 in. wide
(b) 8 ft long, 5 in. deep, and 4 in. wide
(c) 15 ft long, 9 in. deep, and 8 in. wide

20-42 Given a piece of timber 12 ft long, 6 in. deep, and 4 in. wide, find the thickness of a piece of the same material 16 ft long and 8 in. deep that will support twice as much weight as the first piece.

20-43 Given a stick of timber 12 ft long, 5 in. deep, and 3 in. wide, find the thickness of a stick of the same material 14 ft long and 6 in. deep that will support four times as much weight as the first stick.

20-44 Given a stick of timber 12 ft long, 6 in. deep, and 4 in. wide, find the depth of a stick of the same material 20 ft long and 5 in. wide that will support twice as much weight as the first stick.

20-45 The personnel department of a certain company estimates that the value of a man after he becomes 65 years old varies inversely with the square of his age. If a man is worth $49,000 when he is 65 years old and if he lives to be 70 years old, what would be the value of this man to his company at 70?

GEOMETRY
GEOMETRY
GEOMETRY
GEOMETRY
GEOMETRY
GEOMETRY

CHAPTER 21

INTRODUCTION TO GEOMETRY

21-1 INTRODUCTION TO GEOMETRY

Many of the facts of geometry are common knowledge. The word "geometry" comes from two Greek words meaning "earth measurement," and the subject was so called because of its early use in measuring land areas. The oldest traces of a systematic knowledge of geometry are found among the Egyptians and Babylonians as seen in the construction of their pyramids and temples. About the seventh century B.C. the geometry of the Egyptians became known to the Greeks, and in Greece the subject was developed to a remarkable degree. The geometry of Euclid, written about 300 B.C., might be used today as a textbook with very little change.

In the following pages some facts of geometry are discussed and some applications to practical problems are illustrated. The endeavor is to illustrate and clarify the principles and lay a broad foundation for the solution of applied problems. Individual students can select those which are best suited to their needs.

21-2 FUNDAMENTAL DEFINITIONS OF GEOMETRY

Points, lines, and planes are undefined elements in geometry. It is best to think of the statements of these elements as properties of geometry. A *material body*—for example, a block of wood—occupies a definite volume of space. In geometry no attention is given to substance of the material body. It may be iron, stone, wood, or air, or it may be a vacuum. Geometry considers only the space occupied by the substance.

This space is a *geometric solid* or simply a *solid*. If you visualize a brick, and then think only of the volume of space occupied by the brick, you will have an understanding of a geometric solid. A solid has *length, width,* and *thickness*. The solid pictured in Fig. 21-1 has six faces: a top and a bottom, two sides, and two ends.

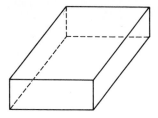

Fig. 21-1. Geometric solid.

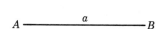

Fig. 21-2. Straight line.

A boundary face of a solid is called a *surface*. A surface has *length* and *width* but no thickness.

The boundary of a surface, or that which separates one surface from an adjoining surface, is called a *line*.† An example would be a boundary line between two states or two farms. A line has *length* only.

That which separates one part of a line from an adjoining part is called a *point*. A point has neither *length, width,* nor *thickness*. A point has *position* only.

A point is identified by placing a capital letter near it. A line is identified by placing letters at its ends or by placing a single letter upon it. Capital letters are usually used at the ends of a line; lowercase are used when a single letter is placed on the line. The line in Fig. 21-2 is identified as "the line *AB*" or "the line *a*."

A *straight line* is a line having the same direction throughout its length. Line *AB*, Fig. 21-2, is a straight line.

A *curved line* is a line that is continually changing in direction. The line *CD* in Fig. 21-3 is a curved line.

A *broken line* is a series of connected straight lines. A broken line, *EFGHI*, is shown in Fig. 21-4. Note that each straight segment of the broken line is identified separately.

If a surface contains two points that can be connected by a straight line lying within the surface, it is called a *plane surface* or a *plane*. A carpenter determines whether or not the surface of a board is a plane by laying the edge of his square (or another straightedge) on the surface in

Fig. 21-3. Curved line.

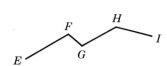

Fig. 21-4. Broken line.

†The word "line" is understood to mean a *finite* part of a line, or a line segment. In this text the word "line" refers to a line segment.

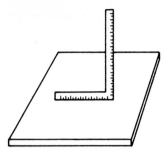

Fig. 21-5. Determining a plane.

several positions and observing whether the edge touches the surface at all points. This test is illustrated in Fig. 21-5.

A *curved surface* is a surface no part of which is a plane surface. The surfaces of a circular pipe and a baseball, for example, are curved surfaces.

21-3 ANGLES

Two straight lines that meet at a point form an *angle*. Figure 21-6 shows an angle formed by the two straight lines *AB* and *BC* meeting at *B*. Point *B* is called the *vertex* of the angle; the two lines are the *sides* of the angle. An angle is identified by reading the letter at the vertex or the letter at the vertex and the letters at the ends of the sides. Letters placed at the vertex and ends of the sides are usually capitals. The angle in Fig. 21-6 can be identified as angle *B* or angle *ABC*.

The symbol ∠ is used instead of the word "angle." The identification of the angle in Fig. 21-6 may then be written ∠*B* or ∠*ABC*.

The size of an angle is independent of the lengths of the two sides. Angle *ABC* in Fig. 21-7 is larger than angle *DEF*. The *size* of an angle is determined by the amount of rotation that would be necessary to move one of the sides onto the other. In Fig. 21-7, the rotation from *BA* to *BC* is greater than the rotation from *ED* to *EF*.

If *BC* is rotated or turned so that point *C* traces a circle, point *C* would sweep or rotate through 360°, as illustrated in Fig. 21-8. The size of an angle is determined by this rotation and is measured by a unit called a "degree." A *degree* is defined as $\frac{1}{360}$ of a circle; a circle includes 360 degrees. The symbol ° placed to the right and above a number is

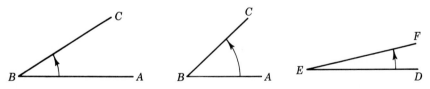

Fig. 21-6. Angle.   **Fig. 21-7.** Angle ABC is larger than angle DEF.

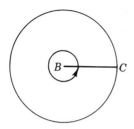

Fig. 21-8. 360° rotation of C.

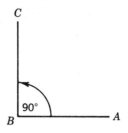

Fig. 21-9. Right (90°) angle.

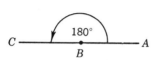

Fig. 21-10. Straight (180°) angle.

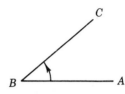

Fig. 21-11. Acute angle.

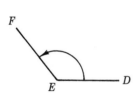

Fig. 21-12. Obtuse angle.

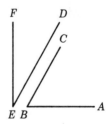

Fig. 21-13. Complementary angles.

usually used instead of the word "degree." Thus, 60 degrees is written 60°.

If the line *BC* rotates from *BA* through one-fourth of a circle, it will, with *BA*, form an angle of 90°, a *right angle*. This is shown in Fig. 21-9.

If the line *BC* rotates through one-half of a circle, it forms, with *AB*, an angle of 180°, a *straight angle,* as shown in Fig. 21-10.

An angle of less than 90° is an *acute angle.* An angle greater than 90° but less than 180° is an *obtuse angle.* An acute angle, *ABC,* is shown in Fig. 21-11; and an obtuse angle, *DEF,* is shown in Fig. 21-12.

Two angles whose sum is 90° are *complementary angles.* Angles *ABC* and *DEF* in Fig. 21-13 are complementary angles because $\angle DEF + \angle ABC = 90°$.

Two angles whose sum is two right angles, or 180°, are *supplementary angles.* Angles *ABC* and *DEF* in Fig. 21-14 are supplementary angles, because $\angle ABC + \angle DEF = 180°$.

When a pair of lines intersect, the opposite angles are *vertical angles,* and the pairs of vertical angles are equal.

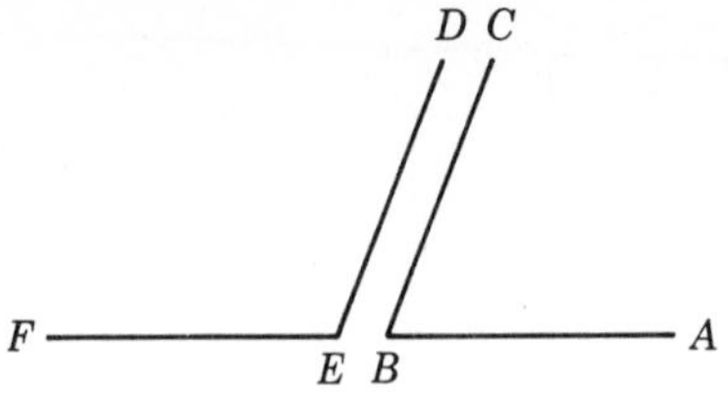

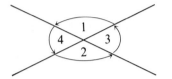

Fig. 21-14. Supplementary angles.

$\angle 1 = \angle 2$ and $\angle 3 = \angle 4$

$\angle 1 + \angle 3 = 180°$; $\angle 3 + \angle 2 = 180°$ etc.

Fig. 21-15. Vertical or opposite angles.

21-4 MEASURING ANGLES WITH A PROTRACTOR

The instrument for measuring angles is the *protractor,* shown in Fig. 21-16. The protractor is a semicircular scale of convenient size. The divisions of the scale are numbered 0° to 180°, beginning at each end. The sum of the two readings at any point is 180°. This method of numbering enables you to measure an angle from either end of the protractor and to lay out an angle in either direction.

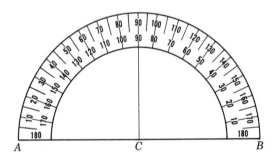

Fig. 21-16. A protractor.

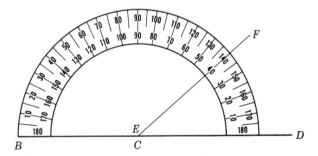

Fig. 21-17. Measuring with a protractor.

To measure an angle with a protractor, place the protractor on the angle to be measured, so that either half of *AB* will fall on one side of the angle and the point *C* will fall on the vertex. The reading on the scale where the second side of the angle crosses the scale part of the protractor is the measure of the angle in degrees. This procedure is illustrated in Fig. 21-17. Angle *DEF* in Fig. 21-17 is an angle of 40°. Angles are generally measured in a counterclockwise direction.

21-5 LAYING OUT ANGLES WITH A PROTRACTOR

To lay out an angle with a protractor, draw one side of the angle, and locate the vertex. Place the side *AB* of the protractor on the side drawn, with the point *C* on the vertex. Locate the reading of the value of the angle required on the scale of the protractor. Connect this point with the vertex.

21-6 RELATIONS OF LINES

Parallel lines are lines in the same plane that are equidistant from each other along their entire lengths. Three sets of parallel lines are shown in Fig. 21-18.

A line that crosses a set of parallel lines is a *transversal*. Fig. 21-19 shows parallel lines intersected by transversal *LM*. Angles *a* and *b* are

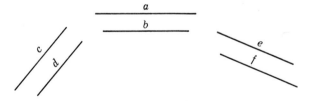

Fig. 21-18. Parallel lines.

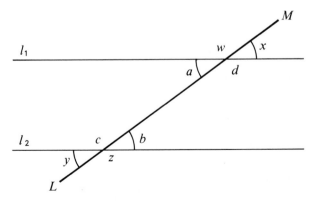

Fig. 21-19. Parallel lines intersected by transversal LM.

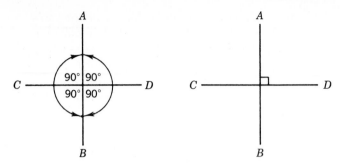

Fig. 21-20. Perpendicular lines.

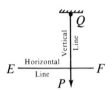

Fig. 21-21. Vertical line (*PQ*) and horizontal line (*EF*).

alternating interior angles as are angles *c* and *d*. Angles *w* and *z* are *alternating exterior angles* as are angles *x* and *y*. The properties of vertical angles apply to interior and exterior angles.

Two lines that form right angles, or angles of 90°, are *perpendicular* to each other. In Fig. 21-20, *AB* is perpendicular to *CD* and *CD* is perpendicular to *AB*. Figure 21-20 indicates the symbol which shows that lines *AB* and *CD* are perpendicular to each other.

A *vertical line,* or *plumb line,* is the line along which a string hangs when suspended at one end and weighted at the other.

A *horizontal line* is a line perpendicular to a vertical line. In Fig. 21-21, the line *PQ* is a vertical line, and line *EF* is a horizontal line.

■ EXERCISES

21-1 Locate points *A*, *B*, and *C* so that line *ABC* is a broken line. Join these three points with three straight lines. Is any one of the three angles formed a right angle? An acute angle? An obtuse angle?

21-2 Draw three curved lines. Does your drawing contain any angles?

21-3 Locate three points so that the three straight lines connecting them will form one right angle and two acute angles.

21-4 Locate three points so that the three straight lines connecting them will form one obtuse angle and two acute angles.

21-5 Draw a broken line *ACB* so that *AC* is vertical and *CB* is horizontal. Is *AC* perpendicular to *CB*?

21-6 Draw any broken line *ACB* so that the angle *ACB* is a right angle. If either of the lines *AC* or *CB* is vertical, is the other horizontal?

21-7 Use a protractor to construct the following angles. 30°, 45°, 60°, 120°, 125°, 1°, 210°, 310°. Label the acute angles. Label the obtuse angles. Pick out two angles whose sum is a right angle. Select two complementary angles. Select two supplementary angles.

21-8 Draw the following: two acute angles; two obtuse angles; three angles larger than two right angles. Use a protractor and measure the size of each of the angles that you have drawn.

21-9 A person driving east turns to drive northeast. Through what angle does she turn?

21-10 A person driving due north turns and drives due west. Through what angle does he turn?

Solve the following by algebra (review the algebra section if needed):

21-11 One angle has twice as many degrees as the other. Find each angle if they are complementary.

21-12 Given two angles such that one is 4 times the other, find each if they are supplementary.

21-13 Find the size of complementary angles if one is 30° larger than the other.

21-14 Find the value of x if $30° − 3\angle x = 2\angle x + 15°$.

12-15 Find the value of x if $4\angle x − 40° = 6\angle x − 100°$.

21-16 Two angles are supplementary. Their difference is 90°. Find the angles.

21-17 Two angles are complementary. Their difference is 20°. Find the angles.

21-7 POLYGONS AND THEIR PROPERTIES

A *polygon* is a plane surface bounded by any number of straight lines. Each of these lines is a *side*. The point where two sides meet is a *vertex*. The distance around the polygon, or the sum of the lengths of the sides, is the *perimeter* of the polygon.

A *regular polygon* is a polygon with the sides all equal in length and all angles equal.

A *triangle* is a polygon of three sides.

A *quadrilateral* is a polygon of four sides.

A *pentagon* is a polygon of five sides.

A *hexagon* is a polygon of six sides.

An *octagon* is a polygon of eight sides.

A *diagonal* is a line joining any two not adjacent vertices in a polygon. A polygon may have one or more diagonals. "Vertices" is the plural form of "vertex."

The chart in Fig. 21-22 illustrates the classification of polygons.

21-8 TRIANGLES AND THEIR PROPERTIES

A *triangle* is a polygon of three sides. A line drawn from any vertex of a triangle perpendicular to the opposite side and ending on it is an *altitude*

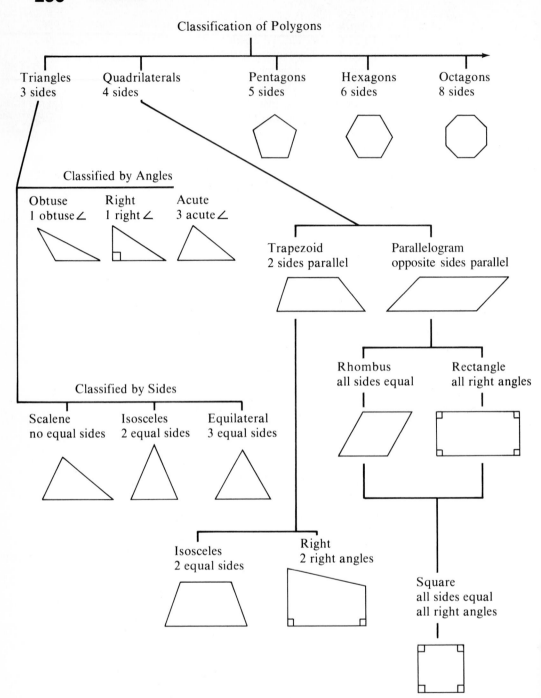

Fig. 21-22. Classification of polygons.

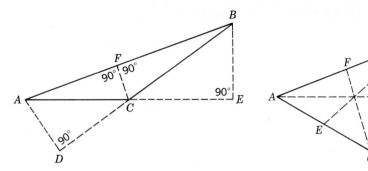

Fig. 21-23. Altitudes of a triangle.   **Fig. 21-24.** Medians of a triangle.

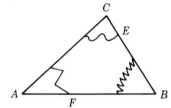   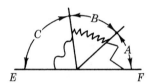

Fig. 21-25. Angles of a triangle.   **Fig. 21-26.** ∠A + ∠B + ∠C = 180°.

of the triangle. Since a triangle has three vertices, it has three altitudes. The altitude may meet the opposite side naturally, as *CF* does in triangle *ABC* in Fig. 21-23, or the opposite side may have to be extended to meet it, as with *AD* and *BE*.

A line drawn from any vertex to the midpoint of the opposite side is called a *median*. Each triangle, then, will have three medians. In Fig. 21-24, *AD*, *CF*, and *BE* are medians.

The word "bisect" always means "to divide into two equal parts." For example, a line that is drawn from the vertex of an angle and divides the angle into two equal angles is the *bisector* of the angle.

The medians *AD*, *CF*, and *BE bisect* the sides *BC*, *AB*, and *AC*. That is, by construction *AE* = *EC*. *AF* = *FB*, and *BD* = *DC*. To bisect means to divide into two equal parts. Here the sides, *AB*, *BC*, and *AC* have been divided into two equal parts by the three medians.

The sum of the three angles of any triangle is 180°. In Fig. 21-26, ∠A + ∠B + ∠C = 180°. This can be shown by measuring the angles of a triangle with a protractor and finding the sum. It can also be shown by drawing any triangle and then tearing off the corners and placing them as shown in Fig. 21-26. The sides will form a straight line *EF*.

If one angle of a triangle is a right angle, the sum of the other two angles is 90°, or a right angle. These other two angles are, therefore, complementary angles and must be acute angles. A triangle containing a right angle is a *right triangle*.

21-9 CONGRUENT TRIANGLES

Two triangles are *congruent* if one can fit exactly over the other. There are several sets of conditions that will make two triangles congruent:

1. If two angles and the side between them in one triangle are equal (respectively) to two angles and the side between them in another triangle, the triangles are congruent. This is known as angle-side-angle congruence.

    Figure 21-27 shows two such triangles. Obviously, if one is placed upon the other, it will fit exactly.

2. If two sides and the angle between them in one triangle are equal (respectively) to two sides and the angle between them in another triangle, the triangles are congruent. This is known as side-angle-side congruence. Figure 21-28 shows two such triangles. Observe that these are congruent triangles. Suppose the two triangles are as shown in Fig. 21-29; are these congruent triangles?

3. If the three sides of one triangle are equal (respectively) to the three sides of another triangle, the triangles are congruent. This is known as side-side-side congruence. Draw two such triangles and show that they can be fitted together exactly.

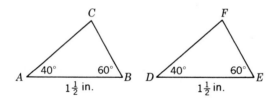

Fig. 21-27. Congruent triangles.

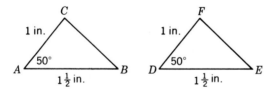

Fig. 21-28. Congruent triangles.

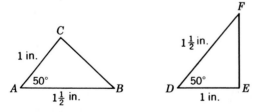

Fig. 21-29. Congruent triangles.

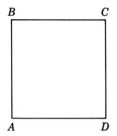

Fig. 21-30. Square.

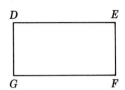

Fig. 21-31. Rectangle.

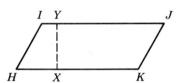

Fig. 21-32. Parallelogram.

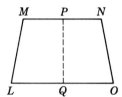

Fig. 21-33. Trapezoid.

Since two congruent triangles can be fitted together exactly, it follows that for each angle in one of two equal triangles there is a corresponding equal angle in the other and that for each side in one of two congruent triangles there is a corresponding equal side in the other. These are *corresponding angles* and *corresponding sides*.

21-10 QUADRILATERALS AND THEIR PROPERTIES

A *quadrilateral* is a polygon having four sides. Several quadrilaterals are shown in Figs. 21-30, 21-31, 21-32, and 21-33.

A *square* (Fig. 21-30) is a four-sided polygon whose four angles are all right angles and whose sides are all equal. The *rectangle* (shown in Fig. 21-31) is a four-sided polygon in which all four angles are right angles and the opposite sides are equal in length. A *parallelogram* (Fig. 21-32) is a four-sided polygon whose opposite sides are equal in length and parallel. Side *HK* is the base of parallelogram *HIJK* in Fig. 21-32. The altitudes are the perpendicular distances between opposite sides, and *XY* is an altitude of parallelogram *HIJK*. A *trapezoid* (Fig. 21-33) is a quadrilateral with two sides parallel. The parallel sides of a trapezoid are called the *bases*. The *altitude* is the perpendicular distance between the bases. In Fig. 21-33, *MN* and *LO* are the bases and *PQ* is the altitude of the trapezoid *LMNO*.

▊ EXERCISES

21-18 Draw three triangles: (a) one with two acute angles; (b) one with three acute angles; and (c) one with an obtuse angle.

21-19 Draw two triangles: (a) one with three acute angles; and (b) one with one obtuse angle. Draw the three altitudes in each triangle.

21-20 Draw two triangles. Draw the three medians in each triangle.

21-21 Draw two triangles. Draw the three bisectors of the angles of each triangle.

21-22 Draw two triangles such that the three angles of one are equal to the three angles of the other but the three sides of one are not equal (respectively) to the three sides of the other.

21-23 Draw two triangles in which the three sides of one are equal to the three sides of the other. What can be said about the three angles of these two triangles?

21-24 Can a triangle have two 90° angles?

21-25 Draw the following regular polygons with ruler and compass: (a) three-sided; (b) four-sided; (c) six-sided; (d) eight-sided.

21-26 Draw the diagonals of each polygon in Exercise 21-25. How many diagonals has each? How many triangles were formed in each case?

21-27 What are the angles of each polygon of Exercise 21-25? What are the perimeters?

21-28 Draw (a) a rectangle; (b) a square; (c) a parallelogram; (d) a trapezoid; (e) a quadrilateral that is not any of these. Draw the altitudes of each. What exception is there?

21-29 Are all the triangles that were formed in each drawing of Exercise 21-26 equal?

21-30 Draw a parallelogram with one side 3 units long and its adjacent side 5 units long. What is its perimeter?

21-31 Draw a parallelogram with adjacent sides of $4x + 2$ and $3x - 1$ units. Find each side if the perimeter is 44 units.

21-32 Draw a regular hexagon with sides 3 units long. Find its diagonals. Find the sum of all its angles.

21-33 What is the sum of all the angles of each drawing in Exercise 21-25? [If n = the number of sides, is the number of degrees in all the angles = $(n - 2)180$?] Note this formula. It will be used again.

21-34 Draw a trapezoid with parallel sides 4 and 6 units long. If the other two sides are each 2 units long, measure to determine the altitude. Measure the acute angles of this trapezoid. Measure the obtuse angles.

POLYGONS

22-1 INTRODUCTION TO POLYGONS

A polygon is a closed plane surface bounded by three or more straight sides and containing three or more angles. The word "polygon" means "many sides." *Closed* means each line connects with another line at each end, and *plane* means it has no depth. Examples of *regular* polygons are the equilateral triangle, the square, the pentagon, and the octagon. (Regular in this case means equal length of sides and equal size of angles.)

22-2 AREA

Area is measured in square units and is the amount of surface of an object or the amount of material it takes to cover the surface. It is measured in square units. The fundamental idea of a unit of area is a square whose sides are each one linear unit. Therefore, a square foot is the area of a square 1 ft on a side; a square centimeter is a square 1 cm on a side; a square yard is a square 1 yd on a side; a square meter is a square 1 m on a side.

22-3 AREA OF A RECTANGLE

The procedure to find the area of a rectangle is illustrated by Fig. 22-1. The rectangle in Fig. 22-1 has a length AD of 5 m and width AB of 4 m. The rectangle is divided into small squares 1 m on a side; so each small square represents 1 m^2. Since there are four rows of squares, each containing 5 m^2, there are $4 \times 5 \text{ m}^2 = 20 \text{ m}^2$ in the rectangle. This method also works if the lengths of the sides are given as fractions or decimals; thus we have the following rule:

Procedure for finding the area of a rectangle:

The area of a rectangle is equal to the product of its length and its width.

$$A = lw$$

295

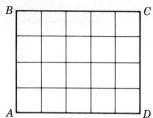

Fig. 22-1. Finding the area of a rectangle.

The length and the width of the rectangle must be expressed in the *same unit of measure* to find their product. The product is in square units of the same dimension as the linear unit of measure. Thus, if the unit of length is the meter, the area will be in square meters.

22-4   AREA OF A PARALLELOGRAM

A parallelogram and a rectangle having the same base and altitude are equal in area. This is illustrated in Fig. 22-2 by rectangle *ABCD* and parallelogram *ABEF*. The altitude *BC* is the same for each, and they have the same base *AB*. Since the portion *BCE* of the parallelogram may be cut off and fitted on *ADF* of the rectangle, it is evident that the parallelogram is equal in area to the rectangle. Therefore, we have the following rule:

Procedure for finding the area of a parallelogram:

The area of a parallelogram is equal to the product of its base and its height, *or* altitude.

$$A = bh$$

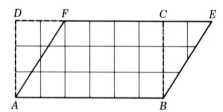

Fig. 22-2. Area of a parallelogram.

22-5   BASIC FORMULAS FOR AREA

A *formula* is a rule expressed in letters, symbols, and constant terms. It is a short way of stating a rule, or an algorithm. If *A* is used as an abbreviation for area, *b* for base, and *h* for height, the rule (formula) for the area of a rectangle or a parallelogram is given in the following format:

$$A = bh$$

where A = area
   b = base
   h = height, or altitude

Since the altitude times the base equals the area, by solving the formula we can obtain the altitude or base for the rectangle or parallelogram. The altitude equals the area divided by the base.

$$h = \frac{A}{b}$$

The base equals the area divided by the altitude.

$$b = \frac{A}{h}$$

22-6   AREA OF A TRIANGLE

If a triangle and a parallelogram have the same base and their altitudes are equal, the area of the triangle is half the area of the parallelogram. This is illustrated in Fig. 22-3 by the parallelogram *ABCD*. The diagonal *BD* divides it into two triangles *ABD* and *BCD*, which are equal in area. Thus the formula for the area of a parallelogram is as follows.

Procedure for finding the area of a triangle:

The area of any triangle is equal to one-half of its base times its height or altitude.

$$A = \frac{bh}{2}$$

It follows that if the area and either base or altitude of a triangle are given, the other dimension (altitude or base) is found by dividing twice the area by the given dimension. When *A* stands for the area, *h* for the height, and *b* for the base, we have the formulas for the height and base of a triangle:

$$h = \frac{2A}{b} \qquad b = \frac{2A}{h}$$

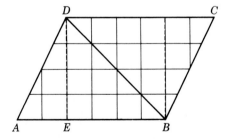

Fig. 22-3. Area of a triangle.

■ EXERCISES

Suggestion: *Sketching the polygons on graph paper may help you to understand the concept of area.*

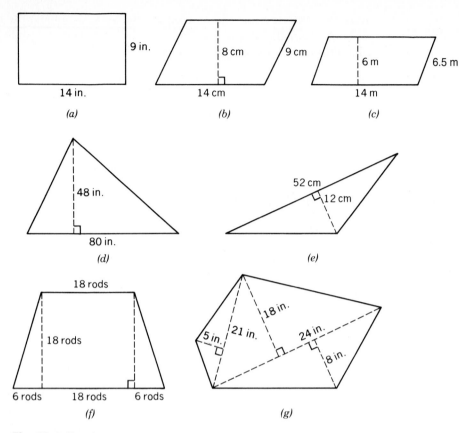

Fig. 22-4. Exercise 22-1.

22-1 Compute the areas of the figures in Fig. 22-4, using the dimensions as given.

22-2 The sides of a parallelogram are 3 ft and 4 ft. Find its perimeter.

22-3 Draw a triangle and calculate its area by drawing the altitude and measuring the altitude and base.

22-4 What measurements are necessary in order that the area of a parallelogram may be found?

22-5 Why does a median of a triangle divide the triangle into two triangles that are equal in area?

22-6 A parallelogram has sides 10 km, 10 km, 20 km, and 20 km long; find its perimeter. A rectangle has the same sides; find its area.

22-7 The two diagonals of a parallelogram bisect each other and divide the parallelogram into four triangles. Are these triangles equal in area? Show why, with a sketch.

22-8 Find the area of the following squares:
(a) One side 12 m   (c) One side 6 ft a in.
(b) One side 6 ft 3 in.   (d) One side a ft b in.

Note: *The dimensions must be converted to one unit of measure before they can be multiplied.*

22-9 Find the side of a square whose area is 225 ft².

22-10 Draw rectangles for the following:
(a) Given $a = 10$ ft and $b = 5$ ft 3 in., find the area in square feet.
(b) Given $a = 4.6$ cm and $b = 6.4$ cm, find the area in square centimeters.
(c) Given $A = 6.25$ in.² and $a = 2.5$ in., find b.
(d) Given $A = 160$ acres and $b = 40$ rods, find a. (1 acre = 160 square rods.)

Note: *Both dimensions must be expressed in the same unit of measure before they can be multiplied.*

22-11 Sketch the following triangles:
(a) Given $h = 2$ ft 6 in. and $b = 6$ ft, find A in square feet. Could the triangle sketched for this problem have various shapes? Why?
(b) Given $h = 20$ m and $b = 5$ m, find A in square meters.
(c) Given $A = 200$ cm² and $b = 2.5$ cm, find h.
(d) Given $A = 116$ ft² and $h = 116$ in., find b.

22-7   AREA OF A TRIANGLE WITH THREE SIDES GIVEN

Sometimes it is necessary to find the area of a triangle when only the lengths of the three sides are given. The formula for finding the area of a triangle in this case is known as Hero's formula. Hero of Alexandria, a great mathematician, derived the formula for surveying about 2000 years ago.

Procedure for finding the area of a triangle given all three sides:

Let a, b, and c stand for the three sides of a triangle, and if s equals one-half the sum of a, b, and c, then the area A of a triangle is given by the formula

$$A = \sqrt{s(s - a)(s - b)(s - c)} \qquad s = \frac{1}{2}(a + b + c)$$

(See Fig. 22-5.) This formula is derived and proved in trigonometry.

Since a formula is a rule stated in symbols, a formula such as Hero's formula may be stated in words as follows: Find half the sum of the three sides. Subtract each side from this half-sum. Take the continued product

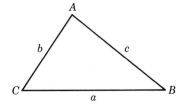

Fig. 22-5. Hero's formula for finding the area of a triangle from the lengths of the sides:
$s = \frac{1}{2}(a + b + c)$
Area $= \sqrt{s(s - a)(s - b)(s - c)}$.

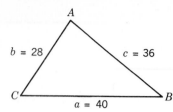

Fig. 22-6. Example of Hero's formula.

of the half-sum and the three differences. The square root of this product is the area of the triangle.

Hero's formula can be illustrated best by an example.

EXAMPLE 22-1 Find the area of a triangle with sides of 40 ft, 28 ft, and 36 ft.

Solution

$$s = \frac{1}{2}(a + b + c)$$

$$s = \frac{1}{2}(40 + 28 + 36) = 52$$

$$A = \sqrt{s(s - a)(s - b)(s - c)}$$
$$= \sqrt{52(52 - 40)(52 - 28)(52 - 36)}$$
$$= \sqrt{52(12)(24)(16)}$$
$$= \sqrt{4 \times 13 \times 4 \times 3 \times 4 \times 3 \times 2 \times 16}$$

Therefore,

$$A = 2 \times 2 \times 2 \times 3 \times 4\sqrt{26} = 96\sqrt{26}$$
$$\text{Area} = 489.506 \text{ ft}^2$$

The area of the triangle with the three sides given can also be found by constructing the triangle to scale. The altitude can then be measured and the area be found, with reasonable accuracy, by taking one-half the product of the base and the altitude; however, this process is time-consuming and in practical applications sometimes impossible. Therefore, Hero's formula should be used to find the area when three sides of a nonright triangle are given.

22-8 AREA OF A TRAPEZOID

A diagonal of a trapezoid divides it into two triangles which have the same altitude and have as bases the two bases of the trapezoid. Thus, in the trapezoid in Fig. 22-7, the diagonal JL divides the trapezoid into two triangles JKL and JLM. The area of $JKL = \frac{1}{2}$ of JK times h, and the area of $JLM = \frac{1}{2}$ of ML times h'. But $h = h'$; hence the sum of the areas of the two triangles $= \frac{1}{2}(JK + ML) \times h$. The area of the trapezoid, then, can be found by finding the sum of the areas of the two triangles.

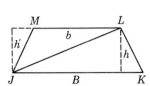

Fig. 22-7. Area of a trapezoid.

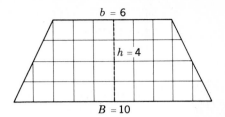

Fig. 22-8. Example 22-2: Find the area of a trapezoid.

Procedure for finding the area of a trapezoid:

The area of a trapezoid equals one-half the sum of the two bases times the height or altitude.

If B and b stand for the two bases and h stands for the height of the trapezoid, the formula is

$$A = \frac{1}{2}(B + b)h$$

EXAMPLE 22-2   Find the area of a trapezoid whose lower base is 10 rods, upper base is 6 rods, and altitude is 4 rods. (See Fig. 22-8.)

Solution $A = \frac{1}{2}(B + b)h$. Substituting the given numbers in the formula,

$$A = \frac{1}{2}(10 + 6)(4)$$
$$= 2(16)$$
$$\text{Area} = 32 \text{ square rods}$$

EXERCISES

Suggestion: Sketch the polygons on graph paper, it will help illustrate the area concept.

22-12   Sketch the following parallelograms and find the parts not given:
(a) Base = 12.5 cm; altitude = 25 cm; area = ?
(b) Altitude = 42 ft; area = 462 ft²; base = ?
(c) Base = 44 m; area = 990 m²; altitude = ?
(d) Altitude = 72 ft 8 in.; area = 2162 ft²; base = ?

22-13   The top of a table is a rectangle 4.5 ft by 6.5 ft. Find its area.

22-14   An oriental rug is a rectangle 18 ft by 12 ft. Find its price at \$20/yd².

22-15   A special tool box 60 cm long, 40 cm wide, and 30 cm deep has six rectangular faces. Find the outside surface area of the box, including the top and bottom.

22-16   A metal cabinet is 60 in. high, 18 in. deep, and 20 in. wide. Find the outside surface area of the cabinet.

22-17 A wardrobe is 2 yd by 24 in. by 2 ft. Find the total area of its surface.

22-18 A utility cabinet is to be painted inside and outside. It has four shelf spaces (three shelves). If it is 50 in. high, 24 in. wide, and 12 in. deep, what is the total surface in *square feet* to be painted? Disregard the thickness of the material and shelves.

22-19 How many paving blocks, each 4 in. by 4 in. by 10 in. placed on their sides, will it take to pave an alley 600 ft long and 12 ft 6 in. wide?

22-20 How many bricks, each 9 in. by $4\frac{1}{2}$ in. by $1\frac{3}{4}$ in., will it take to pave a court 16 ft by 126 ft if the bricks are laid flat? ($A = 9$ in. $\times$ $4\frac{1}{2}$ in.) If laid on edge? ($A = 9$ in. $\times$ $1\frac{3}{4}$ in.)

22-21 What will be the expense of painting the walls and ceiling of a room 12 ft 6 in. by 16 ft and 8 ft 4 in. high at $2.62/yd^2?

22-22 Find the cost of sodding a lawn 36.6 m wide and 46 m long at $3.50/m^2.

22-23 At $3.45/ft^2, find the cost of laying a concrete walk 6 ft wide on two adjacent sides of a corner lot 33 ft by 100 ft.

22-24 Find the area of a trapezoid whose bases are 17 mm and 11 mm and whose altitude is 26 mm.

22-25 The roof of a house is 20 ft long, 18 ft wide at one end, and 10 ft wide at the other end. Find its area.

22-26 The trapezoidal gable of a house is 12 m long at the bottom, 3.5 m long at the top, and 4 m high. Find its area in square feet.

22-27 Find the area of a solar panel 14 ft long, 18 in. wide at one end, and 12 in. wide at the other end.

22-28 Find the area in acres of a wooded lot that is represented on paper as a rectangle $3\frac{3}{4}$ in. by $2\frac{1}{2}$ in. on a scale of $\frac{1}{16}$ in. to the rod.

22-29 A television screen is a rectangle whose base and altitude are in the following ratio:

$$\frac{\text{Base}}{\text{Altitude}} = \frac{13}{11}$$

In each case find the base and altitude: (a) If the picture area is 1287-cm^2; (b) If the picture area is 1557.27 cm^2.

22-30 The most beautiful rectangle, according to artists, has a base and altitude in the following ratio:

$$\text{Base} = \sqrt{5} \times \text{altitude}$$
$$= 2.236 \times \text{altitude}$$

This ratio is known as the Golden Section or Golden Rectangle, and is frequently used as a guide for maintaining proper proportion in art. If the area of such a rectangle is 165 in.^{2}, find its base and altitude.

22-31 A rectangle is 105 cm long and 0.1 m wide. Find its area in square meters.

22-32 Find the number of square feet in the floor of the room shown in Fig. 22-9.

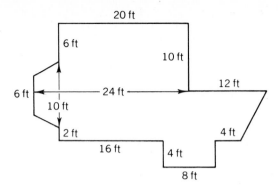

Fig. 22-9. Exercise 22-32.

Suggestion: *Divide the figure into rectangles and trapezoids.*

22-33   Find the area of Fig. 22-10*a*.

22-34   Find the area of Fig. 22-10*b*.

22-35   Find the area of Fig. 22-10*c*.

22-36   Find the surface area of each of seven footings supporting a total load of 168,000 lb if the safe bearing load of a soil is 4000 lb/ft^2.

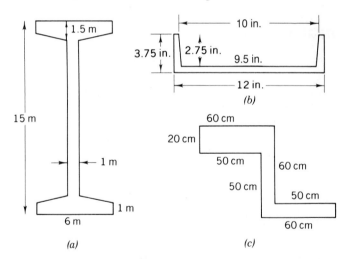

Fig. 22-10. (*a*) Exercise 22-33; (*b*) Exercise 22-34; (*c*) Exercise 22-35.

TRIANGLES

23-1 INTRODUCTION TO TRIANGLES

The triangle has many vital applications in manufacturing and construction. This is because the triangle is the strongest geometric form. Thus the triangle is used in bridge supports, roof design, and a host of other applications.

23-2 THE RIGHT TRIANGLE

A *right triangle* is a special three-sided polygon containing one right angle. The side opposite the right angle is the *hypotenuse*, and the sides of the right angle are the *base* and *altitude*. Many practical problems can be solved by applying the basic right triangle's properties.

In the right triangle shown in Fig. 23-1, $\angle ACB$ is the right angle. Since $\angle ACB$ equals 90°, the sum of $\angle ABC$ and $\angle BAC$ must equal 90°. Because the sum of the three angles of a triangle is 180°,

$$\angle ACB + \angle CAB + \angle ABC = 180°$$

Substituting 90° for $\angle ACB$ and solving,

$$90° + \angle CAB + \angle ABC = 180°$$
$$\angle CAB + \angle ABC = 90°$$

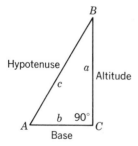

Fig. 23-1. Right triangle.

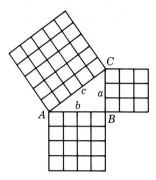

Fig. 23-2. $a^2 + b^2 = c^2$.

It can be seen that if one acute angle of a right triangle is known, the second acute angle can be found.

The relationship between the lengths of the three sides of a right triangle is given as: *The length of the hypotenuse squared is equal to the sum of the base squared plus the altitude squared.* This is illustrated by Fig. 23-2. *AC* is the hypotenuse and is 5 units in length; *AB* is the base and is 4 units in length; *BC* is the altitude and is 3 units in length. *AC* squared is equal to the sum of the squares of *AB* and *BC*. Then $\overline{AC}^2 = \overline{AB}^2 + \overline{BC}^2$. In Fig. 23-2,

$$5^2 = 4^2 + 3^2$$
$$25 = 16 + 9$$
$$25 = 25$$

Note: $\overline{AC}^2$ *means that the length of line AC has been squared.*

This principle may be stated with the lengths of the sides represented by c, b, and a; then $c^2 = b^2 + a^2$. This equation may be solved for c, b, or a; if two of the three terms in the equation are known, the third can be calculated. From the relationship $c^2 = b^2 + a^2$ three formulas may be derived:

$$c = \sqrt{a^2 + b^2} \qquad a = \sqrt{c^2 - b^2} \qquad b = \sqrt{c^2 - a^2}$$

EXAMPLE 23-1   Find the hypotenuse of a right triangle whose base is 14 cm and whose altitude is 16 cm.

Solution   Select the correct formulas, substitute the given values, and solve.

$$c = \sqrt{a^2 + b^2}$$
$$= \sqrt{16^2 + 14^2}$$
$$= \sqrt{256 \times 196}$$
$$= \sqrt{452}$$
$$= 21.26 + \text{cm}$$

The relationship between the length of the hypotenuse and the lengths of the other two sides of a right triangle is perhaps the most famous theorem in geometry. It is known as the *Pythagorean theorem* because it is said to have been proved by Pythagoras, a famous Greek mathematician who lived about 500 B.C.

23-3   THE SQUARE AND ITS DIAGONALS

The diagonal of a square divides a square into two congruent right triangles, as shown in Fig. 23-3. Given that sides *AB*, *BC*, *CD*, and *AD* are equal in length, and the four acute angles formed by the diagonal are each 45°, if *s* is a side and *d* the diagonal, then

$$d^2 = s^2 + s^2 = 2s^2$$
$$d = s\sqrt{2}$$

This may be stated: *The diagonal of a square is equal to the length of the side of the square multiplied by* $\sqrt{2}$. Thus, $d = s\sqrt{2}$.

In Fig. 23-4 both diagonals of the square have been drawn. These diagonals are equal in length, that is, $AC = BD$. In Fig. 23-4, angle $AOD = 90°$, and AO and OD equal one-half the length of the diagonal. In triangle *AOD*,

$$s^2 = \left(\frac{d}{2}\right)^2 + \left(\frac{d}{2}\right)^2$$

$$s^2 = \frac{d^2}{4} + \frac{d^2}{4}$$

$$s^2 = \frac{2d^2}{4}$$

$$\sqrt{s^2} = \sqrt{\frac{d^2}{2}}$$

$$s = \frac{d}{\sqrt{2}}$$

$$s = \frac{d}{(\sqrt{2})}\frac{(\sqrt{2})}{(\sqrt{2})}$$

$$s = \frac{d\sqrt{2}}{2} = 0.707d$$

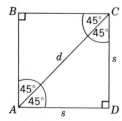

Fig. 23-3. The diagonal of a square divides it into two equal right triangles.

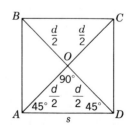

Fig. 23-4. Square with both diagonals drawn.

Thus, the formula can be stated: *The length of side of a square is equal in length to one-half the diagonal multiplied by* $\sqrt{2}$. Thus $s = d\sqrt{2}/2$ or $0.707d$.

■ EXERCISES

23-1 Find the length of a diagonal brace for a rectangular gate 9 ft by 12 ft.

23-2 A city lot is a square whose diagonal is 58 m; find the length of its sides.

23-3 If the edge of a cube is 1 m, find its diagonal.

23-4 Find the diagonal of a cube 9 ft on an edge. This cube is shown in Fig. 23-5.

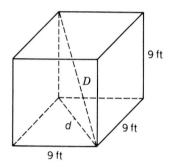

Fig. 23-5. Exercise 23-4.

23-5 If the total surface area of a cube is 12 m², find its diagonal.

23-6 If the diagonal of a cube is $3\sqrt{3}$ ft, find its total area.

23-7 A person swims at a right angle to the bank of a stream at the rate of 3.5 mph. If the current is 7.5 mph, find the rate the person is moving.

Suggestion: *The combined effect of the current and the swimmer's own swimming will cause the swimmer to move on the hypotenuse of a right triangle.*

23-8 A child rows at a right angle to the banks of a river. If the river is 1 km wide and flows at the rate of 4 km/h, how long will it take the child to row across if her rate of rowing is 3 km/h in still water? How many miles will she have rowed when she has crossed?

23-9 One cruise ship goes due north at the rate of 15 mph, and another goes due west at 18 mph. If both start from the same place, how far apart will they be in 6 h?

23-10 The diagonal of a rectangle is 130 cm, and its altitude is 32 cm. Find its area.

23-11 What is the length of the longest line that can be drawn within a rectangular box 12 m by 4 m by 3 m?

23-12 The hypotenuse of a right triangle whose base and altitude are equal is 12 m. Find the length of the base and altitude

23-13 The base of a triangle is 20 ft, and the altitude is 18 ft. What is the side of a square having the same area?

23-14 The area of a rectangular lawn is 5525 m², and the length of one of its sides is 850 dm. Find the length of its diagonal in meters to three decimal places.

23-15   Find the length of the diagonals of the following rooms:
(a) 20 m by 16 m by 12 m
(b) a ft by b ft by c ft
(c) $2a$ ft by $2b$ ft by $2c$ ft

23-16   Find the cost at $150 per M (1000 fbm) of roof boards on a $\frac{1}{3}$-pitch roof of a barn 45 ft by 65 ft, if projections at the ends and eaves are 2 ft. (A $\frac{1}{3}$-pitch roof has a distance from its base to its ridge equal to one-third the width of the building; fbm stands for foot board measure, where 1 fbm is defined as a piece of wood with an area of 1 ft^2 and a thickness of 1 in.

23-17   In fitting a hydraulic pipe to the form $ABCD$ (Fig. 23-6), making a bend of 45°, the technician assumes that $BC = CE + \frac{5}{12}CE$. Find BC by the technician's method, if $CE = 18$ in. What is the correct length of BC, and what is the percentage of error by the technician method?

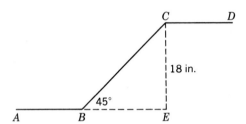

Fig. 23-6. Exercise 23-17.

23-18   To find the diagonal of a square, multiply the side by 10, take away 1 percent of this product, and divide the remainder by 7. Test the accuracy of this rule of thumb.

Solution: For example, take a 25″ square:

$$10 \times 25 = 250$$
$$1\% \text{ of } 250 = \underline{\quad 2.5\quad}$$
$$\text{Remainder} = 247.5$$
$$247.5 \div 7 = 35.357 + \text{in.} = \text{diagonal, by rule}$$

By the Pythagorean formula for the hypotenuse,

$$\text{Diagonal} = \sqrt{25^2 + 25^2} = 35.355 + \text{in.}$$

Hence,

$$\text{Error} = 35.357 \text{ in.} - 35.355 \text{ in.} = 0.002 \text{ in.}$$

and

$$0.002 \div 35.355 = 0.0056\% = \text{percentage of error}$$

It is evident that this rule is both accurate and easy to apply; however, with a hand-held calculator it is easy to find the exact value.

23-19   What is the distance across the corners of a square nut (Fig. 23-7) that is 8.6 cm on a side?

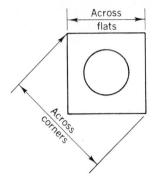

Fig. 23-7. Exercise 23-19.

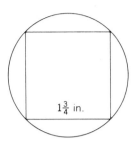

Fig. 23-8. Exercise 23-20.

23-20 What must be the diameter of round stock so that a square bolt head $1\frac{3}{4}$ in. on a side can be milled from it? (Fig. 23-8.)

23-21 A smokestack is held in position by three guy wires that reach the ground 49 ft from the foot of the stack. Find the length of a guy wire if they are fastened to the stack 70 ft from the ground.

23-22 The base of a right triangle is 30 m and the hypotenuse is 6 m longer than the altitude; find the altitude.

23-33 The hypotenuse of a right triangle is 24 m and the base is 8 m longer than the altitude; find the base and altitude.

23-24 The area of a right triangle is 30 ft², and the altitude is 7 ft longer than the base; find the hypotenuse.

23-4 SIMILAR TRIANGLES

Two triangles are *similar triangles* if the corresponding angles of both triangles are equal. Fig. 23-9 illustrates this definition. In the two triangles ABC and $A'B'C'$, $\angle A = \angle A'$, $\angle B = \angle B'$, and $\angle C = \angle C'$. Triangles ABC and $A'B'C'$ are similar triangles.

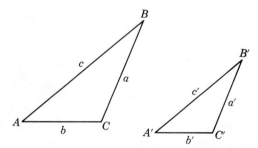

Fig. 23-9. Similar triangles.

Draw two similar triangles so that side AB is three-fourths as long as side $A'B'$, as in Fig. 23-10. Triangles ABC and $A'B'C'$ are similar by construction.

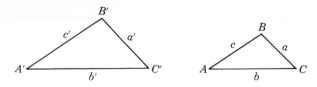

Fig. 23-10. Triangles similar by construction.

Side $AB = \frac{3}{4}A'B'$ by construction. Now measure side $B'C'$ and compare its length with BC. $BC = \frac{3}{4}B'C'$. Measure sides AC and $A'C'$ and compute the ratio; it is $\frac{3}{4}$. *Corresponding sides of similar triangles form a proportion.* That is, in Fig. 23-10,

$$\frac{a}{a'} = \frac{b}{b'} \qquad \frac{a}{c} = \frac{a'}{c'} \qquad \frac{b}{c} = \frac{b'}{c'}$$

EXAMPLE 23-2 Using the principles of similar triangles, find the distance between the points P and Q on opposite banks of a stream (Fig. 23-11), where P is inaccessible.

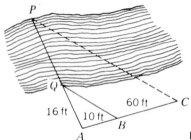

Fig. 23-11. Example 23-2: Finding distance PQ.

Solution Set up a proportion with similar sides of the given triangles.

$$\frac{AC}{AB} = \frac{PA}{QA}$$

$$\frac{70}{10} = \frac{PA}{16}$$

$$\frac{70 \times 16}{10} = PA$$

$$PA = 112 \text{ ft} \qquad QA = 16 \text{ ft}$$

Then distance $PQ = 112 - 16$ or 96 ft.

23-5 PRACTICAL APPLICATION: TAPERS

In machine trades, finding the *taper per foot* is important so that a lathe may be set up to turn a specified taper or design. Figure 23-12 is a sketch of a tapered pin to be turned on a lathe. The total taper is $4\frac{1}{2}$ in. $- 3$ in.,

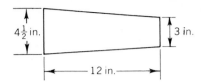

Fig. 23-12. Tapered pin.

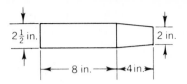

Fig. 23-13. Finding taper per foot.

or $1\frac{1}{2}$ in. Since the pin is 12 in. in length, the taper per foot is $1\frac{1}{2}$ in. The taper for the pin pictured in Fig. 23-13 is $2\frac{1}{2}$ in. $-$ 2 in. $= \frac{1}{2}$ in. The taper is over a length of 4 in., or $\frac{1}{3}$ ft. The taper per foot is, then, $3 \times \frac{1}{2}$ in., or $1\frac{1}{2}$ in.

If l equals the length of the tapered part in feet, t equals the taper in inches, and T equals the taper in inches per foot, the following proportion is true by similar triangles:

$$\frac{l}{1} = \frac{t}{T}$$

The taper for the total length of a part is the taper per foot times the length in feet.

EXAMPLE 23-3 What is the taper per foot of the pin pictured in Fig. 23-14?

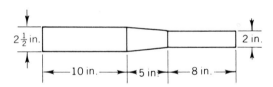

Fig. 23-14. Finding taper per foot.

Solution Substitute into the given proportion and solve.

$$\frac{l}{1} = \frac{t}{T}$$

$$\frac{\frac{5}{12}}{1} = \frac{\frac{1}{2}}{T}$$

$$\tfrac{5}{12}T = (\tfrac{1}{2})1$$

$$(\tfrac{12}{5})\tfrac{5}{12}T = (\tfrac{1}{2})1(\tfrac{12}{5})$$

$$T = 1\tfrac{1}{5} \text{ inches per foot}$$

▨ EXERCISES

23-25 Find the tapers per foot of the following tapers per inch: 0.0013 in.; 0.0260 in.; 0.0473 in.; 0.0758 in.

23-26 The shaft of a golf club is 38 in. long, $\frac{3}{8}$ in. in diameter at the small end, $\frac{7}{8}$ in. in diameter at the large end. What is its taper per foot?

23-27 The heating plant at Illinois Institute of Technology has a smokestack 6 m in diameter at the bottom. It is 60 m high and has a top diameter of 2 m. What is its taper per meter?

23-28 If the barrel of a gun is 30 in. long, 1.5 in. in diameter at the small end, and 2 in. in diameter at the large end, what is its taper per foot?

23-29 The standard pipe thread taper is $\frac{3}{4}$ in. per foot. How much is this per inch?

23-30 Find the taper per decimeter to be used in turning a pulley with a 35.56-cm. face crowned 50 mm.

23-31 If the crowning of a pulley is $\frac{1}{24}$ the width of the face, find the taper per foot to be used in turning a pulley with a 10-in. face.

23-32 The roof truss in Fig. 23-15 shows that the timbers *CB, DF,* and *EG* are perpendicular to *CA*. From the given dimensions find the lengths of *DF, EG, CF, DG,* and *AB*.

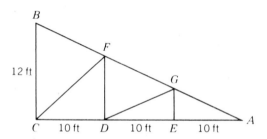

Fig. 23-15. Exercise 23-32.

23-33 In Fig. 23-16, find the distance *BC* across the lake. The given measurements are: *AB* = 200 m, *AD* = 80 m, and *DE* = 60 m. Is it necessary to make right-angle triangles?

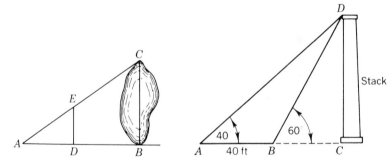

Fig. 23-16. Exercise 23-33.   **Fig. 23-17.** Exercise 23-34.

23-34 Find the height of the smokestack *CD* in Fig. 23-17, when the foot of the stack cannot be reached.

Suggestion: On a level place, measure from A to B in a line with C, and measure angles BAD and CBD. Suppose the line AB is 40 ft and the angles are BAD = 40° and CBD = 60°. Construct a figure to scale and measure the line that corresponds to the smokestack. This problem can be solved very easily by using trigonometry. This type of problem will be covered in the trigonometry section of this text.

23-6 PRACTICAL APPLICATION: THE STEEL SQUARE

One of the most useful tools is a *steel square* or *carpenter's square* (Fig. 23-18). It is made in several sizes; the most common size is that with the longer arm—the *body, blade,* or *stock*—24 in. in length and 2 in. in width, and the shorter arm—the *tongue*—16 or 18 in. in length and $1\frac{1}{2}$ in. in width.

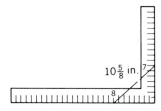

$10\frac{5}{8}$ in.

8

Fig. 23-18. Example 23-4: Using the carpenter's square.

The principles involved in using the square are mainly those having to do with the solution of right-triangle and similar-triangle problems.

EXAMPLE 23-4 Find the length of the hypotenuse of a right triangle if the base is 8 in. and the altitude is 7 in.

Solution Measure the distance from the 8-in. mark on the blade to the 7-in. mark on the tongue. This measures about $10\frac{5}{8}$ in. (Fig. 23-18.) By computation, the hypotenuse = $\sqrt{8^2 + 7^2} = 10.630$ in. The answer obtained by measurement is accurate for practical purposes.

23-7 PRACTICAL APPLICATION: RAFTERS AND ROOFS

The *run* of a rafter is the distance measured on the horizontal from its lower end to a point under its upper end. The *rise* is the distance of the upper end above the lower end. In Fig. 23-19, *AC* is the run and *CB* the rise.

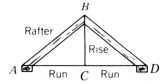

B

Rafter

Rise

A   Run   *C*   Run   *D*

Fig. 23-19. Run (*AC*) and rise (*CB*).

The slant of a roof or roof rafter is usually stated in terms of the relationship of the rise to the run. It is sometimes stated "the rise per foot of run is 6 in. to 1 ft." A second way to express this is to state the *pitch* of the roof; the pitch of a roof is equal to the rise divided by the full width of the building.

EXAMPLE 23-5 Determine the pitch in Fig. 23-19 if $AC = 12$ ft and $BC = 4$ ft.

Solution

$$\text{Pitch} = \frac{\text{rise}}{2 \times \text{run}} = \frac{4}{2 \times 12} = \frac{1}{6}$$

The roof has a $\frac{1}{6}$ pitch.

▪ EXERCISES

23-35 Using a carpenter's square, determine the length of a brace for a run of 6 ft 6 in. and a rise of 4 ft 6 in.

23-36 Find the length of the legs of the sawhorse shown in Fig. 23-20. Determine the length of the brace.

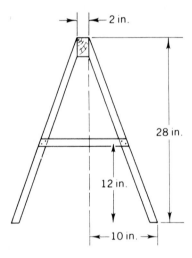

Fig. 23-20. Exercise 23-36.

23-8 ISOSCELES AND EQUILATERAL TRIANGLES

A triangle with two sides equal in length is an *isosceles triangle*. A triangle with all three sides equal in length is an *equilateral triangle*.

An equilateral triangle is also isosceles; the properties of an isosceles triangle apply to an equilateral triangle.

The facts concerning these two triangles can be proven by theorems. You can satisfy yourself that they are true by constructing triangles and measuring the parts.

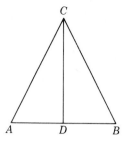

Fig. 23-21. Isosceles triangle.

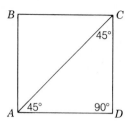

Fig. 23-22. The diagonal of a square divides it into two equal 45° right triangles.

23-9 PROPERTIES OF THE ISOSCELES TRIANGLE

An isosceles triangle is shown in Fig. 23-21. Sides *AC* and *CB* are equal in length. Angles *A* and *B* opposite the equal sides are equal. Line *CD* bisects the vertex angle *C* and is perpendicular to and bisects the base *AB;* thus *AD* = *DB*. Line *CD* also divides the isosceles triangle into two equal right triangles *ACD* and *DCB*. Line *CD* bisects angle *C* and is also a median and the altitude of the triangle.

The diagonal of a square divides the square into two congruent right isosceles triangles (Fig. 23-22). In these isosceles triangles each of the equal angles is 45°. Such a triangle is often called the 45° right triangle.

23-10 PROPERTIES OF THE EQUILATERAL TRIANGLE

In Fig. 23-23, the three sides of triangle *ABC* are equal in length; *ABC* is an equilateral triangle. The angles opposite the equal sides are equal, and each angle = 60° because the sum of the three angles is 180°.

The line drawn from vertex *A* and bisecting angle *A* is perpendicular to, and bisects, the opposite side *BC*. It also divides the equilateral triangle into two congruent right triangles *ADB* and *ADC*. Lines *BE* and *CF* each divide the triangle in the same way. Each of these lines is a *bisector* of an angle, a median, and an altitude of the equilateral triangle.

The point *O* where the three lines *AD*, *EB*, and *CF* intersect is called the *geometric center*, the *centroid*, or the *center of gravity* of the equilateral

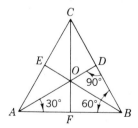

Fig. 23-23. Equilateral triangle.

triangle. It is one-third the distance from one side to the opposite vertex, that is, $DO = \frac{1}{3}DA$; $FO = \frac{1}{3}FC$; and $EO = \frac{1}{3}EB$. It follows that $AO = 2DO$, $BO = 2EO$, and $CO = 2FO$.

Each of the triangles formed when an equilateral triangle is divided into two triangles by an altitude is a right triangle with acute angles of 30° and 60°. Such a triangle is a *30°–60° right triangle*. It can be seen that *in a 30°–60° right triangle, the hypotenuse is twice the shortest side*. Thus in Fig. 23-24, $AB = 2BC$ because BC is one-half a side of the equilateral triangle of which triangle ABC is a part.

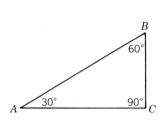

Fig. 23-24. 30°–60° right triangle.

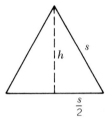

Fig. 23-25. Altitude of a equilateral triangle.

23-11 ALTITUDE AND AREA OF AN EQUILATERAL TRIANGLE

The altitude of an equilateral triangle can be found if a side is given as in Fig. 23-25.

Let s = a side, and let h = the altitude. Then, from the Pythagorean theorem,

$$h = \sqrt{s^2 - \left(\frac{s}{2}\right)^2} = \sqrt{s^2 - \frac{s^2}{4}} = \sqrt{\frac{4s^2}{4} - \frac{s^2}{4}} = \sqrt{\frac{3s^2}{4}} = \frac{s\sqrt{3}}{2}$$

Therefore,

$$h = \frac{s}{2}\sqrt{3} = \frac{s}{2}(1.732) = 0.866s$$

Stated in words: *An altitude of an equilateral triangle equals one-half of a side times $\sqrt{3}$.*

Solving for s,

$$s = \frac{2h}{\sqrt{3}} = \frac{2h\sqrt{3}}{3} = 1.155h$$

Stated in words: *A side of an equilateral triangle equals twice the altitude divided by $\sqrt{3}$ or 1.155h.*

Since the area A is one-half the base times the altitude,

$$A = \frac{s}{2}h = \left(\frac{s}{2}\right)\left(\frac{s}{2}\sqrt{3}\right)$$

Therefore,

$$A = \left(\frac{s}{2}\right)^2 \sqrt{3} = \left(\frac{s^2}{4}\right)\sqrt{3} = \frac{s^2\sqrt{3}}{4} = 0.433s^2$$

Stated in words: *The area of an equilateral triangle equals one-fourth the square of a side times* $\sqrt{3}$ *or* $0.433s^2$.

EXAMPLE 23-6 Find the altitude and area of an equilateral triangle whose sides are 8 in.

Solution Use the given formulas and evaluate:

$$h = \frac{s\sqrt{3}}{2} = \left(\frac{8}{2}\right)(1.732) = (4)(1.732) = 6.928 \text{ in.}$$

$$A = \frac{s^2\sqrt{3}}{4} = \frac{(8)^2}{4}(1.732) = (16)(1.732) = 27.712 \text{ in.}^2$$

23-12 PROPERTIES OF THE REGULAR HEXAGON

A regular hexagon is a six-sided figure with equal sides and equal angles (Fig. 23-26). This polygon is found in many practical applications.

Diagonals drawn in Fig. 23-26 divide the hexagon into six congruent equilateral triangles. The distance from the center O to any vertex equals the length of a side. The area of the regular hexagon is equal to 6 times the area of an equilateral triangle with sides equal to the sides of the hexagon. The altitude ON is the *apothem*—or the radius of the inscribed circle—and may be found by multiplying $AB \times \sqrt{3}/2$ or $0.866\ AB$.†

If s = one side, h = the altitude NO, and A = the area of the regular hexagon, the following formulas may be derived:

$$h = \frac{s\sqrt{3}}{2} = 2h = s\sqrt{3} = 1.732s$$

$$s = \frac{2h}{\sqrt{3}} = s = \frac{2}{3}h\sqrt{3} = 1.155h$$

$$A = 6\left(\frac{s}{2}\right)^2\sqrt{3} = A = \frac{3}{2}s^2\sqrt{3} = 2.598s^2$$

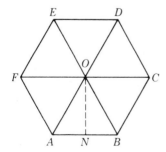

Fig. 23-26. Regular hexagon.

†See Sec. 24-2 for the definition of an *inscribed circle*.

■ EXERCISES

23-37 Sketch the following isosceles triangles and find their geometric centers and the other parts indicated:
(a) The equal sides are 8 ft and the base is 7 ft; find the altitude.
(b) The equal sides are 36 in. and the altitude is 33 in.; find the base.
(c) The sides of an equilateral triangle are 48 km; find the altitude.
(d) The altitude of an equilateral triangle is 12 m; find the length of the sides.

23-38 In Fig. 23-26, given AB = 36m, find ON, the apothem.

23-39 The sides of an equilateral triangle are 3 km in length; find the area.

23-40 In an isosceles triangle the equal sides are 1.5 ft and the base is 1 ft; find the area.

23-41 The area of an equilateral triangle is 21.217 ft^2; find the length of one side.

23-42 The area of an equilateral triangle is 108.996 in.^{2}; find the sides and the altitude.

23-43 Find the total length of the steam pipe $ABCD$ in Fig. 23-27 if FG = 6 ft, CE = 16 in., and angle EBC = 30°.

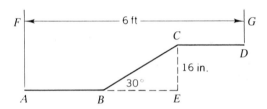

Fig. 23-27. Exercise 23-43.

23-44 Two triangles often used by draftsmen are right triangles: One is a right isosceles triangle with acute angles of 45°, and the second is a right triangle with acute angles of 30° and 60°. These two triangles are shown in Fig. 23-28. If one of the equal sides of the isosceles right triangle is 6 in.,

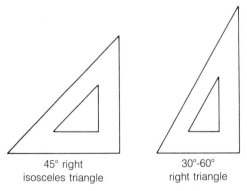

45° right
isosceles triangle

30°-60°
right triangle

Fig. 23-28. Exercise 23-44, 23-45, 23-46.

find the hypotenuse. If the hypotenuse of the 30°–60°–90° triangle is 12, find the two sides.

23-45 If the shortest side of the 30°–60° right triangle (Fig. 23-28) is 4 in., find the other sides.

23-46 Using the triangles in Fig. 23-28, show how to construct the following angles: 15°, 75°, 105°, 120°, 135°, and 150°.

23-47 A hexagonal nut for an $\frac{11}{16}$-in. bolt is $1\frac{1}{2}$ in. across the flats. See Fig. 23-29. Find the diagonal, or the distance across the corners, of such a nut.

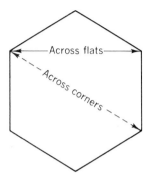

Fig. 23-29. Exercise 23-47.

23-48 What is the distance across the corners of a hexagonal nut that is 20 mm on a side? What is the distance across the flats of this nut?

23-49 In a standard hexagonal bolt nut the distance across the flats is given by the formula $F = 1.5D + \frac{1}{8}$, where F is the distance across the flats and D is the diameter of the body of the bolt. Find the distance across the flats and across the corners of a hexagonal nut for a bolt $1\frac{3}{4}$ in. in diameter.

23-50 To what diameter should a piece of stock be turned so that it can be milled to a hexagon and will be 4.445 cm across the flats?

23-13 PRACTICAL APPLICATION: SCREWS AND SCREW THREADS

Screws are used for fastening. For convenience in entering wood, a wood screw is tapered, but a machine screw is the same diameter over its length and is often fitted with a nut.

The *thread* of a screw is formed by cutting a spiral groove around the body of the screw. The threads on fasteners and those used in machines to communicate motion are made by a cutting process in which a lathe and a single-pointed cutting tool of the proper size are used. Threads can be formed of various sizes and of several shapes, depending on the size and intended use of the screw.

23-14 PITCH AND LEAD OF SCREW THREADS

The *pitch* of a screw thread is the distance from the center of the top of one thread to the center of the top of the next.

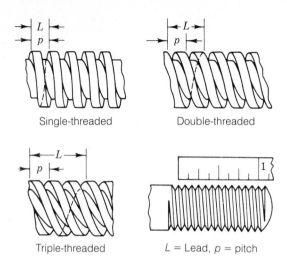

Single-threaded Double-threaded

Triple-threaded   *L* = Lead, *p* = pitch

Fig. 23-30. Pitch and lead.

The *lead* of a screw thread is the distance the screw will move for each complete turn of the screw.

In a *single-threaded screw* there is only one groove cut spirally any number of times around the body of the screw. In a *double-threaded screw* there are two grooves running side by side spirally around the screw. A *triple-threaded screw* has three grooves running around the screw. For a single-threaded screw, the pitch and the lead are equal. For a double-threaded screw, the lead is twice the pitch. For a triple-threaded screw, the lead is 3 times the pitch.

The meaning of *pitch* and *lead* and the relations between them are illustrated in Fig. 23-30. The pitch can be easily determined by placing a scale on the screw, as shown in Fig. 23-30, and counting the number of threads for 1 in. If the screw is single-threaded, the number of threads in 1 in. is the number of times that the screw must be turned around to advance 1 in. This is the pitch. Thus, a 10-pitch single-threaded screw requires 10 turns to advance 1 in.

23-15 KINDS OF SCREW THREADS

In the United States several different kinds of screw threads are in use. Two of the common threads are (1) the *sharp V thread*, or *common V thread* and (2) the *U.S. standard screw thread*. Computations concerning screw threads frequently require finding the depth of the thread to determine the strength of a bolt. In this computation, the equilateral triangle is involved.

23-16 METRIC THREADS

A *metric thread* is like the U.S. standard thread in form, but its dimensions are in millimeters. There are two standard metric threads, the international standard thread and the French standard thread. These two threads differ only slightly.

Because the standard measures are different, a nut with a metric thread will not fit a bolt with a U.S. standard thread, and a separate set of dies is necessary for cutting each kind of thread. As international trade expands, metric threads and metric fasteners are becoming more common.

■ EXERCISES

23-51 A jackscrew is single-threaded and has three threads to an inch. How far does it advance in $\frac{1}{3}$ of a turn? How many turns will advance it 6 in.?

23-52 What is the lead of a single-threaded screw that advances $\frac{1}{2}$ in. in 8 turns? What is the pitch?

23-53 What is the lead of a double-threaded screw that has 18 threads to an inch? What is the pitch?

23-54 According to the Franklin Institute standards for the dimensions of bolts and nuts, a $\frac{5}{8}$-in. bolt has 11 threads per inch. What is the pitch? What is the lead if the bolt is single-threaded? How many full turns of the nut will it take to advance the bolt $2\frac{1}{4}$ in.?

23-55 In a special threaded screw for a screw-power stump puller, the screw is double-threaded with a pitch of $\frac{11}{16}$ in. How many turns of the nut are required to lift the stump $4\frac{1}{3}$ ft?

23-56 The lead screw on the table of a milling machine has a double thread with a pitch of $\frac{1}{4}$ in. How many inches per minute is the feed if the lead screw is making 4 rpm?

23-57 A $4\frac{1}{4}$-in. bolt has $2\frac{7}{8}$ threads per inch. What is the pitch? What is the lead if triple-threaded?

23-58 How many turns must be made with a triple-threaded screw having $4\frac{3}{4}$ threads to the inch to have it advance a distance of 3 in.?

THE CIRCLE

24-1 INTRODUCTION TO THE CIRCLE

The circle has been called the perfect geometric form. It is probably the most important of all geometric forms.

The importance of a geometric form in the study of practical mathematics is determined by the frequency of its application, and the circle occurs more frequently in applied mathematics than any other geometric form. Wires, tanks, pipes, steam boilers, pillars, and many other forms are based on the circle. In this chapter we will consider the more useful facts concerning the circle and some of their applications.

24-2 DEFINITIONS OF THE CIRCLE

A circle is a plane curve with all points the same distance from a fixed point that is the *center;* the common distance of the points on the curve from the center is the *radius.* The region bounded by (enclosed within) the circle is generally referred to as a circle or the area of a circle. To avoid confusion, the curve (of the circle as defined above) is referred to as the *perimeter* or *circumference* of the circle. These terms denote both the curve itself and its length.

A *diameter* of a circle is any line drawn from one side of the circumference to the other and passing through the center. A diameter divides both the area and the circumference into two equal parts called *semicircles.* Any line drawn from the center to the circumference is a *radius.* (*Note:* The plural form of "radius" is "radii.")

Any portion of the circumference is an *arc.* In Fig. 24-1, *BC* and *DME* are arcs. Any two points *B* and *C* on the circumference divide it into two arcs. Unless the two points are on the same diameter, one of the arcs will be smaller than the other; unless otherwise specified, the arc *BC* is understood to mean the smaller of the two arcs. The length of the arc *BC* depends on the radius of the circle and the angle *BOC*.

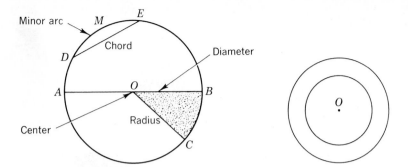

Fig. 24-1. Circle and its parts.

Fig. 24-2. Concentric circles.

A *chord* is a line segment connecting any two points on the circle; in Fig. 24-1, *DE* is a chord. The chord is said to *subtend* its arc. The chord *DE* subtends the arc *DME*. The area bounded by an arc and a chord is a *segment;* in Fig. 24-1, the area *DME* is a segment. The area bounded by two radii and an arc is a *sector*. A sector is an area shaped like a wedge of pie, as is the area *BOC* in Fig. 24-1.

Circles are said to be *concentric* when they have a common center, like those in Fig. 24-2.

A polygon is *inscribed* in a circle when it is inside the circle and has its vertices on the circumference. The circle is then *circumscribed about the polygon*. The polygon *ABCDEF* in Fig. 24-3 is inscribed in the circle.

A line is *tangent* to the circle when it touches but does not cut through the circumference. In Fig. 24-4, *AT* is tangent to the circle at point *T*. The point (here, *T*) where a tangent line touches the circle is the *point of tangency*. A radius drawn at this point is perpendicular to the tangent.

A polygon is *circumscribed about a circle*, or a circle is *inscribed in a polygon*, when the sides of the polygon are all tangent to the circle. In Fig. 24-5, the polygon *ABCDEF* is circumscribed about the circle.

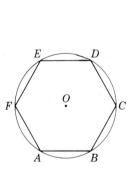

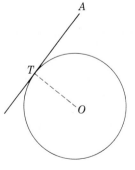

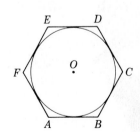

Fig. 24-3. Polygon inscribed in a circle.

Fig. 24-4. Line tangent to a circle.

Fig. 24-5. Polygon circumscribed around a circle.

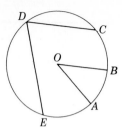

Fig. 24-6. Central angle (*AOB*) and inscribed angle (*CDE*).

A *central angle* is an angle whose vertex is at the center of the circle. In Fig. 24-6, angle *AOB* is a central angle. An *inscribed angle* is an angle with its vertex on the circumference of the circle; in Fig. 24-6, angle *CDE* is an inscribed angle. An inscribed or a central angle is said to *intercept* the arc between its sides. The sides of the angle *AOB* intercept the arc *AB*, and the sides of the angle *CDE* intercept the arc *CE*.

Two circles that are tangent to the same straight line at the same point are *tangent circles*. When both circles are on the same side of the common tangent, they are *tangent internally*. When they are on opposite sides of the common tangent, they are tangent *externally*. In Fig. 24-7, circles with centers *A* and *B* are tangent internally and circles with centers *C* and *D* are tangent externally.

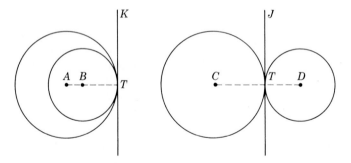

Fig. 24-7. Internally tangent circles (*A* and *B*) and externally tangent circles (*C* and *D*).

24-3 PROPERTIES OF THE CIRCLE

In geometry, facts about circles and their relations to straight lines, angles, and polygons can be proved. These facts are *properties* of the circle. The student should be able to see the truth of these properties and understand them. Become familiar with these facts and satisfy yourself that they are true by making actual drawings and measurements.

Rule: In the same circle or in equal circles, chords that are the same distance from the center are equal.

Rule: A radius drawn to the center of a chord is perpendicular to the chord and bisects the arc that the chord subtends.

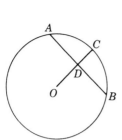

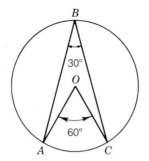

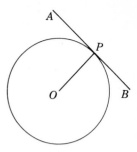

Fig. 24-8. Radius *OC* bisects chord *AB*.

Fig. 24-9. Central angle *AOC* and inscribed angle *ABC* intercept the same arc.

Fig. 24-10. Radius *OP* is drawn to point of contact of tangent *AB*.

In Fig. 24-8, the radius *OC* is drawn through the center of the chord *AB*; it is perpendicular to *AB* and makes arc *AC* = arc *CB* and *AD* = *DB*. This can be shown by measuring the parts in the drawing.

Rule: The central angle which intercepts an arc is double any inscribed angle that intercepts the same arc.

In Fig. 24-9, the central angle *AOC* = 60° and the inscribed angle *ABC* is measured and found to equal 30°. Draw several figures like this, and measure the angles with a protractor; this will give you a better understanding of these properties.

Rule: A radius drawn to the point of contact of a tangent is perpendicular to the tangent.

In Fig. 24-10, *OP* is drawn to the point of contact of tangent *AB*. If the angles are measured, they will be found to be right angles. Thus, the radius is perpendicular to the tangent.

Rule: If two circles are tangent to each other, the straight line joining their centers passes through the point of tangency.

▨ EXERCISES

24-1   Draw circles with a compass, then draw the following:
   (a) A radius
   (b) A diameter
   (c) A tangent
   (d) A chord 5 cm long
   (e) A central angle of 90°
   (f) An inscribed angle of 90°
   (g) A chord subtending an arc of 120°; of 45°
   (h) An inscribed square; a circumscribed square
   (i) An inscribed triangle; a circumscribed triangle
   (j) An inscribed hexagon; a circumscribed hexagon
   (k) An inscribed angle of 60°
   (l) A central angle of 60°

24-2 In a circle of radius 75 mm, draw a radius and then draw four lines perpendicular to this radius so that the four lines divide the radius into three equal parts. Indicate a diameter, a chord, and a tangent in this drawing.

24-3 Draw two concentric circles of radii 2 in. and 1 in. Draw tangents to the smaller circle. Are the chords thus formed all of equal length?

24-4 Draw two circles that are tangent externally. If one of these circles has a radius of 2 in. and the other has a radius of 1 in., how long is the longest line which passes through the point of tangency and is contained within the two circles?

24-4 RELATIONS BETWEEN DIAMETER, RADIUS, AND CIRCUMFERENCE

If the diameter and the circumference of a circle are measured and the length of the circumference is divided by the length of the diameter, the result will be nearly 3.142. Thus, 3.142 is the approximate ratio of the circumference of a circle to its diameter. This ratio cannot be expressed exactly in numbers, because both the diameter and the circumference are measurements which cannot be considered exact. This ratio is represented by the Greek letter π (pi). The value of π to five decimal places is 3.14159. To understand the number π, wrap a string around a round object, measure its length and the diameter of the object, and then divide the length of the diameter into the length of the string. The quotient will equal approximately 3.142, which is considered π. Most hand-held calculators have a special button which will give π accurate to eight places.

Because of this relationship, if the diameter, the radius, or the circumference is given, the other two dimensions can be found.

Rule: *The radius equals one-half the diameter; or, the diameter equals twice the radius.*

$$r = \frac{d}{2} \quad \text{or} \quad d = 2r$$

Rule: *The circumference equals the diameter times π.*

$$C = d\pi$$

Rule: *The diameter equals the circumference divided by π.*

$$d = \frac{C}{\pi}$$

Rule: *The radius equals the circumference divided by 2π.*

$$r = \frac{C}{2\pi}$$

24-5 AREA OF A CIRCLE

The formula for finding the area of a circle when the radius, diameter, or circumference is given can be proved. The following explanation will

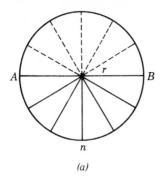
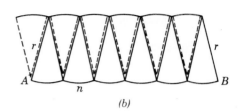

(a)   (b)

Fig. 24-11. Area of a circle.

show the reasonableness of the rule. Consider Fig. 24-11*a*, and suppose that one semicircle is cut, as indicated, from the center nearly to the circumference in each direction. Then the semicircle is spread out as in Fig. 24-11*b*. The length *AB* is the half-circumference. Suppose the other half of the circle is cut in the same way and fitted into the first half, as shown by the dashed lines in Fig. 24-11*b*. If we made a large number of cuts, the figure formed will approximate a rectangle whose length is equal to one-half the circumference and whose width is equal to the radius. Thus the following rule:

Rule: *To find the area of a circle, multiply one-half the circumference by the radius.*

This may also be expressed in either of the following ways:

Rule: *The area of a circle equals π times the square of the radius;* or, *the area of a circle equals one-fourth π times the square of the diameter.*

If A = area, C = circumference, d = diameter, and r = radius, these rules are expressed by the following formulas:

$$A = \frac{Cr}{2}$$

$$A = \pi r^2 \quad \text{or} \quad 3.142r^2$$

$$A = \frac{\pi d^2}{4} \quad \text{or} \quad A = 0.7854d^2$$

The most efficient way to solve problems concerning π is to use a hand-held calculator. If the calculator does not have a π key, use 3.14159 or the formula $A = 0.7854d^2$ when finding the area of a circle. Note the convenient location of 7, 8, 5, and 4 on the calculator.

From the formula $A = \pi r^2$, if the area of the circle is given, the radius equals the square root of the quotient of the area divided by π. Thus we have the formula

$$r = \sqrt{\frac{A}{\pi}}$$

From the formula $A = \pi d^2/4$, we derive

$$d = \sqrt{\frac{A}{0.7854}}$$

EXAMPLE 24-1 Find the radius of a circle whose area is 28 ft².

Solution Using the formula $r = \sqrt{A\pi}$ and substituting the given numbers.

$$r = \sqrt{\frac{28}{3.142}}$$

$$= \sqrt{8.9115}$$

Radius = 2.985 ft

24-6 HISTORICAL NOTE CONCERNING PI

From ancient times down through the centuries, many efforts have been made to find an exact numerical value for π. The people of Babylon and the early Hebrews (I Kings 7:23) used 3 as the value of π. Perhaps the earliest known attempt to find the area of a circle accurately was made by Ahmes of Egypt about 1700 B.C. His method gave a value for π which, in our notation, equals 3.1604.

Archimedes (287–212 B.C.) found that π must be between $3\frac{1}{7}$ and $3\frac{10}{71}$. Rudolph van Ceulen of Holland (1540–1610) found π to 35 decimal places, devoting 15 years to the computation. In 1761, it was proved that π cannot be expressed as a common fraction or as a repeating decimal.

The value of π to 42 decimal places is

$\pi = 3.141\ 592\ 653\ 589\ 793\ 238\ 462\ 643\ 383\ 279\ 502\ 884\ 197\ 169$

In 1949 an Electronic Numerical Integrator and Computer (ENIAC) computed π to 2037 decimal places in 70 hours; in 1958 an IBM 704 calculated it to 10,000 places in 100 minutes; and computers today can be programmed to calculate π to any number of places.

24-7 AREA OF A RING (ANNULUS)

To find the area of a ring (annulus), which is the area between the circumferences of two concentric circles, subtract the area of the small circle from the area of the large circle.

If A and a, R and r equal the areas and the radii of the two circles, and A_r is the area of the ring (annulus) (Fig. 24-12a), then

$$A_r = A - a = \pi R^2 - \pi r^2 = \pi(R^2 - r^2) = \pi(R + r)(R - r)$$

See Fig. 24-12a. This formula $A_r = \pi(R + r)(R - r)$ may be stated in words as follows:

Rule: *To find the area of a ring (annulus), multiply the product of the sum and the difference of the two radii by π.*

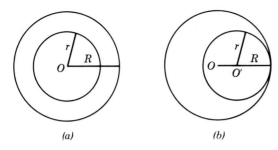

Fig. 24-12. Area of an annulus.

EXAMPLE 24-2 Find the area of a ring of inner diameter 8 in. and outer diameter 12 in.

Solution Using the formula $A_r = \pi(R + r)(R - r)$,

$$A_r = 3.142(6 + 4)(6 - 4),$$

$$= 3.142(10)(2)$$

$$= 3.142(20) = 62.832$$

$$= 62.84 \text{ in.}^2$$

Note that the rule holds even when the circles are not concentric; the circles may be, for example, as in Fig. 24-12*b*.

▨ EXERCISES

24-5 Choose a suitable scale (for example, 1 ft = $\frac{1}{2}$ in.) and draw the following when d = diameter, C = circumference, and A = area. Find the quantities indicated, first estimate the quantities and then compute:
(a) Given d = 2 ft, find C and A.
(b) Given d = 3 ft, find C and A.
(c) Given d = 4.5 m, find C and A.
(d) Given d = 20 ft, find C and A.
(e) Given d = 12 km, find C and A.
(f) Given C = 21.415 93 cm, find d and A.
(g) Given C = 113.0973 ft, find d and A.
(h) Given A = 201.0619 ft^2, find d and C.
(i) Given A = 3.141 593 ft^2, find d and C.

24-6 Find the area of the cross section of a 1-in. rod.

Note: The term "cross section" means the area of the end of the rod formed when a piece is cut squarely off.

24-7 Find the area of the cross section of a pipe with an outer diameter of 40 cm and an inner diameter of 36 cm.

24-8 Find the area of the cross section of a water main whose outer diameter is 18 in. if the cast iron pipe is 1 in. thick.

Use the following proportion for Exercises 24-9 to 24-13:

$$\frac{\text{intercepted arc length}}{\text{circumference}} = \frac{\text{angle at center}}{360}$$

24-9 If the angle is 15° and the radius is 10 in., find the length of the intercepted arc.

24-10 If the angle is 30° and $r = 30$ km, find the intercepted arc length.

24-11 If the angle is 45° and the intercepted arc length $= \pi$ ft, find r.

24-12 If the angle is 270° and the intercepted arc length $= 3\pi$ ft, find r.

24-13 In a circle with a 6-in. radius, a central angle of 30° intercepts what arc length? What arc length is intercepted by an inscribed angle of 90°?

24-14 The diameter of the safety valve in a boiler is 3 in. Find the total pressure tending to raise the valve when the pressure of the steam is 120 lb/in.^{2}.

24-15 If the diameter of a piston is 105 mm, find the total pressure on the piston when the pressure is 0.175 kg/mm^2.

24-16 A circular sheet of steel 2 ft in diameter increases in diameter by 0.005 when the temperature is increased by a certain amount.
(a) Find the increase in the area of the sheet.
(b) Find the percent of increase in area.

24-17 A 15-cm water pipe can carry how many times more than 2-cm pipe?

24-18 Find the ratio of the cross-sectional areas in the following wires.
(a) A 1-in. wire and a 0.5-in. wire
(b) A $\frac{1}{8}$-in. wire and a $\frac{1}{2}$-in. wire
(c) A 2.6-mm wire and a 25-mm wire
(d) An s-in. wire and an S-in. wire

24-19 How many 3-in. steam pipes could open off an 18-in. steam pipe?

24-20 In a steel plate 3 ft by $2\frac{1}{2}$ ft, 26 round holes are cut out, each $1\frac{3}{4}$ in. in diameter. Find the area of steel remaining.

24-21 If the drive wheels of a train engine are 66 in. in diameter, find the rpm required to go 40 mph.

Solution: Find the distance the wheels must travel to go 40 miles and divide by the distance traveled in 1 revolution.

$$\frac{40 \times 5280 \times 12}{60 \times 66 \times 3.1416} = 203.7 \text{ rpm}$$

Note: *Use cancellation and a calculator to solve. The quantities above the line convert 40 miles to inches; the quantities below the line convert from inches per hour to inches per minute times the distance the wheel would go per revolution.*

24-22 The circumference of a drive wheel of a train engine is 16 ft. If a train is to hold a speed of 60 mph, and if there is no slipping, how many revolutions per minute must the wheel make?

24-23 A diesel locomotive wheel 5 ft in diameter made 18,000 revolutions in a distance of 50 mi. What distance was lost due to the slipping of the wheel?

24-24 There are 32 climbing hoops on the circumference of a circular tank weighing 3 lb per linear foot on a tank 15 ft in diameter. Find the weight of the iron hoops.

24-25 The side of a square is 50 cm. Find the following:
(a) The circumference and the area of the inscribed circle.
(b) The circumference and the area of the circumscribed circle.
(c) The area of the ring formed by the two circles in (a) and (b).

24-26 What is the waste in cutting the largest possible circular plate from a piece of sheet steel 17 by 20 in.?

24-27 The minute hand of a tower clock is 2 m long. What distance will the extremity move over in 36 min?

24-28 The area of a square is 49 m². Find the length of the circumference and the area of the circle inscribed in this square.

24-29 Find the size of the largest square timber which can be cut from a log 24 in. in diameter.

24-30 Using 4000 mi as the radius of the earth, find the length in miles of 1° of arc on the equator (the central angle is 1°).

24-31 Using 4000 mi as the radius of the earth, find the length in miles of an arc of 1° (a) on the parallel of 45° north; (b) on the parallel of 60° north.

Suggestion: (See Fig. 24-13.) For the parallel of 45° north, CB is the radius. But CB = OC because the triangle OCB is a right triangle with two equal angles.

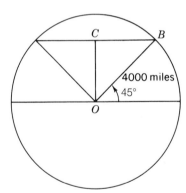

Fig. 24-13. Exercise 24-31.

24-32 A 10-in. pipe is to be branched off into two equal pipes. What must be the diameter of each of these pipes if the two pipes are to equal the 10-in. pipe in cross-sectional area?

24-33 If an automobile wheel has a diameter of 28 in., how many revolutions per minute will the wheel make when the automobile is going 60 mph?

24-34 Find the length in feet of the arc of contact of a belt with a pulley if the pulley is 3 ft 6 in. in diameter and the angle of contact is 210°.

24-35 As in Exercise 24-34, find the length of the arc of contact if the angle of contact is 120° and the diameter of the pulley is 40 cm.

24-8 AREA OF A SECTOR

The area of a sector of a circle equals a fraction of the area of the circle determined by the central angle divided by 360. Thus, if the angle of the sector is 90°, the area of the sector is $\frac{90}{360}$, or $\frac{1}{4}$ the area of the circle.

EXAMPLE 24-3 Find the area of a sector of 60° in a circle of radius 10 in.

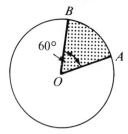

Fig. 24-14. Example 24-3: Finding the area of a sector.

Solution The sector AOB has a central angle of 60°. Its area equals $\frac{60}{360} \times \pi r^2$. If the radius is 10 in., the area of the sector is

$$A = (\tfrac{1}{6})(3.1416)(10^2) = 52.36 \text{ in.}^2$$

If θ (the Greek letter theta) equals the number of degrees in the angle of the sector, the area of the sector is given by the formula

$$A = \left(\frac{\theta}{360}\right)\pi r^2 \quad \text{or} \quad \left(\frac{\theta}{360}\right)(0.7854d^2)$$

24-9 PROPERTIES OF A SEGMENT

In applied problems, it is necessary to find the radius of the circle when we know a chord and the height of a segment, for example, chord AB and the height of the segment DC in Fig. 24-15. Let r = the radius of the circle, h = the height of the segment, and w = the length of the chord. Now note that ODB is a right triangle; the hypotenuse is $OB = r$ and the

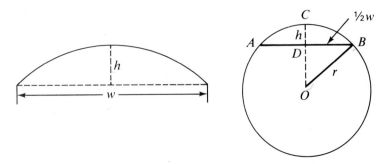

Fig. 24-15. Finding the radius.

other two sides are $OD = r - h$ and $DB = \frac{1}{2}w$. Therefore,

$$\left(\frac{w}{2}\right)^2 + (r - h)^2 = r^2$$

or

$$\left(\frac{w}{2}\right)^2 + r^2 - 2rh + h^2 = r^2$$

Simplifying,

$$\left(\frac{w}{2}\right)^2 + h^2 - 2rh = 0$$

Solving this, in turn, for r and w, and solving the first equation above for h, we obtain the following formulas:

$$r = \frac{\left(\frac{w}{2}\right)^2 + h^2}{2h}$$

$$h = r - \sqrt{r^2 - \left(\frac{w}{2}\right)^2}$$

$$w = 2\sqrt{h(2r - h)}$$

EXAMPLE 24-4 If the chord of the segment of a circle is 5 ft 6 in. (66 in.) and the height of the segment is 10 in., find the radius of the circle (Fig. 24-16).

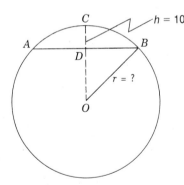

$AB = w = 66''$
$DB = \frac{1}{2}w = 33''$

Fig. 24-16. Example 24-4: Finding the radius.

Solution Substitute the given values into the correct formula and evaluate. Remember to have all units in the same dimension.

$$r = \frac{\left(\frac{w}{2}\right)^2 + h^2}{2h}$$

$$r = \frac{(33)^2 + 10^2}{2 \times 10}$$

Therefore,

$$\text{Radius} = \frac{1089 + 100}{20}$$

$$= 59.45 \text{ in.}$$

■ EXERCISES

24-36 An arched doorway has a width of 6 ft ($w = 6$ ft), and the height of the arch is $1\frac{1}{2}$ ft ($h = 1\frac{1}{2}$ ft). Find the radius of the arch. (See Fig. 24-15.)

24-37 Find the height of a segment cut off from a circle 8 ft in diameter by a 3-ft chord.

24-38 A segment of a circle 6 ft in radius has a height of 3 ft; find the length of the chord. Give the result to the nearest $\frac{1}{8}$ in.

24-39 In a circle of radius 5 m, there is a chord 6.5 m in length. Find the height of the segment.

24-40 A segment of a circle cut off by a chord 4 ft 6 in. in length has a height of 1 ft 10 in. Find the radius of the circle.

24-41 Find the radius of a circle in which a chord of 10 ft has a middle ordinate of 3 in.

24-42 In Fig. 24-17, W is a wall with a round corner where a molding is to be placed. Given the dimensions, find the radius of the circle of which arc ANB is a part.

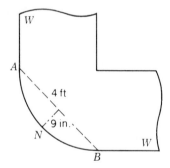

Fig. 24-17. Exercise 24-42.

24-10 AREA OF A SEGMENT

It is evident that the area of segment ABD (the shaded portion) in Fig. 24-14 equals the area of sector AOB minus the area of triangle AOB. Trigonometry is required to find the area of a triangle when we have only two sides and an angle or to find the area of a sector when we have only the chord and the angle is unknown. However, if the angle and the lines in the segment are given, the area of the sector and the triangle can be found. The difference between these areas is the area of the segment.

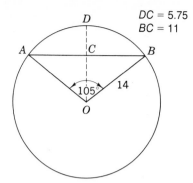

DC = 5.75
BC = 11

Fig. 24-18. Example 24-5: Area of a segment.

EXAMPLE 24-5 In a circle of radius 14 in., find the area of a segment subtending an angle at the center of 105°.

Solution The dimensions are given in Fig. 24-18.

The area of sector $OADB = \frac{105}{360}$ of the area of the circle. Thus

$$\text{Area of segment} = \text{area of sector} - \text{area of triangle}$$
$$\text{Area of sector} = (\tfrac{105}{360})(3.1416)(14)^2 = 179.67 \text{ in.}^2$$
$$\text{Area of triangle } OAB = (\tfrac{1}{2})(22)(8.25) = 90.75 \text{ in.}^2$$
$$\text{Area of segment} = 179.67 \text{ in.}^2 - 90.75 \text{ in.}^2$$
$$= 88.92 \text{ in.}^2$$

Many formulas are used for finding the *approximate* area of a segment. Two of the common formulas are:

$$A = \frac{2hw}{3} + \frac{h^3}{2w}$$

$$A = \frac{4h^2}{3}\sqrt{\frac{2r}{h} - 0.608}$$

In these formulas, A = area of the segment, h = height, w = width, and r = radius of the segment circle.

If the height of the segment is less than one-tenth the radius of the circle, the first formula may be shortened to

$$A = \frac{2hw}{3}$$

When engineers need to find the area of a segment and the height is nearly equal to the radius, they may use an "approximate" method. In Fig. 24-19, to find the area of the segment CnD, find the area of the semicircle AnB and then the area of the part $ACDB$ which is considered a rectangle. The approximate area of the segment is the difference between these.

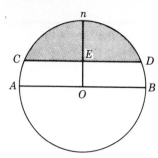

Fig. 24-19. Finding the area of segment *CnD*.

EXAMPLE 24-6   Find the area of the segment whose chord is 10 ft and whose height is 1.5 ft.

Solution   Substitute into the correct formula and evaluate,

$$A = \frac{2hw}{3} + \frac{h^3}{2w}$$

$$= \frac{2}{3}(1.5)(10) + \frac{1.5^3}{2 \times 10} = 10.168 \text{ ft}^2$$

By the second formula,

$$A = \frac{4h^2}{3} \sqrt{\frac{2r}{h} - 0.608}$$

First find *r* by formula:

$$r = \frac{\left(\frac{w}{2}\right)^2 + h^2}{2h}$$

$$r = \frac{5^2 + 1.5^2}{2 \times 1.5} = 9.083$$

$$A = \frac{4}{3}(1.5^2) \sqrt{\frac{2 \times 9.083}{1.5} - 0.608}$$

$$A = 10.175 \text{ ft}^2$$

▨ EXERCISES

24-43   Find the areas of the following sectors of circles (*first draw the figure, then estimate the area, then compute the area*):
(a) Radius = 5 in., angle of sector = 30°
(b) Radius = 10 km, angle of sector = 120°
(c) Radius = 20 in., angle of sector = 225°
(d) Radius = 43.17 ft, arc of sector = 49.79 ft
(e) Radius = 100 m, arc of sector = 100 m

24-44   Find the areas of the following segments of circles:
(a) Chord = 20 m, height = 3 m
(b) Chord = 21 in., height = 1 in.

(c) Radius = 10 ft, height = 9 ft. Use the "approximate method" discussed on the previous page.

24-45 How many gallons of water will go through a circular culvert 6 ft in diameter in 1 h if the depth of the water is 2 ft and the water is moving at 20 ft/min? (1 ft^3 = 7.5 gallons)

24-11 PROPERTIES OF THE ELLIPSE

An *ellipse* is a curved line such that the sum of the distances of any point in it from two fixed points is constant.

In Fig. 24-20, F and F' are the two fixed points. The fixed points are *foci* (singular, *focus*). The point O is the *center* of the ellipse, NA is the *major axis*, and MB is the *minor axis*. OA and OB are the *semiaxes*.

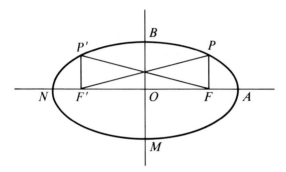

Fig. 24-20. Ellipse and its parts.

The curved line of an ellipse is so traced that the sum of the distances of any point on the ellipse to foci F and F' is a constant.

The following method may be used for drawing an ellipse: Tie the ends of a length of string together to form a loop. The string should be about twice as long as the major axis of the required ellipse. On a piece of cardboard place straight pins at F and F'. The distance between F and F' should be somewhat shorter than the length of the major axis. Place the loop of string so that it encloses F and F'. Place a pencil inside the loop and move the pencil in a curve, keeping the string taut. The pencil will trace an ellipse.

The ellipse is closely related to the circle; sometimes it is called a "flattened circle." For example, if a circular pipe is cut on a slant, the section is a true ellipse.

24-12 USES OF THE ELLIPSE

The ellipse is used in practical applications as well as in theoretical mathematics. Ellipses frequently occur in nature. Kepler (1571–1630) stated that the orbits of the planets are ellipses, and Newton (1642–1727) showed that by the law of gravitation the orbits must be ellipses.

In architecture the elliptic arch is frequently used because of the beauty of its form. Some famous structures have been built in the form of an ellipse; for example, the Colosseum in Rome. The ellipse is also used in bridge structures; many famous stone-arch bridges have elliptical arches. The elliptical structure gives them maximum strength.

In machinery, elliptical gears are often used in situations where changeable rates of motion are desired, as in shapers, planers, and slotters where the cutting speed is less than the return motion.

In the study of electricity and mechanics, the ellipse is also frequently encountered.

24-13 AREA OF AN ELLIPSE

The area of a circle is given by the formula $A = \pi r^2$, which may be written $A = \pi rr$. If a circle is flattened, it will take the form of an ellipse; and the semiaxes of such an ellipse (like OA and OB of Fig. 24-20) will be lengthened and shortened radii. If a stands for AO and b for OB, it can be proved that the area of the ellipse can be found by substituting ab for rr in the formula for the area of the circle. This gives the following formula for the area of an ellipse:

$$A = \pi ab$$

where a = semimajor axis, or $\frac{1}{2}$ major axis
b = semiminor axis, or $\frac{1}{2}$ minor axis

EXAMPLE 24-7 Find the area of an ellipse whose two axes are 30 and 26 ft.

Solution Use the formula $A = \pi ab$; substitute the given values and evaluate.

$$A = (3.1416)(15)(13) = 612.612 \text{ ft}^2$$

24-14 CIRCUMFERENCE OF AN ELLIPSE

Although the area of an ellipse is easily found when the major and minor axes are given, the circumference, or perimeter, of an ellipse is difficult to determine accurately. Various approximate formulas are given for finding the circumference of an ellipse. If the ellipse is very nearly the shape of a circle (that is, if its major and minor axes are nearly equal), then

$$P = \pi(a + b) \tag{1}$$

where P = perimeter or circumference
a = semimajor axis
b = semiminor axis

When the ellipse differs considerably from a circle (when there is considerable difference between the major and minor axes), either of the following formulas may be used:

$$P = \pi[\tfrac{3}{2}(a + b) - \sqrt{ab}] \tag{2}$$

$$P = \pi\sqrt{2(a^2 + b^2)} \tag{3}$$

The result from formula 2 is too small, and the result from formula 3 is too large. However, the mean, or average, of these results is approximately correct.

EXAMPLE 24-8 Find the circumference of an ellipse whose major axis is 18 in. and whose minor axis is 6 in.

Solution Here $a = 9$ and $b = 3$. Using the three formulas for P we obtain

From formula 1: $P = 3.1416(9 + 3) = 37.699$ in.

From formula 2: $P = 3.1416[\tfrac{3}{2}(9 - 3) - \sqrt{9 \times 3}] = 40.225$ in.

From formula 3: $P = 3.1416\sqrt{2(9^2 + 3^2)} = 42.148$ in.

Formula 2 is the best to use when the two axes are not nearly equal.

Note: *The mean of the results obtained with formula 1 and formula 3 is 39.974 in., which is approximately the result obtained from formula 2.*

▨ EXERCISES

24-46 A garden is an ellipse whose major axis is 22 m and whose minor axis is 18 m. Find the area of the garden.

24-47 An elliptic pipe has a major axis of 16 in. and a minor axis of 10 in. Find the diameter of a circular pipe that has the same area of cross section.

24-48 A horizontal steam pipe 8 in. in diameter is to pass through a $\tfrac{1}{2}$-pitch roof. If the direction of the pipe is perpendicular to the sides of the building, what are the dimensions of the ellipse that must be cut in the roof?

Suggestion: *This hole will be an ellipse with a minor axis of 8 in., and the major axis will be $8\sqrt{2}$.*

24-49 A 10-in. smoke pipe is cut off on an angle of 45°. Use the average of the formulas $P = \pi(a + b)$ and $P = \pi[\tfrac{3}{2}(a + b) - \sqrt{ab}]$ to find the circumference of the section.

24-50 The Colosseum in Rome is in the form of an ellipse whose major axis is 620 ft and whose minor axis is 510 ft. Find its area and its circumference.

24-51 Given two joining pipes 12 in. and 8 in. in diameter, find the diameter x of the continuation. (See Fig. 24-21.)

Note: *If the joining pipes have radii of R and r, derive the formula for x as a function of R and r.*

24-52 Determine by using a carpenter's square, how to find the diameter of a circle having the same area as the sum of the areas of any number of given circles.

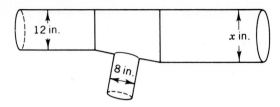

Fig. 24-21. Exercise 24-51.

24-53 If an arc of a circle is equal in length to the radius, what is the value of the central angle that it measures?

Suggestion: $\dfrac{central\ angle}{360°} = \dfrac{intercepted\ arc}{circumference}$

Note: This central angle is a radian. *It will be discussed further in Sec. 26-6.*

24-54 A regular hexagon with a perimeter of 42 m is inscribed in a circle. Find the area of the circle.

24-55 The maximum circumferential velocity of cast-iron flywheels is 80 ft/s. Find the maximum number of revolutions per minute for a cast-iron flywheel 8 ft in diameter.

24-56 An emery wheel may have a circumferential velocity of 5500 ft/min. Find the number of revolutions per second that an emery wheel 9 in. in diameter may make.

24-57 The peripheral speed of an abrasive stone of heavy grain should not exceed 47 ft/s. Find the number of revolutions per minute that an abrasive stone 3 ft in diameter may turn.

24-58 Four of the largest possible equal-sized pipes are enclosed in a square box 18 in. on an edge. What percent of the space do the pipes occupy?

24-59 Three circles are enclosed in an equilateral triangle. If the circles are 10 cm in radius, find the sides of the triangle.

24-60 The strength of a hemp rope is determined by the formula

$$W = 1420A$$

where W = strength, lb
A = area of cross section of the rope, in.^{2}

Determine the diameter of a hemp rope to support a weight of 2400 lb.

24-61 Find the diameter of a single pipe having the same carrying capacity as three pipes of diameters $1\frac{1}{2}$ cm, $2\frac{1}{4}$ cm, and $3\frac{1}{2}$cm.

24-62 The following rule is used by sheet-metal workers. Divide the radius AO (see Fig. 24-22) into four equal parts, and place one of these lengths on the axis from A to C and another from B to D, the ends of two perpendicular diameters. Connect C and D. This gives the side of a square of the same area as the circle. Show the percentage of error if this method is incorrect.

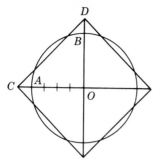

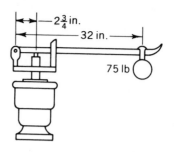

Fig. 24-22. Exercise 24-62.   **Fig. 24-23.** Exercise 24-63.

24-63   The stem of a 4-in. safety valve (see Fig. 24-23) is $2\frac{3}{4}$ in. from the fulcrum. Suppose that the valve will blow when the gage reads 7 lb without any weight on the lever (7 lb/in.² on the valve overcomes the weight of valve and lever). At what pressure would it blow with a weight of 75 lb 32 in. from the fulcrum?

24-64   What weight would be required to allow the valve in Exericse 24-63 to blow off at 80 lb?

24-65   If in Exercise 24-63 the original weight of 34 kg is used, what distance from the fulcrum should it be placed to allow the valve to blow off at 36 kg?

24-66   An approximate formula for determining the number of inscribed equal tangent circles (Fig. 24-24) is

$$N = 0.907\left(\frac{D}{d} - 0.94\right)^2 + 3.7$$

where N = number
D = diameter of enclosing circle
d = diameter of inscribed circles

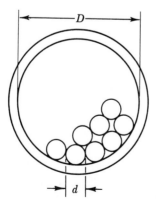

Fig. 24-24. Exercise 24-66.

Apply the formula to determine how many pipes $\frac{1}{2}$ in. in diameter can be placed inside a 5-in. pipe.

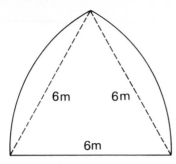

Fig. 24-25. Exercise 24-69.   **Fig. 24-26.** Exercise 24-73.

24-67 How many steel balls 0.4 in. in diameter can rest at the bottom of a closed pipe $2\frac{1}{2}$ in. in diameter?

24-68 From the formula of Exercise 24-66, we can derive

$$D = d\left(0.94 + \sqrt{\frac{N - 3.7}{0.907}}\right)$$

Use the formula to find the size of a pipe or conduit required to hold 50 wires each 0.075 in. in diameter.

24-69 A Gothic arch is formed by two arcs, each one-sixth of a circle. The center of each circle is at the extremity of the width of the arch; that is, the two radii equal each other and the width of the arch. Find the area of an arch of radius 6 m (Fig. 24-25).

24-70 Each of four steam boilers is supplied by a 6-in. steam pipe. These open from a single steam pipe. Find the diameter of the larger pipe that has the same capacity as the four 6-in. pipes.

24-71 A milling cutter $4\frac{1}{2}$ in. in diameter is cutting soft steel at the rate of 45 ft/min. Find the number of revolutions per minute.

24-72 How many turns per second must a drill $\frac{1}{2}$ in. in diameter make so that the outer edge of the lip will have a cutting speed of 35 ft/min?

24-73 Which would occupy the greater portion of the square shown in Fig. 24-26, the four small circles or the large circle?

24-74 The drive wheels on a diesel locomotive make 210 rpm and are 195 cm in diameter. Find the speed of the locomotive in kilometers per hour if 2 percent is allowed for slipping.

24-75 If the total pressure on the piston of a brake cylinder is 8100 lb, what is the diameter of the cylinder if the pressure is 60 lb/in.²?

24-76 Determine the area in square inches of the opening of a cold-air return for a hot-air furnace to supply seven hot-air pipes 9 in. in diameter and one pipe 14 in. in diameter, if the area of the cold-air opening is three-fourths the area of the hot-air pipes?

24-77 Piston rings (Fig. 24-27) for gasoline engines are turned so that they are $1\frac{1}{2}$ percent larger in diameter than the diameter of the cylinder barrel. They then have a piece cut out and are sprung into place. Find the diameter of the original ring for a cylinder 4 in. in diameter. Find the

Fig. 24-27. Exercise 24-77.

length of the piece to be cut out if when sprung into place the ring has a clearance of 0.005 in. between ends. Give dimensions to the nearest ten-thousandths of an inch.

24-78 Find the speed in feet per second of a belt running over a pulley with a diameter of 22 in. making 320 rpm, if 2 percent is allowed for slippage.

24-79 If the greatest and least diameters of an elliptical manhole are 2 ft 7 in. and 2 ft 3 in., find its area.

24-80 Given an elliptical pipe whose longest axis is 400 mm and whose shortest axis is 250 mm, find the diameter of the circular pipe having the same area of cross section.

24-81 What is the horsepower of a gasoline engine with a 6.5-in. stroke and a 4.5-in. piston making 1000 rpm, with an average compression pressure of 400 lb/in.^{2} ($H = PLAN/33,000$)

24-15 REGULAR POLYGONS AND CIRCLES

Finding the dimensions of a regular polygon inscribed in or circumscribed about a given circle or finding the size of a circle that can be inscribed in or circumscribed about a given polygon require additional information or a proper handbook. Such problems are solved by trigonometry, and some of them may be solved by geometry. Handbooks give rules for how computations can be made. In Table 24-1, certain facts about regular polygons are listed. These facts can be applied to regular polygons of any size.

Rule: *To find the area of a regular polygon when the length of one side is given, multiply the square of the side by the number given in column (3) in Table 24-1.*

This rule is an application of the rule that similar areas are in the same ratio as the squares of their like dimensions.

EXAMPLE 24-9 Find the area of a regular pentagon having sides of 7 in.

Solution

$$\text{Area} = (7^2)(1.720\ 477\ 4\ \text{in.}^2) = 84.303\ 39\ \text{in.}^2$$

Rule: *To find the side of a regular polygon when its area is given, divide the area of the polygon by the proper number in column (3). The square root of the quotient is the required side of the polygon.*

TABLE 24-1 REGULAR POLYGONS

No. of sides (1)	Name of polygon (2)	Area when side = 1 (3)	Radius of circumscribed circle		Radius of inscribed circle when side = 1 (6)	Length of sides when radius of circumscribed circle = 1 (7)	Angle at center (8)	Angle between adjacent sides (9)
			When perpendicular from center = 1 (4)	When side = 1 (5)				
3	Triangle	0.4330127	2.	0.5773	0.2887	1.732	120°	60°
4	Square	1.	1.414	0.7071	0.5	1.4142	90°	90°
5	Pentagon	1.7204774	1.238	0.8506	0.6882	1.1756	72°	108°
6	Hexagon	2.5980762	1.156	1.	0.866	1.	60°	120°
7	Heptagon	3.6339124	1.11	1.1524	1.0383	0.8677	51.429°	128.571°
8	Octagon	4.8284271	1.083	1.3066	1.2071	0.7653	45°	135°
9	Nonagon	6.1818242	1.064	1.4619	1.3737	0.684	40°	140°
10	Decagon	7.6942088	1.051	1.618	1.5388	0.618	36°	144°
12	Dodecagon	11.1961524	1.037	1.9319	1.866	0.5176	30°	150°

EXAMPLE 24-10 The area of an octagon is 4376 ft^2; find a side.

Solution

$$\text{Side} = \sqrt{4376 \div 4.828} = \sqrt{906.379} = 30.106 \text{ ft}$$

Rule: To find the radius of the circumscribing circle when a side of the polygon is given, multiply the length of a side by the proper number in column (5).

This can be used to draw a regular polygon when a side is given.

EXAMPLE 24-11 Construct a regular decagon having sides of $2\frac{1}{2}$ in.

Solution The radius of the circumscribing circle = 2.5 × 1.618 in. = 4.045 in.
Construct a circle of this radius with a compass. Then, with the compass open $2\frac{1}{2}$ in., step around the circle; this will divide the circumference into 10 equal parts. Connect these points in succession, and a decagon with 2.5 in. sides will result.

Rule: To find the length of the side of a polygon that can be inscribed in a circle of a given radius, multiply the given radius by the proper number from column (7).

EXAMPLE 24-12 Construct a regular heptagon in a circle with a 3-in. radius.

Solution A side of the polygon = 3 × 0.8677 in. = 2.6 in.
With the compass open 2.6 in., step around the circle, which should thus be divided into 7 equal parts. Connect these points in succession, and the construction is complete.

▪ EXERCISES

Use Table 24-1 to solve Exercises 24-82 to 24-88.

24-82 A round drive shaft is D in. in diameter. The end of this round shaft is to be cut so that it forms a regular polygon whose side is S in. Find S when the regular polygon is:
(a) A triangle, and $D = 4$ in.
(b) A square, and $D = 4$ cm.
(c) A hexagon, and $D = 4$ in.
(d) An octagon, and $D = 4$ cm.

24-83 Tabletops are regular polygons with a side S. Circular covers are to be made for these tables. Find the radius R of these covers if the tabletops are:
(a) Triangular, and $S = 20$ in.
(b) Square, and $S = 30$ in.
(c) Hexagonal, and $S = 40$ cm.

24-84 Find the diameter of the bearings that can be made on the following regular polygon shafts:
(a) Triangular shaft, side $= 1$ in.; side $= s$ in.
(b) Hexagonal shaft, side $= 1$ cm; side $= s$ in.
(c) Decagonal shaft, side $= 1$ in.; side $= s$ in.
(d) Dodecagonal shaft, side $= 1$ in.; side $= s$ in.

24-85 The design in a floor consists of two concentric circles with a square inscribed in each. If the large circle has a radius of 6 ft and the smaller has a radius of 4 ft, find the ratio of the areas of the two squares. What is the ratio of the area of the two circles?

24-86 The area of a regular hexagon inscribed in a circle is $24\sqrt{3}$. Find the area of the circle and the length of the circumference.

24-87 A square end 0.875 in. on a side must be milled on a shaft. What is the diameter to which the shaft should be turned?

24-88 A pipe 10 in. in diameter is connected to a hexagonal pipe of the same area in cross section. Find the edge of the hexagon of the cross section of the hexagonal pipe.

24-16 BELTS, PULLEYS, AND GEARS

The relations between size and speed of driving and driven gears are the same as those of belt pulleys. In calculating for gears, we use the diameter of the pitch circle or the number of teeth.

Technicians should be able to determine quickly and accurately the speed of any shaft or machine and find the size of a pulley needed so that a shaft or machine may run at a desired speed. They should master the principles and formulas and be able to use them. Good students should work many problems on pulley speeds before considering special formulas. This will help them understand the principles and will make them independent of the formulas. This will enable them to derive, or prove, the formulas.

Note: *For a more detailed discussion of questions on belts and belting, see a mechanical engineer's handbook.*

In the study of pulleys and gears, the diameter and the rpm of the driving (power) shaft and the driven (tool) shaft are of fundamental importance. The power may be conveyed from the power shaft to the tool shaft by belts, gears, or chains.

It should be understood the rpm of the tool may be changed either by changing the rpm of the driving shaft (as in speeding up the engine in a car) or by changing the ratio of the diameters of the pulleys on the power shaft and the tool shaft (as in shifting gears in a car).

EXAMPLE 24-13 A driving shaft makes 840 rpm, and has a pulley 6 in. in diameter. If its speed is to be reduced to 320 rpm but the belt speed, or tool speed, is to remain unchanged, what size pulley should be used to replace the one on the driving shaft?

Solution If the 6-in. pulley makes 840 rpm, a point on the belt moves (6)(3.1416)(840 in./min). Then, to make 320 rpm, the pulley must be (6)(3.1416)(840 320 in.) in circumference. Therefore

$$\text{Pulley} = \frac{(6)(3.1416)(840)}{(320)(3.1416)} = 15.75 \text{ in. diameter}$$

EXAMPLE 24-14 In Fig. 24-28, A is the driving pulley and B is the driven pulley. Disregarding slippage, a point on the circumference of B must move as far in 1 min as a point on the circumference of A. Since A makes R rpm, a point on its circumference will move $R\pi D$ units per minute. Similarly, a point on the circumference of B will move $r\pi d$ units per minute. Therefore,

$$(R)(\pi)D = (r)(\pi)d \quad \text{or} \quad (R)(D) = (r)(d) \quad \text{and} \quad r = \frac{(R)(D)}{d}$$

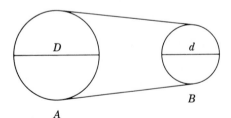

D

d

B

A

Fig. 24-28. Exercise 24-92.

▓ EXERCISES

24-89 A dynamo must make 1700 rpm; it has a pulley 4 in. in diameter. The shaft is to be belted to a driving shaft that makes 500 rpm. What must be the diameter of the pulley on the driving shaft?

24-90 A shaft has on it two pulleys, each 8 in. in diameter. The speed of the shaft is 400 rpm. What must be the size of the pulleys of two machines if, when belted to the shaft, one of them has a speed of 300 rpm and the other a speed of 900 rpm?

24-91 The pulley on the headstock of a lathe is 3 in. in diameter. This is belted to an 8-in.-diameter pulley on a shaft that makes 420 rpm. At what rate will a block of wood placed in the chuck revolve?

24-92 In two connected belt pulleys, or gear wheels, D = diameter of the driving wheel, d = diameter of the driven wheel, R = rpm of the driving wheel, and r = rpm of the driven wheel; find r in terms of D, d, and R.

24-93 The number of revolutions that the governor of a diesel engine is intended to run is given by the builder. If the speed of the governor is 1200 rpm, the size of the governor pulley is 8 in., and the desired speed of the engine is 900 rpm, find the diameter of the pulley to put on the engine shaft to run the governor pulley.

24-17 THE CIRCULAR MIL AND THE SQUARE MIL

Wires and cables used as electrical conductors generally have a circular cross section. Their diameters are usually expressed in mils and their cross-sectional areas in circular mils. The *mil* is a unit of length equal to one one-thousandth of an inch. The *circular mil* (cmil) is a unit of area equal to the area of a circle 1 mil in diameter. The area of a circle is found by the formula $A = \pi d^2/4$, and by definition the area of a circle 1 mil in diameter is 1 circular mil. Then

$$1 \text{ cmil} = \frac{\pi}{4} = 0.7854 \text{ mil}^2$$

If A is the area in circular mils of any circle and d is the diameter in mils, then, since a circle 1 mil in diameter has an area of 1 cmil, we have the proportion

$$\frac{1}{A} = \frac{1^2}{d^2}$$

That is, the areas of the circles are in the same ratio as the squares of their diameters. This proportion gives

$$A = d^2$$

Stated in words, this principle is: *The area of a circle in circular mils is the square of the diameter in mils, or in thousandths of an inch.*

Thus, a 0000-gage B. & S. wire is 0.46 in. = 460 mils in diameter and therefore has an area of $460^2 = 211,600$ cmil.

If the area in circular mils is given, the diameter in mils can be found by taking the square root of the area. That is,

$$d = \sqrt{A}$$

The *square mil* (mil^2) is the area of a square whose side is 1 mil. Since the area of a circle is $0.7854d^2$, it can be seen that 0.7854 mil^2 = 1 cmil.

◼ EXERCISES

24-94 A solid wire has a cross-sectional area of 225,000 cmil. What is its diameter in mils?

24-95 A solid wire having a diameter of 112 mil has a cross-sectional area of how many circular mils?

24-96 How many square mils are there in a bar $\frac{1}{2}$ in. by $\frac{3}{8}$ in. in cross section?

24-97 How many circular mils are equal to 20,000 mil^2?

24-98 Find the diameter in mils and in inches of a circular rod having a cross section of 237,000 cmil.

24-99 What is the area of the cross section in circular mils of 0000-gage B. & S. wire? (0000-gage B. & S. diameter = 0.46 in.)

GEOMETRIC SOLIDS

25-1 INTRODUCTION TO GEOMETRIC SOLIDS

In many occupations, it is necessary to calculate the volume or surface area of geometric solids. Some of these are illustrated in Fig. 25-1.

25-2 PRISMS

A *prism* is a solid whose ends, or *bases,* are parallel congruent polygons, and whose sides, or *faces,* are parallelograms.

Figure 25-2 shows a prism. The bases *ABCD* and *EFGH* are parallel polygons, in this case squares. A prism is called triangular, rectangular, hexagonal, etc., according to the shape of the bases. One side of a base (line AB) is a *base edge.*

The faces of the prism are *ABFE, BCGF,* etc. These faces are parallelograms. If they were rectangles, angle *EAB* would be 90° and the prism would be a *right prism.* A prism that is not a right prism is an *oblique prism.* A side of one of these parallelograms is a *lateral edge.*

The *altitude* of a prism is the distance between the planes of the two bases (*h* in Fig. 25-2). In a right prism, the altitude is the same length as a lateral edge. This is not true for an oblique prism.

25-3 SURFACE AREA OF A PRISM

The lateral area of a prism is the total of the area of the faces. For example, in Fig. 25-2,

$$\text{Lateral area } S = h(AB) + h(BC) + h(CD) + h(DA)$$
$$= h(AB + BC + CD + DA)$$

Rule: *The lateral area of a prism is equal to the perimeter of the base times the altitude.*

25-4 VOLUME OF A PRISM

Consider the right prism in Fig. 25-3. There are as many cubic units in each layer parallel to the base *ABCD* as there are square units in the area

349

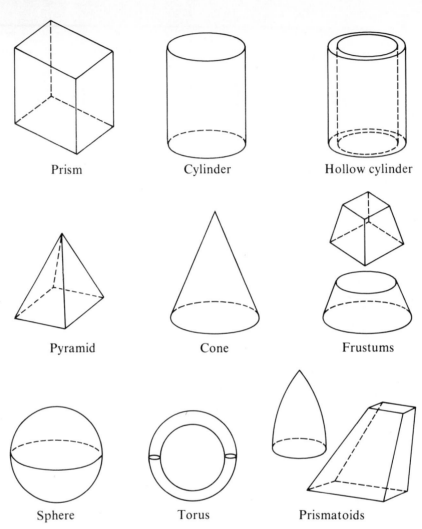

Fig. 25-1. Geometric solids.

Prism

Cylinder

Hollow cylinder

Pyramid

Cone

Frustums

Sphere

Torus

Prismatoids

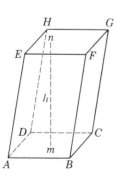

Fig. 25-2. Prism.

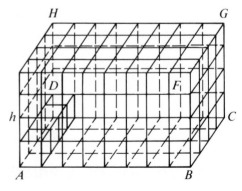

Fig. 25-3. Volume of a prism.

of the base. There are also as many layers of cubic units in the prism as there are linear units in the altitude (*h*). We can, therefore, find the total number of cubic units in the prism by multiplying the area of the base by the altitude. This formula holds true for oblique prisms as well, although the proof of the formula for oblique prisms is a bit more difficult.

$$V = A h$$

Rule: *The volume of a prism equals its base area times its altitude.*

25-5 CYLINDERS

Cylindrical-shaped objects occur regularly in daily life. We see cylinders in wire, pipes, tanks, and other containers, in pillars, and in various parts of machines.

A *right circular cylinder,* or a *cylinder of revolution,* is a solid formed by revolving a rectangle about one of its sides, which is used as an axis. This is shown in Fig. 25-4; the rectangle *OABO'* is revolved about *OO'* (the axis) to form the cylinder.

From this definition, it follows that the two *bases* of a right circular cylinder are circles and the lateral surface is a curved surface. The *axis* of the cylinder is the line (*OO'* in Fig. 25-4) joining the centers of the bases. The axis of a right cylinder is perpendicular to the bases; therefore, it is equal to the *altitude* of the cylinder, which is the perpendicular distance between the two bases. The *cross section* of a right circular cylinder is a section perpendicular to the axis and is a circle.

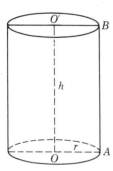

Fig. 25-4. Right circular cylinder.

25-6 VOLUME AND SURFACE AREA OF A CYLINDER

In the previous section on prisms, we determined that the lateral area of a prism was equal to the perimeter of the base times the altitude, that the total area was equal to the lateral area plus the areas of the two bases, and that the volume was equal to the area of the base times the altitude. These relationships also hold true for a cylinder.

The similarity between a cylinder and a prism will become more clear if you think of a prism with an infinite number of sides and compare this prism to a cylinder.

The general formulas for the surface area and the volume of a prism are true for cylinders also. If S = lateral area, V = volume, A = area of base, p = perimeter of base and h = altitude, the rules may be expressed by the following formulas:

$$S = ph = 2\pi r h$$
$$A = \pi r^2$$
$$V = Ah = \pi r^2 h$$

Rule: *The area of the lateral surface of a cylinder equals the circumference of the base times the altitude. The volume of a cylinder equals the area of the base times the altitude.*

25-7 VOLUME OF A HOLLOW CYLINDER

The volume of metal in a pipe (Fig. 25-5) may be found by subtracting the volume of the cylindrical hollow from the volume of the whole cylin-

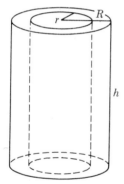

Fig. 25-5. Volume of a hollow cylinder.

der. If R is the outside radius of the cylinder and r the inside radius of the cylinder, then the volume of a hollow cylinder is

$$V = \pi R^2 h - \pi r^2 h = \pi h (R^2 - r^2) = \pi h (R + r)(R - r)$$

25-8 THE PYRAMID

A *pyramid* is a solid whose base is a polygon and whose sides (or *faces*) are triangles with their vertices at a common point. This common point is the *vertex* or *apex* of the pyramid. A pyramid is named according to the shape of its base; thus, a *triangular pyramid* is a pyramid that has a triangular base, a *square pyramid* is one with a square base, etc. The *altitude* of a pyramid is a line drawn from the vertex to the base and perpendicular to the base.

A *right pyramid,* or *regular pyramid,* is a pyramid whose base is a regular polygon and whose sides, or faces, are congruent isosceles triangles. Figure 25-6 shows a regular pyramid with a square base.

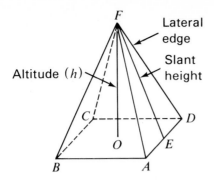

Fig. 25-6. Pyramid.

In a regular pyramid the *axis*—the line drawn from the vertex to the center of the base—is perpendicular to the base. Thus, in a regular pyramid the axis and the altitude are identical. In Fig. 25-6 *OF* is the altitude. The *slant height* of a right pyramid is a line drawn from the vertex to the center of one edge of the base. In Fig. 25-6 *EF* is a slant height. A *lateral edge* is a line in which two faces meet; *BF* in Fig. 25-6 is a lateral edge, as is *AF, DF,* etc.

25-9 SURFACE AREA OF A PYRAMID

The *lateral area* of a right pyramid is the sum of the areas of the triangles forming the faces of the pyramid. In a right pyramid, by definition, these are congruent triangles. Also by definition, the base of a right pyramid is a regular polygon. Therefore the bases of the triangular faces are equal; their altitudes are also equal and are equal to the slant height of the pyramid. We have the following:

Rule: *The lateral area of a right pyramid equals the perimeter of the base times one-half the slant height. The total surface area equals the lateral area plus the area of the base.*

Thus, if s = slant height, p = perimeter of the base, S = lateral area, B = area of base, and A = total surface area,

$$\text{Lateral area} = S = \frac{\text{perimeter} \times \text{slant height}}{2} = \frac{ps}{2}$$
$$\text{Base area} = B = (\text{side})^2$$
$$S = \frac{ps}{2}$$
$$A = S + B = \frac{ps}{2} + B$$

For a regular square pyramid (such as the one in Fig. 25-6), if a = length of a side of the base, we have $p = 4a$ and $B = a^2$. Thus for a regular square pyramid:

$$S = \frac{4as}{2} = 2as$$
$$A = 2as + a^2$$

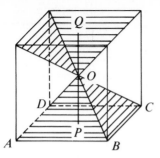

Fig. 25-7. Volume of a pyramid.

25-10 VOLUME OF A PYRAMID

The cube shown in Fig. 25-7 illustrates how to calculate the volume of a pyramid. The opposite vertices of the cube were connected, the lines of connection, meeting at the center O. This divided the cube into six congruent pyramids. The volume of the cube equals the area $ABCD$ times the altitude PQ. The volume of any one of the six pyramids, such as O-$ABCD$, equals one-sixth of the volume of the cube, and therefore the volume of one pyramid equals the area of the base $ABCD$ times one-third of PO. Thus, in this special case (that of a right square pyramid whose altitude equals one-half the length of a side of the base) we see that the volume of the pyramid equals the area of the base times one-third the altitude; in other words, the volume of the pyramid equals one-third the volume of a prism at the same base and altitude. These rules are true for any pyramid.

Rule: *The volume of a pyramid is equal to one-third the volume of a prism of the same base and altitude.*

This means if a prism is cut away to form a pyramid, exactly two-thirds of the original is cut away.

Rule: *The volume of a pyramid equals the area of the base times one-third the altitude.*

This may be written as a formula:

$$V = \frac{Ah}{3}$$

25-11 THE CONE

A *circular cone* is a solid whose base is a circle and whose lateral surface tapers uniformly to a point; that point is the *vertex* or *apex*. The *axis* of the cone is a straight line drawn from the vertex to the center of the base.

A *right circular cone* is a cone whose base is a circle and whose axis is perpendicular to the base. A right circular cone V-AOB is shown in Fig. 25-8. Such a cone can also be defined as a solid formed by a right triangle revolved about one of its legs as an axis; it may therefore be called a *cone of revolution*.

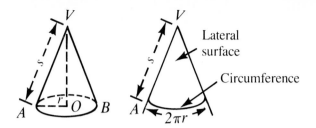

Fig. 25-8. Right circular cone.

The *altitude* of a cone is the perpendicular line from the vertex to the base (*OV* in Fig. 25-8). The *slant height* is the length of a straight line drawn from the vertex to the circumference of the base (*AV* in Fig. 25-8).

25-12 VOLUME AND SURFACE AREA OF A CONE

In an earlier section, we observed a similarity between prisms and cylinders. We now can do the same with the pyramid and the cone.

The lateral area of the pyramid was found to be the circumference of the base times one half the slant height. We also found that the volume of a pyramid was one-third the base area times the altitude. The same relations are true for cones:

Rule: *The lateral area of a right circular cone equals the perimeter of the base times one-half the slant height. The total surface area equals the lateral area plus the area of the base. The volume of a cone equals the area of the base times one-third the altitude.*

$$S = \frac{ps}{2}$$
$$A = \pi r^2$$
$$S + A = \frac{ps}{2} + \pi r^2$$
$$V = \frac{\pi r^2 h}{3}$$

25-13 FRUSTUMS

If the top of a pyramid or a cone is cut off by a plane parallel to the base, the remaining part is called a *frustum* of the pyramid or cone. Figure 25-9 shows a square frustum and a circular frustum.

The *altitude* of a frustum is the length of the perpendicular between the bases. The *slant height* of the frustum of a right pyramid is the shortest line between the perimeters of the two bases (this line would be perpendicular to both perimeters).

Look carefully at Figs. 25-9 and 25-10, you will see that a frustum is the bottom portion of a larger pyramid or cone. The volume of these prisms, therefore, can be obtained by subtracting the volume of the small

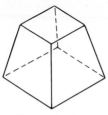

Square frustum

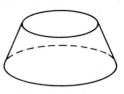

Circular frustum

Fig. 25-9. Frustums.

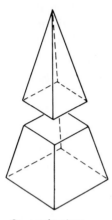

Square frustum
cut from a pyramid

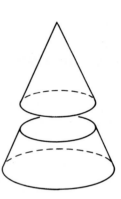

Circular frustum
cut from a cone

Fig. 25-10. Frustums.

pyramid or cone from the volume of the large pyramid or cone. The same applies to the lateral area.

Let S = lateral area, T = total surface area, V = volume, h = altitude, s = slant height, P and p = perimeters of the bases, B and b = base areas, and h' and s' = altitude and slant height of the small pyramid or cone. The altitude and slant height of the large pyramid or cone will be $s' + s$ and $h' + h$, and we will have the following formulas:

$$S = \frac{P(s' + s)}{2} - \frac{ps'}{2} = \frac{Ps + (P - p)s'}{2}$$
$$T = S + B + b$$
$$V = \frac{B(h + h')}{3} - \frac{bh'}{3} = \frac{Bh + (B - b)h'}{3}$$

Note the following proportions:

$$s':(s' + s) = p:P$$
$$h':(h' + h) = p:P$$
$$p^2:P^2 = b:B \quad or \quad p:P \ \sqrt{b}:\sqrt{B}$$

Solving the first proportion for s',

$$Ps' = p(s' + s)$$
$$(P - p)s' = ps$$
$$s' = \frac{p}{P - p}s$$

Substituting this in the formula for S, we get

$$S = \frac{Ps + ps}{2} = \frac{(P + p)s}{2}$$

Solving the second proportion above for h',

$$h' = \frac{p}{P - p}h$$

From the third proportion above,

$$p\sqrt{B} = P\sqrt{b}$$

or

$$P = p\frac{\sqrt{B}}{\sqrt{b}}$$

Hence,

$$h' = \frac{p}{p\sqrt{B}/\sqrt{b} - p}h = \frac{p\sqrt{b}}{p\sqrt{B} - p\sqrt{b}}h = \frac{\sqrt{b}}{\sqrt{B} - \sqrt{b}}h = \frac{\sqrt{b}(\sqrt{B} + \sqrt{b})}{B - b}h$$
$$= \frac{\sqrt{Bb} + b}{B - b}h$$

Substituting this in the formula for V,

$$V = \frac{Bh + (\sqrt{Bb} + b)h}{3} = \frac{(B + b + \sqrt{Bb})h}{3}$$

Thus, the formulas for the lateral area, total surface area, and volume of a frustum are:

$$S = \frac{(P + p)s}{2}$$
$$T = \frac{(P + p)s}{2} + B + b$$
$$V = \frac{(B + b + \sqrt{Bb})h}{3}$$

25-14 THE SPHERE

A *sphere* is a solid bounded by a curved surface every point of which is equally distant from some point; that point is the *center*. A straight line starting at the surface, passing through the center, and ending at the surface is a *diameter*. A line extending from the center to the surface is a *radius*.

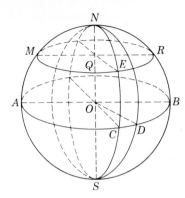

Fig. 25-11. Sphere and its parts.

If a sphere is cut by a plane, the section is a circle. If the section passes through the center of the sphere, it is a *great circle;* if it does not pass through the center, it is a *small circle.* The *circumference of a sphere* is the same as the circumference of a great circle. In Fig. 25-11, circles *ACB* and *NCS* are great circles, and *MER* is a small circle.

25-15 SURFACE AREA OF A SPHERE

The area of the surface of a sphere is equal to 4 times the area of a circle with the same radius.

$$S = 4\pi r^2 = \pi d^2$$

where S = area of the surface of the sphere
 r = radius
 d = diameter

25-16 VOLUME OF A SPHERE

The volume of a sphere is equal to the area of the surface times one-third the radius.

$$V = \frac{Sr}{3} = \frac{4\pi r^3}{3} = \frac{\pi d^3}{6} = 0.5236d^3$$

25-17 SEGMENT AND ZONE OF A SPHERE

The portion of the volume of a sphere included between two parallel planes is a *segment of the sphere.* If both planes cut the surface of the sphere, the segment is a *segment of two bases.* In Fig. 25-12, the segment between planes *ABC* and *DEF* is a segment of two bases. The part of the sphere above *DEF* is a *segment of one base.*

The portion of the surface of the sphere between two parallel planes is a *zone.* The altitude of a segment, or of a zone, is the perpendicular between the parallel planes. The altitude between planes *ABC* and *DEF* in Fig. 25-12 is *OQ.*

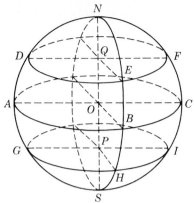

Fig. 25-12. Segments and zones of a sphere.

The area of a zone is equal to the circumference of a great circle of the sphere times the altitude of the zone, or

$$Z = 2\pi r h$$

where Z = area of the zone
h = altitude
r = radius of the sphere

From this formula note that the areas of any two zones on the same sphere or equal spheres are in the same ratio as their altitudes. It follows that the area of any zone is in the same ratio to the surface of the sphere as the altitude of the zone is to the diameter of the sphere.

If a sphere is cut by parallel planes that are equal distances apart, such as the planes in Fig. 25-12, then the zones are equal. On the earth, for example, the parallels 30° north latitude and 30° south latitude are in planes that bisect the radii drawn to the North Pole and South Pole, so that one-half of the surface of the earth is within 30° of the equator.

The volume of a spherical segment (Fig. 25-13) is determined with the following formula:

$$V = \frac{h\pi}{2}(r_1{}^2 + r_2{}^2) + \frac{\pi h^3}{6}$$

where V = volume
h = altitude
r_1 and r_2 = radii of the bases of the segment

If the segment has only one base, one of the radii is 0.

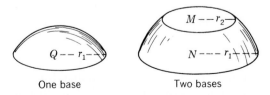

One base   Two bases

Fig. 25-13. Volume of a spherical segment.

EXAMPLE 25-1 Calculate the surface, volume, and weight of a cast-iron ball of radius $12\frac{1}{2}$ in.

Solution

$$S = 4\pi r^2 = (4)(3.1416)(12.5^2)$$
$$= 1963.5 \text{ in.}^2$$
$$V = \frac{Sr}{3}$$
$$= \left(\frac{1963.5}{3}\right)(12.5) = 8181.25 \text{ in.}^3$$

Since 1 in.^{3} of cast iron weighs 0.26 lb, the weight of the cast-iron ball is $8181.25 \times 0.26 = 2127.125$ lb.

EXAMPLE 25-2 The sphere of radius 8 in. shown in Fig. 25-14 is cut by two parallel planes, one passing 2 in. from the center and the other 6 in. from the center. Find the area of the zone and the volume of the segment between the two planes if both planes are on the same side of the center.

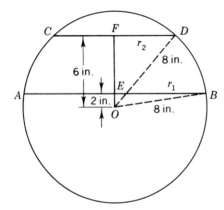

Fig. 25-14. Example 25-2: Finding the area of zone and volume of a segment.

Solution

$$Z = 2\pi rh = (2)(3.1416)(8)(4) = 201.06 \text{ in.}^3$$
$$r_1 = \sqrt{(OB)^2 - (OE)^2} = \sqrt{8^2 - 2^2} = \sqrt{60}$$
$$r^2 = \sqrt{(OD)^2 - (OF)^2} = \sqrt{8^2 - 6^2} = \sqrt{28}$$
$$V = \frac{h\pi}{2}(r_1^2 + r_2^2) + \frac{\pi h^3}{6}$$
$$= \frac{(4)\ 3.1416(60 + 28)}{2} + \frac{(3.1416)\ 4^3}{6}$$
$$= 586.43 \text{ in.}^3$$

25-18 THE TORUS

A ring formed by bending a cylinder into a circular form is called an *anchor ring* or *torus*. A torus is shown in Fig. 25-15. The average length of the rod in such a ring is the circumference of a circle of radius *ON*. Any

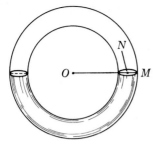

Fig. 25-15. Torus.

cross section of a torus will be a circle. Since the ring is a cylinder in circular form, the area of its surface is the lateral area of the surface of the cylinder that forms the ring and equals $2\pi \times ON \times$ the circumference of a cross section. If $ON = R$, and the radius of the cross section $NM = r$, we have the formula for the area:

$$A = (2\pi R)(2\pi r) = 4\pi^2 Rr$$

The volume is the same as the volume of a cylinder with an altitude equal to the mean circumference of the ring; we therefore have the following formula:

$$V = (2\pi R)(\pi r^2) = 2\pi^2 Rr^2$$

These rules apply to any circular ring.

Rule: *The area of the surface equals the perimeter of the cross section times the circumference drawn through the centroid, or the center of gravity of the cross section. The volume equals the area of the cross section times the circumference drawn through the center of gravity of the cross section.*

Thus the cross section may be a square, a triangle, or any other shape; but it is necessary to be able to find the centroid of the cross section.

To understand the meaning of the term "center of gravity," think of the center of gravity as the balancing point of the area. For instance, a circular piece of tin would balance on a pencil point if it rested with its center on the pencil point. A square piece of tin would also be balanced if its center was placed on the point of support. A triangle could be balanced if it was placed so that the point of support was under the point of the triangle that is the intersection of the medians. For an irregular-shaped piece of thin material, a balancing point can be found by trial. In each case, the balancing point would be the *center of gravity* or the *centroid* of the piece. A solid of any shape also has a center of gravity, although it is not so easily determined. But if a wire, a baseball bat, or a brick was balanced in a horizontal position on your finger, the center of gravity would be a point on the vertical line through the tip of the finger.

25-19 THE PRISMATOID

A *prismatoid* is a solid whose bases are parallel polygons and whose faces are quadrilaterals or triangles. Two special cases are:

(1) One base may be a point, in which case the prismatoid is a pyramid. (2) One base may be a line, in which case the prismatoid is wedge-shaped. The rule for finding the volumes of prismatoids also holds in many cases when the faces have become curved surfaces and the prismatoid has thus become a cone, a frustum of a cone, a cylinder, a sphere, or other various forms. The formula is as follows:

Rule: *To find the volume of a prismatoid, add together the areas of the two bases and four times the area of the cross section midway between them, then multiply the sum by one-sixth the perpendicular distance between the bases.*

The rule may be stated as a formula:

$$V = \frac{h}{6}(B_1 + 4M + B_2)$$

where B_1 and B_2 = areas of the two bases
M = area of the midsection

Figure 25-16 shows some forms that the rule for the volume of a prismatoid applies to. The dimensions of the midsection may be found by actual measurement or computation.

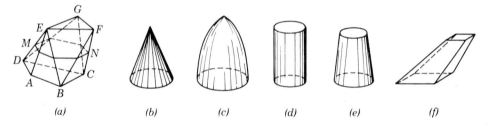

| (a) | (b) | (c) | (d) | (e) | (f) |

Fig. 25-16. For these solids, $V = (B + 4M + B)$.

EXAMPLE 25-3   Find the volume of a frustum of a pyramid in which the bases are regular hexagons 10 in. and 6 in. on a side and in which the altitude of the whole is 18 in. (See Fig. 25-17.)

Solution

$$V = \frac{h}{6}(B_1 + 4M + B_2)$$

$AB = \frac{1}{2}(10 \text{ in.} + 6 \text{ in.}) = 8 \text{ in.}$

Area of lower base = $B_1 = (5^2)(1.732)(6) = 259.8$

Area of upper base = $B_2 = (3^2)(1.732)(6) = 93.528$

$4 \times$ area of midsection = $4M = (4)(4^2)(1.732)(6) = 665.088$

Therefore,

$$V = \frac{18}{6}(259.8 + 665.088 + 93.528)$$

$$= 3055.2 \text{ in.}^3$$

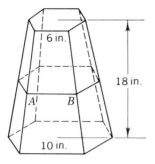

Fig. 25-17. Example 25-3: Finding the volume of a frustum.

25-20 THE FIVE REGULAR POLYHEDRONS

A *polyhedron* is a solid bounded by planes. A *regular polyhedron* is a polyhedron whose faces are all congruent regular polygons. Figure 25-18 shows the regular polyhedrons. The lines of intersection of the planes are the *edges* of the polyhedron; the points of intersection of the edges are the *vertices;* and the portions of the planes between the edges are the *faces*. A line joining any two vertices not in the same face is a *diagonal* of the polyhedron.

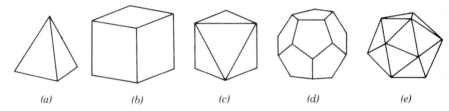

Fig. 25-18. (a) Tetrahedron; (b) hexahedron; (c) octahedron; (d) dodecahedron; (e) icosahedron.

Only five regular convex polyhedrons can be formed. These are:

1. The *regular tetrahedron* has four triangular faces any three of which meet at a vertex. It has four vertices and six edges.
2. The *regular hexahedron,* or *cube,* has six square faces meeting three at a vertex. It has eight vertices and twelve edges.
3. The *regular octahedron* has eight triangular faces meeting four at a vertex. It has six vertices and twelve edges.
4. The *regular dodecahedron* has twelve pentagonal faces meeting three at a vertex. It has twenty vertices and thirty edges.
5. The *regular icosahedron* has twenty triangular faces meeting five at a vertex. It has twelve vertices and thirty edges.

Students will find it very instructive to make models of the five regular convex polyhedrons, as shown in Fig. 25-19. Cut pieces of cardboard

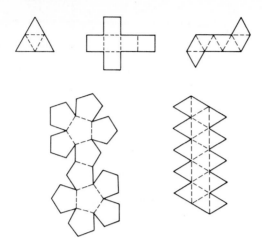

Fig. 25-19. Models of regular polyhedrons.

in the shapes shown and fold them along the dotted lines until the edges come together. To make the folding easier, cut the cardboard halfway through along the dotted lines. Transparent tape or glue can be used at the vertices to hold the edges together.

EXERCISES

25-1 A regular container is 12 cm by 6 cm by 3 cm. Its volume, therefore, is 216 cm^3.

(a) What would the dimensions be if the same amount of tin were used to make a cubical container? What would be the volume?

Note: The total surface area of the rectangular container equals the amount of material available. Exercise 25-1a asks for the dimensions and volume of a cube whose surface area is 252 cm^2.

(b) What would be the dimensions and volume if the same amount of material were used to make a spherical container?

(c) What would the volume be for a pyramid with a 9 cm by 9 cm base using this same amount of material?

(d) What would be the volume of a cone with a base diameter of 9 cm, using this same amount of material?

25-2 A manufacturer wants to minimize the cost of the package containing soap powder. Each container should hold 108 in.^{3} of soap powder, and the cost of the package is directly proportional to the amount of plastic used.

(a) What is the most economical package? What are the dimensions of this package?

(b) If you were to use a prism, what would be the most economical dimensions?

Note: Look at the two prisms in Fig. 25-20 and at the ratio of volume to surface area. Make a general statement as to what type of geometric solids maximize volume in relation to surface area?

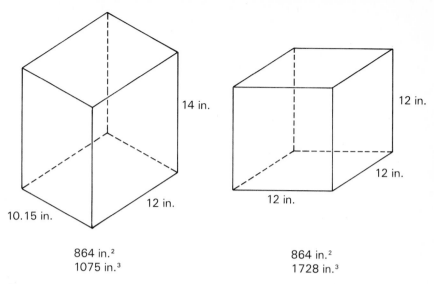

14 in.

12 in.

10.15 in.

12 in.

12 in.

12 in.

12 in.

864 in.²
1075 in.³

864 in.²
1728 in.³

Fig. 25-20. Exercise 25-2.

25-3 A rectangular piece of sheet iron 1000 cm² in area is bent to form a cylinder 31.83 cm in diameter. Find the height and the volume of this cylinder.

25-4 A wire $\frac{1}{16}$ in. in diameter was made from 1 ft³ of copper. Estimate the length of this wire in miles, and then find its precise length. (5280 ft = 1 mi)

25-5 The height of a right circular cylinder is 6 cm, and its entire area is 1884 cm². Find the radius of the base.

25-6 A pail is marked 1 gal. It is 6 in. in diameter at the bottom and 8 in. in diameter at the top. The pail is open at the top. Find the number of square inches of tin required to make the pail.

25-7 A common brick is 2 in. by 4 in. by 8 in. Find the number of bricks in a pile $8\frac{1}{2}$ ft by 4 ft by 10 ft. (Use cancellation.)

25-8 The lining of the bottom and sides of a cubical vessel requires 180 ft² of zinc. How many cubic feet of water will this vessel hold?

25-9 A boxcar that is $36\frac{2}{3}$ ft long and $8\frac{1}{6}$ ft wide (inside measurements) can be filled with wheat to a height of $5\frac{1}{2}$ ft. Find how many bushels of wheat it will hold if $\frac{3}{4}$ ft³ = 1 bushel.

25-10 A sun-tan-oil bottle is a rectangular solid whose inside measurements are 5.5 in. by 3 in. by 2 in. It is marked 1 pint. The bottom of this bottle curves inward. If there are 231 in.³ to 1 gallon, how many cubic inches of oil are displayed by this curved bottom?

25-11 A swimming pool is 40 ft long by 20 ft wide. It is 3 ft deep at one end and slopes uniformly to a depth of 10 ft at the other end. How many gallons does it hold when it is full if 7.5 gallons = 1 ft³?

25-12 How many cubic yards of soil will it take to fill in a lot 50 ft by 100 ft if it is to be raised 3 ft in the rear end and gradually sloped to the front, where it is to be 1½ ft deep?

Hint: *The vertical cross section of this lot, taken the long way, is a trapezoid having parallel sides of lengths 3 ft and 1½ ft and an altitude of 100 ft.*

25-13 A cube has a 10-in. edge. If the surface area of this cube were used to form a rectangular solid x in. by x in. by $2x$ in., how much volume would be lost?

25-14 Two different soap containers which are rectangular solids 9 in. by $3\frac{1}{4}$ in. by $6\frac{1}{4}$ in. and $8\frac{3}{8}$ in. by $3\frac{1}{8}$ in. by $6\frac{5}{8}$ in. are sold at the same price. Which is the better buy if value depends only on volume?

25-15 How many rectangular solids 3 in. by 4 in. by 9 in. will fill a box 4 ft by 6 ft. by 8 ft?

25-16 Find the number of cubic yards of crushed rock needed to make a road 1 mi in length whose cross section is to be as shown in Fig. 25-21.

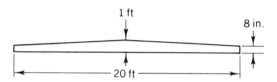

Fig. 25-21. Exercise 25-16.

Solution: The area of the vertical cross section can be found by considering it as two trapezoids, each having parallel sides of 8 in. and 1 ft and an altitude of 10 ft.

$$\text{Area of cross section} = 2\left(\frac{h}{2}\right)(B_1 + b_2)l = (\tfrac{2}{3} \div 1)10 = 16\tfrac{2}{3} \text{ ft}^2$$
$$\text{Rock, ft}^3 = (5280)(16\tfrac{2}{3}) = 88{,}000$$
$$\text{Rock, yd}^3 = \frac{88{,}000}{27} = 3259.259$$

25-17 What length must be cut from a bar of steel 0.5 cm by 1.25 cm in cross section to make 1 cm^3?

25-18 One cm^3 of steel weighs 0.008 kg. An I beam has a cross section as shown in Fig. 25-22 and a length of 6 m. Find its weight.

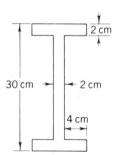

Fig. 25-22. Exercise 25-18.

25-19 If there are as many square feet in the surface area of a cube as there are cubic feet in its volume, find the length of its edge.

25-20 A copper rod 1 in. in diameter and 8 in. long is made into a wire of uniform diameter 200 ft long. Find the diameter of the wire.

25-21 Determine the ratio of the lateral areas of the rod and the wire in Exercise 25-20.

25-22 Using 4.211 ft^3 = 1 barrel, find the number of barrels in the following cylindrical tanks:
(a) h = 20 ft and r = 15 ft
(b) h = 9 ft and r = 4 ft

25-23 A 2$\frac{1}{4}$-in. diameter cylindrical container with an altitude of 3$\frac{5}{8}$ in. is marked 6 ounces. How much would 1 quart of this powder weight? (Use 231 in.^{3} = 1 gallon, 4 quarts = 1 gallon, and 32 ounces = 1 quart.)

25-24 A container is 60 cm in diameter and 60 cm high. Another can is 50 cm in diameter and holds the same amount as the first container; find its height.

25-25 The bottom part of a container is a cube with an edge of 2 in. and the top part is a cylinder with a diameter of 2 in. and a height of 2 in. Find the volume and total surface area of this container.

25-26 The external diameter of a hollow cast-iron pipe is 45 cm, and its internal diameter is 25 cm. Calculate its weight if the length is 6 m and cast iron has a density of 7.48 g cm^3.

25-27 Water is flowing at the rate of 10 mph through a pipe 16 in. in diameter into a rectangular reservoir 197 yd long and 87 yd wide. Calculate the time it will take to raise the surface 3 in. (10 mph = 10 × 5280 60 = 880 ft/min is the number of cubic feet of water that will flow through the 16-in. pipe in 1 min.) Then find the number of cubic feet required to fill the reservoir 3 in. The quotient of dividing the required number of cubic feet by the flow per minute is the time in minutes.

25-28 In a table giving weights and sizes of square nuts for bolts, a nut 2 in. square and 1$\frac{1}{4}$ in. thick with a hole 1$\frac{1}{16}$ in. in diameter has a given weight of 1.042 lb. How many cubic inches does it contain?

25-29 Find the length of steel wire in a coil if the diameter of the wire is 0.0625 cm and its weight is 22.5 kg. (Use a density of 8.0 g/cm^3.)

25-30 Find the weight of a cylinder of lead 1 ft long and 1 in. in diameter (1 in.^{3} of lead weighs 0.412 lb).

25-31 Find the weight of 360 m of lead pipe with an inside diameter of 1.09 cm and an outer diameter of 3.09 cm if lead has a density of 11.85 g/cm^3.

25-32 Down the middle of a four-lane highway there is a safety guard made of concrete. The base of a cross section of the guard is a trapezoid and the top is a semicircle. The lower base of the trapezoid is 18 in.; the upper base, which is also the diameter of the semicircle, is 6 in.; and the distance between the bases is 12 in. How many cubic yards are there in 1 mi of the safety guard?

25-33 A wrought iron washer 3$\frac{1}{2}$ cm in diameter with a hole 1$\frac{1}{2}$ cm in diameter is $\frac{5}{32}$ cm thick. Find the number of washers this size in a keg of 100 kg (1 cm³ wrought iron = 0.008 kg).

25-34 How many cubic inches of metal are there in a length of pipe 10 ft long if the inside diameter is 5 in. and the metal is $\frac{1}{4}$ in. thick?

25-35 A conduit made of concrete has a cross section as shown in Fig. 25-23. How many cubic yards of concrete are used in making 500 yd of this conduit?

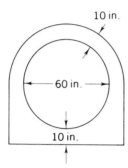

10 in.

60 in.

10 in.

Fig. 25-23. Exercise 25-35.

25-36 A round steel bar $\frac{3}{8}$ in. in diameter has an area of 0.1104 in.² and a weight of 0.376 lb per linear foot, according to a table giving weights and cross-sectional areas of steel bars. Verify this if the density of steel is 489.6 lb/ft².

25-37 A cylindrical water tank holds 10 barrels. It has a radius of 3 ft. Find its height (1 barrel = 4.211 ft³).

25-38 Water is being pumped through a 6-in. pipe. If the flow is 3 ft/s, how many barrels are being pumped per hour?

25-39 To test the flow of water through a $\frac{1}{4}$-in. circular nozzle, a flow of 60 gallons was recorded in 10 min. Find the rate in feet per second of the flow of water if the stream is supposed to be as large as the nozzle.

25-40 Find the pull per square inch necessary to break a rod $2\frac{1}{2}$ in. in diameter that breaks with a load of 270,000 lb.

25-41 The rain that falls on a flat roof 22 ft by 36 ft is conducted to a cylindrical cistern 8 ft in diameter. How much rainfall would it take to fill the cistern to a depth of $7\frac{1}{2}$ ft?

25-42 If a wrought-iron bar 2 in. by $1\frac{1}{4}$ in. in cross section breaks under a load of 125,000 lb, what load will break a wrought-iron rod $2\frac{1}{2}$ in. in diameter?

25-43 A wrought-iron cylindrical rod 2000 ft long and $1\frac{1}{2}$ in. in diameter is suspended vertically from its upper end. What is the total pull at the upper end and the pull per square inch of cross section? (Use a density of 0.28 lb/in.³)

25-44 Find the length in miles (in hundredths) of a 1.5-in. diameter wrought iron rod supported vertically at its upper end that will just break under its own weight.

25-45 At $3.50 per pound, find the cost of the copper in a sheet-copper cylindrical tank 3 ft high and 1½ ft in diameter, open at the top and weighing 2 lb per square foot.

25-46 A rectangular block of wood whose dimensions are a, b, and c has a circular hole d in. in diameter bored through it, perpendicular to the faces with dimensions a and b. Write a formula for the volume V of the remaining wood.

25-47 A rectangular block of wood has the dimensions a, b, and c. Each of its four edges of length b is rounded in the form of a quarter of a circle of radius r. Find the volume V of the remaining block.

25-48 An asphalt-cooling cylinder is 4 ft in diameter and 9 ft long and makes 4 rpm. The hot asphalt covers the surface to a depth of ⅛ in. How many pounds of asphalt will be cooled in 1 h if the specific gravity of asphalt is 0.9? Assume that an entire change of asphalt passes over the cylinder during each revolution.

25-49 If a tank 5 ft in diameter and 10 ft deep holds 10,000 lb of soybean oil, what will be the depth of a tank of 2000-lb capacity if its diameter is 3 ft? If this tank has a jacket around it on the bottom and sides 3 in. from the surface of the tank (as shown in Fig. 25-24), how many gallons of water will the space between the jacket and the tank hold?

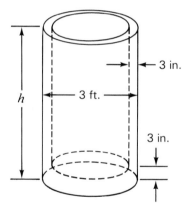

Fig. 25-24. Exercise 25-49.

25-50 Find the volume of a gasoline tank whose bottom is in the form of a rectangle with a semicircle at each end. The rectangle is 10 cm by 14 cm and the semicircles are on the smaller dimensions. The depth of the gasoline tank is 16 cm.

25-51 In making the pattern of a special oil change container to hold 5 quarts, it is decided to have the bottom an ellipse with axes of 10 in. and 7 in. Find the height of the container.

25-52 How long a piece of plastic will it take to make the body of the above container?

25-53 Find the amount of sheet metal in an elliptical tank whose base has axes of 42 in. and 32 in. and whose height is 40 in. The tank is to have a bottom but no top. Find the number of gallons the tank will hold.

25-54   In drilling in soft steel, a $1\frac{9}{16}$-in. twist drill makes 37 rpm with a feed of $\frac{1}{80}$ in. Find the number of cubic inches cut away in $3\frac{1}{2}$ min.

25-55   A $\frac{3}{8}$-in. twist drill makes 310 rpm with a feed of $\frac{1}{125}$ in. Find the volume cut away in $3\frac{1}{4}$ min.

25-56   (a) to (c) Find the number of pounds of cast iron removed per hour by a shaper under the conditions in Table 25-1 (consider the cutting as if on a plane). (0.26 lb $= 1$ in^3.)

TABLE 25-1   EXERCISE 25-56

	Speed per Minute, ft	Depth of Cut, in.	Breadth of Cut, in.
(a)	37.90	0.125	0.015
(b)	25.82	0.015	0.125
(c)	25.27	0.048	0.048

25-57   The flanges at the joining of two ends of flanged steam pipes 9 in. in inside diameter are bolted together by 12 bolts $\frac{3}{4}$ in. in diameter. If the pressure in the pipes is 200 lb/in.^{2}, find what each bolt must hold. How much is this per square inch of cross section of the bolts? Suppose the bolts have 10-pitch U.S. Standard thread. This will make the root diameter 0.620 in. (Fig. 25-25).

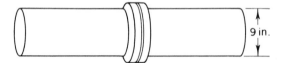

Fig. 25-25. Exercise 25-57.

25-58   From Exercise 25-57, if the steam pipe is 18 in. in diameter, and if the same pull per square inch of cross section of each bolt is allowed, find the number of bolts $1\frac{1}{8}$ in. in diameter at a joint of the pipe. (A $1\frac{1}{8}$-in. bolt with 7-pitch U.S. Standard thread is 0.940 in. in diameter at the root of the thread.)

25-59   The following rule is often used to find the heating surface of any number of equal tubes in a steam boiler: Multiply the number of tubes by the diameter of one tube in inches, multiply this product by the length of a tube in feet and then by 0.2618. The final product is the number of square feet of heating surface. Using this rule, find the heating surface of sixty-six 3-in. tubes each 18 ft long?

25-60   The cylinder of a pump is 6 in. in diameter, the length of stroke is 8 in., and the number of strokes per minute is 160. Find the flow in gallons per minute.

25-61   What is the weight of a cast-iron pipe 3 m long, 60 cm in outer diameter, and 2.5 cm thick? (Use 0.0075 kg/cm^3.)

25-62 The Cleveland Twist Drill Company records a test in which a $1\frac{1}{4}$-in. "Paragon" high-speed drill removed 70.56 in.^{3} of cast iron per minute. The penetration per minute was $57\frac{1}{2}$ in.; the feed was $\frac{1}{10}$ in. How many rpm would this require?

25-63 A tank car with a cylindrical tank 8 ft in diameter and 34 ft long will hold how many gallons? What weight of oil will it hold if the specific gravity of oil is 0.94?

25-64 Find the height of a cylindrical tank having a diameter of 30 in. and holding 4 barrels.

25-65 A cylindrical tank 22 ft long and 6 ft in diameter rests on its side in a horizontal position. Find the number of gallons of oil that it will hold when the depth of the oil is 8 in. and when the depth of the oil is 2 ft 6 in.

25-66 Because the body of a bolt is greater in diameter than the threaded part, the two parts will not stretch uniformly when the bolt is under strain. For this reason the bolt is most likely to break where the threaded part joins the other part. To overcome this, a hole is sometimes drilled from the center of the head to the beginning of the threaded part. This hole is made of such size that the cross-sectional area of the body is the same as that at the root of the thread.

  Find the diameter of the hole to be drilled in the following bolt, in accordance with the preceding explanation: diameter of bolt = $\frac{3}{4}$ in., with 10 U.S. standard threads to 1 in. (See Table 4.)

25-67 A cone 12 in. in altitude with a circular base 8 in. in diameter has a hole 2 in. in diameter bored through the center from apex to base. Find the volume of the part remaining.

Hint: *The portion cut away, as shown in Fig. 25-26, consists of a cylinder 9 in. in altitude and a cone 3 in. in altitude. The height of the small cone can be found from the similar triangles AOP and BO'P in which this proportion holds:*

$$\frac{AO}{BO'} = \frac{OP}{O'B}$$

or

$$\frac{4}{1} = \frac{12}{O'B}$$
$$4\,O'B = 12$$
$$O'B = 3$$

25-68 When the right circular cylinder of greatest volume possible is inscribed in a given right circular cone, the altitude of the cylinder is equal to one-third of that of the cone. Find the volume of the largest right circular cylinder that can be inscribed in a right circular cone with $d = 12$ cm and $h = 9$ cm. (See Exercise 25-67.)

25-69 Find the weight of a green fir log 215 ft long, 4 ft 6 in. in diameter at one end, and 20 in. in diameter at the other end. The specific gravity of fir is 0.78.

25-70 Find the weight of a tapered brick stack of 3 m inside diameter, with a

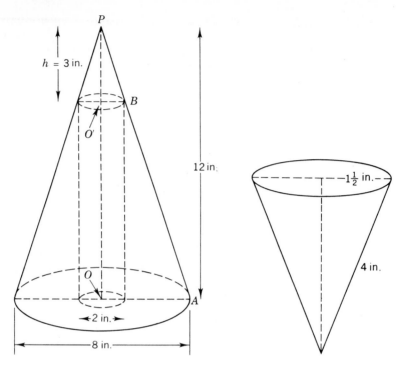

Fig. 25-26. Exercise 25-67.   **Fig. 25-27.** Exercise 25-74.

wall 1.2 m thick at the base, 0.45 m at the top, and 52.5 m high. A brick has a density of 1860 kg/m³.

25-71 It is known that the most economical proportions for a conical tent of a given capacity are $h = r\sqrt{2}$. Find the least amount of canvas to make a conical tent with $V = (1000\pi\sqrt{2})/3$ ft³.

25-72 A cast-iron driver in the form of a frustum of a square pyramid is used in a pile-driving machine. Find the weight of the driver if it is 40 cm square at the bottom, 17.5 cm square at the top, and 20 cm thick. The density of cast iron is 0.0075 kg/cm³.

25-73 A cast-iron cone pulley is 34 in. long. The diameter of one end is 12 in., and that of the other end is 5 in. A circular hole 2 in. in diameter extends the length of the pulley. Find the weight of the pulley. The density of cast iron is 0.26 lb/in.³

25-74 Determine how to cut a pattern to make a tin cone of the dimensions shown in Fig. 25-27.

Hint: *The radius of the sector is 4 in. Angle θ of the sector is the same part of 360° that the circumference of the cone is of a circumference of radius 4 in.*

$$\frac{8\pi}{3\pi} = \frac{360°}{\theta}$$
$$8\pi(\theta) = (3\pi)(360°)$$
$$\theta = 135°$$

The pattern is a sector of 135° in a circle of 4-in. radius.

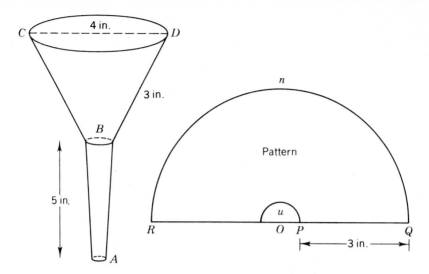

Fig. 25-28. Exercise 25-75.

25-75 Determine how to draw a pattern for the upper part of a funnel with dimensions as shown in Fig. 25-28. The diameter at B is 1 in. and the diameter at A is $\frac{1}{2}$ in.

Suggestion: *The length OP = x and may be determined as follows:*

$$\frac{x}{x+3} = \frac{1}{4}$$
$$4x = x + 3$$
$$x = 1$$

The radius to use is therefore 4 in.

To determine the central angle θ, by the same reasoning as in the solution of the previous exercise, we have

$$\frac{\text{arc } Qn\,R}{8\pi} = \frac{\theta}{360°}$$

Arc $QnR = 4\pi$, which, substituted in the proportion, gives $\theta = 180°$.

25-76 Determine how to draw a pattern for the lower part of the funnel shown in Fig. 25-28.

25-77 How many square inches of material would be necessary to make the funnel in Fig. 25-28?

25-78 The decorations of the top part of a gatepost is a sphere 8 in. in diameter. If this sphere is cut from an 8-in. cube, what part of the volume is wasted?

25-79 The bottom part of a container is a hemisphere with a 4-ft diameter. The top part is a right circular cone in which $r = h = 2$ ft. How many gallons will this container hold? Use 7.5 gallons = 1 ft^3.

25-80 How many square feet of paint surface are there in 20 of the containers of Exercise 25-79? Only the outside is to be painted.

25-81 Find the weight of 10 lead balls 8 cm in diameter. The density of lead is 0.01 kg/cm^3.

25-82 The nose portion of a small plane is a hemisphere with a 4-ft diameter. A hole with a 6-in. radius passes through the nose. Find the area of the zone left.

25-83 A cast-iron ball 3 in. in diameter is covered with a coating of ice 1 in. thick. Find the weight of the ice.

25-84 A hollow copper sphere used as a float weighs 10 oz and is 5 in. in diameter. How heavy a weight will it support in water?

25-85 How many acres of land are there on the surface of the earth if one-fourth of the surface is land and the earth's radius is 4000 mi?

25-86 The radius of the earth is 3960 mi, and the radius of the moon is 1080 mi. Find the ratio of their surface areas.

25-87 Find the volume of a cylinder 2 m in diameter and 2 m in altitude; of a sphere 2 m in diameter; and of a cone 2 m in diameter and 2 m in altitude. Compare the three volumes, showing they are in the ratio 3 : 2 : 1; that is, the volume of the sphere is two-thirds of the cylinder, and the volume of the cone is one-third that of the cylinder.

25-88 A ball of lead 2 in. in diameter is pounded into a circular sheet 0.01 in. thick. How large in diameter is the sheet?

25-89 A cylindrical water tank with hemispherical ends is 1.8 m in total length and 45 cm in diameter. Find its capacity in gallons.

25-90 For both a cube and a sphere, volume = 1 ft^3. Find the ratio of their areas.

Note: Of all solids with the same given volume, the sphere has the least surface area.

25-91 A silo is a cylinder with a hemispherical roof. If the diameter of the silo equals its height, find its total area. (Let diameter = d ft.)

Note: This is the most economical proportion for building a silo (that is, it takes the least material to build for a maximum volume).

25-92 In Exercise 25-91, the floor of the silo is twice as thick as the walls and roof. For the most economical shape the height of the cylinder should be equal to its diameter. Find the number of board feet of lumber to build this silo if the floor is 2 in. thick and the radius equals r in. (1 fbm = 1 ft^2 of 1-in.-thick lumber.)

25-93 A circular flower bed in a park is 25 ft in diameter and is raised 2 ft 6 in. in the center, making a spherical segment. How many loads of soil did it take to build it up if one load is $1\frac{1}{2}$ yd^2?

25-94 Find the percent of error in using the following rule: To find the weight of a cast-iron ball, multiply the cube of the diameter in inches by 0.1377; the product is the weight in pounds.

25-95 The cross section of a wedding ring is a circle $\frac{1}{16}$ in. in radius. The inner radius of this torus is $\frac{3}{8}$ in. If the ring is solid gold, find its weight. (The density of gold is 0.695 lb/in.^{3})

25-96 If the torus in Fig. 25-15 is a solid cork float with *ON* equal to 1 ft and *NM* equal to 4 in., find its weight. (The density of cork is 15 lb/ft^3.)

25-97 If 1000 floats like the one in Exercise 25-96 are to be painted, find the number of square feet of painted surface.

25-98 The cross section of the rim of a cast-iron flywheel of an oil well engine is a rectangle 6 in. by 8 in., the shorter dimension being in the diameter of the wheel. The wheel is 8 ft in outer diameter. Find the volume of the rim and its weight. (The density of cast iron is 0.26 lb/in.^{3})

25-99 Find the weight of a cast-iron water main 12 ft in length, 2 ft in outer diameter, and 1 in. thick.

25-100 Find the area of the surface and the volume of a ring whose outer diameter is 10 in. and which is made of round iron 1 in. in diameter. What is its weight at a density of 0.28 lb/in.^{3}?

25-101 An anchor ring 18 in. in outer diameter, of $1\frac{1}{4}$-in. round iron, has the same volume as a bar $1\frac{1}{4}$ in. by $1\frac{1}{2}$ in. in cross section. How long is the bar?

25-102 Find the weight of a torus of cast iron whose outer diameter is 3 ft, the iron being circular in cross section and 6 in. in diameter. (Use a density of 450 lb/ft^3.)

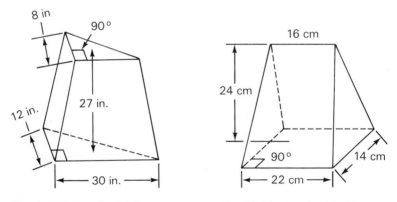

Fig. 25-29. Exercise 25-104.　　　　**Fig. 25-30.** Exercise 25-105.

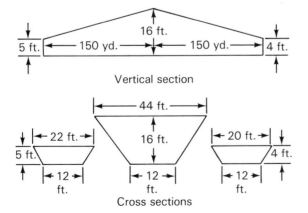

Fig. 25-31. Exercise 25-106.

25-103   Using the formula for the volume of a prismatoid, find the volumes of the following:

(a) A cube with an edge equal to x

(b) A right circular cylinder with radius r and height h

(c) A right circular cone with radius r and altitude h

(d) A hemisphere with radius r

25-104   Find the volume of the solid shown in Fig. 25-29.

25-105   Find the volume of the solid shown in Fig. 25-30.

25-106   A railroad cut through a hill has the dimensions given in Fig. 25-31, which shows the vertical section and three cross sections. Find the volume of the earth removed, in cubic yards.

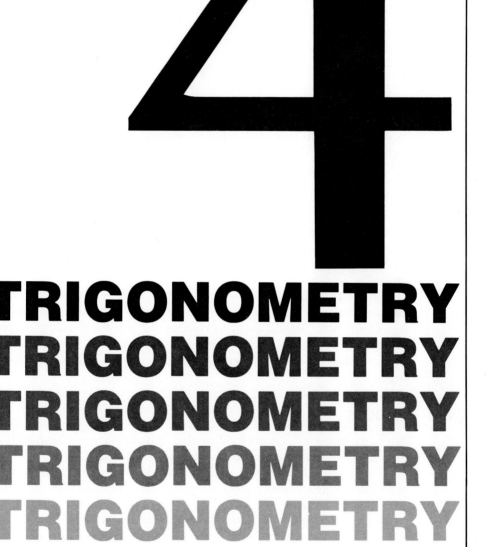

4

TRIGONOMETRY
TRIGONOMETRY
TRIGONOMETRY
TRIGONOMETRY
TRIGONOMETRY
TRIGONOMETRY

INTRODUCTION TO TRIGONOMETRY

26-1 INTRODUCTION TO TRIGONOMETRY

As you progress in your study of mathematics, you should be aware that each mathematical subject is based on the preceding subject. With additional mathematical knowledge, more problems can be solved more easily.

This section is devoted to the study of trigonometry. In trigonometry, you will use arithmetic, algebra, and geometry; in fact, some of the errors you will make in trigonometry will actually be errors in arithmetic, algebra, or geometry. Do not let this disturb you; correcting these errors will increase your understanding of the subjects you have already studied.

The word "trigonometry" is derived from two Greek words: *trigonon*, meaning "triangle"; and *metria*, meaning "measurement." The derivation of the word would seem to confine the subject to triangles, but the measurement of triangles is only part of the subject. Trigonometry includes many other problem-solving procedures involving angles.

26-2 ANGLES

Since trigonometry is a study of angles, it is necessary to have a clear conception of the nature and measurement of angles.

Trigonometry works with angles of any size and with *negative* as well as positive angles. Thus a more comprehensive definition of an angle becomes necessary.

If a line is turned about a fixed point on the line and kept in the same plane, it is said to *generate*, or *sweep out*, an angle. The hand of a clock is an example of a line that is revolving and generating an angle.

The *size* of an angle is determined by the *amount of turning* made by the line.

If the line turns in a *counterclockwise direction*—that is, opposite in direction to the hands of a clock—the angle described is a *positive angle*. If the line turns in a *clockwise direction*, the angle described is a *negative angle*.

The position of the line at the start is the *initial line* or *side,* and the final position is the *terminal line* or *side.* A circular arrow which is drawn between the initial line and the terminal line and which has the arrow on the terminal line shows the *sense*—the direction of turning—and the *size* of the angle. In Fig. 26-1*a*, line *OX* can be imagined as pinned at *O* and

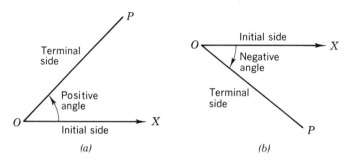

Fig. 26-1. (*a*) Positive angle; (*b*) negative angle.

turning in a counterclockwise direction to the position *OP*. The angle described is *positive* and is read "angle *XOP*." Notice that the initial line is read first.

In Fig. 26-1*b*, line *OX* is turning in a clockwise direction; it thus describes a *negative* angle, *XOP*.

It is evident that, according to the idea of an angle given here, an angle can be of any value whatever, positive or negative. Thus, an angle of 467° is one complete turn (360°) and 107°. Such an angle is shown in Fig. 26-2*a*. An angle of −229° is a turn of 229° in the negative (clockwise) direction; Fig. 26-2*b* shows such an angle. An angle of 720° is two complete turns of the initial line. An angle of 3760° is ten complete turns and 160° more.

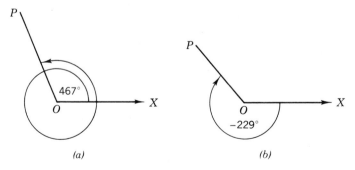

Fig. 26-2. (*a*) 467° angle; (*b*) −229° angle.

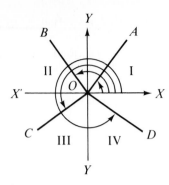

Fig. 26-3. Quadrants.

26-3 QUADRANTS

Angles are located in any of four quadrants. Figure 26-3 illustrates this: two lines, $X'X$ and $Y'Y$, are drawn at right angles to each other, forming four 90° angles. The area bounded by XOY is called the *first quadrant*, YOX' the *second quadrant*, $X'OY'$ the *third quadrant*, and $Y'OX$ the *fourth quadrant*. The positive part of the X axis, OX, is taken as the initial side of an angle, the angle is said to be in the first quadrant if its terminal side lies between OX and OY. It does not matter how many complete revolutions are made. Thus angles of 40°, 400°, and 760° all lie in the first quadrant.

Similarly, if the terminal side of an angle lies between OY and OX', the angle is said to be in the second quadrant. If the terminal side lies between OX' and OY', the angle lies in the third quadrant. If the terminal side of an angle is located between OY' and OX, the angle is in the fourth quadrant. In Fig. 26-3, angle XOA is in the first quadrant, XOB is in the second, XOC is in the third, and XOD is in the fourth.

If the terminal side falls on OX, OY, OX', or OY', the angle lies between two quadrants. This will be true for angles of 0°, 90°, 180°, 270°, 360°, 450°, and so on.

When line OX in Fig. 26-3 is rotated one-fourth of a complete turn, a *right angle* is formed; since the rotation is one-fourth of a complete revolution, the right angle is equal to 90°. Angle XOY in Fig. 26-3 is a right angle. When line OX in Fig. 26-3 is turned or rotated through one-half a turn or circle, a straight angle is formed. A straight angle is equal to two right angles, or 180°. Angle XOX' in Fig. 26-3 is a straight angle. If OX is rotated three-fourths of a complete turn, the angle formed is equal to three right angles, or 270°.

▦ EXERCISES

In the following exercises, show the sense and size of these angles by using circular arrows. Use the protractor to draw the angles.

26-1 Draw eight positive angles, one in each quadrant and one between each quadrant.

26-2 Draw eight negative angles, one in each quadrant and one between each quadrant.

26-3 Draw the following angles and tell in which quadrant or between which quadrants each angle's terminal side lies: 65°, 240°, 374°, 180°, 375°, 210°, 15°, 790°.

26-4 Draw the terminal side of 60°. Give four other positive angles which have the same terminal side.

26-4 MEASUREMENT OF ANGLES

Note that the size of an angle is determined by the amount of rotation and *not* by the length of the sides of the angle. For example, angle *ABC* is larger than angle *DEF* in Fig. 26-4 because the terminal side *BC* has been rotated further from *BA* than the terminal side *EF* from *ED*.

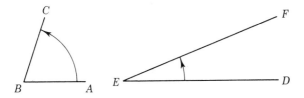

Fig. 26-4. ∠*ABC* is larger than ∠*DEF*.

There are two units for measuring this amount of rotation: the degree and the radian.

26-5 DEGREE MEASUREMENT OF ANGLES

One basic unit for measuring angles is the *degree* (symbolized °). A degree is defined as $\frac{1}{360}$ of a revolution (Fig. 26-5).

Fig. 26-5. The degree.

For more accurate measurement, a degree can be subdivided into smaller parts. One method is to use minutes and seconds as parts of a degree. A *minute* (symbolized ') is defined as $\frac{1}{60}$ of a degree. A smaller subdivision is the *second* (symbolized "). A second is defined as $\frac{1}{60}$ of a minute. Thus,

$$1 \text{ circle} = 360°$$
$$1° = 60'$$
$$1' = 60''$$

Another method for subdividing degrees is by using decimal parts. A decimal number accurate to the nearest tenth would subdivide the degree into 10 equal parts. Likewise, a decimal number accurate to the nearest one-hundredth would subdivide the degree into 100 equal parts.

The benefit of either system is largely historical. The degree-minute-second system has been the standard designation for many industries. Older equipment used in these industries may still use the degree-minute-second increments.

The decimal degree notation is much easier to work with and is the standard notation in electronic hand-held calculators. Most industries now use the decimal degree notation. For these reasons, this book will use the decimal degree system.

You may need to translate a measurement in one system to a figure in the other. This is done by recalling: a minute is defined as $\frac{1}{60}$ of a degree, and a second is defined as $\frac{1}{60}$ of a minute, or $\frac{1}{3600}$ of a degree.

To convert minutes to decimal parts of degrees, you must multiply the minutes by 1°/60'. Thus:

$$30' = (\overset{1}{\cancel{30}}) \left(\frac{1°}{\underset{2}{\cancel{60'}}} \right) = 0.5°$$

As a second is $\frac{1}{60}$ of a minute, you can convert from seconds to minutes in the same way, and then convert from minutes to decimal parts of a degree. However, it is easier to think of a second as $\frac{1}{3600}$ of a degree. Thus:

$$30'' = (\overset{1}{\cancel{30''}}) \left(\frac{1°}{\underset{120}{\cancel{3600''}}} \right) = 0.0083°$$

EXAMPLE 26-1 Convert 35° 15′ 55″ to decimal degrees.

Solution Multiply minutes by 1°/60″; seconds by 1°/3600″; then combine.

$$35° \, 15' \, 55'' = 35° + \left(\overset{1}{\cancel{15'}} \times \frac{1°}{\underset{4}{\cancel{60'}}} \right) + \left(55'' \times \frac{1°}{3600'} \right)$$

$$= 35° + 0.25° + 0.0153°$$

$$= 35.2653°$$

To convert from decimal degrees to degrees, minutes, and seconds, first multiply the decimal part by 60 to obtain the number of minutes, and then multiply the decimal part of the number of minutes by 60 to obtain the number of seconds. Thus:

$$35.2653° = 35° + \left(.2653 \times \frac{60'}{1°} \right) = 35° + 15.9180'$$

$$= 35° + 15' + \left(.9180' \times \frac{60''}{1'} \right) = 35° \, 15' \, 55''$$

EXERCISES

26-5 Convert the following to decimal degrees:
(a) 15′   (e) 50″
(b) 30′   (f) 17′ 27″
(c) 45′   (g) 29′ 15″
(d) 15″   (h) 7° 18′ 24″

26-6 Convert the following to degrees, minutes, and seconds:
(a) 0.25°   (e) 0.174°
(b) 0.75°   (f) 3.758°
(c) 0.60°   (g) 2.777°
(d) 0.305°   (h) 3.5139°

26-6 RADIAN MEASUREMENT OF ANGLES

Another unit for measuring the amount of rotation of an angle is the
radian (rad or *r*). The radian is defined as the angle created by an arc
equal in length to the radius of the circle. This definition is illustrated in
Fig. 26-6, where the length of arc *AB* is equal to the radius *OA*. The angle
created, *AOB*, is equal to 1 rad.

EXAMPLE 26-2 Draw an angle of 2π rad.

Solution

1. Rotate a line until an arc of 2π rad has been generated.

2. But by definition, the arc generated by 1 rad is equal in length to the radius
of the generating arm.

3. Therefore, an angle of 2π rad is the same as an angle having an arc
length of $2\pi r$. In other words, a 2π-rad angle is a complete circle (Fig.
26-7).

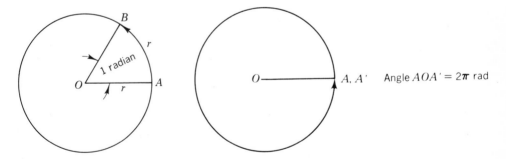

Fig. 26-6. The radian.   **Fig. 26-7.** $\angle AOA' = 2\pi$ rad.

EXAMPLE 26-3 Draw an angle of $\pi/2$ rad.

Solution

1. A complete circle has an angle of 2π rad, we want to make an angle of
$\pi/2$ rad.

2. Given these two facts, we can set up the following proportion for x, the angle we want.

$$\frac{\text{complete revolution}}{2\pi} = \frac{x}{\dfrac{\pi}{2}}$$

$$\frac{\pi}{2} \text{ rad} = \frac{1}{4} \text{ complete revolution, or a right angle.}$$

▨ EXERCISE

26-7 Draw the following angles; determine which quadrant or between which quadrants each angle's terminal side lies.

(a) π rad

(b) 2π rad

(c) $\dfrac{\pi}{2}$ rad

(d) $\dfrac{3}{2}\pi$ rad

(e) $\dfrac{\pi}{6}$ rad

(f) 3.1416 rad

(g) $\dfrac{\pi}{4}$ rad

(h) 1.047 rad

26-7 THE RELATION BETWEEN THE RADIAN AND THE DEGREE

A full revolution equals 2π radians and also equals $360°$. Therefore, $360° = 2\pi$ rad and

$$1 = \frac{360°}{2\pi}$$

$$1 \text{ rad} = \frac{\overset{180°}{\cancel{360°}}}{\underset{1}{\cancel{2\pi}}} = \frac{180°}{\pi} = 57.295\ 78°$$

$$1° = \frac{\overset{1}{\cancel{2\pi}} \text{ rad}}{\underset{180}{\cancel{360}}} = \frac{\pi \text{ rad}}{180} = 0.017\ 453 \text{ rad}$$

Use these two formulas to convert any angle from degrees to radians or from radians to degrees.

To convert radians to degrees, multiply the number of radians by $180/\pi$, or $57.29578\ldots$.

To convert degrees to radians, multiply the number of degrees by $\pi/180$, or $0.017\ 453\ldots$.

EXAMPLE 26-4 Convert 2.5 rad to degrees.

Solution Multiply 2.5 rad by 57.29578°.

$$1 \text{ rad} = \frac{180°}{\pi} = 57.295\ 78°$$

$$2.5 \text{ rad} = 2.5 \times 57.295\ 78°$$
$$= 143.2394°$$

EXAMPLE 26-5 Convert 143.2394° to radians.

Solution Multiply 143.2394° by 0.017453 rad.

$$1° = \frac{180}{\pi} \text{ rad} = 0.017\ 453 \text{ rad}$$

$$143.2394° = 143.2394 \times 0.017\ 453 \text{ rad}$$
$$= 2.5 \text{ rad}$$

Table 26-1 shows the degree and radian equivalents for common first-quadrant angles. Listed in the left-hand column are the conventional degree measurements for six angles. Listed in the center column are the radian measurements, in fractional multiples of π, for the same angles. The right-hand column shows radians to four decimal places.

TABLE 26-1 RELATIONSHIP BETWEEN RADIAN AND DEGREE ANGLE MEASUREMENTS

Degrees	Radians as Multiples of π	Radians as Decimals, to Four Places
15°	$\dfrac{\pi}{12}$	0.2618
30°	$\dfrac{\pi}{6}$	0.5236
45°	$\dfrac{\pi}{4}$	0.7854
60°	$\dfrac{\pi}{3}$	1.0472
75°	$\dfrac{5\pi}{12}$	1.3090
90°	$\dfrac{\pi}{2}$	1.5708

EXERCISES

26-8 Express the following angles in radians: (a) 18°, (b) 20°, (c) 30°, (d) 36°, (e) 45°, (f) 60°, (g) 72°, (h) 90°, (i) 120°, (j) 135°, (k) 180°, (l) 210°, (m) 225°, (n) 270°, (o) 300°, (p) 315°, (q) 330°, (r) 360°, (s) 540°, (t) 720°, (u) 1080°.

26-9 Express the following angles in degrees: (a) 0.3142 rad, (b) 0.1111π rad, (c) 0.5236 rad, (d) 0.2π rad, (e) 0.7854 rad, (f) 0.3333π rad, (g) 1.2566 rad, (h) 0.5π rad, (i) 2.0944 rad, (j) 0.75 rad, (k) 3.1416 rad, (l) 1.1667π rad, (m) 3.9270 rad, (n) 1.5π rad, (o) 5.2360 rad, (p) 1.75π rad, (q) 5.7596 rad, (r) 2.0π rad, (s) 9.4248 rad, (t) 6.0π rad, (u) 18.8496 rad.

26-10 Given a circle with a radius of 1 unit, what is the arc length generated by the following angles?

(a) 47°

(b) 0.8203 rad

(c) 60°

(d) 1 rad

(e) 90°

(f) 1.5708 rad

(g) 23°

(h) 0.4630 rad

(i) 20°

(j) 2π rad

(k) 53°

(l) $\dfrac{\pi}{4}$ rad

26-11 Express the angles of the following regular polygons in both degrees and radians: (a) three sides; (b) four sides; (c) five sides; (d) six sides; (e) eight sides; (f) twelve sides.

26-12 If you walk completely around a circle, through how many π radians do you turn? If the radius of the circle is 10 m, how far do you walk?

26-13 In 3 h, through how many radians has the minute hand of a clock turned? The hour hand? The second hand?

26-14 How many radians are in each of the angles of a right triangle if one of the acute angles is 36.7833°?

26-15 How many degrees are there in each of the angles of an isosceles triangle if the angle at the vertex is $\frac{1}{6}\pi$ rad?

26-16 Two of the angles of a triangle are $\frac{2}{3}$ and $\frac{2}{5}$ rad. Find the number of radians and degrees in the third angle.

TRIGONOMETRIC FUNCTIONS

27-1 THE TRIGONOMETRIC FUNCTIONS

Calculations involving trigonometry are based on certain ratios, for example, the comparison of one side of a right triangle to another side. The six ratios are the sine, cosine, tangent, cotangent, secant and cosecant. These are *trigonometric functions*. Understanding the definitions of the six functions will make solving trigonometry problems much easier.

27-2 THE SIX FUNCTIONS

If an acute angle *XOA* is drawn (Fig. 27-1), and from *P* (any point on *OA*) a perpendicular *QP* is drawn to *OX*, a right triangle *QOP* is formed. The ratio of the length of *QP* to *OP* is the *sine* (pronounced "sign") of angle θ, and is usually written sin θ. The symbol θ, is a Greek letter called "theta," it is used because it is different from traditional letters. Thus, the ratio *QP/OP* is the sine of angle θ (written sin θ), or *QP/OP* = sin θ. The numerical value of sin θ can be determined for all angles from 0 to any number of degrees.

It is possible to determine by a scale drawing the approximate value of the sine of any angle. For example, to determine the value of sin θ

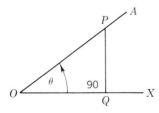

Fig. 27-1. *QP/OP* = sin θ.

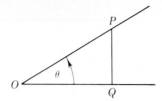

Fig. 27-2. Determining sin θ when θ = 30°.

when $\theta = 30°$, construct an angle of 30° (Fig. 27-2), measure *OP* any distance, and draw *PQ* perpendicular to *OQ*.

Measure the lengths of *OP* and *PQ* and substitute these two lengths in the formula $QP/OP = \sin \theta$.

In Fig. 27-2, *OP* = 4 cm and *PQ* = 2 cm. Then

$$\sin 30° = \frac{QP}{OP} = \frac{2}{4} = 0.500$$

The sin 30° is *always* equal to 0.500 regardless of the size of the triangle. This ratio is a constant. You can show the validity of this statement by constructing several triangles and determining this ratio for sin 30° for each triangle constructed.

If θ has the same value in each of two or more right triangles constructed, the triangles are similar. Since corresponding sides of similar triangles are proportional, the ratio of *QP* to *OP* will therefore have a constant value for each value of θ.

The six ratios defined in terms of triangle *OQP* in Fig. 27-2 are:

The ratio *QP/OP* is the *sine* of angle θ (sin θ).

The ratio *OQ/OP* is the *cosine* of angle θ (cos θ).

The ratio *QP/OQ* is the *tangent* of angle θ (tan θ).

The ratio *OQ/QP* is the *cotangent* of angle θ (cot θ).

The ratio *OP/OQ* is the *secant* of angle θ (sec θ).

The ratio *OP/QP* is the *cosecant* of angle θ (csc θ).

The last three functions, cot θ, sec θ, and csc θ, are reciprocals of tan θ, cos θ, and sin θ, respectively. It is necessary to memorize only the first three ratios and remember that the last three are reciprocals.

27-3   FUNCTIONS FOR ANY ANGLE

Draw any four angles, one in each quadrant, as shown in Fig. 27-3. Here the four angles formed are XOP_1, XOP_2, XOP_3, and XOP_4. On the terminal sides of these four angles, locate the points P_1, P_2, P_3, and P_4 at any convenient distances from *O*, and draw perpendiculars to the *X* axis. Let the distances from *O* along the terminal side of any angle to the points chosen (P_1, P_2, P_3, or P_4) be known as *r*; these distances are *always positive*.

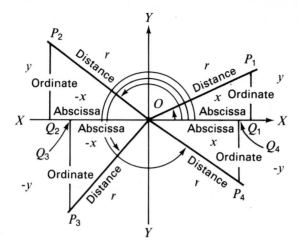

Fig. 27-3. Trigonometric ratios.

The perpendiculars to the X axis are ordinates of the four points P_1, P_2, P_3, and P_4. The ordinates Q_1P_1 and Q_2P_2 are *positive* lengths, since they extend above the X axis. The ordinates Q_3P_3 and Q_4P_4 are *negative*, since they extend below the X axis.

Distances OQ_1 and OQ_4 are *positive* because they are to the right of the Y axis. Distances OQ_2 and OQ_3 are *negative* because they extend to the left of the Y axis.

The ratios for any of the four angles drawn, where θ is at the origin, are:

$$\sin \theta = \frac{y}{r} \qquad \csc \theta = \frac{r}{y}$$

$$\cos \theta = \frac{x}{r} \qquad \sec \theta = \frac{r}{x}$$

$$\tan \theta = \frac{y}{x} \qquad \cot \theta = \frac{x}{y}$$

Note that the following relation can be developed from these six functions:

$$\tan \theta = \frac{\sin \theta}{\cos \theta} \qquad \tan \theta = \frac{1}{\cot \theta}$$

$$\cot \theta = \frac{\cos \theta}{\sin \theta} \qquad \cot \theta = \frac{1}{\tan \theta}$$

$$\sec \theta = \frac{1}{\cos \theta} \qquad \cos \theta = \frac{1}{\sec \theta}$$

$$\csc \theta = \frac{1}{\sin \theta} \qquad \sin \theta = \frac{1}{\csc \theta}$$

$$(\sin \theta)^2 - (\cos \theta)^2 = 1$$

From these relations it can be seen that if any one of the six trigonometric functions is known for an angle, the values of the other five functions for that angle can be determined.

In formulas containing powers of trigonometric functions, the exponent is placed before the angle: $(\sin \theta)^2$ is written $\sin^2 \theta$. Thus, the last formula above is usually written

$$\sin^2 \theta + \cos^2 \theta = 1$$

▨ EXERCISES

Note: *Remember that these trigonometric ratios are real numbers and hence are positive, negative, or zero, depending on the quadrant in which the terminal side of the angle lies.*

27-1 In each of the following, give the quadrant or quadrants in which the terminal side of the angles lies, and draw the function.
(a) When $\sin \theta$ is positive
(b) When $\sin \theta$ is negative
(c) When $\sin \theta$ is 0
(d) When $\cos \theta$ is positive
(e) When $\cos \theta$ is negative
(f) When $\cos \theta$ is 0
(g) When $\tan \theta$ is (1) positive, (2) negative, (3) 0
(h) When all functions are positive
(i) When all functions are negative

27-2 What two functions are positive in the second quadrant?

27-3 What four functions are negative in the fourth quadrant?

27-4 In what quadrant is $\sin \theta$ positive and $\cos \theta$ negative?

27-5 In what quadrant is $\sin \theta$ negative and $\cos \theta$ positive?

27-6 In what quadrant is $\cos \theta$ positive and $\tan \theta$ negative?

27-7 In what quadrant is $\sec \theta$ negative and $\csc \theta$ positive?

27-8 In what quadrant is $\cot \theta$ positive and $\sec \theta$ positive?

27-9 When the terminal side lies along the X axis, what functions are 0? What functions are unity? What functions are not defined?

Note: *Division by 0 is undefined.*

27-10 When the terminal side lies along the Y axis, what functions are 0? What functions are unity? What functions are not defined?

27-11 What functions are equal in value when the terminal side lies along the 45° line? The 225° line?

27-12 Give the algebraic signs of each of the trigonometric functions of the following angles (draw the graph):
(a) 30°   (d) 315°   (g) −45°   (j) $\frac{1}{4}\pi$
(b) 120°   (e) 420°   (h) −150°   (k) $-\frac{5}{8}\pi$
(c) 225°   (f) 750°   (i) −300°   (l) $\frac{2}{3}\pi$

27-13 Is there an angle whose tangent is positive and whose cotangent is negative? Is there an angle whose secant is positive and whose cosine is negative? Is there an angle whose secant is positive and whose cosecant is negative?

27-14 Draw a pair of coordinate axes and locate an angle of 60°. Locate a point on the terminal side 2 in. from the origin, and find the abscissa and ordinate of this point. Finally, tabulate the trigonometric functions of 60°.

Suggestion: *Here the distance is 2, the abscissa is 1, and the ordinate is $\sqrt{3}$.*

27-4 COMPUTATION OF TRIGONOMETRIC FUNCTIONS

Two right triangles have definite relations between sides and angles: the right isosceles triangle with acute angles of 45°, and the right triangle with acute angles of 30° and 60°. The trigonometric functions for angles of 30°, 45°, and 60° can be calculated, since the relations between the sides of these two triangles are known or can be determined.

27-5 TRIGONOMETRIC FUNCTIONS OF 30°

In a 30°-60° right triangle, the hypotenuse is twice the shorter side. In any right triangle the hypotenuse squared is equal to the sum of the squares of the other two sides.

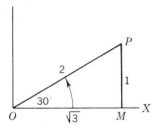

Fig. 27-4. Computing functions of a 30° angle.

To compute the trigonometric functions for an angle of 30°, draw angle $XOP = 30°$ (Fig. 27-4). Locate P two units from O and draw PM perpendicular to OX. The distance PM will then be one unit. To determine the length of OM:

$$
\begin{aligned}
(OP)^2 &= (PM)^2 + (OM)^2 \\
2^2 &= 1^2 + (OM)^2 \\
4 &= 1 + (OM)^2 \\
3 &= (OM)^2 \\
\sqrt{3} &= OM
\end{aligned}
$$

Since r, x, and y are now known ($r = 2$, $x = \sqrt{3}$, and $y = 1$) it is possible to calculate the six trigonometric ratios for an angle of 30°:

$$\sin 30° = \frac{y}{r} = \frac{1}{2} = 0.5 \qquad\qquad \csc 30° = \frac{r}{y} = \frac{2}{1} = 2$$

$$\cos 30° = \frac{x}{r} = \frac{\sqrt{3}}{2} = 0.866 \qquad\qquad \sec 30° = \frac{r}{x} = \frac{2}{\sqrt{3}} = \frac{2}{3}\sqrt{3} = 1.1547$$

$$\tan 30° = \frac{y}{x} = \frac{1}{\sqrt{3}} = \frac{\sqrt{3}}{3} = 0.577 \qquad \cot 30° = \frac{x}{y} = \frac{\sqrt{3}}{1} = \sqrt{3} = 1.732$$

27-6 TRIGONOMETRIC FUNCTIONS OF 45°

In a 45° right triangle the sides are equal in length. The hypotenuse squared of a right triangle is equal to the sum of the squares of the other two sides; it is therefore possible to determine the value of all three sides of a 45° right triangle.

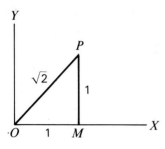

Fig. 27-5. Computing functions of a 45° angle.

To find the functions of an angle of 45°, draw $XOP = 45°$, as in Fig. 27-5. Locate M one unit distance from O and construct MP perpendicular to OX. Then MP will be one unit, equal in length to OM. Calculate OP:

$$(OP)^2 = (OM)^2 + (MP)^2$$
$$(OP)^2 = 1^2 + 1^2$$
$$(OP)^2 = 2$$
$$OP = \sqrt{2}$$

Since r, x, and y are now known, it is possible to calculate the six trigonometric ratios for an angle of 45°.

$$\sin 45° = \frac{y}{r} = \frac{1}{\sqrt{2}} = \frac{1}{2}\sqrt{2} = 0.707 \qquad \csc 45° = \frac{r}{y} = \frac{\sqrt{2}}{1} = \sqrt{2} = 1.414$$

$$\cos 45° = \frac{x}{r} = \frac{1}{\sqrt{2}} = \frac{1}{2}\sqrt{2} = 0.707 \qquad \sec 45° = \frac{r}{x} = \frac{\sqrt{2}}{1} = \sqrt{2} = 1.414$$

$$\tan 45° = \frac{y}{x} = \frac{1}{1} = 1 \qquad\qquad \cot 45° = \frac{x}{y} = \frac{1}{1} = 1$$

27-7 TRIGONOMETRIC FUNCTIONS OF OTHER ANGLES

It is possible to determine the values of the trigonometric functions of angles of 0°, 60°, 45°, 90°, 135°, 120°, and so on, by using the procedures from the last two sections. But this method can be used only where the lengths of the sides of the triangle are known or can be determined.

The method used to calculate the values of the trigonometric functions of angles other than those listed above is beyond the scope of this text and will therefore not be explained. Tables and calculators will be used to determine these values.

27-8 APPLICATIONS TO RIGHT TRIANGLES

The abscissa, the ordinate, and the distance from O of any point in the terminal side form an acute right triangle where the given angle θ is one of the acute angles. Because of the numerous applications of the right triangle in trigonometry, the definitions of the trigonometric functions will be stated with reference to the right triangle.

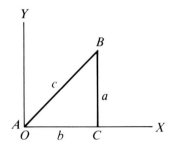

Fig. 27-6. Functions of a right angle.

Draw a right triangle ABC, as in Fig. 27-6, with the vertex A at the origin O and with AC on OX. The lowercase letters a, b, and c represent the lengths of the three sides of the right triangle: The side opposite angle A is labeled a; the side opposite angle B is labeled b; and the side opposite angle C is labeled c.

By definition, the six trigonometric functions of angle A are:

$$\sin A = \frac{\text{ordinate}}{\text{distance}} = \frac{a}{c} = \frac{\text{side opposite}}{\text{hypotenuse}}$$

$$\cos A = \frac{\text{abscissa}}{\text{distance}} = \frac{b}{c} = \frac{\text{side adjacent}}{\text{hypotenuse}}$$

$$\tan A = \frac{\text{ordinate}}{\text{abscissa}} = \frac{a}{b} = \frac{\text{side opposite}}{\text{side adjacent}}$$

$$\cot A = \frac{\text{abscissa}}{\text{ordinate}} = \frac{b}{a} = \frac{\text{side adjacent}}{\text{side opposite}}$$

$$\sec A = \frac{\text{distance}}{\text{abscissa}} = \frac{c}{b} = \frac{\text{hypotenuse}}{\text{side adjacent}}$$

$$\csc A = \frac{\text{distance}}{\text{ordinate}} = \frac{c}{a} = \frac{\text{hypotenuse}}{\text{side opposite}}$$

The expression $\sin A$ = side opposite/hypotenuse may be read "the sine of an acute angle in a right triangle is equal to the length of the side opposite the angle divided by the length of the hypotenuse." The $\sin B = b/c$, since b is the side opposite the acute angle B and c is the hypotenuse.

The six trigonometric functions for angle B of Fig. 27-6 are:

$$\sin B = \frac{\text{side opposite}}{\text{hypotenuse}} = \frac{b}{c}$$

$$\cos B = \frac{\text{side adjacent}}{\text{hypotenuse}} = \frac{a}{c}$$

$$\tan B = \frac{\text{side opposite}}{\text{side adjacent}} = \frac{b}{a}$$

$$\cot B = \frac{\text{side adjacent}}{\text{side opposite}} = \frac{a}{b}$$

$$\sec B = \frac{\text{hypotenuse}}{\text{side adjacent}} = \frac{c}{a}$$

$$\csc B = \frac{\text{hypotenuse}}{\text{side opposite}} = \frac{c}{b}$$

The six functions for these two acute angles A and B appear in Table 27-1. Note that $\sin A = a/c$ and $\cos B = a/c$. It follows that $\sin A = \cos B$. Angles A and B are complementary, and their sum is equal to 90°. Also $\cos A = b/c$ and $\sin B = b/c$; thus $\cos A = \sin B$; and so on for the other four comparisons.

The sine, cosine, tangent, cotangent, secant, and cosecant of an acute angle are, respectively, the cosine, sine, cotangent, tangent, cosecant, and secant of the complement of that angle.

The following examples illustrate this statement:

$$\cos 75° = \sin (90 - 75)° = \sin 15°$$
$$\tan 80° = \cot (90 - 80)° = \cot 10°$$

TABLE 27-1 FUNCTIONS OF COMPLEMENTARY ANGLES

Angle A	Angle B
$\sin A = \dfrac{a}{c}$	$= \cos B$
$\cos A = \dfrac{b}{c}$	$= \sin B$
$\tan A = \dfrac{a}{b}$	$= \cot B$
$\cot A = \dfrac{b}{a}$	$= \tan B$
$\sec A = \dfrac{c}{b}$	$= \csc B$
$\csc A = \dfrac{c}{a}$	$= \sec B$

27-9 DETERMINING TRIGONOMETRIC FUNCTIONS BY CONSTRUCTION AND MEASUREMENT

It is possible to determine the approximate value of the functions of angles by construction and measurement. The approximate values of the sine, cosine, and tangent of 40° will be determined here to illustrate this method.

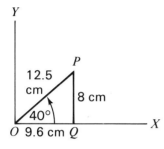

Fig. 27-7. Determining functions by construction and measurement.

Draw angle $XOP = 40°$ (Fig. 27-7). Take any convenient distance OP on the terminal side and draw the perpendicular QP. Measure the lengths of the sides of the right triangle—OP, PQ, and OQ.

$$\text{The sine of } 40° = \frac{\text{side opposite}}{\text{hypotenuse}} = \frac{PQ}{OP} = \frac{8 \text{ cm}}{12.5 \text{ cm}} = 0.64.$$

$$\text{The cosine of } 40° = \frac{\text{side adjacent}}{\text{hypotenuse}} = \frac{OQ}{OP} = \frac{9.6 \text{ cm}}{12.5 \text{ cm}} = 0.77.$$

$$\text{The tangent of } 40° = \frac{\text{side opposite}}{\text{side adjacent}} = \frac{PQ}{OQ} = \frac{8 \text{ cm}}{9.6 \text{ cm}} = 0.83.$$

Determine by construction and measurement the sine, cosine, and tangent of the angles listed in Table 27-2. After completing the table,

TABLE 27-2

Angle	sin	cos	tan	Angle	sin	cos	tan
10°				50°			
15°				55°			
20°				60°			
25°				65°			
30°				70°			
35°				75°			
40°				80°			
45°				85°			

compare the values of the sine and cosine of 10° and 80°, 15° and 75°, 20° and 70°, and so on. The sine of the one angle and the cosine of the other should be the same; that is, sin 10° = cos 80°, etc. As stated earlier, any function of an angle is the cofunction of the complement of that angle. For example, from the completed Table 27-2, determine the value of the cotangent of 5°.

It follows that any function of an angle larger than 45° is a function of an angle less than 45°. If a table is made, then, for the functions of all angles from 0° to 45°, it can be used for finding the functions of the angles from 45° to 90° as well.

■ EXERCISES

27-15 Give orally the six trigonometric ratios of each of the acute angles of the right triangles in Fig. 27-8.

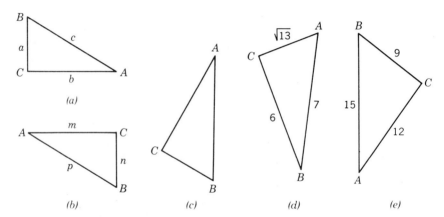

Fig. 27-8. Exercise 27-15.

27-16 In a right triangle find a if $\sin A = \frac{2}{5}$ and $c = 3.45$.

Solution: Substituting in the formula $\sin A = a/c$, we obtain $\dfrac{2}{5} = \dfrac{a}{3.45}$, solving for a, $a = 1.38$.

27-17 In a right triangle find a if $\sin A = \frac{3}{4}$ and $c = 1$.

27-18 In a right triangle find b if $\cos A = \frac{2}{3}$ and $c = 2$.

27-19 In a right triangle find a if $\tan A = 2$ and $b = 4$.

27-20 In a right triangle find c if $\sin A = \frac{5}{16}$ and $a = 8$.

27-21 In a right triangle find b if $\sin B = \frac{3}{7}$ and $c = 16$.

27-22 In a right triangle find a and c if $\sin B = \frac{2}{3}$ and $b = 48$.

27-23 In a right triangle find a and b if $\sin A = 0.245$ and $c = 100$.

27-24 Express the following functions as functions of the complements of these angles: sin 60°; sin 20°; cos 45°; tan 70°; sec 27°; csc 42°.

27-25 If $\sin \theta = \cos 60°$, find θ.

27-26 If $\tan 40° = \cot \theta$, find θ.

27-27 If $\tan \theta = \cot 3\theta$, find θ.

27-28 Express each of the following functions as functions of angles less than 45°: $\cos 70°$; $\sin 80°$; $\cot 82.43°$; $\tan 61.7093°$.

TRIGONOMETRIC TABLES

28-1 INTRODUCTION TO TRIGONOMETRIC TABLES

In the previous chapter, definitions of the six trigonometric (trig) functions were given. The values of these functions (ratios) were calculated for angles of 30° and 45°. A method for determining the approximate values of functions of an angle of 40° was demonstrated. Another explanation of the trigonometric ratios, to reemphasize, is that the values of the functions and cofunctions of complementary angles are equal.

Trigonometric ratios are abstract numbers and in general cannot be expressed exactly as decimals. The values for these ratios or functions are arranged in tables similar to tables of logarithms. The functions may be carried to any number of decimal places. The greater the number of decimal places in a table, the more accurate the computations in which the table is used.

A good hand-held calculator is the most efficient way to determine the value of trig functions. The calculator is instantaneous and more accurate than tables, generally to eight places. A calculator will give the ratio of any size angle including the sign of the function. The procedure to obtain the trig values will vary according to the calculator. This process will be explained in the particular calculator manual.

A calculator is recommended in the solution of trigonometry problems. However, being able to determine the trig values from tables will provide greater understanding and appreciation of trigonometry. The procedures are covered in this chapter.

28-2 DETERMINING VALUES OF TRIGONOMETRIC FUNCTIONS FROM A TABLE

Trigonometric tables are used to obtain values of trigonometric functions of various angles. For example, Tables 10, 11, and 12 in the appendix list the values of sine, cosine, and tangent for angles up to 90°.

When the angle is in terms of whole degrees, find the appropriate table, locate the angle in the lefthand column, and read off the value in the column headed 0.0.

EXAMPLE 28-1   Find tan 5°.

Solution

1. Go to Table 12.
2. In the lefthand column, find 5°.
3. In the second column (0.0), read tan 5°.

$$\tan 5° = 0.087\ 49$$

When the angle includes tenths of a degree, find the appropriate table, locate the whole degrees in the lefthand column, and read the value in the column headed by the appropriate tenth of a degree.

EXAMPLE 28-2   Find cos 14.8°.

Solution

1. Go to Table 11.
2. In the lefthand column, find 14°.
3. Read across until you find the column headed 0.8. This is the value of the cos of 14.8°.

$$\cos 14.8° = 0.966\ 82$$

When an angle includes hundredths of a degree, you must interpolate as you did with logarithms.

EXAMPLE 28-3   Find the sine of 16.46°.

Solution

1. Turn to Table 10.
2. Find 16° in the lefthand column.
3. Read across until you reach the columns headed 0.4 and 0.5.
4. The sine of 16.46° will lie between these two figures: sin 16.4° = 0.282 34 and sin 16.5° = 0.284 02.

$$\frac{0.06}{0.10} = \frac{x}{0.001\ 68}$$

$$x = \frac{0.001\ 68 \times 0.06}{0.10} = 0.001\ 01$$

$$\begin{aligned} \sin 16.46° &= \sin 16.4° + x \\ &= 0.282\ 34 + 0.001\ 01 \\ &= 0.283\ 35 \end{aligned}$$

Note: *To check this interpolation, make certain that the value is between the sine values of 16.4° and 16.5°.*

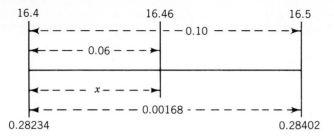

Fig. 28-1. Interpolation for Example 28-3.

To find tangent values, use Table 12 just as you used the sine tables. Table 11 is for natural cosines. The format of the table is like the table for sines; however, the cosine values are of *decreasing* value with increasing angles.

For example,

Angle	Sine Value	Cosine Value
0.0°	0.00000	1.00000
44.9°	0.70587	.70834

With cosine values, you must subtract when you interpolate.

EXAMPLE 28-4 Find cos 26.49°.

Solution

1. Go to Table 11.
2. Find 26° in the lefthand column.
3. Read across this line until you reach the columns headed 0.4 and 0.5.
4. The cosine of 26.49° will be between these values: cos 26.4° = 0.895 71 and cos 26.5° = 0.894 93.

$$\frac{0.09}{0.10} = \frac{x}{-0.000\ 78}$$

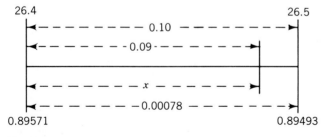

Fig. 28-2. Interpolation for Example 28-4.

$$x = \frac{-0.000\ 78 \times 0.09}{0.10} = -0.000\ 70$$

$$\cos 26.49 = \cos 26.4 + x$$
$$= 0.895\ 71 - 0.000\ 70$$
$$= 0.895\ 01$$

Note: *To check this interpolation, make certain that the value is between the cosine values for 26.4° and 26.5°.*

EXERCISES

28-1 Find the sine values of the following angles:

(a) 3.5°   (e) 78.5°   (i) 57.7°
(b) 7.8°   (f) 45.0°   (j) 60.0°
(c) 25.1°   (g) 90.0°   (k) 30.0°
(d) 33.3°   (h) 40.5°

28-2 Find the cosine values of the angles in Exercise 28-1.

28-3 Find the tangent values of the angles in Exercise 28-1.

28-3 FINDING THE ANGLE CORRESPONDING TO A VALUE OF A TRIGONOMETRIC FUNCTION

To find the angle corresponding to a given value of a trigonometric function, you must reverse the steps outlined above.

EXAMPLE 28-5 Find x when $\sin x = 0.332\ 16$

Solution

1. Go to Table 10.
2. Go down the second column (0.0) until you come to a number *larger* than or equal to 0.332 16. In this case, you will be at 20°, 0.342 02.
3. Go back up one line and read across until you find a number equal to or greater than 0.332 16.
4. In this case, 0.332 16 is in the table, at the intersection of the 19° line and the 0.4° column. $x = 19.4°$.

EXAMPLE 28-6 Find x when $\sin x = 0.332\ 50$.

Solution

1. Follow steps 1 to 3 outlined above.
2. 0.332 50 does not occur in the table. However, $\sin 19.4° = 0.332\ 16$ and $\sin 19.5° = 0.333\ 81$. Therefore, x must be between 19.4° and 19.5°.
3. Interpolate for x.

$$x = 19.4° + y$$

$$\frac{y}{0.000\ 34} = \frac{0.1°}{0.001\ 65}$$

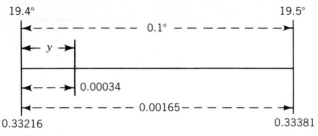

Fig. 28-3. Interpolation for Example 28-6.

$$y = \frac{0.1° \, (0.000\ 34)}{0.001\ 65}$$

$$= 0.020\ 61°$$
$$x = 19.4° + 0.02°$$
$$= 19.42°$$

Note: *To check this result, make certain that 19.42° is between 19.4° and 19.5°.*

EXAMPLE 28-7 Find x when $\cos x = 0.954\ 50$.

Solution

1. Turn to Table 11.

2. Read down the second column until you reach a number *smaller* than or equal to 0.954 50. (Remember, the table of cosine values is descending in magnitude.) This will be the 18° line.

3. Go up one line and read across until you reach a number which is less than or equal to 0.954 50.

4. 0.954 50 falls between the numbers in the 17° line in the columns headed by 0.3 and 0.4. Therefore, x must be between 17.3° and 17.4°.

5. Interpolate as before.

$$x = 17.3° + y$$

$$\frac{y}{-0.000\ 26} = \frac{0.1°}{-0.000\ 52}$$

$$y = \frac{0.1° \ (-0.000\ 26)}{-0.000\ 52}$$

$$y = 0.05°$$
$$x = 17.3° + 0.05° = 17.35°$$

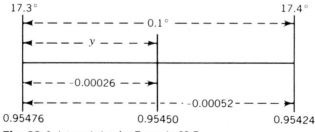

Fig. 28-4. Interpolation for Example 28-7.

■ EXERCISES

Find the angle for each of the following trigonometric values:

28-4	$\sin x = 0.587\ 79$	**28-16**	$\cos x = 0.433\ 09$
28-5	$\sin x = 0.832\ 92$	**28-17**	$\cos x = 0.923\ 06$
28-6	$\sin x = 0.500\ 00$	**28-18**	$\cos x = 0.301\ 00$
28-7	$\sin x = 0.707\ 11$	**28-19**	$\cos x = 0.801\ 08$
28-8	$\sin x = 0.881\ 75$	**28-20**	$\tan x = 0.577\ 35$
28-9	$\sin x = 0.450\ 00$	**28-21**	$\tan x = 1.0000$
28-10	$\sin x = 0.733\ 00$	**28-22**	$\tan x = 1.732\ 05$
28-11	$\sin x = 0.620\ 00$	**28-23**	$\tan x = 0.582\ 01$
28-12	$\cos x = 0.707\ 11$	**28-24**	$\tan x = 0.291\ 03$
28-13	$\cos x = 0.866\ 03$	**28-25**	$\tan x = 0.415\ 21$
28-14	$\cos x = 0.500\ 00$	**28-26**	$\tan x = 1.629\ 78$
28-15	$\cos x = 0.346\ 94$	**28-27**	$\tan x = 0.791\ 19$

28-4 EVALUATION OF FORMULAS

Formulas in numerous areas of business and industry contain trigonometric functions. When formulas include powers of trigonometric functions, the exponent is placed before the symbol for the angle. Thus, $\sin^2 30°$ means the square of $\sin 30°$ or $(\sin 30°)^2$.

EXAMPLE 28-8 Given $R = \dfrac{3a\ \sin^2(x/3)}{4}$, find R if $a = 4$ and $x = 66°$.

Solution

$$R = \frac{3a\ \sin^2(x/3)}{4}$$

$$= \frac{(3)\ \overset{1}{\cancel{4}}\ \sin^2(66°/3)}{\underset{1}{\cancel{4}}}$$

$$= (3)\ \sin^2(22°)$$
$$= (3)(.37461)^2$$
$$= (3)(.1403326521)$$
$$= 0.420997956$$
$$= 0.421$$

■ EXERCISES

Note: *Use a scientific calculator for the most efficient solution of these exercises.*

28-28 Given $R = 0.75a\ \sin^2 \frac{1}{3}x$, find R if $a = 4$ and $x = 66.5°$.

28-29 Given $R = \dfrac{2a}{\cos^3 0.5x}$, find R if $a = 16$ and $x = 16.3°$.

28-30 Given $R = 3a \sin x \cos x$, find R if $a = 2$ and $x = 15°$.

28-31 Given $R = \dfrac{a(5 - 4 \cos x)^{1.5}}{9 - 6 \cos x}$, find R if $a = 100$ and $x = 18°$.

28-32 Given $R = \dfrac{a(1 - e^2)(1 - 2e \cos x + e^2)^{1.5}}{(1 - e \cos x)^3}$, find R if $a = 5$, $e = 0.4$, and $x = 4.6°$.

28-33 Find the numerical value of $r^{2/3}(s^2 - t^2) \tan \theta$, where $r = 25.2$, $s = 90$, $t = 49.6$, and $\theta = 31.8°$.

28-34 Given $x = \dfrac{\tan 1.3788}{\sqrt{3} + \frac{4}{3}\pi}$, find x to four decimal places.

Suggestion: *Notice that the angle is expressed in radians; it is first necessary to convert 1.3788 rad to degrees to find the tangent.*

28-35 If two sides a and b and the included angle C of a triangle are given, then the third side c is given by the formula

$$c = \sqrt{a^2 + b^2 - 2ab \cos C}$$

Find c if $a = 748$, $b = 375$, and $C = 63.35°$.

28-36 In constructing lenses, the formula

$$n = \frac{\sin [(A + D)/2]}{\sin (A/2)}$$

is used. Find n if $A = 42.3°$ and $D = 34.87°$.

28-37 To determine the height of an object (such as the height of an airplane), the formula

$$h = \frac{s}{(\cot A - \cot B)}$$

may be used. Find h if $A = 69.40°$, $B = 80°$, and $s = 1$ km.

28-38 Find the value of each of the following:
(a) $\sin^2 33° + \cos^2 33°$
(b) $\sin^2 41° + \cos^2 41°$
(c) $\sin^2 68.10° + \cos^2 68.10°$
(d) $\sin^2 83.40° + \cos^2 83.40°$

Compare the results and state your conclusions.

28-39 The velocity v of a body sliding a distance s down a smooth plane inclined at an angle ϕ from the horizontal is given by the formula

$$v = \sqrt{2gs \sin \phi}$$

where $g = 9.81$ m/s². See Fig. 28-5. Find v when $s = 50$ cm and $\phi = 27°$.

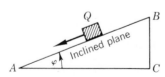

Fig. 28-5. Exercise 28-39.

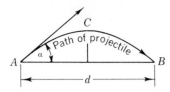

Fig. 28-6. Exercise 28-40.

28-40 If the resistance of the air is disregarded, the distance a projectile will travel along a horizontal plane is given by the formula

$$d = \frac{v^2 \sin 2\alpha}{g}$$

where v = velocity at which the body is projected, m/s
$\qquad \alpha$ = angle that the initial direction makes with the horizontal
$\qquad d$ = distance along the horizontal

The value of g may be taken as 9.81 m/s². This is shown in Fig. 28-6. Find d if v is 240 m/s and α is 5°. Using the same velocity, find d when α is 20°, 30°, 40°, and 45°. Notice that the distance is greatest when the angle is 45°.

28-41 Disregarding the resistance of the air, the highest point reached by a projectile is given by the formula

$$y = \frac{v^2 \sin^2 \alpha}{2g}$$

Find the greatest height above the starting point reached by a projectile having an initial velocity of 2000 ft/s when α is 5°, 10°, 20°, 30°, 45°, 60°, and 90°.

28-42 The height y reached by a projectile when it has traversed a horizontal distance x, if it has been projected with a velocity v in a direction making an angle α with the horizontal, is given by the following formula:

$$y = x \tan \alpha - \frac{gx^2}{2v^2 \cos^2 \alpha}$$

Find y when x = 1000 yd, v = 2000 ft/s, α = 5°, and g = 32 ft/s².

28-43 If the resistance of the air is disregarded, the greatest horizontal distance a missile will travel is found by making the initial direction at an angle of 45° from the horizontal. Find the greatest horizontal distance a missile having an initial velocity of 2200 ft/s can reach.

28-44 If F is the force required to move a weight W up a plane inclined from the horizontal at an angle α, and μ is the coefficient of friction, then

$$F = W\left(\frac{\sin \alpha + \mu \cos \alpha}{\cos \alpha - \mu \sin \alpha}\right)$$

Calculate F if W = 90 kg, α = 30°, and μ = 0.2.

28-45 To compute the illumination on a surface that is not perpendicular to the

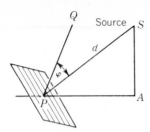

Fig. 28-7. Exercise 28-45.

rays of light from a source of light (Fig. 28-7), the following formula is used:

$$E = \frac{I}{d^2} \cos \phi$$

when E = illumination at the point on the surface in lumens per square foot (lm/ft^2)

I = luminous intensity of the source in candelas (cd)

d = distance from the source of light in ft

ϕ = angle between the incident ray and a line perpendicular to the surface

Solve the above formula for d and I.

28-46 From the formulas found in Exercise 28-45, compute the following:
(a) E when $I = 50$ cd, $d = 10$ ft, and $\phi = 75°$
(b) d when $I = 60$ cd, $E = 0.25$ ft, and $\phi = 65°$
(c) I when $E = 4$ lm/ft^2, $d = 8$ ft, and $\phi = 45°$

28-47 What do the formulas in Exercise 28-45 become if $\phi = 0°$, that is, if the rays are perpendicular to the surface?

28-48 To compute the illumination on a horizontal surface from a source of light at a given vertical distance from the surface (see Fig. 28-8), the following formula is used:

$$E_h = \frac{I}{h^2} \cos^3 \phi$$

where E_h = illumination at a point on the horizontal surface, lm/ft^2

I = luminous intensity of the source, cd

h = vertical distance from the horizontal surface to the source of light, ft

ϕ = angle between the incident ray and a vertical line

Solve the above formula for h and I and obtain the following formulas:

$$h = \sqrt{\frac{I \cos^3 \phi}{E_h}} \qquad I = \frac{E_h h^2}{\cos^3 \phi}$$

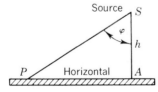

Fig. 28-8. Exercise 28-48.

28-49 Using the previous formulas, compute the following:
(a) E_h when I = 250 cd, h = 12 ft, and ϕ = 55°
(b) h when I = 100 cd, E_h = 65 lm/ft², and ϕ = 12°
(c) I when E_h = 0.85 lm/ft², h = 8 ft, and ϕ = 37°

28-50 The perimeter P and the area A of a regular polygon of n sides inscribed in a circle of radius R is given by these formulas:

$$P = 2nR \sin\left(\frac{180°}{n}\right)$$

$$A = nR^2 \sin\left(\frac{180°}{n}\right) \cos\left(\frac{180°}{n}\right)$$

Find the perimeter and area of a regular polygon inscribed in a circle of unit radius in the following instances:
(a) n = 3
(b) n = 12
(c) n = 100 (Compare the results with the circumference and area of the circle.)

28-5 TRIGONOMETRIC FUNCTIONS OF ANGLES GREATER THAN 90°

Angles may be of any magnitude—such as 127°, 350°, or −130°. A table could be developed for these different angles, but it would be many pages—much larger than the tables included in this text. More important, it would be impossible to set up a table that would include all the possible angles.

It is possible, however, to develop formulas that make the present tables usable for angles of any size and for negative angles.

Note: *A scientific calculator will give the value of the function, including the sign, of any size angle, including negative angles. However, being able to determine the functions from tables of angles greater than 90 degrees and negative angles will provide greater understanding and appreciation of trigonometry. The procedures are covered in sections 28-6 to 28-11.*

28-6 TRIGONOMETRIC FUNCTIONS OF AN ANGLE IN THE SECOND QUADRANT

The angle 90° + θ, where θ is an acute angle, is an angle in the second quadrant.

In Fig. 28-9, θ is any acute angle. Locate P on the terminal side of angle θ. Locate P' on the terminal side of angle 90° + θ so that $OP = OP'$. Triangles OPM and $OP'M'$ are right triangles by construction; that is, $P'M'$ and PM were drawn at right angles to $M'OM$. Angle $M'P'O$ = angle POM.

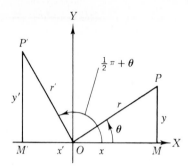

Fig. 28-9. Functions of an angle in the second quadrant.

Note: *Angle P'OP is a right angle by construction. Angle M'OP' = angle OPM. Then right triangles M'P'O and MPO are equal and y' = x, x' = −y, and r' = r.*

$$\sin(90° + \theta) = \frac{\text{ordinate}}{\text{distance}} = \frac{y'}{r'}$$

Since $y' = x$ and $r = r'$,

$$\sin(90° + \theta) = \frac{x}{r}$$

And since $\dfrac{x}{r}$ is cos θ.

$$\sin(90° + \theta) = \cos\theta$$

The five remaining functions are as follows:

$$\cos(90° + \theta) = \frac{x'}{r'} = \frac{-y}{r} = -\frac{y}{r} = -\sin\theta$$

$$\tan(90° + \theta) = \frac{y'}{x'} = \frac{x}{-y} = -\frac{x}{y} = -\cot\theta$$

$$\cot(90° + \theta) = \frac{x'}{y'} = \frac{-y}{x} = -\frac{y}{x} = -\tan\theta$$

$$\sec(90° + \theta) = \frac{r'}{x'} = \frac{r}{-y} = -\frac{r}{y} = -\csc\theta$$

$$\csc(90° + \theta) = \frac{r'}{y'} = \frac{r}{x} = \sec\theta$$

Note that in each of the formulas above, the function to the right is the cofunction of the one to the left. That is, sin (90° + θ) = cos θ, for example. Applications of this principle are:

$$\sin 130° = \sin(90° + 40°) = \cos 40° = 0.766\ 04$$
$$\sin 130° = 0.766\ 04$$

$$\cot 110° = \cot(90° + 20°) = -\tan 20° = -0.363\ 97$$
$$\cot 110° = -0.363\ 97$$

It is possible to determine the values of the functions for all angle values from 90° to 180°, using the tables in this text.

28-7   TRIGONOMETRIC FUNCTIONS OF AN ANGLE IN THE THIRD QUADRANT

All six trigonometric functions have been included in explanations and demonstrations up to this point. Only three functions will be discussed in the remaining sections of this chapter. Remember that the three remaining functions are reciprocals of the three given functions.

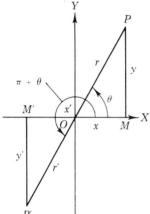

Fig. 28-10. Functions of an angle in the third quadrant.

In Fig. 28-10, let θ be an acute angle. Construct $180° + \theta$; take $OP' = OP$; represent the other parts as shown.

Then $x' = -x$; $y' = -y$; and $r' = r$; and we have

$$\sin (180° + \theta) = \frac{y'}{r'} = \frac{-y}{r} = -\frac{y}{r} = -\sin \theta$$

$$\cos (180° + \theta) = \frac{x'}{r'} = \frac{-x}{r} = -\frac{x}{r} = -\cos \theta$$

$$\tan (180° + \theta) = \frac{y'}{x'} = \frac{-y}{-x} = \frac{y}{x} = \tan \theta$$

Notice that in the formulas above, in each line the function at the end is the same function as at the beginning.

EXAMPLE 28-9 $\tan 230° = \tan (180° + 50°) = \tan 50°$

EXAMPLE 28-10 $\cos 205° = \cos (180° + 25°) = -\cos 25°$

28-8   TRIGONOMETRIC FUNCTIONS OF AN ANGLE IN THE FOURTH QUADRANT

Angle $270° + \theta$, where θ is an acute angle, is an angle in the fourth quadrant. In Fig. 28-11, let θ be an acute angle. Construct $270° + \theta$; take $OP' = OP$; and represent the other parts as shown.

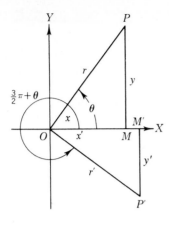

Fig. 28-11. Functions of an angle in the fourth quadrant.

Then $y' = -x$; $x' = y$; and $r' = r$; and we have

$$\sin (270° + \theta) = \frac{y'}{r'} = \frac{-x}{r} = -\frac{x}{r} = -\cos \theta$$

$$\cos (270° + \theta) = \frac{x'}{r'} = \frac{y}{r} = \sin \theta$$

$$\tan (270° + \theta) = \frac{y'}{x'} = \frac{-x}{y} = -\frac{x}{y} = -\cot \theta = -\frac{1}{\tan \theta}$$

Notice that in the formulas above in each line the function at the end is the cofunction of the one at the beginning.

EXAMPLE 28-11 $\cos 310° = \cos (270° + 40°) = \sin 40°$

28-9 FUNCTIONS OF A NEGATIVE ANGLE

Sometimes it is necessary to find the function of a negative angle. The following formulas are proved for θ if θ is an acute angle, but they can be shown to hold for any value of $-\theta$.

Construct θ and $-\theta$ as in Fig. 28-12; take $OP' = OP$; and represent the other parts as shown. Then $x' = x$; $y' = -y$; and $r' = r$; we have

$$\sin (-\theta) = \frac{y'}{r'} = \frac{-y'}{r} = -\frac{y}{r} = -\sin \theta$$

$$\cos (-\theta) = \frac{x'}{r'} = \frac{x}{r} = \cos \theta$$

$$\tan (-\theta) = \frac{y'}{x'} = \frac{-y}{x} = -\frac{y}{x} = -\tan \theta$$

Notice that the functions of the negative angle are the same as those of the positive angle, but of the opposite sign—except the cosine, which is of the same sign.

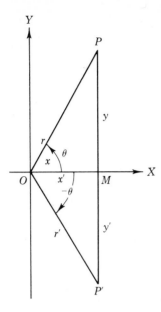

Fig. 28-12. Functions of a negative angle.

EXAMPLE 28-12 $\sin(-40°) = -\sin 40°$

EXAMPLE 28-13 $\cos(-145°) = \cos 145°$

EXAMPLE 28-14 $\tan(-240°) = -\tan 240°$

28-10 SUMMARY OF THE REDUCTION FORMULAS

In the second quadrant, we could have used $180° - \theta$; in the third quadrant, $270° - \theta$; and in the fourth quadrant, $360° - \theta$. The functions of these angles, together with those of $90° - \theta$ and the ones derived, are arranged in Table 28-1 for reference.

Although the proofs of all the formulas in Table 28-1 were based on the assumption that θ is an acute angle, they are true for all values of θ and can be used for any value of θ in the same way as when θ is an acute angle.

28-11 FUNCTIONS OF AN ANGLE GREATER THAN 360°

Any angle α greater than 360° has the same trigonometric functions as α (the Greek letter alpha) minus an integral multiple of 360°, because α and $(\alpha - n)(360°)$ have the same initial and terminal sides. That is, the functions of α equal the same functions of $(\alpha - n)(360°)$, where n is an integer. In other words, a function of an angle that is larger than 360° can be found by dividing the angle by 360° and finding the required function of the remainder.

TABLE 28-1 REDUCTION FORMULAS

$\sin(90° - \theta) = \cos\theta$	$\sin(90° + \theta) = \cos\theta$
$\cos(90° - \theta) = \sin\theta$	$\cos(90° + \theta) = -\sin\theta$
$\tan(90° - \theta) = \cot\theta = 1/\tan\theta$	$\tan(90° + \theta) = -\cot\theta = -1/\tan\theta$
$\sin(180° - \theta) = \sin\theta$	$\sin(180° + \theta) = -\sin\theta$
$\cos(180° - \theta) = -\cos\theta$	$\cos(180° + \theta) = -\cos\theta$
$\tan(180° - \theta) = -\tan\theta$	$\tan(180° + \theta) = \tan\theta$
$\sin(270° - \theta) = -\cos\theta$	$\sin(270° + \theta) = -\cos\theta$
$\cos(270° - \theta) = -\sin\theta$	$\cos(270° + \theta) = \sin\theta$
$\tan(270° - \theta) = \cot\theta = 1/\tan\theta$	$\tan(270° + \theta) = -\cot\theta = -1/\tan\theta$
$\sin(360° - \theta) = -\sin\theta$	$\sin(-\theta) = -\sin\theta$
$\cos(360° - \theta) = \cos\theta$	$\cos(-\theta) = \cos\theta$
$\tan(360° - \theta) = -\tan\theta$	$\tan(-\theta) = -\tan\theta$

EXAMPLE 28-15 Find $\cos 1240°$.

Solution

$$1240° = 3(360°) + 160°$$

Therefore,

$$\cos 1240° = \cos 160°$$
$$\cos 160° = \cos(90° + 70°) = -\sin 70° = -0.939\,69$$

EXAMPLE 28-16 Find $\tan(-684°)$.

Solution

$$\tan(-684°) = -\tan 684° = -\tan 324° = 0.72654$$

Procedure to find the sine, cosine, or tangent of any angle:

1. If the angle is negative, convert the angle to a positive angle by the method explained in Sec. 28-9.
2. If the angle is greater than 360°, convert to an angle of less than 360° by subtracting an appropriate integral multiple of 360° (Sec. 28-11).
3. If the angle is greater than 90°, convert to an angle of less than 90° by using the appropriate formula in Table 28-1. This may result in a change of sign (steps 1 and 2); in addition, if the angle (after completing steps 1 and 2) is in the second or fourth quadrant, it will result in a change from the function originally to be evaluated to its cofunction.

Note: *Study carefully the double algebraic signs illustrated in Example 28-16.*

To find the cotangent, secant, or cosecant of any angle, remember that these functions are the reciprocals of the tangent, cosine, and sine, respectively.

EXERCISES

28-51 Using the tables of natural functions, find the sine, cosine, and tangent of each of the following angles.

Note: *It will be necessary to apply the formulas of Table 28-1 or the summary given in Sec. 28-11.*

(a)	140°	(g)	460°
(b)	170°	(h)	1220°
(c)	190°	(i)	3890°
(d)	250°	(j)	−1190°
(e)	280°	(k)	−915°
(f)	340°	(l)	−1420.66°

28-52 Find the angle A, less than 90°, for which the following will be true:

(a) $\sin 20° = \sin A$

(b) $\cos 130° = \cos A$

(c) $\tan 250° = \tan A$

(d) $\sin 20° = \cos A$

(e) $\cos 130° = \sin A$

(f) $\tan 250° = \cot A$

RIGHT TRIANGLES

29-1 INTRODUCTION TO RIGHT TRIANGLES

In the previous chapters, you studied trigonometric functions, trigonometric tables, and the relations between functions. The purpose of this chapter is to apply this knowledge to the solution of trigonometry problems. Some of the problems presented will be fairly routine, to help you develop skill in using the facts learned; some will be verbal problems which will give you the opportunity to apply the facts.

Finding the height of a missile, the direction of a road, or the width of a river are examples of problems which can be solved by trigonometry. For simple solutions, these and similar problems are visualized as right-triangle problems.

Using a scientific calculator is the most efficient way to solve right triangles.

29-2 PROPERTIES OF THE RIGHT TRIANGLE

A right triangle is a three-sided figure with one right angle. A right triangle is shown in Fig. 29-1; *ABC* is a right triangle with angle *ACB* the

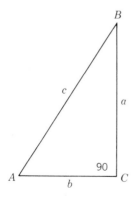

Fig. 29-1. Right triangle.

right angle. It is conventional to label the three angles of a triangle with capital letters (*A*, *B*, and *C*), and the three sides with the lowercase letters (*a*, *b*, and *c*). Side *a* is opposite angle *A;* side *b* is opposite angle *B;* and side *c* is opposite angle *C*.

If angle *C* ($\angle C$) is equal to 90°, then $\angle A + \angle B = 90°$ because $\angle A + \angle B + \angle C$ must equal 180°. In a right-angle triangle, if one acute angle is known, the other acute angle can be determined by subtraction. For example, if $\angle A = 60°$, then $\angle B = 30°$, because $\angle A + \angle B$ must equal 90°.

Remember that $c^2 = a^2 + b^2$: in a right triangle the hypotenuse squared is equal to the sum of the squares of the other two sides.

There are six facts that completely describe a right triangle: three angles and three sides. To determine all six facts by computation, at least three must be known, and one of these three must be a side.

29-3 SOLVING RIGHT-TRIANGLE PROBLEMS

When solving right-triangle problems, you should draw the triangle, list the known facts, and then select an appropriate equation or equations to determine the remaining facts.

Several examples will illustrate the procedure.

EXAMPLE 29-1 In the right triangle *ABC* in Fig. 29-2, given $\angle A = 32.20°$ and $b = 10$ ft. find *B*, *a*, and *c*.

Solution In Fig. 29-2,

$$\sin A = \frac{a}{c} \qquad \cos A = \frac{b}{c} \qquad \tan A = \frac{a}{b}$$

Find:

$$\angle B = ?$$
$$c = ?$$
$$a = ?$$

Since $\angle A + \angle B = 90°$ and $A = 32.20°$, *B* can be found. $32.20° + \angle B = 90°$; by subtraction $\angle B = 57.80°$.

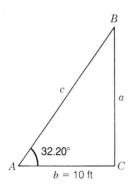

Fig. 29-2. Example 29-1: Find *B*, *a*, *c*.

In the formula $\tan A = a/b$, two facts are known—angle A and side b; the third fact, side a, can be calculated by substituting the known values for $\tan A$ and b:

$$\tan A = \frac{a}{b}$$

$$\tan 32.20° = \frac{a}{10}$$

$$10(\tan 32.20) = a$$
$$10(0.629\ 73) = a$$
$$6.30 \text{ ft} = a$$

In the formula $\cos A = b/c$, two facts are known—side b and angle A; the third fact, side c, can therefore be calculated:

$$\cos A = \frac{b}{c}$$

$$\cos 32.20° = \frac{10}{c}$$

$$0.846\ 19 = \frac{10}{c}$$

$$c = \frac{10}{0.846\ 19} = 11.82 \text{ ft}$$

The three unknown values have been determined:

$$\angle B = 57.80°$$
$$c = 11.82 \text{ ft}$$
$$a = 6.30 \text{ ft}$$

It is possible to check the accuracy of these three answers. The first check is for the lengths of the three sides:

$$c^2 = a^2 + b^2$$
$$(11.82)^2 \stackrel{?}{=} (6.30)^2 + (10)^2$$
$$139.71 = 139.69$$

These are approximately equal, thus the lengths of the three sides are accepted as correct. If an eight-digit calculator had been used to determine tan and cos, the two numbers would be more nearly equal.

The second check, or test, is for the acute angle B:

$$\sin B = \frac{\text{side opposite}}{\text{hypotenuse}}$$

$$\sin B = \frac{b}{c} = \frac{10}{11.82} = 0.846\ 02 = \sin 57.80° \text{ (approximately)}$$

The subtraction is checked, and the answers calculated for angle B and sides c and a are accepted as correct.

The steps in the solution may be summarized as follows:

Procedure for solving right-triangle problems:

   1. Draw the triangle.
   2. List the known facts.
   3. List the facts to be calculated.
   4. Determine the formula to be used.
   5. Solve the formulas.
   6. Check the calculations.

Note: *It is possible to check the results by constructing the triangle with a protractor and rule.*

EXAMPLE 29-2 Given $A = 67.4°$ and $c = 23.47$ m, find B, a, and b. (Fig. 29-3.)

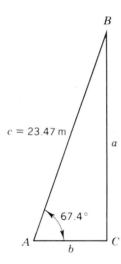

Fig. 29-3. Example 29-2: Construction.

Solution Select the formulas:

   1. $A + B = 90°$; therefore, $B = 90° - A$

   2. $\sin A = \dfrac{a}{c}$; therefore, $a = c \sin A$

   3. $\cos A = \dfrac{b}{c}$; therefore, $b = c \cos A$

Substitute the given values and evaluate.

(1)
$$B = 90° - A$$
$$= 90° - 67.4°$$
$$= 22.6°$$

(2)
$$a = c \sin A$$
$$= 23.47 \sin 67.4°$$
$$= 23.47 \ (0.923 \ 21)$$
$$= 21.67 \text{ m}$$

(3)
$$b = c \cos A$$
$$= 23.47 \cos 67.4°$$
$$= 23.47 \ (0.384 \ 30)$$
$$= 9.02 \ m$$

EXAMPLE 29-3 Given $c = 10.86$ m and $a = 7.27$ m, find b, A, and B (Fig. 29-4).

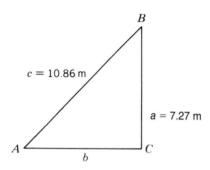

Fig. 29-4. Example 29-3: Find b, A, B.

Solution Select the ratios and formulas to be used.

1. $\sin A = \dfrac{a}{c} = \dfrac{7.27}{10.86}$ or $\cos B = \dfrac{a}{c} = \dfrac{7.27}{10.86}$

2. $b = \sqrt{c^2 - a^2}$ or $b = c \cos A$

Substitute the given quantities and evaluate.

(1)
$$\sin A = \frac{a}{c} = \frac{7.27}{10.86}$$

$$\sin A = 0.669 \ 43$$
$$A = 42.02°$$ and $B = 90° - A = 47.98°$

Or

$$\cos B = 0.669 \ 43$$
$$B = 47.98°$$ and $A = 90° - B = 42.02°$

Note: *In this illustration, one calculation determined angles A and B.*

(2)
$$b = c \cos A$$
$$= 10.86 \cos 42.02°$$
$$= 10.86 \, (0.742\ 91)$$
$$= 8.07 \text{ m}$$

▓ EXERCISES

Solve the following exercises for a triangle where $C = 90°$.

29-1 Given $A = 68.13°$, $c = 200$ cm; find B, a, and b.

29-2 Given $B = 66.30°$, $b = 575$ ft; find A, a, and c.

29-3 Given $A = 13.41°$, $a = 992$ m; find B, b, and c.

29-4 Given $B = 52.4°$, $b = 4$ ft; find A, a, and c.

29-5 Given $A = 53.30°$, $c = 30.69$ ft; find B, a, and b.

29-6 Given $A = 63°$, $c = 43$ m; find B, a, and b.

29-7 Given $B = 36.45°$, $a = 1758$ ft; find A, b, and c.

29-8 Given $B = 85.25°$, $a = 637$ ft; find A, b, and c.

29-9 Given $A = 86°$, $a = 0.0008$ km; find B, b, and c.

29-10 Given $A = 21.8°$, $a = 73$ ft; find B, b, and c.

29-11 Given $a = 2$ cm, $b = 2$ cm; find A, B, and c.

29-12 Given $c = 8$ ft, $b = 4$ ft; find A, B, and a.

29-13 Given $a = 8.49$ cm, $c = 9.35$ cm; find A, B, and b.

29-14 Given $b = 16.926$ ft, $a = 13,690$ ft; find A, B, and c.

29-15 Given $a = 2.19$ cm, $c = 91.92$ cm; find A, B, and b.

29-16 Given $c = 2194$ ft, $b = 1312.7$ ft; find A, B, and a.

29-17 Given $b = \sqrt[3]{2}$ ft, $c = \sqrt{3}$ ft; find A, B, and a.

29-4 VERBAL RIGHT-TRIANGLE PROBLEMS

The problems and solutions shown so far were mechanical and routine. Verbal or word problems are more difficult; your ability to solve a verbal problem is a better opportunity to increase your understanding of trigonometry.

The solutions of several verbal problems follow; notice that the first step in each case is a sketch of the problem. Many of the exercises will be very difficult or even impossible for you to solve unless you make a sketch of the problem.

EXAMPLE 29-4 If a person 1.7 m tall stands on level ground, and if the elevation of the sun is 60°, how long will the shadow be?

Solution Figure 29-5 is a sketch of this problem; Fig. 29-6 is the same sketch with the angles and sides labeled.

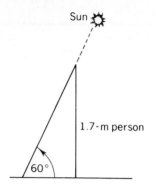

Fig. 29-5. Example 29-4: Sketch.

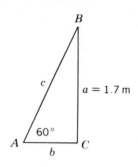

Fig. 29-6. Example 29-4: Angles and sides.

Two facts are known: $A = 60°$ and $a = 1.7$ m; b (the length of the shadow) must be calculated.

$$\tan A = \frac{a}{b}$$

$$b = \frac{a}{\tan A}$$

$$b = \frac{1.7 \text{ m}}{\tan 60°} = \frac{1.7 \text{ m}}{1.732\ 05}$$

$$b = 0.98 \text{ m}$$

This problem can be checked by computing the value of c using $c^2 = b^2 + c^2$ and by calculating angle B using $\angle A + \angle B = 90°$. It can also be checked by constructing the triangle to scale.

EXAMPLE 29-5 What angle does a rafter make with the horizontal if it has a rise of 6 ft in a run of 12 ft?

Solution Figure 29-7 is a sketch of this problem, and Fig. 29-8 is the same sketch with the sides and angles conventionally labeled. To find A, the angle of the rafter with the horizontal,

$$\tan A = \frac{a}{b} = \frac{6}{12} = 0.5000$$

$$A = 26.57°$$

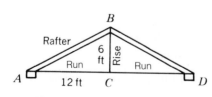

Fig. 29-7. Example 29-5: Sketch.

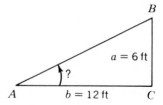

Fig. 29-8. Example 29-5: Angles and sides.

EXAMPLE 29-6 To determine the width of a river, a tree is observed standing directly across, on the opposite bank. The angle of elevation to the top of the tree as seen from the bank on the side of the river was 32°; at 150 ft back from this point, the angle of elevation of the top of the tree was 21°. Find the width of the river.

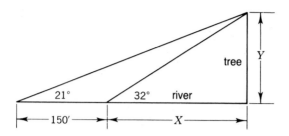

Fig. 29-9. Example 29-6.

Solution Figure 29-9 is the sketch for this problem. Set up two equations with two unknowns—the height of the tree (Y) and the width of the river (X):

$$\tan 32° = \frac{Y}{X} \tag{1}$$

$$\tan 21° = \frac{Y}{150 + X} \tag{2}$$

This is an application of the definition of the tangent:

$$\tan A = \frac{\text{side opposite}}{\text{side adjacent}}$$

The sketch not only helps us understand the problem but also—more important—directs our attention to the proper solution, here an application of the tangent function.

Thus with two equations with two unknowns, it is possible to solve the problem. The simultaneous equations are solved by methods covered in the algebra section.

$$\tan 32° = \frac{Y}{X} \tag{1}$$

$$\tan 21° = \frac{Y}{150 + X} \tag{2}$$

Solve equation (1) for Y:

$$X \tan 32 = Y$$

Substitute this value for Y in equation (2):

$$\tan 21° = \frac{X \tan 32°}{150 + X}$$

Substitute the tangent values and solve the equation for X:

$$0.383\ 86 = \frac{0.624\ 87X}{150 + X}$$

$$0.383\ 86(150 + X) = 0.624\ 87X$$
$$57.58 + 0.383\ 86X = 0.624\ 87X$$
$$0.241\ 01X = 57.58$$
$$X = 238.9\ \text{ft}$$

The height of the tree could also be determined.

The three examples given illustrate the value of sketching a problem and then applying trigonometric definitions to its solution. You should also have observed the application of principles learned in algebra.

■ EXERCISES

29-18 What is the inclination from the vertical of the face of a wall having a batter of $\frac{1}{8}$? A "batter of $\frac{1}{8}$" means an inclination 1 ft from the vertical in a rise of 8 ft.

29-19 What is the angle of inclination of a stairway from the floor if the steps have a tread of 8 in. and a rise of $6\frac{1}{2}$ in.?

29-20 Find the angle between the rafters and the horizontal in roofs of the following pitches: $\frac{2}{3}$; $\frac{1}{2}$; $\frac{1}{3}$; $\frac{1}{4}$.

Note: *The "pitch" of a roof is the ratio of the rise of the rafters to twice the run, or, in a V-shape roof (Fig. 29-10), the ratio of the distance from the plate to the ridge to the width of the building.*

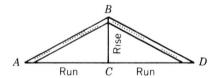

Fig. 29-10. Exercise 29-23.

29-21 A building 183 m tall stands on level ground, and the angle of elevation of the top is observed to be 5°. Find the distance from the point of observation to the foot of the building.

29-22 If the building in Exercise 29-21 is the Empire State Building, which is 380 m high, solve the problem.

29-23 The bottom of a picture on the wall is level with the eye of the observer. If the picture is 8 ft from top to bottom, and if the elevation of the top is 20°, how far is the observer's eye from the bottom of the picture?

29-24 One of the equal sides of an isosceles triangle is 4.2 in., and one of the base angles is 27.15°. Find the altitude and the base.

29-25 The base of an isosceles triangle is 60 cm and the vertex angle is 58.30°. Find the equal sides and the base angles.

29-26 If it is 1000 yd to the foot of an object 1 yd high, what angle does this object subtend? This is approximately a mil, which is used as the unit of angle measurement in artillery practice.

Note: Use the tangent function.

29-27 A mountain road has a rise of 1 mi for every 6 mi along the road. What is the angle of rise?

29-28 A grade of 1 percent in a roadbed is a rise of 1 ft in a horizontal distance of 100 ft, and a proportionate rise for other grades. What is the angle of slope of a roadbed that has a grade of 5 percent? Of a roadbed with a grade of 0.25 percent?

29-29 A road rises 400 m in a horizontal distance of 2600 m. Find the percent of grade and the inclination of the roadbed from the horizontal.

29-30 A person whose eyes are 5 ft 6 in. above the ground is on a level with, and 150 ft distant from, the foot of a flagstaff 72 ft high. What angle does the person's line of sight make with the horizontal line when the person is looking at the top of the staff?

29-31 In surveying, measurements were taken as shown in Fig. 29-11. Find the distance on a straight line from A to E.

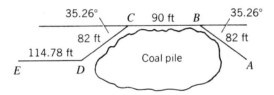

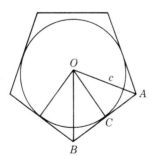

Fig. 29-11. Exercise 29-31.

29-32 The side of a regular pentagon (five-sided figure) is 12 in. Find the radius (apothem) of the inscribed circle and the area of the pentagon.

Suggestion: Draw the pentagon and inscribe the circle as in Fig. 29-12. Angle AOB = 72°. Triangle AOB is isosceles. We have tan 36° = 6/r. Therefore, r = 6 ÷ tan 36°. Could this problem be solved by geometry?

Fig. 29-12. Exercise 29-32.

29-33 Find a side of the regular octagon circumscribed about a circle 120 cm in diameter.

29-34 A building stands on level ground. From a point A on the ground, the elevation of the second-floor window sill, which is 20 ft from the ground, is 40°. When viewed from A, the angle of elevation of the top of the building is 70°. Find the height of the building.

29-35 An observation balloon was attached to a point A on the ground. On a level with A, and in the same straight line, points B and C were chosen so that BC equaled 9 m. From points B and C the angles of elevation of the balloon were 40° and 30°, respectively. Find the height of the balloon.

Suggestion: *Form two equations in two unknowns.*

29-36 Find the shorter altitude and the area of a parallelogram whose sides are 10 m and 25 m and whose acute angle between sides is 75°.

29-37 Two points C and B are on opposite banks of a river. A line AC at right angles to CB is measured 40 rods long, and the angle CAB is measured and found to be 41.40°. Find the width of the stream.

29-38 A ship is sailing due east at 16 mph. A lighthouse is observed due south at 8:30 A.M. At 9:45 A.M. the bearing of the same lighthouse is S 38.5° W. Find the distance from the ship to the lighthouse at the time of the first observation.

29-39 In order to locate accurately holes that are to be drilled in a piece of work, the piece is clamped to the table of a milling machine. The table is so constructed that it can be moved in two directions at right angles to each other. In order to drill five holes accurately spaced at the vertices of a regular pentagon inscribed in a circle of radius 1 in., the lengths of OF, FB, OG, and GC in Fig. 29-13 must be determined. Find these to the nearest 0.001 in.

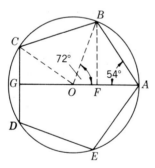
Fig. 29-13. Exercise 29-39.

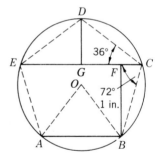
Fig. 29-14. Exercise 29-40.

29-40 Locate the holes in the same piece by determining the lengths of AB, BF, FC, EC, EG, and GD in Fig. 29-14 to the nearest 0.001 in.

29-41 Locate the centers of the holes B and C in Fig. 29-15 by finding the distance each is to the right of and above the center O. The radius of the circle is 1.5 in. Compute correctly to three decimal places.

29-42 To drill holes accurately at A, B, and C in Fig. 29-16, it is necessary to determine AC and CB. Find them to the nearest 0.001 in.

29-43 For drilling holes at A, B, and C in Fig. 29-17, determine lengths AD, DC, and DB to the nearest 0.001 cm.

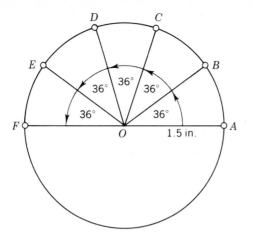

Fig. 29-15. Exercise 29-41.

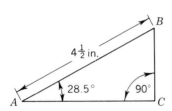

Fig. 29-16. Exercise 29-42.

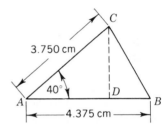

Fig. 29-17. Exercise 29-43.

29-44 One side of a regular pentagon inscribed in a circle is 250 mm. Find the radius of the circle.

29-45 One side of a regular octagon inscribed in a circle is 25 cm. Find the radius of the circle.

29-46 A man surveying a mine measures a length $AB = 220$ ft due east with a dip of 6.15°, then a length $BC = 325$ ft due south with a dip 10.45°. How much deeper is C than A?

29-47 In the side of a hill that slopes upward at an angle of 32° a tunnel is bored sloping downward at an angle of 12.15° from the horizontal. How far below the surface of the hill is a point 115 ft down the tunnel?

29-48 The angle of elevation of a balloon from a point due south of it is 60°; from another point 1 mi due west of the first point, the angle of elevation is 45°. Find the height of the balloon.

29-49 From the top of a mountain 1050 ft high, two buildings are seen on a level plane and in a direct line from the foot of the mountain. The angle of depression of the first is 35°, and that of the second is 24°. Find the distance between the buildings.

29-50 A chord of 2 ft is in a circle of radius 3 ft. Find the length of the arc subtended by the chord and the number of degrees in it. (See Fig. 29-18.)

29-51 Find the angle between the diagonal of a cube and one of the diagonals of a face that meets it.

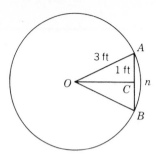

Fig. 29-18. Exercise 29-50.

29-52 What size target at 300 m from the eye subtends the same angle as a target 2 m in diameter at 1000 m? Find the angle that it subtends.

29-53 An observation balloon has an angle of elevation of 42.50° from point A on level ground and an elevation of 36.75° from point B directly above A. If AB is 66 ft, find the height of the balloon and the distance from A to a point directly under the balloon.

29-54 At a certain point the angle of elevation of a mountain peak is 44.50°, at a distance of 3 miles farther away in the same horizontal plane, its angle of elevation is 29.75°. Find the distance of the top of the mountain above the horizontal plane and the horizontal distance from the first point of observation to the peak.

29-55 At a certain point A the angle of elevation of a mountain peak is α; at point B, a miles farther away in the same horizontal plane, its angle of elevation is β. If h represents the distance of the peak above the plane and x represents the horizontal distance of the peak from A, derive the following formulas:

$$h = \frac{a \tan \alpha \tan \beta}{\tan \alpha - \tan \beta}$$

$$x = \frac{a \tan \beta}{\tan \alpha - \tan \beta}$$

Note: *In working with these formulas, it is most convenient to use natural functions.*

29-56 Find the height of a tree if the angle of elevation of its top changes from 35° to 64.5° when you walk toward it 150 ft in a horizontal line through its base.

29-57 A person walking on a level plain toward a tower observes that at a certain point the angle of elevation of the top of the tower is 28° and that after she has walked 78 m directly toward the tower, the angle of elevation of the top is 48°. Find the height of the tower if the point of observation each time is 2 m above the ground.

29-5 VECTORS

A *vector* has direction and magnitude, and it is a line connecting two points, one of which is identified as the beginning point and the other as

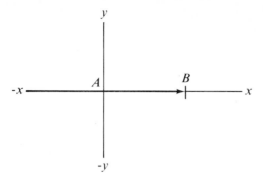

Fig. 29-19. 50-lb force.

the endpoint. In a diagram, vectors are drawn with an arrowhead at the endpoint.

Vectors are used to represent velocities, forces, and other quantities that have direction as well as magnitude. For example, the vector $\overrightarrow{AB}$ in Fig. 29-19 represents, to scale, a 50-lb force acting horizontally and toward the right. If the arrowhead were placed at A instead of at B, we would have the vector $\overrightarrow{BA}$ representing a force of equal magnitude but acting in a direction opposite to that of $\overrightarrow{AB}$.

Two vectors having the same length and direction are considered identical. That is, if $\overrightarrow{AB}$ and $\overrightarrow{BC}$ are of equal length, parallel, and pointing in the same direction, then $\overrightarrow{AB} = \overrightarrow{BC}$. For example, if A and B are two points and C is a point midway between A and B, then $\overrightarrow{AC} = \overrightarrow{CB}$, but $\overrightarrow{AC} = \overrightarrow{BC}$. Thus the beginning point of a vector may be placed anywhere; what is important is not the location but the *length and* the *direction*.

To add two vectors, place the beginning point of the second vector at the endpoint of the first, then draw a vector from the beginning point of the first vector to the endpoint of the second vector. For example, in Fig. 29-20, $\overrightarrow{AB}$ and $\overrightarrow{BC}$ represent, respectively, a 50-lb force and a 20-lb force, both acting in the direction of the positive X axis. The vector $\overrightarrow{AC}$ represents the sum of these two forces: a 70-lb force acting in the same direction.

In a diagram, vectors are generally drawn with the beginning point at the point where the force is exerted; thus, although vectors may be

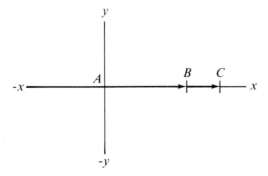

Fig. 29-20. Addition of forces.

moved for purposes of addition, it is important to remember that forces can only be added if they are acting at the same point (or on the same object). For example, in Fig. 29-20 it is assumed that the force $\overrightarrow{BC}$ acts on the point or object represented by A and that its starting point has been moved to B merely for the purpose of performing the addition.

When two vectors are added, the result does not depend on the order of addition: $\overrightarrow{AB} + \overrightarrow{AC} = \overrightarrow{AC} + \overrightarrow{AB}$. This is obvious in the case of two vectors pointing in the same or in opposite directions; such as when the angle between the two vectors is $0°$ or $180°$. When this angle is not $0°$ or $180°$, if $\overrightarrow{AC}$ is moved so that its starting point is B and if D represents the endpoint of the moved vector, if $\overrightarrow{AC} = \overrightarrow{BD}$, then A, B, C, and D will be the vertices of a parallelogram. Hence, $\overrightarrow{CD} = \overrightarrow{AB}$, and therefore $\overrightarrow{AB} + \overrightarrow{AC} = \overrightarrow{AB} + \overrightarrow{BD} = \overrightarrow{AD}$ and $\overrightarrow{AC} + \overrightarrow{AB} = \overrightarrow{AC} + \overrightarrow{CD} = \overrightarrow{AD}$. Thus, another way of adding the two vectors $\overrightarrow{AB}$ and $\overrightarrow{AC}$ is to draw the parallelogram two of whose sides are AB and AC; the vector represented by the diagonal starting at A will then be the sum.

The sum of two or more vectors is also called the *resultant* vector; if the vectors represent forces or velocities, the sum is called the resultant or net force or velocity. In Fig. 29-21, the vector $\overrightarrow{AE}$ represents the resultant of the four forces pictured.

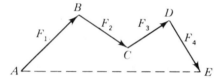

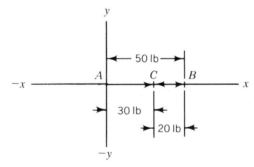

Fig. 29-21. Resultant forces.

The negative of a vector is a vector of the same magnitude that is pointing in the opposite direction: $-\overrightarrow{AB} = \overrightarrow{BA}$.

To subtract one vector from another, add the negative of the vector to be subtracted to the other vector. For example, in Fig. 29-22, $\overrightarrow{AB} - \overrightarrow{CB} = \overrightarrow{AB} + \overrightarrow{BC} = \overrightarrow{AC}$.

Fig. 29-22. Subtraction of forces.

Sometimes it is necessary to resolve or convert a vector into its components, that is, to represent the given vector as a sum of two vectors

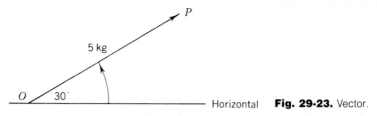

Horizontal   **Fig. 29-23.** Vector.

such that one of them is parallel to a given line or plane and the other perpendicular to it. For example, in Fig. 29-23, the vector $\overrightarrow{OP}$ represents a 5-kg force acting at an angle of 30° from the horizontal. In Fig. 29-24, the same force is represented as the sum (or resultant) of two forces; $\overrightarrow{OQ}$, the horizontal component, and $\overrightarrow{OR}$, the vertical component. The magni-

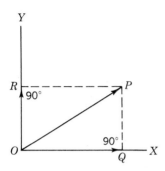

Fig. 29-24. Vectors.

tudes of these components may be determined by making an accurate drawing to scale and measuring the lengths OQ and OR. But a more accurate answer can be obtained by calculating the magnitudes of the component forces with the use of trigonometry.

In Fig. 29-24,

$$\frac{OQ}{OP} \cos 30°$$

$$OQ = OP \cos 30° = 5(0.866\ 03) = 4.33 \text{ kg}$$

and

$$\frac{OR}{OP} = \sin 30°$$

$$OR = OP \sin 30° = 5(0.5) = 2.5 \text{ kg}$$

The components of a vector are also referred to as projections; in Fig. 29-24, for example, $\overrightarrow{OQ}$ is the projection of $\overrightarrow{OP}$ onto the X axis or onto the horizontal plane.

Two examples will illustrate further.

EXAMPLE 29-7 A car is moving up an incline, making an angle of 35° with the horizontal, at the rate of 26 ft/s. What is its horizontal velocity? Its vertical velocity?

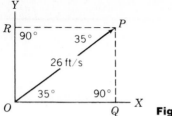

Fig. 29-25. Example 29-7.

Solution   See Fig. 29-25. Here $\vec{OQ}$ is the horizontal velocity and $\vec{OR}$ the vertical velocity. Then

$$\frac{OQ}{OP} = \cos 35° \qquad\qquad OQ = OP \cos 35°$$

$$OQ = 26(0.819\ 15) \qquad OQ = 21.30 \text{ ft/s}$$

$$\frac{OR}{OP} = \sin 35° \qquad\qquad OR = OP \sin 35°$$

$$OR = 26(0.573\ 58) \qquad OR = 14.91 \text{ ft/s}$$

EXAMPLE 29-8   A weight W is resting on a rough horizontal surface as shown in Fig. 29-26. Assume that a pull of 40 lb is acting on the weight at an angle of 20° with the horizontal. What is the horizontal pull on the weight? The vertical pull?

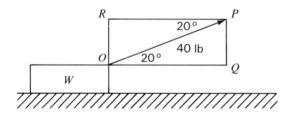

Fig. 29-26. Example 29-8.

Solution   Here the horizontal pull will be $\vec{OQ}$ and the vertical pull $\vec{OR}$. Then

$$\frac{OQ}{OP} = \cos 20° \qquad\qquad OQ = OP \cos 20°$$

$$OQ = 40\ \ (0.939\ 69) \qquad OQ = 37.58 \text{ lb}$$

$$\frac{OR}{OP} = \sin 20° \qquad\qquad OR = OP \sin 20°$$

$$OR = 40\ \ (0.342\ 02) \qquad OR = 13.68 \text{ lb}$$

Sometimes it is necessary to find the component of a vector in a direction other than the horizontal or vertical. For example, if a billiard ball is placed on an inclined plane and then released, the force of gravity acting on the ball will be equal to its weight in magnitude and will point downward in a vertical direction. We can think of this force as having two components: one along the inclined plane and the other perpendicular to the plane. Since the ball is constrained by the plane on which it has been placed, the force perpendicular to the plane will have no effect and the effective force on the ball will be the component along the plane, or the perpendicular projection of the force onto the plane.

▨ EXERCISES

29-58 Two lines OX and OY are perpendicular at O. A line segment PQ is 10 cm long and lies in the plane of OX and OY. If PQ makes an angle of 67° with OX, find its projection on OX. Find its projection on OY.

29-59 In Exercise 29-58 find the projection of PQ on a line that makes an angle of 45° with OX. [Projection $= 10 \cos (67° - 45°)$.]

29-60 A car is moving north at the rate of 60 mph. How fast is it moving east? How fast is it moving southeast?

29-61 A golf ball is hit at an angle of 15° with the horizontal with a force of p lb. What is the force that drives it forward? What force causes it to rise if gravity is neglected?

29-62 The eastward and northward components of the velocity of a ship are, respectively, 8.8 km/h and 17 km/h. Find the direction and the rate at which the ship is sailing.

29-63 A roof is inclined at an angle of 33.5°. The wind strikes this horizontally with a force of 820 kg. Find the pressure perpendicular to the roof.

29-64 A roof 20 ft by 25 ft, inclined at an angle of 27.4° with the horizontal, will shelter how large an area?

29-65 A force of 140 kg is acting on a body lying on a horizontal plane, in a direction that makes an angle of 20° with the horizontal. What is the force tending to lift the body from the plane, and what is the force tending to move the body along the plane?

29-66 A body weighing 45 lb rests on a horizontal table and is acted on by a force of 50 lb acting at an angle of 25.25° with the surface of the table. What is the pressure on the table?

29-67 A body weighing 75 lb rests on a horizontal table and is acted on by a force of 100 lb acting at an angle of $-36.5°$ with the surface of the table. What is the pressure on the table?

29-68 The horizontal and vertical components of a force are, respectively, 245.8 lb and 325.6 lb. What is the magnitude of the force, and what angle does its line of action make with the horizontal?

29-69 The horizontal and vertical components of a force are, respectively, 125.5 and -189.6 lb. What is the magnitude of the force, and what angle does its line of action make with the horizontal?

29-70 A river runs directly south at 4 mph. A person starts at the west bank and rows directly across at the rate of 3 mph. In what direction does the boat move?

29-71 A ferryboat at a point on one bank of a river $\frac{1}{2}$ mi wide must reach a point directly across the river. If the river flows 3.5 mph and the ferryboat can steam 7.6 mph, in what direction should the boat be pointed?

29-72 Two people are lifting a motorcycle by means of ropes running over a pulley and acting in the same vertical plane. One person pulls 38 kg in a direction 23° from the vertical, and the other pulls 56 kg in a direction 42° from the vertical. Determine the weight of the stone.

29-73 Two forces of 240 lb and 180 lb act in the same vertical plane on a heavy body, the first at an angle of 40° with the horizontal and the second at an angle of 65°. Find the total force tending to move the body horizontally. Find the total force tending to lift the body vertically.

GRAPHICAL REPRESENTATION OF TRIGONOMETRIC FUNCTIONS

30-1 INTRODUCTION TO GRAPHING TRIGONOMETRIC FUNCTIONS

In the previous chapters you studied the definitions of the several trigonometric functions and the relationships between these functions. You have had some practice in applying the definitions learned, and in making these applications, you have used a scientific calculator or a table of trigonometric functions.

The values of the functions considered vary in a definite way, as you have noticed. Three of the functions increase in value with an increase in the size of an angle, up to and including 90°. Three of the functions decrease in value as an angle increases, up to and including 90°.

The purpose of this chapter is to picture these relationships. A scientific calculator will provide fast, accurate values for the trig functions and will aid in the graphing of the functions.

30-2 THE SINE CURVE

Inspect the table of trigonometric functions; note that the value of the sine is 0 for an angle of 0 degrees and increases to a maximum value of 1 for an angle of 90°. The values of the sine for angles greater than 90°, up to and including 360°, can be computed for angles of 120°, 135°, 150°, 180°, 210°, 225°, 240°, 270°, 300°, 315°, 330°, and 360° by methods explained in Chapter 28. When these sine values have been read or computed, the *sine curve* can be plotted; the resulting graph is shown in Fig. 30-1. The values for the sine (as found in the table or computed) were plotted on the *Y* axis, and the angle values were plotted on the *X* axis.

For angles greater than 360°, the form of the sine curve repeats itself. This is also true of the curve if negative angle values are plotted. Figure 30-2 shows the curve for angles over 360° and for negative angles.

The sine wave is useful for picturing the voltage developed by an

433

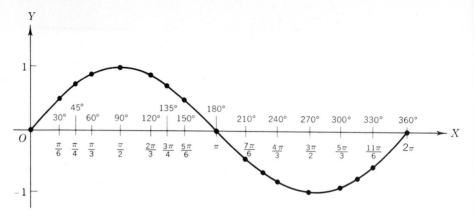

Fig. 30-1. Sine curve.

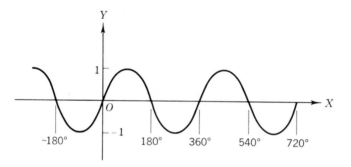

Fig. 30-2. Sine curve.

alternating-current (ac) generator. The voltage is 0 at the beginning of each cycle and then increases to a maximum, then decreases to 0, then it reverses direction and increases to a maximum, then decreases to 0. The complete rotation through 360 electrical-time degrees is a *cycle*. The electricity found in most homes is a 60-cycle current. That is, the flow of electricity reverses 60 times per second. If the alternating current is 25 cycles, it is possible to notice a flicker in lights as the voltage varies.

This same sine wave is used to show the amperes flowing in an alternating-current circuit. In this case the Y values are amperes.

30-3 THE COSINE CURVE

Figure 30-3 is a graph of $y = \cos x$; this is the *cosine wave* or *cosine curve*. This curve was developed using the same method that was used to plot $y = \sin x$.

Figure 30-4 shows the graph of $y = \sin x$ and $y = \cos x$. The unbroken line is $y = \sin x$; the broken line is $y = \cos x$. Notice that the two curves are identical in shape or form over the range of 360°. Observe also

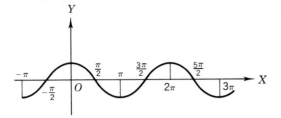

Fig. 30-3. Cosine curve ($y = \cos x$).

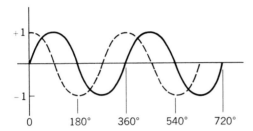

Fig. 30-4. Sine and cosine curves.

that if either curve is shifted 90°, the two curves will coincide, or become the same curve.

30-4 CURVES FOR THE TANGENT, COTANGENT, SECANT, AND COSECANT

Figure 30-5 shows the graphs of $y = \tan x$ and $y = \cot x$. The graph for $y = \tan x$ was drawn using values read from the table; the values of $\cot x$ were calculated using the formula $\cot x = 1/(\tan x)$. The curves of $\tan x$ and $\cot x$ in Fig. 30-5 are not continuous, since the value of $\tan 90°$, infinity, cannot be plotted. The value of $\cot 0°$ is also infinity, and this value too cannot be plotted.

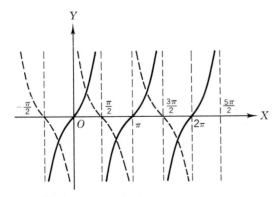

Fig. 30-5. Tangent and cotangent curves.

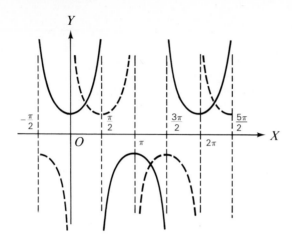

Fig. 30-6. Secant and cosecant graphs.

Figure 30-6 shows the graphs of $y = \sec x$ and $y = \csc x$. Values for sec and csc are not given in the table, but these can be easily calculated from table values of sin and cos, since $\csc x = 1/(\sin x)$ and $\sec x = 1/(\cos x)$. Values were calculated using these relationships, and the two curves shown were drawn.

Notice that these two curves are discontinuous, since some values of sec x and csc x are infinite and cannot be plotted.

▓ EXERCISES

30-1 Plot $y = \sin \theta$.

30-2 Plot $y = \cos \theta$. Use the same scales as in Exercise 30-1.

30-3 Compare the curves plotted in Exercises 30-1 and 30-2. Are they identical? What must you do to match these two curves?

30-4 Plot $y = \tan \theta$.

30-5 Plot $y = \cot \theta$.

30-6 Compare the curves plotted in Exercises 30-4 and 30-5. Are they identical? What must you do to match these two curves?

30-7 Plot $y = \sec \theta$.

30-8 Plot $y = \csc \theta$.

30-9 Compare the curves plotted in Exercises 30-7 and 30-8. Are they identical? What must you do to match these two curves?

30-10 Using the curves constructed in Exercises 30-1, 30-2, 30-4, 30-5, 30-7, and 30-8, answer these questions.
(a) What is the maximum value of the sine? Of the cosine?
(b) What is the minimum value of the sine? Of the cosine?
(c) What are the maximum and minimum values of the tangent and the cotangent?

(d) What are the maximum and minimum values of the secant and cosecant?

30-11 Draw $y = \sin x + \cos x$.

Suggestion: Plot $y = \sin x$ and $y = \cos x$ on the same set of axes. Find the y points for the curve $y = \sin x + \cos x$ by adding the y values (ordinates) for the various values of x.

OBLIQUE TRIANGLES

31·1 INTRODUCTION TO OBLIQUE TRIANGLES

In Chapter 29 several procedures were illustrated for solving right-triangle problems. This chapter will teach several methods for solving problems involving oblique triangles.

An *oblique* triangle is a three-sided figure, including three angles none of which are right angles. To solve an oblique-triangle problem, three of the six parts must be known, and one of the known parts must be a side. Many different triangles can be constructed if only the three angles are given. For example, triangle *ABC* in Fig. 31-1 has all three 60° angles. An infinite number of triangles can be constructed with all angles equal to 60°, all with sides of different lengths. Therefore, one side must be known if one and only one triangle (*a unique* triangle) is to be constructed.

A unique triangle is defined if:

Three sides are given

Two sides and the included angle are given

Two angles and the included side are given

Two angles and the side opposite one of them are given

Try constructing a triangle when two sides and the angle opposite one of the sides are given. That is, construct triangle *ABC* given *a*, *c*, and

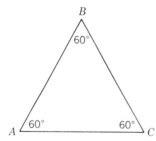

Fig. 31-1. Equilateral triangle.

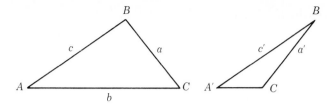

Fig. 31-2. $a = a'$, $c = c'$, and $\angle A = \angle A'$.

$\angle A$. In the triangles of Fig. 31-2 $a = a'$, $c = c'$ and $\angle A = \angle A'$. Obviously, two sides and an angle opposite one of them define more than one triangle fulfilling the original, or given, conditions. In an applied problem it may be possible to eliminate one of two triangles, or answers, as impossible and to accept the remaining triangle as the solution.

31-2 THE LAW OF SINES

The law of sines is: *In any triangle, the sides are proportional to the sine function of the opposite angles.* This law involves a number of relations or formulas. The derivations of these formulas follow.

Draw triangles ABC as in Fig. 31-3. The altitudes of the two triangles must be the same; that is, $h = h$. Then, in either triangle,

$$\frac{h}{c} = \sin A \qquad \frac{h}{a} = \sin C$$
$$h = c \sin A \qquad h = a \sin C$$

Being $h = h$, we have:

$$c \sin A = a \sin C$$

Divide both members of the equation by $\sin A \sin C$:

$$\frac{c}{\sin C} = \frac{a}{\sin A}$$

In the same way, and by constructing an altitude from C to AB,

$$\frac{a}{\sin A} = \frac{b}{\sin B}$$

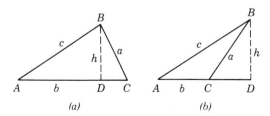

(a)   *(b)*

Fig. 31-3. h in (a) equals h in (b).

Therefore, the law of sines:

$$\frac{a}{\sin A} = \frac{b}{\sin B} = \frac{c}{\sin C}$$

EXAMPLE 31-1   In triangle ABC in Fig. 31-4, given $a = 50$ m, $B = 36°$, and $C = 86°$; find b, c, and A.

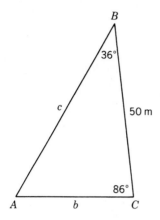

B

36°

c

50 m

86°

A   b   C   **Fig. 31-4.** Example 31-1: Find b, c and $\angle A$.

Solution   First solve for the unknown angle.

$$A + B + C = 180°$$
$$A + 36° + 86° = 180°$$
$$A = 58°$$

Then use two of the ratios from the law of sines, substitute the given values, and solve.

$$\frac{a}{\sin A} = \frac{b}{\sin B}$$

$$\frac{50}{\sin 58°} = \frac{b}{\sin 36°}$$

$$b = \frac{50(\sin 36°)}{\sin 58°}$$

$$= \frac{50(0.587\ 79)}{0.848\ 05} = \frac{29.3895}{0.848\ 05} = 34.66 \text{ m}$$

$$\frac{a}{\sin A} = \frac{c}{\sin C}$$

$$\frac{50}{\sin 58°} = \frac{c}{\sin 86°}$$

$$c = \frac{50 \sin 86°}{\sin 58°}$$

$$= \frac{50(0.997\ 56)}{0.848\ 05} = \frac{49.878}{0.848\ 05} = 58.81 \text{ m}$$

Then

$$c = 58.81 \text{ m}$$
$$b = 34.66 \text{ m}$$
$$A = 58°$$

EXAMPLE 31-2 In triangle ABC in Fig. 31-5, given $a = 450$ cm, $A = 10.33°$, and $B = 47.67°$; find b, c, and C.

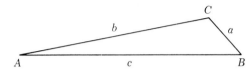

Fig. 31-5. Example 31-2: Find b, c and $\angle C$.

Solution First solve for the unknown angle.

$$A + B - C = 180°$$
$$C = 180° - 10.33° - 47.67°$$
$$= 122°$$

Substitute the given values into two of the law-of-sine ratios and solve.

$$\frac{a}{\sin A} = \frac{b}{\sin B}$$

$$b = \frac{a \sin B}{\sin A} = \frac{450(0.739\ 28)}{0.179\ 32}$$

$$= 1855 \text{ cm}$$

$$\frac{a}{\sin A} = \frac{c}{\sin C}$$

$$c = \frac{a \sin C}{\sin A}$$

$$= \frac{450(0.848\ 05)}{0.179\ 32}$$

$$= 2128 \text{ cm}$$

31-3 MOLLWEIDE'S EQUATIONS

Two convenient equations for checking the accuracy of numerical solutions like the preceding ones are known as "Mollweide's equations." These equations are:

$$\frac{a - b}{c} = \frac{\sin\left[\frac{1}{2}(A - B)\right]}{\cos\left(\frac{1}{2}C\right)} \tag{1}$$

$$\frac{a + b}{c} = \frac{\cos\left[\frac{1}{2}(A - B)\right]}{\sin\left(\frac{1}{2}C\right)} \tag{2}$$

Note: *All six elements of a triangle are included in each formula: three angles and three sides.*

The answers for Example 31-1 in Sec. 31-2 are checked using formula 1.

$$\frac{a-b}{c} = \frac{\sin\left[\frac{1}{2}(A-B)\right]}{\cos\left(\frac{1}{2}C\right)} \tag{1}$$

From Example 31-1,

$$a = 50 \text{ m} \qquad A = 58°$$
$$b = 34.66 \text{ m} \qquad B = 36°$$
$$c = 58.81 \text{ m} \qquad C = 86°$$

Then,

$$\frac{50-34.66}{58.81} \overset{?}{=} \frac{\sin\left[\frac{1}{2}(58°-36°)\right]}{\cos\left[\frac{1}{2}(86°)\right]}$$

$$\frac{15.34}{58.81} \overset{?}{=} \frac{\sin\left[\frac{1}{2}(22°)\right]}{\cos\left[\frac{1}{2}(86°)\right]}$$

$$\frac{15.34}{58.81} \overset{?}{=} \frac{\sin 11°}{\cos 43°}$$

$$0.26 = 0.26$$

The answers for Example 31-2 are checked using formula 2.

$$\frac{a+b}{c} = \frac{\cos\left[\frac{1}{2}(A-B)\right]}{\sin\left(\frac{1}{2}C\right)} \tag{2}$$

From Example 31-2,

$$a = 450 \text{ cm} \qquad A = 10.33°$$
$$b = 1855 \text{ cm} \qquad B = 47.67°$$
$$c = 2128 \text{ cm} \qquad C = 122°$$

$$\frac{450-1855}{2128} \overset{?}{=} \frac{\cos\left[\frac{1}{2}(10.33°-47.67°)\right]}{\sin\left[\frac{1}{2}(122°)\right]}$$

$$\frac{2305}{2128} \overset{?}{=} \frac{\cos\left[\frac{1}{2}(-37.34°)\right]}{\sin 61°}$$

$$\frac{2305}{2128} \overset{?}{=} \frac{\cos(-18.67°)}{\sin 61°}$$

$$1.08 = 1.08$$

Another method to check is to construct the triangle using the three given and the three calculated facts. A construction test is an approximate check only, but if you are solving a problem on the job it should be made in addition to checks by formula. As a matter of fact, it is best in any event to construct a triangle using the three given facts, for reasons which will be given in the next section.

31-4   CONSTRUCTION TESTS

A triangle should be sketched before solving a problem to decide: (1) if a solution is possible; (2) if there are two solutions; and (3) if there is one and only one solution.

EXAMPLE 31-3   Given $a = 12$, $c = 20$, $A = 62°$, determine b, B, and C.

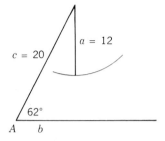

Fig. 31-6. Example 31-3: Determine a triangle.

Solution   See Fig. 31-6.

1. Lay out b to the right of A, of any convenient length. Since the length of b is unknown, this is an indefinite length.

2. Draw an angle of 62° at A.

3. Lay out, on the terminal side of angle A, the length of side c, 20 units.

4. Draw side a, opposite angle A. With the compass point on the end of side c, swing an arc of length 12 units.

5. Since the arc drawn in step 4 does not intercept b, it is not long enough to complete a triangle, and a solution is impossible.

6. If the arc drawn in step 4 is tangent to b, then ABC is a right triangle and angle $C = 90°$.

7. If the arc drawn in step 4 intercepts b in two places, then two solutions are possible.

8. If the arc drawn in step 4 intercepts b in one place only, there is one and only one solution.

Construction of the triangle, then, provides you with an approximate check on the answers to be calculated and, in addition, gives you a clue to the number of possible solutions.

31-5   THE LAW OF COSINES

The law of cosines is: *In any triangle, the square of any side equals the squares of the other two sides minus twice the product of these two sides and the cosine of the angle between them.* The development of three formulas expressing this law follows.

Draw any two triangles ABC with $h = h$, as in Fig. 31-7a and b. In either

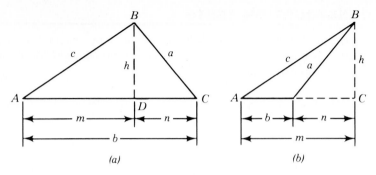

Fig. 31-7. h in (a) equals h in (b).

triangle,

$$a^2 = h^2 + n^2$$

In Fig. 31-7a,

$$n = b - m \qquad m = c \cos A$$
$$n = b - c \cos A$$

In Fig. 31-7b,

$$n = m - b \qquad m = c \cos A$$
$$n = c \cos A - b$$

In either case, $n^2 = b^2 - 2bc \cos A + c^2 \cos^2 A$; $h/c = \sin A$; $h = c \sin A$; and $h^2 = c^2 \sin^2 A$. Substituting these values for n^2 and h^2 in the original statement,

$$a^2 = h^2 + n^2$$
$$a^2 = c^2 \sin^2 A + b^2 - 2bc \cos A + c^2 \cos^2 A$$
$$a^2 = b^2 + c^2 \sin^2 A - 2bc \cos A + c^2 \cos^2 A$$
$$a^2 = b^2 + c^2(\sin^2 A + \cos^2 A) - 2bc \cos A$$

Since $\sin^2 A + \cos^2 A = 1$, then

$$a^2 = b^2 + c^2 - 2bc \cos A$$
$$a = \sqrt{b^2 + c^2 - 2bc \cos A} \tag{1}$$

Similarly,

$$b^2 = a^2 + c^2 - 2ac \cos B$$
$$b = \sqrt{a^2 + c^2 - 2ac \cos B} \tag{2}$$
$$c^2 = a^2 + b^2 - 2ab \cos C$$
$$c = \sqrt{a^2 + b^2 - 2ab \cos C} \tag{3}$$

Studying these three statements of the law of cosines indicates that the formulas are most appropriate when two sides and an angle between the two sides are given. In fact, they cannot be used unless two sides are given.

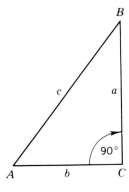

Fig. 31-8. $C = 90°$; $c = \sqrt{a^2 + b^2}$.

Note: *It is interesting to observe that in triangle ABC in Fig. 31-8, if C = 90°, then the formula 3, c = $\sqrt{a^2+b^2-2ab \cos C}$, reduces to c = $\sqrt{a^2+b^2}$, since cos 90° = 0. This is, of course, the Pythagorean theorem.*

It is possible to solve all three formulas so that they can be used to determine the three angles of a triangle when the three sides are given. This solution is illustrated for formula 3:

$$c^2 = a^2 + b^2 - 2ab \cos C$$
$$2ab \cos C + c^2 = a^2 + b^2$$
$$2ab \cos C = a^2 + b^2 - c^2$$
$$\cos C = \frac{a^2 + b^2 - c^2}{2ab}$$

Similarly,

$$\cos A = \frac{b^2 + c^2 - a^2}{2bc}$$

$$\cos B = \frac{a^2 + c^2 - b^2}{2ac}$$

EXAMPLE 31-4 Given $a = 10$, $b = 12$, and $c = 15$, find A, B, and C. See Fig. 31-9.

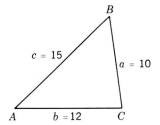

Fig. 31-9. Example 31-4: Find A, B, and C.

Solution Substitute the given values and evaluate.

$$\cos A = \frac{b^2 + c^2 - a^2}{2bc}$$

$$\cos A = \frac{12^2 + 15^2 - 10^2}{(2)(12)(15)} = 0.747\ 22$$

$$A = 41.65°$$

$$\cos B = \frac{a^2 + c^2 - b^2}{2ac}$$

$$\cos B = \frac{10^2 + 15^2 - 12^2}{(2)(10)(15)} = 0.603\ 33$$

$$B = 52.89°$$

$$\cos C = \frac{a^2 + b^2 - c^2}{2ab}$$

$$\cos C = \frac{10^2 + 12^2 - 15^2}{(2)(10)(12)} = 0.079\ 17$$

$$C = 85.46°$$

Check:
$$A + B + C \overset{?}{=} 180°$$
$$41.65° + 52.89° + 85.46° = 180°$$

EXAMPLE 31-5 Given $a = 56.7$, $b = 45.2$, and $C = 47.75°$, find c, A, and B. See Fig. 31-10.

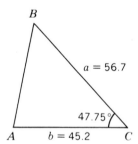

A $b = 45.2$ C **Fig. 31-10.** Example 31-5: Find c, A, and B.

Solution Since a, b, and C are known, use the law of cosines formula solved for c.

$$c = \sqrt{a^2+b^2-2ab \cos C}$$
$$= \sqrt{56.7^2+45.2^2-2(56.7)(45.2) \cos 47.75°}$$
$$= 42.56$$

Then use the law of sines to solve for angles A and B.

$$\frac{a}{\sin A} = \frac{c}{\sin C}$$

$$\sin A = \frac{a \sin C}{c}$$

$$\sin A = \frac{56.7(0.740\ 22)}{42.56} = 0.986\ 15$$

$$A = 80.45°$$

$$\frac{b}{\sin B} = \frac{c}{\sin C}$$

$$\sin B = \frac{b \sin C}{c}$$

$$\sin B = \frac{45.2(0.740\ 22)}{42.56} = 0.786\ 13$$

$$B = 51.83°$$

Check:

$$A + B + C \stackrel{?}{=} 180°$$
$$47.75° + 80.45° + 51.83° \stackrel{?}{=} 180°$$
$$180.03° = 180°$$

The check above is only a partial check; it is best to use one of Mollweide's formulas for a complete test:

$$\frac{a - b}{c} = \frac{\sin\left[\frac{1}{2}(A - B)\right]}{\cos\left(\frac{1}{2}C\right)}$$

$$\frac{56.7 - 45.2}{42.56} \stackrel{?}{=} \frac{\sin\left[\frac{1}{2}(80.45° - 51.83°)\right]}{\cos\left[\frac{1}{2}(47.75°)\right]}$$

$$\frac{11.50}{42.56} \stackrel{?}{=} \frac{\sin\left[\frac{1}{2}(28.62°)\right]}{\cos\left[\frac{1}{2}(47.75°)\right]}$$

$$0.27 = 0.27$$

The check using one of Mollweide's formulas is a better check of the answers because it uses all six elements of the triangle.

EXAMPLE 31-6 Two highways intersect at point A at an angle of 43.5°. From A to B along one highway is a straightaway of 16 km; from A to C, along another straightaway, it is 24 km. A third straightaway connects B and C. How long will it take a car averaging 80 km/h to go from A to B to C to A?

Solution Sketch the problem (Fig. 31-11).

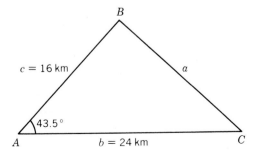

Fig. 31-11. Example 31-6: Sketch.

Distance BC, or a, must be calculated so that the entire distance traveled can be determined. By the law of cosines:

$$a = \sqrt{b^2+c^2-2\ bc\ \cos A}$$
$$= \sqrt{24^2+16^2-2(24)(16)(0.725\ 37)}$$
$$= \sqrt{576 + 256 - (768)(0.725\ 37)}$$
$$= \sqrt{832 - 557.1}$$
$$= \sqrt{274.9}$$
$$= 16.58 \text{ km}$$

Then

$$\text{Total distance traveled} = (16 + 24 + 16.58) \text{ km}$$
$$= 56.58 \text{ km}$$

Since rate = 80 km/h,

$$\text{Travel time} = \frac{d}{r} = \frac{56.58 \text{ km}}{80 \text{ km/h}} = 0.707 \text{ h} = 42 \text{ min}$$

Check 1: By construction, in Fig. 31-12 the distance from B to C and angle B were determined by measurement. The measured distance, 16.5 km, checks the value calculated, 16.58 km.

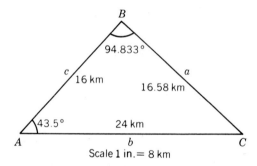

Fig. 31-12. Example 31-6: Check by construction.

Check 2: By calculation, using the law of sines:

$$\frac{b}{\sin B} = \frac{a}{\sin A} \qquad \sin B = \frac{b \sin A}{a}$$

$$\sin B = \frac{24(0.688\ 35)}{16.58} = 0.996\ 41$$

Since sin B is positive, angle B may have two values, 85.14° or 94.86°. This is true because the sines of supplementary angles are equal. Since the check indicated that B is greater than 90°, you should reject the answer 85.14° and accept 94.86°. If you have difficulty accepting this statement, construct two angles of 85.14° and 94.86° and set up equations for the value of the two sines.

$$C = 180° - A - B$$
$$= 180° - 43.5° - 94.86°$$
$$= 41.64°$$

$$\frac{a + b}{c} = \frac{\cos \left[\frac{1}{2}(A - B)\right]}{\sin \left(\frac{1}{2}C\right)}$$

$$a = 16.58 \text{ km} \quad A = 43.5°$$
$$b = 24 \text{ km} \quad B = 94.86°$$
$$c = 16 \text{ km} \quad C = 41.64°$$

$$\frac{a + b}{c} \overset{?}{=} \frac{\cos [\frac{1}{2}(A - B)]}{\sin (\frac{1}{2}C)}$$

$$\frac{16.58 + 24}{16} \overset{?}{=} \frac{\cos \frac{1}{2}(43.5° - 94.86°)}{\sin \frac{1}{2}(41.64°)}$$

$$2.536 \overset{?}{=} \frac{\cos(-25.68°)}{\sin 20.82°}$$

$$2.536 \overset{?}{=} \frac{0.901\ 23}{0.355\ 44}$$

$$2.536 = 2.535$$

Granted, the check is longer than the problem; however, as much trigonometry is learned by completing the check as in solving the original problem. In these examples, the value of an accurate sketch of the problem has been demonstrated.

31-6 AREA OF A TRIANGLE

The area of a triangle is equal to one-half the product of the base and altitude, or

$$A = \tfrac{1}{2}bh$$

where A = area, b = base, and h = altitude.

The following formulas may be used to compute the area of a triangle:

$$A = \tfrac{1}{2}bc \sin A = \tfrac{1}{2}ac \sin B = \tfrac{1}{2}ab \sin C$$

and

$$A = \sqrt{s(s - a)(s - b)(s - c)}$$

where $s = (a + b + c)/2$.

EXAMPLE 31-7 Find the area of the triangle in Fig. 31-13.

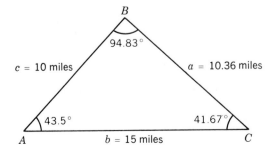

Fig. 31-13. Find the area.

Solution

$$A = \tfrac{1}{2}bc \sin A$$
$$= \tfrac{1}{2}(15)(10)(0.688\ 35)$$
$$= 51.6\ \text{mi}^2$$

or

$$A = \sqrt{s(s - a)(s - b)(s - c)}$$

$$s = \frac{a + b + c}{2} = \frac{35.36}{2} = 17.68$$

$$A = \sqrt{(17.68)(17.68 - 10.36)(17.68 - 15)(17.68 - 10)}$$
$$= \sqrt{(17.68)(7.32)(2.68)(7.68)}$$
$$= 51.6\ \text{mi}^2$$

In the process of solving oblique triangles, the facts given or known should be listed first. The values to be found should be indicated. Then a sketch of the triangle should be made. Consider the facts known and the given quantities, the formula or formulas should be selected, and then the unknown values should be calculated. Three checks, or tests, should be made before the values calculated are accepted: one, check the answers by measurement of a constructed triangle; two, apply Mollweide's equations; three, check the answer or answers against the original conditions given in the problem—the answer should agree with the given situation.

▌ EXERCISES

31-1 Solve the following:
(a) Given $A = 33°$, $B = 72.5°$, $a = 10$, find C, c, b.
(b) Given $A = 10.2°$, $B = 46.6°$, $a = 50$, find C, b, c.
(c) Given $A = 12.82°$, $B = 141.98°$, $a = 82$, find C, b, c, K.
(d) Given $B = 77°$, $C = 65°$, $b = 99.9$, find A, a, c, K.
(e) Given $A = 99.92°$, $C = 35.67°$, $a = 80.4$, find B, b, c, K.

31-2 Solve the following:
(a) Given $a = 840$, $b = 485$, $A = 21.5°$, find B, C, c.
(b) Given $A = 51.54°$, $a = 91.06$, $b = 77.04$, find B, C, c.
(c) Given $B = 16.09°$, $a = 75$ ft, $b = 29$ ft, find the difference between the areas of the two corresponding triangles.

31-3 Solve the following:
(a) Given $a = 4$, $c = 6$, $b = 60°$, find b, A, C.
(b) Given $a = 17$, $b = 12$, $C = 59.1°$, find A, B, c.
(c) Given that the two sides of a triangle are each equal to 6 and the included angle is $120°$, find the third side.

31-4 Solve the following, and check using the formula $A + B + C = 180°$:
(a) Given $a = 4$, $b = 6$, $c = 8$, find A, B, C.
(b) Given $a = 19$, $b = 34$, $c = 49$, find A, B, C.

(c) Given $a = 51$, $b = 65$, $c = 20$, find A, B, C.

(d) Given $a = 40$ ft, $b = 13$ ft, $c = 37$ ft; find A, B, C.

31-5 Find the area of a triangle with sides of 12.5 ft and 17.05 ft and the included angle of 106.215°.

31-6 Find the area of a triangle whose three sides are 46.45, 27.3, and 32.75 ft long.

31-7 To find the distance AB through a swamp, shown in Fig. 31-14, the following data were measured: $a = 748$ rods, $b = 375$ rods, and $C = 63.21°$. Compute the distance AB.

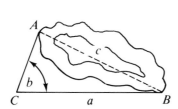

Fig. 31-14. Exercise 31-7.

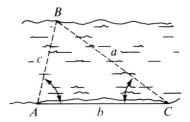

Fig. 31-15. Exercise 31-8.

31-8 Compute the inaccessible distance AB in Fig. 31-15 from the measured data: $b = 1113.8$ ft, $A = 78.31°$, and $C = 47.23°$.

31-9 Two points P and Q in Fig. 31-16 are on opposite sides of a stream and invisible from each other because of an island in the stream. A straight line AB is run through Q, and the following measurements are taken: $AQ = 824$ ft, $QB = 662$ ft, angle $QAP = 42.20°$, and angle $QBP = 57.27°$. Compute QP.

31-10 Two headlands P and Q are separated by water. In order to find the distance between them, a third point A is chosen from which both P and Q are visible, and the following measurements are made: $AP = 1140$ ft, $AQ = 1846$ ft, and angle $PAQ = 58.30°$. Find the distance PQ.

31-11 A vertical TV tower casts a shadow 80 ft long on a hillside that slopes at an angle of 10° from the horizontal. If the angle of elevation of the sun is 49°, find the length of the tower.

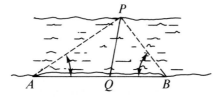

Fig. 31-16. Exercise 31-9.

31-12 A triangular lot in a suburban area was offered for sale at $5000 per acre. The sides of the lot were 5 rods, 9 rods, and 10 rods. Find the sale price of the lot (160 square rods = 1 acre).

31-13 A classic painting hangs on a wall so that from a point on the floor the angle of elevation of the top is 60° and the angle of elevation of the bottom is 30°. Prove that the distance from the top to the bottom of the tapestry (measured vertically) is twice the vertical distance from the floor to the bottom of the tapestry.

31-14 An army officer wants to know the distance from a gun emplacement at A to a trench at D (Fig. 31-17). She can measure the distance a and the angles α and β. Find the distance AD. Find the distances x and y, and then check the result found for AD by showing that $x^2 + y^2 = \overline{AD}^2$.

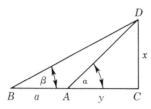

Fig. 31-17. Exercise 31-14.

Solution:

$$\tan \beta = \frac{x}{a + y} \qquad \tan \alpha = \frac{x}{y}$$

$$y \, (\tan \alpha) = x$$

$$\tan \beta = \frac{y \, (\tan \alpha)}{a + y}$$

$$(a + y) \tan \beta = y \, (\tan \alpha)$$

$$a \tan \beta + y \tan \beta = y \tan \alpha$$
$$a \tan \beta = y \tan \alpha - y \tan \beta$$
$$a \tan \beta = y \, (\tan \alpha - \tan \beta)$$

$$\frac{a \tan \beta}{\tan \alpha - \tan \beta} = y$$

$$y = \frac{a \tan \beta}{\tan \alpha - \tan \beta}$$

31-15 Two scouts stationed on the opposite sides of an observation balloon observe its angles of elevation as 44.33° and 36.67°. They find that the distance between them is 215 m. What is the height of the balloon?

31-16 To find the distance between two inaccessible points A and B, a base line $CD = 245$ m is measured in the same plane as A and B, and the angles $DCA = 106°$, $DCB = 39°$, $CDB = 122°$, and $CDA = 41°$ are measured. Find the distance AB.

31-17 Show that the area of any quadrilateral is equal to one-half the product of its diagonals and the sine of the included angle.

31-18 From a point on a horizontal plane, the angle of elevation of the top of a hill is 23.77° and a tower 14 m high standing on the top of a hill subtends an angle of 5.27°. Find the height of the hill.

31-19 Two observers at A and B, 500 m apart on a horizontal plane, observe an airplane at the same moment. The angle of elevation of the aircraft as seen at A is 68.97°. In the horizontal plane, the projections of the observers' lines of sight form angles with the line AB of 43.45° at A and 23.75° at B. Find the height of the aircraft.

31-20 B is 68 km from A in a direction of N 68° W, and C is 94 km from A in a direction N 17° E. What is the position of C relative to B?

31-21 From a point 90 m above the level of a lake and at some distance from one side, an observer finds the angles of depression of the two ends of the lake to be 4.25° and 3.5°. The angle between the two lines of sight is 58.75°. Find the length of the lake.

31·7   DIAMETER OF A CIRCLE CIRCUMSCRIBED ABOUT A TRIANGLE

Draw a triangle with a circumscribed circle, as shown in Fig. 31-18. The sides of the triangle are a, b, and c, and R is the radius of the circle. Draw OE perpendicular to AC, and draw the radii OA, OB, and OC. Then OE

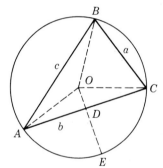

Fig. 31-18. Triangle with a circumscribed circle.

bisects AC and angle AOD equals angle B. In triangle AOD, $AD/AO = \sin AOD$, $AD = \frac{1}{2}b$, $AO = R$, and angle AOD = angle B. Substituting,

$$\text{Diameter} = \frac{b}{\sin B}.$$

$$\frac{\frac{1}{2}b}{R} = \sin B \qquad \text{or} \qquad R = \frac{\frac{1}{2}b}{\sin B}$$

$$2R = \frac{b}{\sin B}$$

The length $2R$ is equal to the diameter of the circle; the diameter of a circle circumscribed about a triangle, then, is equal to the ratio of any side to the sine of the opposite angle. See Section 24-3.

31·8   RADIUS OF A CIRCLE INSCRIBED IN A TRIANGLE

Draw a triangle with an inscribed circle, as shown in Fig. 31-19. Draw the radii OD, OE, and OF and the three lines OA, OB, and OC. The area of

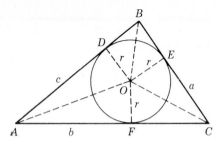

Fig. 31-19. Circle inscribed in a triangle.

triangle *ABC* is equal to the sum of the areas of the three triangles *BOC*, *AOC*, and *AOB*; or

$$\text{Radius} = \frac{\sqrt{(s-a)(s-b)(s-c)}}{s}.$$

Area ABC = area BOC − area AOC + area AOB

Area $BOC = \frac{1}{2}ra$ area $AOC = \frac{1}{2}rb$ area $AOB = \frac{1}{2}rc$

Area $ABC = \frac{1}{2}ra + \frac{1}{2}rb + \frac{1}{2}rc$

A (area of a triangle) $= \sqrt{s(s-a)(s-b)(s-c)}$

Then

$$\sqrt{s(s-a)(s-b)(s-c)} = \frac{1}{2}ra + \frac{1}{2}rb + \frac{1}{2}rc$$
$$= \frac{1}{2}r(a+b+c)$$

Recall that

$$S = \frac{1}{2}(a+b+c)$$
$$= rs$$
$$\sqrt{s(s-a)(s-b)(s-c)} = rs$$
$$\frac{\sqrt{s(s-a)(s-b)(s-c)}}{s} = r$$
$$\sqrt{\frac{(s-a)(s-b)(s-c)}{s}} = r$$

or

$$r = \sqrt{\frac{(s-a)(s-b)(s-c)}{s}}$$

■ EXERCISES

31-22 An art design consists of an equilateral triangle with 20-cm sides, the circle circumscribed about this triangle, and the circle inscribed in this triangle. Find the radii of the two circles.

31-23 A piece of tin is triangular with sides of 15, 18, and 25. Find the radius of the largest circular piece that can be cut from this triangle.

31-24 A midget automobile track is the arc of a circle that must pass through the vertices of a triangle with sides of 15 m, 30 m, and 40 m. Find the radius of this circular track.

Suggestion: *To find an angle, use*

$$\sin \tfrac{1}{2}A = \sqrt{\frac{(s-b)(s-c)}{bc}}$$

31-9 RESULTANT OF FORCES

Two forces are acting on the machine part pictured in Fig. 31-20. It is necessary to compute the magnitude and the direction of the total force in order to design the pin that fits the hole at P.

The two forces are represented by the vectors $\overrightarrow{PQ}$ and $\overrightarrow{PS}$.

Construct the parallelogram $PSRQ$ by drawing QR parallel to PS and SR parallel to PQ.

From Sec. 29-5 you will remember that PR, the diagonal of the parallelogram, is the resultant of the two forces. The resultant of any num-

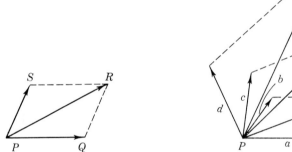

Fig. 31-20. Resultant of forces. **Fig. 31-21.** Vector diagram.

ber of forces is a single force that will produce the same effect as the combined effect of all the forces; it is the result of all of these vectors.

If $\overrightarrow{PQ}$ and $\overrightarrow{PS}$ were constructed to scale, then the resultant force could be determined by measuring PR. The angle of the resultant could also be determined by measurement with a protractor. Answers obtained in this way will be only as accurate as the construction.

It is possible to combine a number of forces into one resultant force and show this in a *vector diagram*. Figure 31-21 is a vector diagram. It illustrates the application of a vector diagram to a simple machine design problem. Vector diagrams are also used in electronic problems, but there the "forces" may be IR drop, voltage, or capacitance. The principle—the application of trigonometry—is the same; the vectors simply carry different labels.

In Fig. 31-21, r_1 is the resultant of forces a and b; r_2 is the resultant of c and r_1; r_3 is the resultant of d and r_2. One force has replaced four forces: a, b, c, and d. This reduction is of great advantage in machine design problems.

31·10 COMPUTATIONS OF A RESULTANT

A resultant can be determined with the application of the proper trigonometric formula. In Fig. 31-22 determine the resultant $\overrightarrow{PR}$.

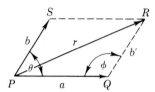

Fig. 31-22. Find resultant $\overrightarrow{PR}$.

Angle θ is the angle between the two forces. Angle ϕ plus angle θ equals 180°; thus if θ is known, ϕ can be determined by subtraction. Sides b and b' are equal, since they are opposite sides of a parallelogram. Side b, or PS, represents the strength of one force, and a or PQ, represents the strength of the second force. Applying the cosine law,

$$r = \sqrt{a^2 + b'^2 - 2ab' \cos \phi}$$

And since $\phi = 180° - \theta$ and $\cos \phi = \cos(180° - \theta) = -\cos \theta$, then

$$r = \sqrt{a^2 + b^2 + 2ab \cos \theta}$$

The angle between the resultant and either force can be found by the law of sines.

Velocities can be combined in the same way.

EXAMPLE 31-8 A force $\mathbf{F_1}$ of 214 kg and a second force $\mathbf{F_2}$ of 236 kg are acting on a balance that registers 426 kg. Find the angle between the forces $\mathbf{F_1}$ and $\mathbf{F_2}$ (Fig. 31-23).

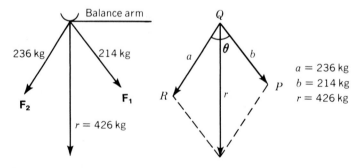

Fig. 31·23. Example 31-8.

Solution In the parallelogram of forces,

$$QR = PS \qquad QP = RS \qquad \text{and} \qquad r = \sqrt{a^2 + b^2 + 2\,ab\,\cos\theta}$$

$$r^2 = a^2 + b^2 + 2\,ab\,\cos\theta$$

$$\cos\theta = \frac{r^2 - a^2 - b^2}{2\,ab}$$

$$= \frac{(426)^2 - (236)^2 - (214)^2}{(2)(236)(214)}$$

$$= \frac{181{,}476 - 55{,}696 - 45{,}796}{101{,}008}$$

$$= \frac{79{,}984}{101{,}008}$$

$$= 0.791\ 86$$

$$\theta = 37.64°$$

■ EXERCISES

31-25 Find the magnitude and direction of the resultant force **F** if a force **F₁** of 18 kg and a force **F₂** of 28 kg each act on an object in the following ways:
(a) In the same direction
(b) In opposite directions
(c) At an angle of 90°
(d) At an angle of 30°

31-26 A car is stuck in a snowbank. A force of 227 kg acting toward the north and another force of 318 kg acting toward the northeast are just sufficient to move the car. Find the resultant force, in both magnitude and direction, that is moving the car.

31-27 If in Exercise 31-26 one of the forces is 186 kg and the other is 145 kg and if they are acting at an angle of 51.62° find the resultant force, both in magnitude and direction, that is moving the car.

31-28 A force **F₁** of 58 kg and another force **F₂** that is unknown are acting on a spring balance that registers 91 kg. If the angle between **F₁** and the direction that the balance is noted to move is 18.4° find the magnitude and direction of **F₂**.

31-29 Two equal forces **F₁** and **F₂** act at an angle of 135° on an object. Find the magnitude and direction of these two forces.

Determine by means of a vector diagram the magnitude and direction of the resultant for the following forces.

31-30 **F₁** = 25 at an angle of 45°, and **F₂** = 60 at an angle of 60°

31-31 **F₁** = 125 at an angle of 100°, and **F₂** = 70 at an angle of 70°

31-32 **F₁** = 100 at an angle of 10°, **F₂** = 90 at an angle of 30°, and **F₃** = 200 at an angle of 45°

31-33 **F₁** = 225 at an angle of 60°, and **F₂** = −70 at an angle of 60°

31-34 **F₁** = 200 at an angle of 45°, and **F₂** = −100 at an angle of 30°

31-11 THE POWER FACTOR

The formula $P = EI \cos \theta$ may be used to compute the power consumed in an alternating-current (ac) circuit. This power in watts is equal to the product of the electrical pressure in volts, the current in amperes and the $\cos \theta$. The angle θ is a measure of how much the voltage and current are out of phase. In many circuits the current and voltage are out of phase, and the amount they are out of phase (angle θ) is measured in degrees. If this phase angle is 0 then $\cos \theta$ is equal to 1 and the formula reduces to $P = EI$. The cosine of angle θ is known as the *power factor;* it can be calculated by measuring the true watts and dividing by the apparent watts.

$$\text{Power factor} = \cos \theta = \frac{\text{true watts}}{\text{apparent watts}} = \frac{P}{EI} \quad \text{or} \quad P = EI \cos \theta$$

If the cosine of θ is known, then θ can be established with a table of trigonometric functions or with a calculator.

EXAMPLE 31-9 Calculate the power in a 115-V alternating-current circuit if the current flowing is 15 A and $\theta = 10°$.

Solution

$$P = EI \cos \theta$$
$$= (115)(15)(0.984\ 81)$$
$$= 1698.8 \text{ W}$$

EXAMPLE 31-10 Determine the power factor if the true watts in a circuit are 1500 and this circuit is drawing $10.\overline{3}$A at 150 V.

Solution

$$P = EI \cos \theta$$
$$1500 = (150)(10.\overline{3}) \cos \theta$$

$$\frac{1500}{150(10.\overline{3})} = \cos \theta$$

$$0.967\ 74 = \cos \theta = \text{power factor} \quad \text{or} \quad \theta = 14.59°$$

It is possible for a "flow" of alternating current to be 90° out of phase; in this case the power available would be 0, since $\cos 90° = 0$. It is reasonable to assume that in most circuits the power factor should be near 1 and the angle θ as near 0 as possible; thus the current and voltage should be in phase.

▮ EXERCISES

31-35 An ac circuit is drawing 20 A at 230 V. If the power factor is 0.90, what is the power in watts?

31-36 What is the power factor if true watts are 200 and apparent watts are 225?

31-37   What is angle θ in problem 31-36?

31-38   An ac motor is drawing 10 A at 440 V. A wattmeter reading is 4200 W. What is cos θ? What is θ?

31-39   Angle θ is 5°; voltage is 220 V; and current is 10 A. Find the power in watts.

31-40   In an ac circuit, P is 500 W, E is 250 V, I is 3 A, and θ is 0°. Do these sound like reasonable values?

31-41   What is the power in watts if θ is 6.5°, voltage is 440 V, and current is 12 A?

31-12 REFRACTION

The path of a ray of light in a single homogeneous medium, such as air, is a straight line. But when a ray of light strikes a polished surface, such as a mirror, it is reflected according to the well-known law that *the angle of incidence is equal to the angle of reflection.*

In Fig. 31-24, the light ray SQ strikes the polished surface and is reflected in the direction QR. Line PQ is perpendicular to the surface at Q. Angle SQP, or angle i, is the angle of incidence; and angle RQP, or angle r, is the angle of reflection. Angles SQP and RQP are equal.

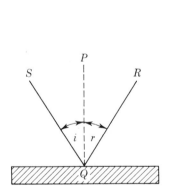

Fig. 31-24. Reflection of light.

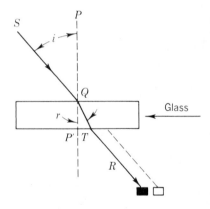

Fig. 31-25. Refraction of light.

When a ray of light passes from one transparent medium to another that is more or less dense, its direction is changed; that is, the ray of light is bent or refracted. This occurs because the light ray travels at a different speed in the more or less dense second medium.

In Fig. 31-25, a ray of light SQ passing through air meets the surface of a piece of glass at Q and is refracted toward the normal, or perpendicular, QP'. The ray of light continues in the direction QT until it meets the other surface of the glass at T. At T it is again refracted, but this time away from the perpendicular, and passed into the air in the direction TR. If the two surfaces of the glass are parallel, then SQ and TR are parallel. Lines QP and QP' are perpendicular to the surface at Q. Angle SQP

(angle i) is the angle of incidence; angle $P'QT$ (angle r) is the angle of refraction. In this figure angle i and angle r are not equal.

It can be seen from Fig. 31-25 that an object (black square) viewed through the glass at an angle will not appear in its true position: it will appear to be at the white square. You can easily demonstrate this with a pencil, a small object, and a large bowl of water. Place the object on the bottom of the bowl of water and then try to touch it with the pencil.

It has been found by experiment that, for a given kind of glass, the ratio

$$\frac{\sin i}{\sin r} = u$$

is a constant whatever the angle of incidence may be. This means that, for a certain kind of glass, if the angle of incidence is changed, then the angle of refraction also changes in such a way that the ratio of the sines is a constant. For a ray of light passing from air to crown glass, this ratio is very nearly $\frac{3}{2}$; for a ray of light passing from air to water, it is $\frac{4}{3}$.

The value of the ratio $(\sin i)/(\sin r) = u$ is the *index of refraction* of the glass with respect to air.

It follows that the index of refraction of air with respect to glass is the reciprocal of that of glass with respect to air. That is, if the index of refraction of glass with respect to air is u, then the index of refraction of air with respect to glass is $1/u$. The statement is true for any two transparent substances.

When a ray of light passes into a medium less dense—from glass to air—the ray of light is bent or refracted away from the perpendicular. It then follows that if the angle of incidence is great enough, the ray will not pass through the glass but will be *internally reflected*. This is illustrated in Fig. 31-26. Here ray A passes from glass to air without refracting, since the angle is 90°. Ray B is refracted, or bent away from the perpendicular. Ray C is at the critical angle; the bent or refracted ray travels along the surface of the glass. Ray D is beyond the critical angle and is reflected back into the glass.

Note: *The principle involved here is used in designing binoculars.*

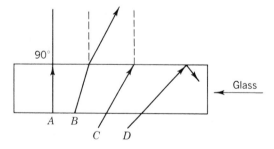

Fig. 31-26. Internal reflection.

■ EXERCISES

31-42 If $u = 1.167$, find the angles of refraction in the following:
(a) Given: The angle of incidence is 18.5°.

Solution:

$$\sin r = \frac{\sin i}{u} = \frac{\sin 18.5°}{1.167}$$

Therefore,

$$\sin r = 0.271\ 90$$
$$r = 15\ 78°$$

(b) Given: The angle of incidence is 18.5°.
(c) Given: the angle of incidence is 37°.

31-43 The sine of the critical angle is equal to the reciprocal of the index of refraction. Find u (the index of refraction) for the following:
(a) For water; the critical angle is 48.47°.
(b) For crown glass; the critical angle is 41.17°.
(c) For diamond; the critical angle is 24.43°.

Note: For jewels with regular facets, the smaller the critical angle, the larger the proportion of the light incident on it that is internally reflected. Hence the brilliancy of the diamond.

31-44 The eye is 25 in. in front of a mirror, and an object appears to be 20 in. behind the mirror. The line of sight makes an angle of 32.5° with the mirror. Find the distance and direction of the object from the eye.

31-45 A ray of light travels a path *ABCD* (Fig. 31-27) in passing through a plate glass *MN* 0.525 in. thick. What is the displacement *CE* if the ray strikes the glass at an angle *ABP* of 43.25° and the index of refraction is $\frac{3}{2}$?

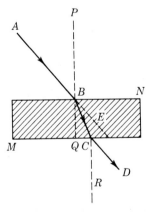

Fig. 31-27. Exercise 31-45.

Solution: Here $u = \frac{3}{2} = 1.5$, and $i = 43.25°$. Therefore,

$$\sin r = \frac{\sin 43.25°}{1.5} = 0.456\ 79 \qquad \text{and} \qquad r = 27.18°$$

$$BC = \frac{BQ}{\cos r} = \frac{0.525}{\cos 27.18°} = 0.590 \text{ in.}$$
$$\text{Angle } CBE = 43.25° - 27.18° = 16.07°$$
$$CE = BC \sin CBE = 0.590 \sin 16.07° = 0.163 \text{ in.}$$

31-46 A source of light is under water. What is the greatest angle a ray of light from this source can make with the normal and pass into the air? For any greater angle, the ray is totally reflected.

Suggestion: *As this ray passes into the air, it is refracted away from the normal. When the angle with the normal in air is 90°, the ray will not pass out of the water. When the ray in the water makes any greater angle with the normal than is necessary to make the angle in the air 90°, the ray is totally reflected at the surface of the water.*

$$\frac{\sin i}{\sin 90°} = \frac{3}{4}$$

31-47 If the eye is at a point under water, what is the greatest angle from the zenith that a star can appear to be?

31-48 A straight rod is partly immersed in water. The image in the water appears to be inclined at an angle of 40° with the surface. Find the inclination of the rod to the surface of the water if the index of refraction is $\frac{4}{3}$.

31-13 THE SINE BAR

A sine bar may be used for angular measurements when the accuracy required is 0.1° or less. This measuring instrument is a steel bar whose ends are machined so that they can receive cylindrical plugs of equal diameter. The sine bar is usually 5 or 10 in. in length and is the hypotenuse of a right triangle. It is usually used with a surface plate, since a true flat surface is needed. Figure 31-28 shows a 5-in. sine bar in place. Here

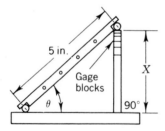

Fig. 31-28. Sine bar.

the angle to be measured or laid out is θ; the vertical distance X is measured with gage blocks. The sine bar can be used to determine either angle θ or the distance X for any value of angle θ with the aid of a table of natural functions.

EXAMPLE 31-11 What is X when θ is 25.3°?

Solution

$$\sin \theta = \frac{X}{5}$$

$$\sin 25.3° = \frac{X}{5}$$

$$X = 5 \sin 25.3° = 5(0.427\ 36)$$
$$= 2.137 \text{ in.}$$

Gage blocks are then used to measure or lay out 2.137 in.

EXAMPLE 31-12 What is θ if X is 1.750 in.?

Solution

$$\sin \theta = \frac{X}{5}$$

$$= \frac{1750}{5}$$
$$= 0.3500$$
$$\theta = 20.5°$$

Since the length of the sine bar is a constant, a special table can be developed which makes it unnecessary to perform calculations to determine either X or θ. Machine-shop handbooks usually include such a table for 5-in. sine bars.

Now available are 100- and 200-mm sine bars. Although these sine bars are not in general use, they are available and will replace the 5- and 10-in. bars. All the principles concerning the 5- and 10-in. bars apply to the metric sine bars.

▉ EXERCISES

Note: *Make a sketch for each exercise before you solve it. Assume a 5-in. sine bar for solving all exercises.*

31-49 Calculate X if θ is 10.5°.

31-50 Calculate θ if X is 2.5 in.

31-51 Calculate X if θ is 35.3°.

31-52 Calculate X if θ is 47.2°.

31-53 Construct a table for a 5-in. sine bar listing these values for θ:

(a) 5°	(d) 20°	(g) 35°	(j) 50°	(m) 65°	(p) 80°
(b) 10°	(e) 25°	(h) 40°	(k) 55°	(n) 70°	(q) 85°
(c) 15°	(f) 30°	(i) 45°	(l) 60°	(o) 75°	

31-54 Would it be possible to construct a similar table for a 10-in. sine bar by multiplying all the values in Exercise 31-53 by 2 (sine 2 × 5 = 10)? Check your answer by calculating several values.

31-55 Rewrite the basic sine-bar equation for a sine bar 10 in. in length.

31-14 AREAS OF SECTORS AND SEGMENTS OF A CIRCLE

The calculations for the areas of sectors and segments of a circle were discussed in the geometry section of this book. We now study them in the context of trigonometry.

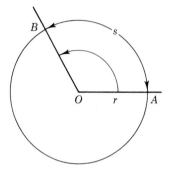

Fig. 31-29. θ (radians) $= \dfrac{s}{r}$.

From the definition of the radian, it is obvious that the number of radians in an angle at the center of a circle can be found by dividing the length of the arc by the length of the radius. Thus, in Fig. 31-29,

$$\text{Number of radians in an angle} = \frac{\text{arc}}{\text{radius}}$$

$$\text{Angle } AOB \text{ (in radians)} = \frac{\text{arc } AB}{\text{radius } OA} = \frac{s}{r}$$

$$\theta = \frac{s}{r}$$
$$s = r\theta$$
$$r = \frac{s}{\theta}$$

As an illustration, find the area of the segment AnX in Fig. 31-30 (the cross-hatched area).

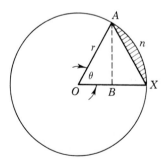

Fig. 31-30. Find the area of segment AnX.

The area of sector $AOnX$ minus the area of triangle AOX will equal the area of segment AnX. Let A = area of the segment, P = area of the sector, and T = area of the triangle. Then

$$A = P - T$$

The area of a triangle is equal to $\frac{1}{2}bh$; therefore $T = \frac{1}{2}(OX)(AB)$. Since $OX = r$ (the radius) and $AB = r \sin \theta$, substituting these values, the area of triangle $AOX = T = \frac{1}{2}(r)(r \sin \theta) = \frac{1}{2}r^2 \sin \theta$.

The area of a sector is equal to one-half the arc times r; or $P = \frac{1}{2}$ arc $\times r$. Arc $s = r\theta$; substituting this value for the arc in $P = \frac{1}{2}$ arc $\times r$, the equation becomes

$$P = \frac{1}{2} \times r\theta \times r = \frac{1}{2}r^2\theta$$

Substituting the values for P and T in our original equation $A = P - T$,

$$A = \frac{1}{2}r^2\theta - \frac{1}{2}r^2 \sin \theta = \frac{1}{2}r^2(\theta - \sin \theta)$$

Note: *Remember that θ is in radians. The example below illustrates this point.*

EXAMPLE 31-13 Find the area of the segment in a circle of radius 16 in. which has a central angle of 75.5°.

Solution Change the given angle from degrees to radians:

$$\theta = 78.5° = 78.5(0.017\ 453) = 1.3701 \text{ rad}$$
$$\sin 78.5° = 0.979\ 92$$

Substituting in the formula,

$$A = \frac{1}{2}r^2(\theta - \sin \theta)$$

$$= \frac{1}{2}(16^2)(1.3701 - 0.9799) = 49.94$$
$$= 49.94 \text{ in.}^2$$

■ EXERCISES

31-56 Find the area of a sector in a circle of radius 244 cm with a central angle of 60°. Find the area of the segment.

31-57 Find the area of the segment whose chord is 1220 mm in a circle 3050 mm in diameter.

31-58 Find the area of a segment whose chord is 366 cm and whose height is 60 cm.

31-59 In a circle with a radius of 152 cm, find the area of a segment having an angle of 63.25°. Find the length of the chord and the height of the segment; take two-thirds of their product; and compare the result with the area found.

31-60 A cylindrical tank resting in a horizontal position is filled with water to within 10 in. of the top. Find the number of cubic feet of water in the tank. The tank is 10 ft long and 4 ft in diameter.

31-61 If an angle of 126° at the center has an arc of 226 ft, find the radius of the circle.

Solution: Use the formula $r = s/\theta$. Remember to convert $126°$ to radians.

$$\theta = 126(0.017453) = 2.199$$
$$r = 226(2.199) = 102.77$$

Therefore, the radius is 102.77 ft.

31-62 A flywheel 20 ft in diameter has an angular velocity of 3π per second. Find the rim velocity.

Solution: Given $\theta R = S$, we can divide both sides of the equation by time (T).

$$\theta R = S$$
$$\frac{\theta R}{T} = \frac{S}{T}$$

$\dfrac{\theta}{T}$ is a measure of angular velocity (rad/s)

$\dfrac{S}{T}$ is also a measure of velocity (ft/s)

31-63 The circumferential speed generally advised by makers of emery wheels is 5,500 ft/min. Find the angular velocity per second in radians for a 10-in. wheel.

31-64 Solve similarly for the velocities of the following:
(a) Ohio grindstones, advised speed, 2,500 ft/min
(b) Huron grindstones, advised speed, 3,500 ft/min
(c) Wood, leather-covered polishing wheels, 7,000 ft/min
(d) Walrus-hide polishing wheels, 8,000 ft/min
(e) Rag wheels, 7,000 ft/min
(f) Hairbrush wheels, 12,000 ft/min

31-65 A flywheel with a 4-ft radius is revolving counterclockwise with a circumferential velocity of 75 ft/s. Find the angular velocity in radians per second.

31-66 Find the radius of a circle in which an arc of 20 ft measures an angle of 2.3 rad at the center. In this circle, find the angle at the center measured by an arc of 3 ft 8 in.

31-67 Find the angular velocity per minute of the minute hand of a watch. Express in degrees and radians.

31-68 A train of cars is going at the rate of 15 mph on a curve of 600-ft radius; find its angular velocity in radians per minute.

31-69 A flywheel 22 ft in diameter is revolving with an angular velocity of 9 rad/s. Find the rate per minute at which a point on the circumference is traveling.

31-70 Find the length of arc that, at the distance of 1 mi, will subtend an angle of $0.17°$ at the eye; find the length of arc that subtends an angle of $0.0003°$ in the same conditions.

31-71 The radius of the earth's orbit, which is about 92,700,000 mi, subtends at the star Sirius an angle of about $0.0001°$. Find the approximate distance of Sirius from the earth.

31-15 RAILROAD AND HIGHWAY CURVES

Imagine yourself driving a car and approaching a cloverleaf intersection. Depending upon where you look at the curve, your head may turn up to 90° from the path your car is traveling.

In theory, any curvature should be measured by the rate at which the road departs from its own tangent. However, such a measurement is very hard to make without special equipment.

In the United States, it is customary to express the curvature of railroads and highways in terms of a central angle:

Rule: *The degree D of a highway curve is given by the central angle in degrees subtending an arc of 100 ft along the curve. Measurement of railroad curves is done with a 100-ft chord instead of the arc.*

EXAMPLE 31-14 Find the radius R of a 1° railroad curve.

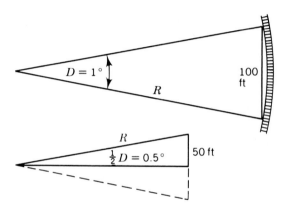

Fig. 31-31. Example 31-14.

Solution Study Fig. 31-31. Use the trig function $\sin \theta = \dfrac{opposite\ side}{hypotenuse}$.

$$\frac{50\ ft}{R} = \sin 0.5°$$

$$R = \frac{50}{\sin 0.5°}$$
$$= 5730\ ft$$

A 1° railroad curve can be laid out by drawing an arc with a radius of 5730 ft.

Note: *It does not matter how long the arc is (or how large the central angle is); the curve is still defined as a 1° railroad curve, since a 1° central angle would subtend a chord of 100 ft.*

EXAMPLE 31-15 Can $R_1° = 5730$ ft be used as a constant in laying out a 5° railroad curve?

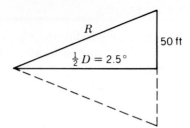

Fig. 31-32. Example 31-15.

Solution See Fig. 31-32 and use the sine function to solve.

$$\sin 2.5° = \frac{50}{R}$$

$$R = \frac{50}{\sin 2.5°} = \frac{50}{0.043\ 62} = 1146$$
$$R_{5°} = 1146$$
$$R_{1°} = 5730$$
$$5R_{5°} = 5730$$

For curves up to 7°, $R_{1°} = 5730$ ft can be used as a constant. That is,

$$R_x = \frac{5730\ \text{ft}}{x}$$

Highway curves differ slightly in that the degree of curvature is defined as the central of a 100-ft arc, rather than as a 100-ft chord. Because of this, the central degree measurement must be converted to radians.

EXAMPLE 31-16 Find the radius R of a 1° highway curve.

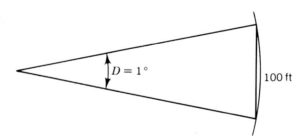

Fig. 31-33. Example 31-16.

Solution Fig. 31-33 illustrates a 1° highway curve. Use the formula to determine the radius.

$$R = \frac{S}{\dfrac{D°2\pi\ \text{rad}}{360°}}$$

$$= \frac{100}{0.01745}$$
$$= 5731\ \text{ft}$$

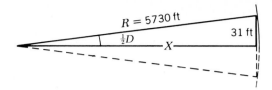

Fig. 31-34. Rule of thumb for railroad curves.

An interesting rule of thumb used by railroads is that the distance between the railroad track and the midpoint of a 62-ft chord indicates in inches the curvature in degrees. That is, if you stretch a 62-ft string as a chord on a 1° curve, the midpoint would be 1 in. away from the track.

To prove this, we can use the fact that a 1° curve has a radius of 5730 ft. In Fig. 31-34, we solve for x and find the distance between the track and the string by subtracting x from 5730 ft.

$$\sin \tfrac{1}{2}D = \frac{31}{5730} = 0.005\,41$$

$$\tfrac{1}{2}D = 0.309\,98°$$

$$\cos \tfrac{1}{2}D = \frac{x}{5730}$$

$$x = 5730 \cos \tfrac{1}{2}D$$
$$= 5730\,(0.99999)$$
$$= 5729.916$$
$$5730 - x = 5730 - 5729.916$$
$$= 0.084 \text{ ft}$$
$$= 1.008 \text{ in.}$$

This rule of thumb works very well for approximations of any curve normally encountered in rail or road construction.

■ EXERCISES

31-72 A train is traveling on a curve of $\tfrac{1}{2}$-mi radius at the rate of 30 mph. Through what angle does it turn in 45 s? Express the answer in both radians and degrees.

Solution:

$$\tfrac{1}{2}D = \frac{s}{R} = \frac{(\text{mi/h})(\text{h/min})(\text{min/s})}{\text{miles}}$$

$$= \frac{30(\tfrac{1}{60})(\tfrac{1}{60})(45)}{\tfrac{1}{2}}$$

31-73 A train on the main line is passing through a 0.5° curve (the radius is 11,460 ft) at 75 mph. If the central angle of the total curve is 10°, how long will it take for the engine to pass through the curve?

31-74 A train is $\tfrac{1}{4}$ mi long and is traveling at 45 mph. How long will some portion of the train be in the curve in Exercise 31-73?

31-75 The maximum curve allowed on a particular main-line railroad is 1.75°. How much track will be needed to change a train's direction by 7°?

31-76 A particular railroad curve is 825 ft long and has a radius of 882 ft. What is the approximate degree of curvature of the track? What is the central angle of this curve (how much will a train's direction be changed)?

31-77 An interstate thruway allows a maximum degree of curvature of 0.30°. What will be the minimum allowable curve radius?

31-78 On a highway, the degree of curvature is inversely proportional to the square of the designed speed ($DV^2 = K$). Fill in the degree of curvature and the radius for the following speeds: (a) 30 mph (b) 40 mph (c) 45 mph (d) 55 mph (e) 70 mph.

Solution:

Design Speed, MPH	Maximum Curvature, Degrees	Minimum Curve Radius
30	21°	273
40		
45		
55		
70		

$$K = DV^2$$
$$= (21)(30)^2 = 21(900)$$
$$= 18,900$$

31-16 DISTANCE AND DIP OF THE HORIZON

Occasionally it is necessary to determine the distance to the horizon from an elevated position. The elevated position may be in an airplane or the crow's nest of a ship, or on the top of a tower, a hill, or a mountain.

In Fig. 31-35, O is the center of the earth, r is the radius of the earth's

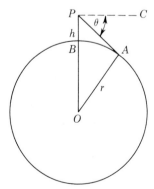

Fig. 31-35. Find distance *PA*.

surface, and h is the height of the point P above the surface. To calculate the distance PA, then,

$$(PA)^2 = (OP)^2 - (OA)^2$$

Since $OP = OB + h$, $OB = r$, and $OA = r$,

$$OP = r + h$$
$$(PA)^2 = (r + h)^2 - (r)^2$$
$$(PA)^2 = r^2 + 2rh + h^2 - r^2$$
$$(PA)^2 = 2rh + h^2$$
$$PA = \sqrt{2rh + h^2}$$

For almost all elevated positions h is very small compared with the product $2rh$. Therefore, for most problems,

$$PA = \sqrt{2rh}$$

In the formula above, PA, r, and h are in the same units. To change this equation where h represents the height of the elevated position in feet and r and PA are expressed in miles, let r (the radius) = 3960 mi. Then

$$PA = \sqrt{2\,(3960)\left(\frac{h}{5280}\right)} = \sqrt{\frac{3}{2}h} = \sqrt{1.5h} \text{ mi}$$

That is, the distance to the horizon here is approximately equal to the square foot of 1.5 times the height of the elevated position in feet.

This last formula would be appropriate for most problems, but $PA = \sqrt{2rh}$ or $PA = \sqrt{2rh + h^2}$ should be used to determine the distance to the horizon when h is in miles (for example, if a spacecraft were involved).

Angle θ, or APC, in Fig. 31-35 is the *dip* of the horizon; this angle can be calculated by application of a trigonometric formula:

$$\tan APC = \tan \theta = \frac{PA}{r}$$

▇ EXERCISES

31-79 A cliff 2000 ft high is on the seashore; how far away is the horizon? What is the dip of the horizon?

31-80 Find the greatest distance at which the lamp of a lighthouse can be seen from the deck of a ship if the lamp is 85 ft above the surface of the water and the deck of the ship is 30 ft above the surface.

31-81 Find the radius of your horizon if you are 400 m above the surface of the earth. What is the radius if you are 5 km above the surface of the earth?

31-82 How high above the earth must you be to see a point on the surface 40 mi away?

31-83 Two lighthouses, one 95 ft high and the other 80 ft high, are just visible from each other over the water. Find how far apart they are.

31-84 An astronaut is in a capsule 200 km above the surface of the earth. How far is it to the horizon from the capsule? What is the error if the formula $PA = \sqrt{2rh}$ is used to determine this distance?

31·17 SLOPE OF A LINE MAKING A GIVEN ANGLE WITH THE LINE OF GREATEST SLOPE

When a roadbed is laid out on a steep hill or grade, it is often necessary to determine the direction of the road so that a given grade will not be exceeded.

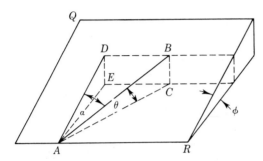

Fig. 31-36. Find angle θ.

In Fig. 31-36 the plane RAQ is inclined at an angle ϕ with the horizontal. AD is a line of greatest slope in plane RAQ and is also at an angle ϕ with the horizontal. Line AB is in plane RAQ but at an angle α with AD. Obviously the angle between AB and the horizontal is less than angle ϕ; line AB is at an angle θ with the horizontal.

To determine angle θ,

$$\sin \theta = \frac{BC}{AB}$$

$$\cos \alpha = \frac{AD}{AB} \quad \text{or} \quad AB = \frac{AD}{\cos \alpha}$$

$$\sin \phi = \frac{DE}{AD} \quad \text{or} \quad DE \quad \text{or} \quad BC = AD \sin \phi$$

Substituting $AD \sin \phi$ for BC and $AD/(\cos \alpha)$ for AB,

$$\sin \theta = \frac{BC}{AB} = \frac{AD \sin \phi}{AD/(\cos \alpha)}$$

$$= \overset{1}{\cancel{AD}} \sin \phi \left(\frac{\cos \alpha}{\underset{1}{\cancel{AD}}} \right)$$

$$= \sin \phi \cos \alpha$$

▨ EXERCISES

31-85 The slope of a roof is 30°. Find the angle θ that is the inclination to the horizontal of a line AB drawn in the roof and making an angle of 35° with the line of greatest slope.

31-86   A hill has a slope of 32°. A path leads up it, making an angle of 45° with the line of greatest slope. Find the slope of the path.

31-87   Two roofs have their ridges at right angles, and each is inclined to the horizontal at an angle of 30°. Find the inclination of their line of intersection to the horizontal.

31-88   A mountainside has a slope of 30°. A road ascending the mountain is to be built, having a grade of 6 percent. Find the angle that the road will make with the line of greatest slope.

31-89   A road making an angle of 7° with the horizontal is on a mountainside that makes an angle of 20° with the horizontal. Find the angle that the road makes with the line of greatest slope.

31-90   A road of 6 percent grade is on a mountainside having a slope of 35 percent. Find the angle that the road makes with the line of greatest slope.

APPENDIX

Table 1   Summary of Formulas

PLANE FIGURES

Circle	$C = 2\pi r = \pi d$
	$A = \pi r^2 = \dfrac{\pi d^2}{4}$
Sector	$A = \dfrac{\theta \pi r^2}{360} = \dfrac{\theta \pi d^2}{1440} \qquad \theta$ in degrees
	$A = \frac{1}{2}\, \text{arc} \cdot r \qquad$ arc in radians
Segment	$r = \dfrac{(w/2)^2 + h^2}{2h}$
	$h = r - \sqrt{r^2 - \left(\dfrac{w}{2}\right)^2}$
	$w = 2\sqrt{h(2r - h)}$
	$A = \dfrac{2hw}{3} + \dfrac{h^3}{2w} = \dfrac{4h^2}{3}\sqrt{\dfrac{2r}{h} - 0.608}$
Ellipse	The mean of
	$P = \pi\left[\dfrac{3}{2}(a + b) - \sqrt{ab}\right] \qquad$ and
	$P = \pi\sqrt{2(a^2 + b^2)}$
	$A = \pi ab$
Parallelogram	$A = bh$
Rectangle	$A = bh$
Ring	$A = \pi(R^2 - r^2) = \pi(R + r)(R - r)$
Tapers	$l{:}1 = t{:}T$
	$x = \dfrac{D - d}{2} \times \dfrac{L}{l}$
Threads	$D_1 = D - \dfrac{1.732}{N} \qquad$ sharp V threads
	$D_1 = D - \dfrac{1.299}{N} \qquad$ U.S. Standard threads
Trapezoid	$A = \dfrac{h}{2}(B + b)$

474

Table 1 Summary of Formulas (Continued)

Triangle $\qquad A = \frac{1}{2}bh$

$A = \sqrt{s(s-a)(s-b)(s-c)}$ $\qquad$ where $s = \dfrac{a+b+c}{2}$

Right triangle $\qquad c = \sqrt{a^2 + b^2}$

$a = \sqrt{c^2 - b^2}$

$b = \sqrt{c^2 - a^2}$

SOLID FIGURES

Cone $\qquad S = \frac{1}{2}ps$
$T = \frac{1}{2}ps + A$
$V = \frac{1}{3}Ah$

Cube $\qquad S = 4a^2$
$T = 6a^2$
$V = a^3$

Cylinder $\qquad S = 2\pi rh$
$T = 2\pi rh + 2\pi r^2$
$V = \pi r^2 h$

Frustum $\qquad S = \frac{1}{2}(P + p)s$
$T = \frac{1}{2}(P + p)s + B + b$
$V = \dfrac{h}{3}(B + b + \sqrt{Bb})$

Frustum of cone $\qquad V = \dfrac{h\pi}{3}(R^2 + r^2 + Rr) = \dfrac{\pi h}{12}(D^2 + d^2 + Dd)$

Prismatoid $\qquad V = \dfrac{h}{6}(B_1 + 4M + B_2)$

Prism $\qquad S = ph$
$T = ph + 2A$
$V = Ah$

Pyramid $\qquad S = \frac{1}{2}ps$
$T = \frac{1}{2}ps + A$
$V = \frac{1}{3}Ah$

Ring $\qquad A = 4\pi^2 Rr$
$V = 2\pi^2 Rr^2$

Sphere $\qquad T = 4\pi r^2$
$V = \frac{4}{3}\pi r^3$

Segment $\qquad V = \frac{1}{2}\pi h(r_1^2 + r_2^2) + \frac{1}{6}\pi h^3$
Zone $\qquad S = 2\pi rh$

Table 1   Summary of Formulas (Continued)

TRIGONOMETRIC FORMULAS

$\sin^2 \theta + \cos^2 \theta = 1$

$\sec^2 \theta = 1 + \tan^2 \theta$

$\csc^2 \theta = 1 + \cot^2 \theta$

$$\sin \theta = \frac{1}{\csc \theta} \quad \text{or} \quad \csc \theta = \frac{1}{\sin \theta}$$

$$\cos \theta = \frac{1}{\sec \theta} \quad \text{or} \quad \sec \theta = \frac{1}{\cos \theta}$$

$$\tan \theta = \frac{1}{\cot \theta} \quad \text{or} \quad \cot \theta = \frac{1}{\tan \theta}$$

$$\tan \theta = \frac{\sin \theta}{\cos \theta}$$

$$\cot \theta = \frac{\cos \theta}{\sin \theta}$$

$\sin (A + B) = \sin A \cos B + \cos A \sin B$

$\sin (A - B) = \sin A \cos B - \cos A \sin B$

$\cos (A + B) = \cos A \cos B - \sin A \sin B$

$\cos (A - B) = \cos A \cos B + \sin A \sin B$

$$\tan (A + B) = \frac{\tan A + \tan B}{1 - \tan A \tan B}$$

$$\tan (A - B) = \frac{\tan A - \tan B}{1 + \tan A \tan B}$$

$\sin 2A = 2 \sin A \cos A$

$\cos 2A = \cos^2 A - \sin^2 A = 1 - 2 \sin^2 A = 2 \cos^2 A - 1$

$$\tan 2A = \frac{2 \tan A}{1 - \tan^2 A}$$

$$\sin A = \pm \sqrt{\frac{1 - \cos 2A}{2}}$$

$$\tan A = \pm \sqrt{\frac{1 - \cos 2A}{1 + \cos 2A}} = \frac{\sin 2A}{1 + \cos 2A} = \frac{1 - \cos 2A}{\sin 2A}$$

Table 1 Summary of Formulas (Continued)

$\sin A \cos B = \frac{1}{2} \sin (A + B) + \frac{1}{2} \sin (A - B)$
$\cos A \sin B = \frac{1}{2} \sin (A + B) - \frac{1}{2} \sin (A - B)$
$\cos A \cos B = \frac{1}{2} \cos (A + B) + \frac{1}{2} \cos (A - B)$
$\sin A \sin B = -\frac{1}{2} \cos (A + B) + \frac{1}{2} \cos (A - B)$
$\sin A + \sin B = 2 \sin \frac{1}{2}(A + B) \cos \frac{1}{2}(A - B)$
$\sin A - \sin B = 2 \cos \frac{1}{2}(A + B) \sin \frac{1}{2}(A - B)$
$\cos A + \cos B = 2 \cos \frac{1}{2}(A + B) \cos \frac{1}{2}(A - B)$
$\cos A - \cos B = -2 \sin \frac{1}{2}(A + B) \sin \frac{1}{2}(A - B)$

$$\frac{a}{\sin A} = \frac{b}{\sin B} = \frac{c}{\sin C}$$

$a^2 = b^2 + c^2 - 2bc \cos A$
$b^2 = a^2 + c^2 - 2ac \cos B$
$c^2 = a^2 + b^2 - 2ab \cos C$

$$\cos A = \frac{b^2 + c^2 - a^2}{2bc}$$

$$\cos B = \frac{a^2 + c^2 - b^2}{2ac}$$

$$\cos C = \frac{a^2 + b^2 - c^2}{2ab}$$

MOLLWEIDE'S EQUATIONS FOR CHECKING TRIANGLES

$$\frac{a - b}{c} = \frac{\sin \frac{1}{2}(A - B)}{\cos \frac{1}{2}C}$$

$$\frac{a + b}{c} = \frac{\cos \frac{1}{2}(A - B)}{\sin \frac{1}{2}C}$$

Table 2 Useful Equivalents

1 ft³ water weighs 62.5 lb (approximately) = 1000 oz

1 gallon water weighs $8\frac{1}{3}$ lb (approximately)

1 atmosphere pressure = 14.7 lb/in.² = 2116 lb/ft²

1 atmosphere pressure = 760 mm mercury

A column of water 2.3 ft high = a pressure of 1 lb/in.²

1 gal = 231 in.³ (by act of Congress)

1 ft³ = $7\frac{1}{2}$ gallons (approximately) or, better, 7.48 gallons

1 ft³ = $\frac{4}{5}$ bushel (approximately)

1 barrel = 4.221 − ft³ (approximately)

1 bushel = 2,150.42 in.³ (by act of Congress) = 1.24446 − ft³

1 bushel = $\frac{5}{4}$ ft³ (approximately)

1 perch = $24\frac{3}{4}$ ft³ (but usually taken as 25 ft³)

1 in. = 25.4000 mm (exactly)

1 ft = 30.4801 cm

1 m = 39.37 in. (by act of Congress)

1 lb (avoirdupois) = 7000 grains (by act of Congress)

1 lb (troy or apothecaries) = 5760 grains

1 g = 15.432 grains

1 kg = 2.204 62 lb (avoirdupois)

1 liter = 1.056 68 quart (liquid) = 0.908 08 quart (dry)

1 quart (liquid) = 946.358 cm³ = 0.946 358 liter or dm³

1 quart (dry) = 1101.228 cm³ = 1.101 228 liters or dm³

π = 3.141 592 653 589 79+ = 3.1416 = $\frac{355}{113}$ = $3\frac{1}{7}$ (all approximate)

1 rad = 57°17′44.8″ = 57.295 779 5°+

1° = 0.017 453 29+ rad

Base of Napierian logarithms = e = 2.718 281 828 . . .

$\log_{10} e$ = 0.434 294 48 . . .

$\log_e 10$ = 2.302 585 09 . . .

1 horsepower-second = 550 ft-lb

1 horsepower-minute = 33,000 ft-lb

$\sqrt{2}$ = 1.414 213 6

$\sqrt{3}$ = 1.732 050 8

$\sqrt{5}$ = 2.236 068 0

$\sqrt{6}$ = 2.449 489 7

$\sqrt[3]{2}$ = 1.259 921 0

$\sqrt[3]{3}$ = 1.442 249 6

Table 3   Decimal and Fractional Equivalents of Parts of an Inch

8THS AND 16THS	32NDS	64THS	
8ths			
1 = 0.125	1 = 0.03125	1 = 0.015625	33 = 0.515625
2 = 0.250	3 = 0.09375	3 = 0.046875	35 = 0.546875
3 = 0.375	5 = 0.15625	5 = 0.078125	37 = 0.578125
4 = 0.500	7 = 0.21875	7 = 0.109375	39 = 0.609375
5 = 0.625	9 = 0.28125	9 = 0.140625	41 = 0.640625
6 = 0.750	11 = 0.34375	11 = 0.171875	43 = 0.671875
7 = 0.875	13 = 0.40625	13 = 0.203125	45 = 0.703125
16ths	15 = 0.46875	15 = 0.234375	47 = 0.734375
1 = 0.0625	17 = 0.53125	17 = 0.265625	49 = 0.765625
3 = 0.1875	19 = 0.59375	19 = 0.296875	51 = 0.796875
5 = 0.3125	21 = 0.65625	21 = 0.328125	53 = 0.828125
7 = 0.4375	23 = 0.71875	23 = 0.359375	55 = 0.859375
9 = 0.5625	25 = 0.78125	25 = 0.390625	57 = 0.890625
11 = 0.6875	27 = 0.84375	27 = 0.421875	59 = 0.921875
13 = 0.8125	29 = 0.90625	29 = 0.453125	61 = 0.953125
15 = 0.9375	31 = 0.96875	31 = 0.484375	63 = 0.984375

Table 4   U.S. Standard and Sharp V Threads

DIAMETER OF SCREW, INCHES	THREADS PER INCH	DEPTH U. S. S.	DEPTH SHARP V	ROOT DIA. U. S. S.	ROOT DIA. SHARP V
$\frac{1}{4}$	20	0.03247	0.04330	0.1850	0.1634
$\frac{5}{16}$	18	0.03608	0.04811	0.2403	0.2163
$\frac{3}{8}$	16	0.04059	0.05412	0.2936	0.2668
$\frac{7}{16}$	14	0.04639	0.06178	0.3447	0.3139
$\frac{1}{2}$	13	0.04996	0.06661	0.4001	0.3668
$\frac{9}{16}$	12	0.05412	0.07216	0.4542	0.4182
$\frac{5}{8}$	11	0.05905	0.07873	0.5069	0.4675
$\frac{3}{4}$	10	0.06495	0.08660	0.6201	0.5768
$\frac{7}{8}$	9	0.07216	0.09622	0.7307	0.6826
1	8	0.08119	0.10825	0.8376	0.7835
$1\frac{1}{8}$	7	0.09277	0.12371	0.9394	0.8776
$1\frac{1}{4}$	7	0.09277	0.12371	1.0644	1.0026
$1\frac{3}{8}$	6	0.10825	0.14433	1.1585	1.0863
$1\frac{1}{2}$	6	0.10825	0.14433	1.2835	1.2113
$1\frac{5}{8}$	$5\frac{1}{2}$	0.11809	0.15745	1.3888	1.3101
$1\frac{3}{4}$	5	0.12990	0.17325	1.4902	1.4035
$1\frac{7}{8}$	5	0.12990	0.17325	1.6152	1.5285
2	$4\frac{1}{2}$	0.14433	0.19244	1.7113	1.6151
$2\frac{1}{4}$	$4\frac{1}{2}$	0.14433	0.19244	1.9613	1.8651
$2\frac{1}{2}$	4	0.16238	0.21650	2.1752	2.0670
$2\frac{3}{4}$	4	0.16238	0.21650	2.4252	2.3170
3	$3\frac{1}{2}$	0.18557	0.24742	2.6288	2.5052
$3\frac{1}{4}$	$3\frac{1}{2}$	0.18557	0.24742	2.8788	2.7552
$3\frac{1}{2}$	$3\frac{1}{4}$	0.19985	0.26647	3.1003	2.9671
$3\frac{3}{4}$	3	0.21666	0.28866	3.3167	3.1727
4	3	0.21666	0.28866	3.5667	3.4227
$4\frac{1}{4}$	$2\frac{7}{8}$	0.2259	0.3012	3.7982	3.6476
$4\frac{1}{2}$	$2\frac{3}{4}$	0.2362	0.3149	4.0276	3.8712
$4\frac{3}{4}$	$2\frac{5}{8}$	0.2474	0.3299	4.2551	4.0901
5	$2\frac{1}{2}$	0.2598	0.3465	4.4804	4.3070
$5\frac{1}{4}$	$2\frac{1}{2}$	0.2598	0.3465	4.7304	4.5500
$5\frac{1}{2}$	$2\frac{3}{8}$	0.2735	0.3647	4.9530	4.7707
$5\frac{3}{4}$	$2\frac{3}{8}$	0.2735	0.3647	5.2030	5.0207
6	$2\frac{1}{4}$	0.2887	0.3849	5.4226	5.2302

Table 5 Standard Gages for Wire and Sheet Metals (Diameter or Thickness Given in Decimals of an Inch)

NUMBER OF GAGE	BIRMINGHAM WIRE GAGE	AMERICAN, BROWN AND SHARP (B. & S.)	UNITED STATES STANDARD PLATE IRON STEEL	BRITISH IMPERIAL	AMERICAN STEEL AND WIRE CO.
0000000	—	—	0.5	0.5	
000000	—	—	0.46875	0.464	
00000	—	—	0.4375	0.432	
0000	0.454	0.46	0.40625	0.400	0.3938
000	0.425	0.409642	0.375	0.372	0.3625
00	0.380	0.364796	0.34375	0.348	0.3310
0	0.340	0.324861	0.3125	0.324	0.3065
1	0.300	0.289297	0.28125	0.300	0.2830
2	0.284	0.257627	0.265625	0.276	0.2625
3	0.259	0.229423	0.25	0.252	0.2437
4	0.238	0.204307	0.234375	0.232	0.2253
5	0.220	0.181940	0.21875	0.212	0.2070
6	0.203	0.162023	0.203125	0.192	0.1920
7	0.180	0.144285	0.1875	0.176	0.1770
8	0.165	0.128490	0.171875	0.160	0.1620
9	0.148	0.114423	0.15625	0.144	0.1483
10	0.134	0.101897	0.140625	0.128	0.1350
11	0.120	0.090742	0.125	0.116	0.1205
12	0.109	0.080808	0.109375	0.104	0.1055
13	0.095	0.071962	0.09375	0.092	0.0915
14	0.083	0.064084	0.078125	0.080	0.0800
15	0.072	0.057068	0.0703125	0.072	0.0720
16	0.065	0.050821	0.0625	0.064	0.0625
17	0.058	0.045257	0.05625	0.056	0.0540
18	0.049	0.040303	0.05	0.048	0.0475
19	0.042	0.035890	0.04375	0.040	0.0410
20	0.035	0.031961	0.0375	0.036	0.0348

Table 5 Standard Gages for Wire and Sheet Metals (Diameter or Thickness Given in Decimals of an Inch) (Continued)

NUMBER OF GAGE	BIRMINGHAM WIRE GAGE	AMERICAN, BROWN AND SHARP (B. & S.)	UNITED STATES STANDARD PLATE IRON STEEL	BRITISH IMPERIAL	AMERICAN STEEL AND WIRE CO.
21	0.032	0.028462	0.034375	0.032	0.03175
22	0.028	0.025346	0.03125	0.028	0.0286
23	0.025	0.022572	0.028125	0.024	0.0258
24	0.022	0.020101	0.025	0.022	0.0230
25	0.020	0.017900	0.021875	0.020	0.0204
26	0.018	0.015941	0.01875	0.018	0.0181
27	0.016	0.014195	0.0171875	0.0164	0.0173
28	0.014	0.012641	0.015625	0.0148	0.0162
29	0.013	0.011257	0.0140625	0.0136	0.0150
30	0.012	0.010025	0.0125	0.0124	0.0140
31	0.010	0.008928	0.0109375	0.0116	0.0132
32	0.009	0.007950	0.01015625	0.0108	0.0128
33	0.008	0.007080	0.009375	0.0100	0.0118
34	0.007	0.006305	0.00859375	0.0092	0.0104
35	0.005	0.005615	0.0078125	0.0084	0.0095
36	0.004	0.005000	0.00703125	0.0076	0.0090
37	—	0.004453	0.006640625	0.0068	
38	—	0.003965	0.00625	0.0060	
39	—	0.003531	—	0.0052	
40	—	0.003144	—	0.0048	

Table 6 Specific Gravities and Densities of Substances

NAME OF SUBSTANCE	POUNDS PER CUBIC INCH	POUNDS PER CUBIC FOOT	SPECIFIC GRAVITY
Air	—	0.0795	—
Aluminum	—	162	2.6
Anthracite coal, broken	—	52 to 60	—
Antimony	—	418	6.7
Asphaltum	—	87.3	1.4
Beech wood	—	46	0.73
Birch wood	—	41	0.65
Brass, cast (copper and zinc)	—	506	8.1
Brass, rolled	—	525	8.4
Brick, common	—	125	—
Brick, pressed	—	150	—
Chalk	—	156	2.5
Coal, bituminous, broken	—	47 to 56	—
Coke, loose	—	23 to 32	—
Corundum	—	—	3.9
Copper, cast	—	542	8.6 to 8.8
Copper, rolled	0.319	555	8.8 to 9
Cork	—	15	0.24
Ebony wood	—	76	1.23
Elm wood	—	35	0.56
Flint	—	162	2.6
Glass	—	186	2.5 to 3.45
Gold	0.695	—	19.3
Granite	—	170	2.56 to 2.88
Hickory wood	—	53	0.85
Ice	—	57.5	0.92
Iron, cast	0.26	450	6.7 to 7.4
Iron, wrought	0.28	480	7.69
Lead	0.412	712	11.42
Marble	—	168.7	2.7
Maple wood	—	49	0.79
Mercury	0.49	—	13.6
Nickel	0.318	—	8.8
Oak wood, red	—	46	0.73 to 0.75
Pine wood, white	—	28	0.45
Pine wood, yellow	—	38	0.61
Platinum	—	—	21.5
Quartz	—	165	2.65
Silver	0.379	655	10.5
Steel	0.29	490	7.85
Tin	—	459	7.2 to 7.5
Zinc	—	438	6.8 to 7.2
Water, distilled, at 32° F	—	62.417	—
Water, distilled, at 62° F	—	62.355	1
Water, distilled, at 212° F	—	59.7	—

Table 6   Specific Gravities and Densities of Substances (Continued)

SPECIFIC GRAVITIES REFERRED TO AIR

Air................ 1
Oxygen 1.11
Hydrogen .... .07
Chlorine gas 2.44

SPECIFIC GRAVITIES REFERRED TO HYDROGEN

Hydrogen 1
Air................ 14.53
Oxygen 15.95
Coal gas 6

Table 7   Strength of Materials

MATERIAL	ULTIMATE TENSILE STRENGTH, POUNDS PER SQUARE INCH°	ULTIMATE COMPRESSIVE STRENGTH, POUNDS PER SQUARE INCH†	COEFFICIENT OF LINEAR EXPANSION, FOR 1°F‡
Hard steel	100,000	120,000	0.0000065
Structural steel	60,000	60,000	0.0000065
Wrought iron.......	50,000	50,000	0.0000067
Cast iron..........	20,000	90,000	0.0000062
Copper	30,000	—	0.0000089
Timber, with grain ..	8,000 to 25,000	4,000 to 12,000	0.0000028
Concrete..........	300	3,000	0.0000055
Granite	—	11,000	0.0000050
Brick	—	3,000	0.0000050

° In the column of ultimate tensile strengths are given the pulls necessary to break a rod of 1 in.² cross section of the given material.

† In the column of ultimate compressive strengths are given the weights necessary to cause a support of 1 in.² cross section to give way under the pressure.

‡ In the column of coefficient of linear expansion are given the fractional parts of their length that bars of the different materials will increase when the temperature rises 1°F.

Table 8 Square Roots

N	0	1	2	3	4	5	6	7	8	9	1	2	3	4	5	6	7	8	9
1.0	1.000	1.005	1.010	1.015	1.020	1.025	1.030	1.034	1.039	1.044	0	1	1	2	2	3	3	4	4
1.1	1.049	1.054	1.058	1.063	1.068	1.072	1.077	1.082	1.086	1.091	0	1	1	2	2	3	3	4	4
1.2	1.095	1.100	1.105	1.109	1.114	1.118	1.122	1.127	1.131	1.136	0	1	1	2	2	3	3	4	4
1.3	1.140	1.145	1.149	1.153	1.158	1.162	1.166	1.170	1.175	1.179	0	1	1	2	2	3	3	3	4
1.4	1.183	1.187	1.192	1.196	1.200	1.204	1.208	1.212	1.217	1.221	0	1	1	2	2	2	3	3	4
1.5	1.225	1.229	1.233	1.237	1.241	1.245	1.249	1.253	1.257	1.261	0	1	1	2	2	2	3	3	4
1.6	1.265	1.269	1.273	1.277	1.281	1.285	1.288	1.292	1.296	1.300	0	1	1	2	2	2	3	3	4
1.7	1.304	1.308	1.311	1.315	1.319	1.323	1.327	1.330	1.334	1.338	0	1	1	2	2	2	3	3	3
1.8	1.342	1.345	1.349	1.353	1.356	1.360	1.364	1.367	1.371	1.375	0	1	1	1	2	2	3	3	3
1.9	1.378	1.382	1.386	1.389	1.393	1.396	1.400	1.404	1.407	1.411	0	1	1	1	2	2	3	3	3
2.0	1.414	1.418	1.421	1.425	1.428	1.432	1.435	1.439	1.442	1.446	0	1	1	1	2	2	2	3	3
2.1	1.449	1.453	1.456	1.459	1.463	1.466	1.470	1.473	1.476	1.480	0	1	1	1	2	2	2	3	3
2.2	1.483	1.487	1.490	1.493	1.497	1.500	1.503	1.507	1.510	1.513	0	1	1	1	2	2	2	3	3
2.3	1.517	1.520	1.523	1.526	1.530	1.533	1.536	1.539	1.543	1.546	0	1	1	1	2	2	2	3	3
2.4	1.549	1.552	1.556	1.559	1.562	1.565	1.568	1.572	1.575	1.578	0	1	1	1	2	2	2	3	3
2.5	1.581	1.584	1.587	1.591	1.594	1.597	1.600	1.603	1.606	1.609	0	1	1	1	2	2	2	3	3
2.6	1.612	1.616	1.619	1.622	1.625	1.628	1.631	1.634	1.637	1.640	0	1	1	1	2	2	2	2	3
2.7	1.643	1.646	1.649	1.652	1.655	1.658	1.661	1.664	1.667	1.670	0	1	1	1	2	2	2	2	3
2.8	1.673	1.676	1.679	1.682	1.685	1.688	1.691	1.694	1.697	1.700	0	1	1	1	1	2	2	2	3
2.9	1.703	1.706	1.709	1.712	1.715	1.718	1.720	1.723	1.726	1.729	0	1	1	1	1	2	2	2	3
3.0	1.732	1.735	1.738	1.741	1.744	1.746	1.749	1.752	1.755	1.758	0	1	1	1	1	2	2	2	3
3.1	1.761	1.764	1.766	1.769	1.772	1.775	1.778	1.780	1.783	1.786	0	1	1	1	1	2	2	2	3
3.2	1.789	1.792	1.794	1.797	1.800	1.803	1.806	1.808	1.811	1.814	0	1	1	1	1	2	2	2	2
3.3	1.817	1.819	1.822	1.825	1.828	1.830	1.833	1.836	1.838	1.841	0	1	1	1	1	2	2	2	2
3.4	1.844	1.847	1.849	1.852	1.855	1.857	1.860	1.863	1.865	1.868	0	1	1	1	1	2	2	2	2
3.5	1.871	1.873	1.876	1.879	1.881	1.884	1.887	1.889	1.892	1.895	0	1	1	1	1	2	2	2	2
3.6	1.897	1.900	1.903	1.905	1.908	1.910	1.913	1.916	1.918	1.921	0	1	1	1	1	1	1	2	2
3.7	1.924	1.926	1.929	1.931	1.934	1.936	1.939	1.942	1.944	1.947	0	1	1	1	1	2	2	2	2
3.8	1.949	1.952	1.954	1.957	1.960	1.962	1.965	1.967	1.970	1.972	0	1	1	1	1	2	2	2	2
3.9	1.975	1.977	1.980	1.982	1.985	1.987	1.990	1.992	1.995	1.997	0	1	1	1	1	2	2	2	2
4.0	2.000	2.002	2.005	2.007	2.010	2.012	2.015	2.017	2.020	2.022	0	0	1	1	1	1	2	2	2
4.1	2.025	2.027	2.030	2.032	2.035	2.037	2.040	2.042	2.045	2.047	0	0	1	1	1	1	2	2	2
4.2	2.049	2.052	2.054	2.057	2.059	2.062	2.064	2.066	2.069	2.071	0	0	1	1	1	1	2	2	2
4.3	2.074	2.076	2.078	2.081	2.083	2.086	2.088	2.090	2.093	2.095	0	0	1	1	1	1	2	2	2
4.4	2.098	2.100	2.102	2.105	2.107	2.110	2.112	2.114	2.117	2.119	0	0	1	1	1	1	2	2	2
4.5	2.121	2.124	2.126	2.128	2.131	2.133	2.135	2.138	2.140	2.142	0	0	1	1	1	1	2	2	2
4.6	2.145	2.147	2.149	2.152	2.154	2.156	2.159	2.161	2.163	2.166	0	0	1	1	1	1	2	2	2
4.7	2.168	2.170	2.173	2.175	2.177	2.179	2.182	2.184	2.186	2.189	0	0	1	1	1	1	2	2	2
4.8	2.191	2.193	2.195	2.198	2.200	2.202	2.205	2.207	2.209	2.211	0	0	1	1	1	1	2	2	2
4.9	2.214	2.216	2.218	2.220	2.223	2.225	2.227	2.229	2.232	2.234	0	0	1	1	1	1	2	2	2
5.0	2.236	2.238	2.241	2.243	2.245	2.247	2.249	2.252	2.254	2.256	0	0	1	1	1	1	2	2	2
5.1	2.258	2.261	2.263	2.265	2.267	2.269	2.272	2.274	2.276	2.278	0	0	1	1	1	1	2	2	2
5.2	2.280	2.283	2.285	2.287	2.289	2.291	2.293	2.296	2.298	2.300	0	0	1	1	1	1	2	2	2
5.3	2.302	2.304	2.307	2.309	2.311	2.313	2.315	2.317	2.319	2.322	0	0	1	1	1	1	2	2	2
5.4	2.324	2.326	2.328	2.330	2.332	2.335	2.337	2.339	2.341	2.343	0	0	1	1	1	1	1	2	2

Table 8 Square Roots (Continued)

N	0	1	2	3	4	5	6	7	8	9	1	2	3	4	5	6	7	8	9
5.5	2.345	2.347	2.349	2.352	2.354	2.356	2.358	2.360	2.362	2.364	0	0	1	1	1	1	1	2	2
5.6	2.366	2.369	2.371	2.373	2.375	2.377	2.379	2.381	2.383	2.385	0	0	1	1	1	1	1	2	2
5.7	2.387	2.390	2.392	2.394	2.396	2.398	2.400	2.402	2.404	2.406	0	0	1	1	1	1	1	2	2
5.8	2.408	2.410	2.412	2.415	2.417	2.419	2.421	2.423	2.425	2.427	0	0	1	1	1	1	1	2	2
5.9	2.429	2.431	2.433	2.435	2.437	2.439	2.441	2.443	2.445	2.447	0	0	1	1	1	1	1	2	2
6.0	2.449	2.452	2.454	2.456	2.458	2.460	2.462	2.464	2.466	2.468	0	0	1	1	1	1	1	2	2
6.1	2.470	2.472	2.474	2.476	2.478	2.480	2.482	2.484	2.486	2.488	0	0	1	1	1	1	1	2	2
6.2	2.490	2.492	2.494	2.496	2.498	2.500	2.502	2.504	2.506	2.508	0	0	1	1	1	1	1	2	2
6.3	2.510	2.512	2.514	2.516	2.518	2.520	2.522	2.524	2.526	2.528	0	0	1	1	1	1	1	2	2
6.4	2.530	2.532	2.534	2.536	2.538	2.540	2.542	2.544	2.546	2.548	0	0	1	1	1	1	1	2	2
6.5	2.550	2.551	2.553	2.555	2.557	2.559	2.561	2.563	2.565	2.567	0	0	1	1	1	1	1	2	2
6.6	2.569	2.571	2.573	2.575	2.577	2.579	2.581	2.583	2.585	2.587	0	0	1	1	1	1	1	2	2
6.7	2.588	2.590	2.592	2.594	2.596	2.598	2.600	2.602	2.604	2.606	0	0	1	1	1	1	1	2	2
6.8	2.608	2.610	2.612	2.613	2.615	2.617	2.619	2.621	2.623	2.625	0	0	1	1	1	1	1	2	2
6.9	2.627	2.629	2.631	2.632	2.634	2.636	2.638	2.640	2.642	2.644	0	0	1	1	1	1	1	2	2
7.0	2.646	2.648	2.650	2.651	2.653	2.655	2.657	2.659	2.661	2.663	0	0	1	1	1	1	1	2	2
7.1	2.665	2.666	2.668	2.670	2.672	2.674	2.676	2.678	2.680	2.681	0	0	1	1	1	1	1	1	2
7.2	2.683	2.685	2.687	2.689	2.691	2.693	2.694	2.696	2.698	2.700	0	0	1	1	1	1	1	1	2
7.3	2.702	2.704	2.706	2.707	2.709	2.711	2.713	2.715	2.717	2.718	0	0	1	1	1	1	1	1	2
7.4	2.720	2.722	2.724	2.726	2.728	2.729	2.731	2.733	2.735	2.737	0	0	1	1	1	1	1	1	2
7.5	2.793	2.740	2.742	2.744	2.746	2.748	2.750	2.751	2.753	2.755	0	0	1	1	1	1	1	1	2
7.6	2.757	2.759	2.760	2.762	2.764	2.766	2.768	2.769	2.771	2.773	0	0	1	1	1	1	1	1	2
7.7	2.775	2.777	2.778	2.780	2.782	2.784	2.786	2.787	2.789	2.791	0	0	1	1	1	1	1	1	2
7.8	2.793	2.795	2.796	2.798	2.800	2.802	2.804	2.805	2.807	2.809	0	0	1	1	1	1	1	1	2
7.9	2.811	2.812	2.814	2.816	2.818	2.820	2.821	2.823	2.825	2.827	0	0	1	1	1	1	1	1	2
8.0	2.828	2.830	2.832	2.834	2.835	2.837	2.839	2.841	2.843	2.844	0	0	1	1	1	1	1	1	2
8.1	2.846	2.848	2.850	2.851	2.853	2.855	2.857	2.858	2.860	2.862	0	0	1	1	1	1	1	1	2
8.2	2.864	2.865	2.867	2.869	2.871	2.872	2.874	2.876	2.877	2.879	0	0	1	1	1	1	1	1	2
8.3	2.881	2.883	2.884	2.886	2.888	2.890	2.891	2.893	2.895	2.897	0	0	1	1	1	1	1	1	2
8.4	2.898	2.900	2.902	2.903	2.905	2.907	2.909	2.910	2.912	2.914	0	0	1	1	1	1	1	1	2
8.5	2.915	2.917	2.919	2.921	2.922	2.924	2.926	2.927	2.929	2.931	0	0	1	1	1	1	1	1	2
8.6	2.933	2.934	2.936	2.938	2.939	2.941	2.943	2.944	2.946	2.948	0	0	1	1	1	1	1	1	2
8.7	2.950	2.951	2.953	2.955	2.956	2.958	2.960	2.961	2.963	2.965	0	0	1	1	1	1	1	1	2
8.8	2.966	2.968	2.970	2.972	2.973	2.975	2.977	2.978	2.980	2.982	0	0	1	1	1	1	1	1	2
8.9	2.983	2.985	2.987	2.988	2.990	2.992	2.993	2.995	2.997	2.998	0	0	1	1	1	1	1	1	2
9.0	3.000	3.002	3.003	3.005	3.007	3.008	3.010	3.012	3.013	3.015	0	0	0	1	1	1	1	1	1
9.1	3.017	3.018	3.020	3.022	3.023	3.025	3.027	3.028	3.030	3.032	0	0	0	1	1	1	1	1	1
9.2	3.033	3.035	3.036	3.038	3.040	3.041	3.043	3.045	3.046	3.048	0	0	0	1	1	1	1	1	1
9.3	3.050	3.051	3.053	3.055	3.056	3.058	3.059	3.061	3.063	3.064	0	0	0	1	1	1	1	1	1
9.4	3.066	3.068	3.069	3.071	3.072	3.074	3.076	3.077	3.079	3.081	0	0	0	1	1	1	1	1	1
9.5	3.082	3.084	3.085	3.087	3.089	3.090	3.092	3.094	3.095	3.097	0	0	0	1	1	1	1	1	1
9.6	3.098	3.100	3.102	3.103	3.105	3.106	3.108	3.110	3.111	3.113	0	0	0	1	1	1	1	1	1
9.7	3.114	3.116	3.118	3.119	3.121	3.122	3.124	3.126	3.127	3.129	0	0	0	1	1	1	1	1	1
9.8	3.130	3.132	3.134	3.135	3.137	3.138	3.140	3.142	3.143	3.145	0	0	0	1	1	1	1	1	1
9.9	3.146	3.148	3.150	3.151	3.153	3.154	3.156	3.158	3.159	3.161	0	0	0	1	1	1	1	1	1

Table 8 Square Roots (Continued)

N	0	1	2	3	4	5	6	7	8	9	1	2	3	4	5	6	7	8	9
10	3.162	3.178	3.194	3.209	3.225	3.240	3.256	3.271	3.286	3.302	2	3	5	6	8	9	11	12	14
11	3.317	3.332	3.347	3.362	3.376	3.391	3.406	3.421	3.435	3.450	1	3	4	6	7	9	10	12	13
12	3.464	3.479	3.493	3.507	3.521	3.536	3.550	3.564	3.578	3.592	1	3	4	6	7	8	10	11	13
13	3.606	3.619	3.633	3.647	3.661	3.674	3.688	3.701	3.715	3.728	1	3	4	5	7	8	10	11	12
14	3.742	3.755	3.768	3.782	3.795	3.808	3.821	3.834	3.847	3.860	1	3	4	5	7	8	9	11	12
15	3.873	3.886	3.899	3.912	3.924	3.937	3.950	3.962	3.975	3.987	1	3	4	5	6	8	9	10	11
16	4.000	4.012	4.025	4.037	4.050	4.062	4.074	4.087	4.099	4.111	1	2	4	5	6	7	9	10	11
17	4.123	4.135	4.147	4.159	4.171	4.183	4.195	4.207	4.219	4.231	1	2	4	5	6	7	8	10	11
18	4.243	4.254	4.266	4.278	4.290	4.301	4.313	4.324	4.336	4.347	1	2	3	5	6	7	8	9	10
19	4.359	4.370	4.382	4.393	4.405	4.416	4.427	4.438	4.450	4.461	1	2	3	5	6	7	8	9	10
20	4.472	4.483	4.494	4.506	4.517	5.528	4.539	4.550	4.561	4.572	1	2	3	4	6	7	8	9	10
21	4.583	4.593	4.604	4.615	4.626	4.637	4.648	4.658	4.669	4.680	1	2	3	4	5	6	8	9	10
22	4.690	4.701	4.712	4.722	4.733	4.743	4.754	4.764	4.775	4.785	1	2	3	4	5	6	7	8	9
23	4.796	4.806	4.817	4.827	4.837	4.848	4.858	4.868	4.879	4.889	1	2	3	4	5	6	7	8	9
24	4.899	4.909	4.919	4.930	4.940	4.950	4.960	4.970	4.980	4.990	1	2	3	4	5	6	7	8	9
25	5.000	5.010	5.020	5.030	5.040	5.050	5.060	5.070	5.079	5.089	1	2	3	4	5	6	7	8	9
26	5.099	5.109	5.119	5.128	5.138	5.148	5.158	5.167	5.177	5.187	1	2	3	4	5	6	7	8	9
27	5.196	5.206	5.215	5.225	5.235	5.244	5.254	5.263	5.273	5.282	1	2	3	4	5	6	7	8	9
28	5.292	5.301	5.310	5.320	5.329	5.339	5.348	5.357	5.367	5.376	1	2	3	4	5	6	7	7	8
29	5.385	5.394	5.404	5.413	5.422	5.431	5.441	5.450	5.459	5.468	1	2	3	4	5	5	6	7	8
30	5.477	5.486	5.495	5.505	5.514	5.523	5.532	5.541	5.550	5.559	1	2	3	4	4	5	6	7	8
31	5.568	5.577	5.586	5.595	5.604	5.612	5.621	5.630	5.639	5.648	1	2	3	3	4	5	6	7	8
32	5.657	5.666	5.675	5.683	5.692	5.701	5.710	5.718	5.727	5.736	1	2	3	3	4	5	6	7	8
33	5.745	5.753	5.762	5.771	5.779	5.788	5.797	5.805	5.814	5.822	1	2	3	3	4	5	6	7	8
34	5.831	5.840	5.848	5.857	5.865	5.874	5.882	5.891	5.899	5.908	1	2	3	3	4	5	6	7	8
35	5.916	5.925	5.933	5.941	5.950	5.958	5.967	5.975	5.983	5.992	1	2	2	3	4	5	6	7	8
36	6.000	6.008	6.017	6.025	6.033	6.042	6.050	6.058	6.066	6.075	1	2	2	3	4	5	6	7	7
37	6.083	6.091	6.099	6.107	6.116	6.124	6.132	6.140	6.148	6.156	1	2	2	3	4	5	6	7	7
38	6.164	6.173	6.181	6.189	6.197	6.205	6.213	6.221	6.229	6.237	1	2	2	3	4	5	6	6	7
39	6.245	6.253	6.261	6.269	6.277	6.285	6.293	6.301	6.309	6.317	1	2	2	3	4	5	6	6	7
40	6.325	6.332	6.340	6.348	6.356	6.364	6.372	6.380	6.387	6.395	1	2	2	3	4	5	6	6	7
41	6.403	6.411	6.419	6.427	6.434	6.442	6.450	6.458	6.465	6.473	1	2	2	3	4	5	5	6	7
42	6.481	6.488	6.496	6.504	6.512	6.519	6.527	6.535	6.542	6.550	1	2	2	3	4	5	5	6	7
43	6.557	6.565	6.573	6.580	6.588	6.595	6.603	6.611	6.618	6.626	1	2	2	3	4	5	5	6	7
44	6.633	6.641	6.648	6.656	6.663	6.671	6.678	6.686	6.693	6.701	1	2	2	3	4	5	5	6	7
45	6.708	6.716	6.723	6.731	6.738	6.745	6.753	6.760	6.768	6.775	1	1	2	3	4	4	5	6	7
46	6.782	6.790	6.797	6.804	6.812	6.819	6.826	6.834	6.841	6.848	1	1	2	3	4	4	5	6	7
47	6.856	6.863	6.870	6.877	6.885	6.892	6.899	6.907	6.914	6.921	1	1	2	3	4	4	5	6	7
48	6.928	6.935	6.943	6.950	6.957	6.964	6.971	6.979	6.986	6.993	1	1	2	3	4	4	5	6	6
49	7.000	7.007	7.014	7.021	7.029	7.036	7.043	7.050	7.057	7.064	1	1	2	3	4	4	5	6	6
50	7.071	7.078	7.085	7.092	7.099	7.106	7.113	7.120	7.127	7.134	1	1	2	3	4	4	5	6	6
51	7.141	7.148	7.155	7.162	7.169	7.176	7.183	7.190	7.197	7.204	1	1	2	3	4	4	5	6	6
52	7.211	7.218	7.225	7.232	7.239	7.246	7.253	7.259	7.266	7.273	1	1	2	3	3	4	5	6	6
53	7.280	7.287	7.294	7.301	7.308	7.314	7.321	7.328	7.335	7.342	1	1	2	3	3	4	5	5	6
54	7.348	7.355	7.362	7.369	7.376	7.382	7.389	7.396	7.403	7.409	1	1	2	3	3	4	5	5	6

Table 8 Square Roots (Continued)

N	0	1	2	3	4	5	6	7	8	9	1	2	3	4	5	6	7	8	9
55	7.416	7.423	7.430	7.436	7.443	7.450	7.457	7.463	7.470	7.477	1	1	2	3	3	4	5	5	6
56	7.483	7.490	7.497	7.503	7.510	7.517	7.523	7.530	7.537	7.543	1	1	2	3	3	4	5	5	6
57	7.550	7.556	7.563	7.570	7.576	7.583	7.589	7.596	7.603	7.609	1	1	2	3	3	4	5	5	6
58	7.616	7.622	7.629	7.635	7.642	7.649	7.655	7.662	7.668	7.675	1	1	2	3	3	4	5	5	6
59	7.681	7.688	7.694	7.701	7.707	7.714	7.720	7.727	7.733	7.740	1	1	2	3	3	4	4	5	6
60	7.746	7.752	7.759	7.765	7.772	7.778	7.785	7.791	7.797	7.804	1	1	2	3	3	4	4	5	6
61	7.810	7.817	7.823	7.829	7.836	7.842	7.849	7.855	7.861	7.868	1	1	2	3	3	4	4	5	6
62	7.874	7.880	7.887	7.893	7.899	7.906	7.912	7.918	7.925	7.931	1	1	2	3	3	4	4	5	6
63	7.937	7.944	7.950	7.956	7.962	7.969	7.975	7.981	7.987	7.994	1	1	2	3	3	4	4	5	6
64	8.000	8.006	8.012	8.019	8.025	8.031	8.037	8.044	8.050	8.056	1	1	2	2	3	4	4	5	6
65	8.062	8.068	8.075	8.081	8.087	8.093	8.099	8.106	8.112	8.118	1	1	2	2	3	4	4	5	5
66	8.124	8.130	8.136	8.142	8.149	8.155	8.161	8.167	8.173	8.179	1	1	2	2	3	4	4	5	5
67	8.185	8.191	8.198	8.204	8.210	8.216	8.222	8.228	8.234	8.240	1	1	2	2	3	4	4	5	5
68	8.246	8.252	8.258	8.264	8.270	8.276	8.283	8.289	8.295	8.301	1	1	2	2	3	4	4	5	5
69	8.307	8.313	8.319	8.325	8.331	8.337	8.343	8.340	8.355	8.361	1	1	2	2	3	4	4	5	5
70	8.367	8.373	8.379	8.385	8.390	8.396	8.402	8.408	8.414	8.420	1	1	2	2	3	4	4	5	5
71	8.426	8.432	8.438	8.444	8.450	8.456	8.462	8.468	8.473	8.479	1	1	2	2	3	4	4	5	5
72	8.485	8.491	8.497	8.503	8.509	8.515	8.521	8.526	8.532	8.538	1	1	2	2	3	3	4	5	5
73	8.544	8.550	8.556	8.562	8.567	8.573	8.579	8.585	8.591	8.597	1	1	2	2	3	3	4	5	5
74	8.602	8.608	8.614	8.620	8.626	8.631	8.637	8.643	8.649	8.654	1	1	2	2	3	3	4	5	5
75	8.660	8.666	8.672	8.678	8.683	8.689	8.695	8.701	8.706	8.712	1	1	2	2	3	3	4	5	5
76	8.718	8.724	8.729	8.735	8.741	8.746	8.752	8.758	8.764	8:769	1	1	2	2	3	3	4	5	5
77	8.775	8.781	8.786	8.792	8.798	8.803	8.809	8.815	8.820	8.826	1	1	2	2	3	3	4	4	5
78	8.832	8.837	8.843	8.849	8.854	8.860	8.866	8.871	8.877	8.883	1	1	2	2	3	3	4	4	5
79	8.888	8.894	8.899	8.905	8.911	8.916	8.922	8.927	8.933	8.939	1	1	2	2	3	3	4	4	5
80	8.944	8.950	8.955	8.961	8.967	8.972	8.978	8.983	8.989	8.994	1	1	2	2	3	3	4	4	5
81	9.000	9.006	9.011	9.017	9.022	9.028	9.033	9.039	9.044	9.050	1	1	2	2	3	3	4	4	5
82	9.055	9.061	9.066	9.072	9.077	9.083	9.088	9.094	9.099	9.105	1	1	2	2	3	3	4	4	5
83	9.110	9.116	9.121	9.127	9.132	9.138	9.143	9.149	9.154	9.160	1	1	2	2	3	3	4	4	5
84	9.165	9.171	9.176	9.182	9.187	9.192	9.198	9.203	9.209	9.214	1	1	2	2	3	3	4	4	5
85	9.220	9.225	9.230	9.236	9.241	9.247	9.252	9.257	9.263	9.268	1	1	2	2	3	3	4	4	5
86	9.274	9.279	9.284	9.290	9.295	9.301	9.306	9.311	9.317	9.322	1	1	2	2	3	3	4	4	5
87	9.327	9.333	9.338	9.343	9.349	9.354	9.359	9.365	9.370	9.375	I	1	2	2	3	3	4	4	5
88	9.381	9.386	9.391	9.397	9.402	9.407	9.413	9.418	9.423	9.429	1	1	2	2	3	3	4	4	5
89	9.434	9.439	9.445	9.450	9.455	9.460	9.466	9.471	9.476	9.482	1	1	2	2	3	3	4	4	5
90	9.487	9.492	9.497	9.503	9.508	9.513	9.518	9.524	9.529	9.534	1	1	2	2	3	3	4	4	5
91	9.539	9.545	9.550	9.555	9.560	9.566	9.571	9.576	9.581	9.586	1	1	2	2	3	3	4	4	5
92	9.592	9.597	9.602	9.607	9.612	9.618	9.623	9.628	9.633	9.638	1	1	2	2	3	3	4	4	5
93	9.644	9.649	9.654	9.659	9.664	9.670	9.675	9.680	9.685	9.690	1	1	2	2	3	3	4	4	5
94	9.695	9.701	9.706	9.711	9.716	9.721	9.726	9.731	9.737	9.742	1	1	2	2	3	3	4	4	5
95	9.747	9.752	9.757	9.762	9.767	9.772	9.778	9.783	9.788	9.793	1	1	2	2	3	3	4	4	5
96	9.798	9.803	9.808	9.813	9.818	9.823	9.829	9.834	9.839	9.844	1	1	2	2	3	3	4	4	5
97	9.849	9.854	9.859	9.864	9.869	9.874	9.879	9.884	9.889	9.894	1	1	2	2	3	3	4	4	5
98	9.899	9.905	9.910	9.915	9.920	9.925	9.930	9.935	9.940	9.945	1	1	1	2	2	3	3	4	4
99	9.950	9.955	9.960	9.965	9.970	9.975	9.980	9.985	9.990	9.995	0	1	1	2	2	3	3	4	4

Table 9   Common Logarithms

	0.00	0.01	0.02	0.03	0.04	0.05	0.06	0.07	0.08	0.09
1.0	.00000	.00432	.00860	.01284	.01703	.02119	.02531	.02938	.03342	.03743
1.1	.04139	.04532	.04922	.05308	.05690	.06070	.06446	.06819	.07188	.07555
1.2	.07918	.08279	.08636	.08991	.09342	.09691	.10037	.10380	.10721	.11059
1.3	.11394	.11727	.12057	.12385	.12710	.13033	.13354	.13672	.13988	.14301
1.4	.14613	.14922	.15229	.15534	.15836	.16137	.16435	.16732	.17026	.17319
1.5	.17609	.17898	.18184	.18469	.18752	.19033	.19312	.19590	.19866	.20140
1.6	.20412	.20683	.20952	.21219	.21484	.21748	.22011	.22272	.22531	.22789
1.7	.23045	.23300	.23553	.23805	.24055	.24304	.24551	.24797	.25042	.25285
1.8	.25527	.25768	.26007	.26245	.26482	.26717	.26951	.27184	.27416	.27646
1.9	.27875	.28103	.28330	.28556	.28780	.29003	.29226	.29447	.29667	.29885
2.0	.30103	.30320	.30535	.30750	.30963	.31175	.31387	.31597	.31806	.32015
2.1	.32222	.32428	.32634	.32838	.33041	.33244	.33445	.33646	.33846	.34044
2.2	.34242	.34439	.34635	.34830	.35025	.35218	.35411	.35603	.35793	.35984
2.3	.36173	.36361	.36549	.36736	.36922	.37107	.37291	.37475	.37658	.37840
2.4	.38021	.38202	.38382	.38561	.38739	.38917	.39094	.39270	.39445	.39620
2.5	.39794	.39967	.40140	.40312	.40483	.40654	.40824	.40993	.41162	.41330
2.6	.41497	.41664	.41830	.41996	.42160	.42325	.42488	.42651	.42813	.42975
2.7	.43136	.43297	.43457	.43616	.43775	.43933	.44091	.44248	.44404	.44560
2.8	.44716	.44871	.45025	.45179	.45332	.45484	.45637	.45788	.45939	.46090
2.9	.46240	.46389	.46538	.46687	.46835	.46982	.47129	.47276	.47422	.47567
3.0	.47712	.47857	.48001	.48144	.48287	.48430	.48572	.48714	.48855	.48996
3.1	.49136	.49276	.49415	.49554	.49693	.49831	.49969	.50106	.50243	.50379
3.2	.50515	.50651	.50786	.50920	.51055	.51188	.51322	.51455	.51587	.51720
3.3	.51851	.51983	.52114	.52244	.52375	.52504	.52634	.52763	.52892	.53020
3.4	.53148	.53275	.53403	.53529	.53656	.53782	.53908	.54033	.54158	.54283
3.5	.54407	.54531	.54654	.54777	.54900	.55023	.55145	.55267	.55388	.55509
3.6	.55630	.55751	.55871	.55991	.56110	.56229	.56348	.56467	.56585	.56703
3.7	.56820	.56937	.57054	.57171	.57287	.57403	.57519	.57634	.57749	.57864
3.8	.57978	.58092	.58206	.58320	.58433	.58546	.58659	.58771	.58883	.58995
3.9	.59106	.59218	.59329	.59439	.59550	.59660	.59770	.59879	.59988	.60097
4.0	.60206	.60314	.60423	.60531	.60638	.60746	.60853	.60959	.61066	.61172
4.1	.61278	.61384	.61490	.61595	.61700	.61805	.61909	.62014	.62118	.62221
4.2	.62325	.62428	.62531	.62634	.62737	.62839	.62941	.63043	.63144	.63246
4.3	.63347	.63448	.63548	.63649	.63749	.63849	.63949	.64048	.64147	.64246
4.4	.64345	.64444	.64542	.64640	.64738	.64836	.64933	.65031	.65128	.65225
4.5	.65321	.65418	.65514	.65610	.65706	.65801	.65896	.65992	.66087	.66181
4.6	.66276	.66370	.66464	.66558	.66652	.66745	.66839	.66932	.67025	.67117
4.7	.67210	.67302	.67394	.67486	.67578	.67669	.67761	.67852	.67943	.68034
4.8	.68124	.68215	.68305	.68395	.68485	.68574	.68664	.68753	.68842	.68931
4.9	.69020	.69108	.69197	.69285	.69373	.69461	.69548	.69636	.69723	.69810
5.0	.69897	.69984	.70070	.70157	.70243	.70329	.70415	.70501	.70586	.70672
5.1	.70757	.70842	.70927	.71012	.71096	.71181	.71265	.71349	.71433	.71517
5.2	.71600	.71684	.71767	.71850	.71933	.72016	.72099	.72181	.72263	.72346
5.3	.72428	.72509	.72591	.72673	.72754	.72835	.72916	.72997	.73078	.73159
5.4	.73239	.73320	.73400	.73480	.73560	.73640	.73719	.73799	.73878	.73957
5.5	.74036	.74115	.74194	.74273	.74351	.74429	.74507	.74586	.74663	.74741

Table 9 Common Logarithms (Continued)

	0.00	0.01	0.02	0.03	0.04	0.05	0.06	0.07	0.08	0.09
5.6	.74819	.74896	.74974	.75051	.75128	.75205	.75282	.75358	.75435	.75511
5.7	.75587	.75664	.75740	.75815	.75891	.75967	.76042	.76118	.76193	.76268
5.8	.76343	.76418	.76492	.76567	.76641	.76716	.76790	.76864	.76938	.77012
5.9	.77085	.77159	.77232	.77305	.77379	.77452	.77525	.77597	.77670	.77743
6.0	.77815	.77887	.77960	.78032	.78104	.78176	.78247	.78319	.78390	.78462
6.1	.78533	.78604	.78675	.78746	.78817	.78888	.78958	.79029	.79099	.79169
6.2	.79239	.79309	.79379	.79449	.79518	.79588	.79657	.79727	.79796	.79865
6.3	.79934	.80003	.80072	.80140	.80209	.80277	.80346	.80414	.80482	.80550
6.4	.80618	.80686	.80754	.80821	.80889	.80956	.81023	.81090	.81158	.81224
6.5	.81291	.81358	.81425	.81491	.81558	.81624	.81690	.81757	.81823	.81889
6.6	.81954	.82020	.82086	.82151	.82217	.82282	.82347	.82413	.82478	.82543
6.7	.82607	.82672	.82737	.82802	.82866	.82930	.82995	.83059	.83123	.83187
6.8	.83251	.83315	.83378	.83442	.83506	.83569	.83632	.83696	.83759	.83822
6.9	.83885	.83948	.84011	.84073	.84136	.84198	.84261	.84323	.84386	.84448
7.0	.84510	.84572	.84634	.84696	.84757	.84819	.84880	.84942	.85003	.85065
7.1	.85126	.85187	.85248	.85309	.85370	.85431	.85491	.85552	.85612	.85673
7.2	.85733	.85794	.85854	.85914	.85974	.86034	.86094	.86153	.86213	.86273
7.3	.86332	.86392	.86451	.86510	.86570	.86629	.86688	.86747	.86806	.86864
7.4	.86923	.86982	.87040	.87099	.87157	.87216	.87274	.87332	.87390	.87448
7.5	.87506	.87564	.87622	.87679	.87737	.87795	.87852	.87910	.87967	.88024
7.6	.88081	.88138	.88195	.88252	.88309	.88366	.88423	.88480	.88536	.88593
7.7	.88649	.88705	.88762	.88818	.88874	.88930	.88986	.89042	.89098	.89154
7.8	.89209	.89265	.89321	.89376	.89432	.89487	.89542	.89597	.89653	.89708
7.9	.89763	.89818	.89873	.89927	.89982	.90037	.90091	.90146	.90200	.90255
8.0	.90309	.90363	.90417	.90472	.90526	.90580	.90634	.90687	.90741	.90795
8.1	.90849	.90902	.90956	.91009	.91062	.91116	.91169	.91222	.91275	.91328
8.2	.91381	.91434	.91487	.91540	.91593	.91645	.91698	.91751	.91803	.91855
8.3	.91908	.91960	.92012	.92065	.92117	.92169	.92221	.92273	.92324	.92376
8.4	.92428	.92480	.92531	.92583	.92634	.92686	.92737	.92788	.92840	.92891
8.5	.92942	.92993	.93044	.93095	.93146	.93197	.93247	.93298	.93349	.93399
8.6	.93450	.93500	.93551	.93601	.93651	.93702	.93752	.93802	.93852	.93902
8.7	.93952	.94002	.94052	.94101	.94151	.94201	.94250	.94300	.94349	.94399
8.8	.94448	.94498	.94547	.94596	.94645	.94694	.94743	.94792	.94841	.94890
8.9	.94939	.94988	.95036	.95085	.95134	.95182	.95231	.95279	.95328	.95376
9.0	.95424	.95472	.95521	.95569	.95617	.95665	.95713	.95761	.95809	.95856
9.1	.95904	.95952	.95999	.96047	.96095	.96142	.96190	.96237	.96284	.96332
9.2	.96379	.96426	.96473	.96520	.96567	.96614	.96661	.96708	.96755	.96802
9.3	.96848	.96895	.96942	.96988	.97035	.97081	.97128	.97174	.97220	.97267
9.4	.97313	.97359	.97405	.97451	.97497	.97543	.97589	.97635	.97681	.97727
9.5	.97772	.97818	.97864	.97909	.97955	.98000	.98046	.98091	.98137	.98182
9.6	.98227	.98272	.98318	.98363	.98408	.98453	.98498	.98543	.98588	.98632
9.7	.98677	.98722	.98767	.98811	.98856	.98900	.98945	.98989	.99034	.99078
9.8	.99123	.99167	.99211	.99255	.99300	.99344	.99388	.99432	.99476	.99520
9.9	.99564	.99607	.99651	.99695	.99739	.99782	.99826	.99870	.99913	.99957

Table 10   Natural Sines

	0.0	0.1	0.2	0.3	0.4	0.5	0.6	0.7	0.8	0.9
0	0.00000	0.00175	0.00349	0.00524	0.00698	0.00873	0.01047	0.01222	0.01396	0.01571
1	0.01745	0.01920	0.02094	0.02269	0.02443	0.02618	0.02792	0.02967	0.03141	0.03316
2	0.03490	0.03664	0.03839	0.04013	0.04188	0.04362	0.04536	0.04711	0.04885	0.05059
3	0.05234	0.05408	0.05582	0.05756	0.05931	0.06105	0.06279	0.06453	0.06627	0.06802
4	0.06976	0.07150	0.07324	0.07498	0.07672	0.07846	0.08020	0.08194	0.08368	0.08542
5	0.08716	0.08889	0.09063	0.09237	0.09411	0.09585	0.09758	0.09932	0.10106	0.10279
6	0.10453	0.10626	0.10800	0.10973	0.11147	0.11320	0.11494	0.11667	0.11840	0.12014
7	0.12187	0.12360	0.12533	0.12706	0.12880	0.13053	0.13226	0.13399	0.13572	0.13744
8	0.13917	0.14090	0.14263	0.14436	0.14608	0.14781	0.14954	0.15126	0.15299	0.15471
9	0.15643	0.15816	0.15988	0.16160	0.16333	0.16505	0.16677	0.16849	0.17021	0.17193
10	0.17365	0.17537	0.17708	0.17880	0.18052	0.18224	0.18395	0.18567	0.18738	0.18910
11	0.19081	0.19252	0.19423	0.19595	0.19766	0.19937	0.20108	0.20279	0.20450	0.20620
12	0.20791	0.20962	0.21132	0.21303	0.21474	0.21644	0.21814	0.21985	0.22155	0.22325
13	0.22495	0.22665	0.22835	0.23005	0.23175	0.23345	0.23514	0.23684	0.23853	0.24023
14	0.24192	0.24362	0.24531	0.24700	0.24869	0.25038	0.25207	0.25376	0.25545	0.25713
15	0.25882	0.26050	0.26219	0.26387	0.26556	0.26724	0.26892	0.27060	0.27228	0.27396
16	0.27564	0.27731	0.27899	0.28067	0.28234	0.28402	0.28569	0.28736	0.28903	0.29070
17	0.29237	0.29404	0.29571	0.29737	0.29904	0.30071	0.30237	0.30403	0.30570	0.30736
18	0.30902	0.31068	0.31233	0.31399	0.31565	0.31730	0.31896	0.32061	0.32227	0.32392
19	0.32557	0.32722	0.32887	0.33051	0.33216	0.33381	0.33545	0.33710	0.33874	0.34038
20	0.34202	0.34366	0.34530	0.34694	0.34857	0.35021	0.35184	0.35347	0.35511	0.35674
21	0.35837	0.36000	0.36162	0.36325	0.36488	0.36650	0.36812	0.36975	0.37137	0.37299
22	0.37461	0.37622	0.37784	0.37946	0.38107	0.38268	0.38430	0.38591	0.38752	0.38912
23	0.39073	0.39234	0.39394	0.39555	0.39715	0.39875	0.40035	0.40195	0.40355	0.40514
24	0.40674	0.40833	0.40992	0.41151	0.41310	0.41469	0.41628	0.41787	0.41945	0.42104
25	0.42262	0.42420	0.42578	0.42736	0.42894	0.43051	0.43209	0.43366	0.43523	0.43680
26	0.43837	0.43994	0.44151	0.44307	0.44464	0.44620	0.44776	0.44932	0.45088	0.45243
27	0.45399	0.45554	0.45710	0.45865	0.46020	0.46175	0.46330	0.46484	0.46639	0.46793
28	0.46947	0.47101	0.47255	0.47409	0.47562	0.47716	0.47869	0.48022	0.48175	0.48328
29	0.48481	0.48634	0.48786	0.48938	0.49090	0.49242	0.49394	0.49546	0.49697	0.49849
30	0.50000	0.50151	0.50302	0.50453	0.50603	0.50754	0.50904	0.51054	0.51204	0.51354
31	0.51504	0.51653	0.51803	0.51952	0.52101	0.52250	0.52399	0.52547	0.52696	0.52844
32	0.52992	0.53140	0.53288	0.53435	0.53583	0.53730	0.53877	0.54024	0.54171	0.54317
33	0.54464	0.54610	0.54756	0.54902	0.55048	0.55194	0.55339	0.55484	0.55630	0.55775
34	0.55919	0.56064	0.56208	0.56353	0.56497	0.56641	0.56784	0.56928	0.57071	0.57215
35	0.57358	0.57501	0.57643	0.57786	0.57928	0.58070	0.58212	0.58354	0.58496	0.58637
36	0.58779	0.58920	0.59061	0.59201	0.59342	0.59482	0.59622	0.59763	0.59902	0.60042
37	0.60182	0.60321	0.60460	0.60599	0.60738	0.60876	0.61015	0.61153	0.61291	0.61429
38	0.61566	0.61704	0.61841	0.61978	0.62115	0.62251	0.62388	0.62524	0.62660	0.62796
39	0.62932	0.63068	0.63203	0.63338	0.63473	0.63608	0.63742	0.63877	0.64011	0.64145
40	0.64279	0.64412	0.64546	0.64679	0.64812	0.64945	0.65077	0.65210	0.65342	0.65474
41	0.65606	0.65738	0.65869	0.66000	0.66131	0.66262	0.66393	0.66523	0.66653	0.66783
42	0.66913	0.67043	0.67172	0.67301	0.67430	0.67559	0.67688	0.67816	0.67944	0.68072
43	0.68200	0.68327	0.68455	0.68582	0.68709	0.68835	0.68962	0.69088	0.69214	0.69340
44	0.69466	0.69591	0.69717	0.69842	0.69966	0.70091	0.70215	0.70339	0.70463	0.70587

Table 10 Natural Sines (Continued)

	0.0	0.1	0.2	0.3	0.4	0.5	0.6	0.7	0.8	0.9
45	0.70711	0.70834	0.70957	0.71080	0.71203	0.71325	0.71447	0.71569	0.71691	0.71813
46	0.71934	0.72055	0.72176	0.72297	0.72417	0.72537	0.72657	0.72777	0.72897	0.73016
47	0.73135	0.73254	0.73373	0.73491	0.73610	0.73728	0.73846	0.73963	0.74080	0.74198
48	0.74314	0.74431	0.74548	0.74664	0.74780	0.74896	0.75011	0.75126	0.75241	0.75356
49	0.75471	0.75585	0.75700	0.75813	0.75927	0.76041	0.76154	0.76267	0.76380	0.76492
50	0.76604	0.76717	0.76828	0.76940	0.77051	0.77162	0.77273	0.77384	0.77494	0.77605
51	0.77715	0.77824	0.77934	0.78043	0.78152	0.78261	0.78369	0.78478	0.78586	0.78694
52	0.78801	0.78908	0.79016	0.79122	0.79229	0.79335	0.79441	0.79547	0.79653	0.79758
53	0.79864	0.79968	0.80073	0.80178	0.80282	0.80386	0.80489	0.80593	0.80696	0.80799
54	0.80902	0.81004	0.81106	0.81208	0.81310	0.81412	0.81513	0.81614	0.81714	0.81815
55	0.81915	0.82015	0.82115	0.82214	0.82314	0.82413	0.82511	0.82610	0.82708	0.82806
56	0.82904	0.83001	0.83098	0.83195	0.83292	0.83389	0.83485	0.83581	0.83676	0.83772
57	0.83867	0.83962	0.84057	0.84151	0.84245	0.84339	0.84433	0.84526	0.84619	0.84712
58	0.84805	0.84897	0.84989	0.85081	0.85173	0.85264	0.85355	0.85446	0.85536	0.85627
59	0.85717	0.85806	0.85896	0.85985	0.86074	0.86163	0.86251	0.86340	0.86427	0.86515
60	0.86603	0.86690	0.86777	0.86863	0.86949	0.87036	0.87121	0.87207	0.87292	0.87377
61	0.87462	0.87546	0.87631	0.87715	0.87798	0.87882	0.87965	0.88048	0.88130	0.88213
62	0.88295	0.88377	0.88458	0.88539	0.88620	0.88701	0.88782	0.88862	0.88942	0.89021
63	0.89101	0.89180	0.89259	0.89337	0.89415	0.89493	0.89571	0.89649	0.89726	0.89803
64	0.89879	0.89956	0.90032	0.90108	0.90183	0.90259	0.90334	0.90408	0.90483	0.90557
65	0.90631	0.90704	0.90778	0.90851	0.90924	0.90996	0.91068	0.91140	0.91212	0.91283
66	0.91355	0.91425	0.91496	0.91566	0.91636	0.91706	0.91775	0.91845	0.91914	0.91982
67	0.92050	0.92119	0.92186	0.92254	0.92321	0.92388	0.92455	0.92521	0.92587	0.92653
68	0.92718	0.92784	0.92849	0.92913	0.92978	0.93042	0.93106	0.93169	0.93232	0.93295
69	0.93358	0.93420	0.93483	0.93544	0.93606	0.93667	0.93728	0.93789	0.93849	0.93909
70	0.93969	0.94029	0.94088	0.94147	0.94206	0.94264	0.94322	0.94380	0.94438	0.94495
71	0.94552	0.94609	0.94665	0.94721	0.94777	0.94832	0.94888	0.94943	0.94997	0.95052
72	0.95106	0.95159	0.95213	0.95266	0.95319	0.95372	0.95424	0.95476	0.95528	0.95579
73	0.95630	0.95681	0.95732	0.95782	0.95832	0.95882	0.95931	0.95981	0.96029	0.96078
74	0.96126	0.96174	0.96222	0.96269	0.96316	0.96363	0.96410	0.96456	0.96502	0.96547
75	0.96593	0.96638	0.96682	0.96727	0.96771	0.96815	0.96858	0.96902	0.96945	0.96987
76	0.97030	0.97072	0.97113	0.97155	0.97196	0.97237	0.97278	0.97318	0.97358	0.97398
77	0.97437	0.97476	0.97515	0.97553	0.97592	0.97630	0.97667	0.97705	0.97742	0.97778
78	0.97815	0.97851	0.97887	0.97922	0.97958	0.97992	0.98027	0.98061	0.98096	0.98129
79	0.98163	0.98196	0.98229	0.98261	0.98294	0.98325	0.98357	0.98389	0.98420	0.98450
80	0.98481	0.98511	0.98541	0.98570	0.98600	0.98629	0.98657	0.98686	0.98714	0.98741
81	0.98769	0.98796	0.98823	0.98849	0.98876	0.98902	0.98927	0.98953	0.98978	0.99002
82	0.99027	0.99051	0.99075	0.99098	0.99122	0.99144	0.99167	0.99189	0.99211	0.99233
83	0.99255	0.99276	0.99297	0.99317	0.99337	0.99357	0.99377	0.99396	0.99415	0.99434
84	0.99452	0.99470	0.99488	0.99506	0.99523	0.99540	0.99556	0.99572	0.99588	0.99604
85	0.99619	0.99635	0.99649	0.99664	0.99678	0.99692	0.99705	0.99719	0.99731	0.99744
86	0.99756	0.99768	0.99780	0.99792	0.99803	0.99813	0.99824	0.99834	0.99844	0.99854
87	0.99863	0.99872	0.99881	0.99889	0.99897	0.99905	0.99912	0.99919	0.99926	0.99933
88	0.99939	0.99945	0.99951	0.99956	0.99961	0.99966	0.99970	0.99974	0.99978	0.99982
89	0.99985	0.99988	0.99990	0.99993	0.99995	0.99996	0.99998	0.99999	0.99999	1.00000

Table 11 Natural Cosines

	0.0	0.1	0.2	0.3	0.4	0.5	0.6	0.7	0.8	0.9
0	1.00000	1.00000	0.99999	0.99999	0.99998	0.99996	0.99995	0.99993	0.99990	0.99988
1	0.99985	0.99982	0.99978	0.99974	0.99970	0.99966	0.99961	0.99956	0.99951	0.99945
2	0.99939	0.99933	0.99926	0.99919	0.99912	0.99905	0.99897	0.99889	0.99881	0.99872
3	0.99863	0.99854	0.99844	0.99834	0.99824	0.99813	0.99803	0.99792	0.99780	0.99768
4	0.99756	0.99744	0.99731	0.99719	0.99705	0.99692	0.99678	0.99664	0.99649	0.99635
5	0.99619	0.99604	0.99588	0.99572	0.99556	0.99540	0.99523	0.99506	0.99488	0.99470
6	0.99452	0.99434	0.99415	0.99396	0.99377	0.99357	0.99337	0.99317	0.99297	0.99276
7	0.99255	0.99233	0.99211	0.99189	0.99167	0.99144	0.99122	0.99098	0.99075	0.99051
8	0.99027	0.99002	0.98978	0.98953	0.98927	0.98902	0.98876	0.98849	0.98823	0.98796
9	0.98769	0.98741	0.98714	0.98686	0.98657	0.98629	0.98600	0.98570	0.98541	0.98511
10	0.98481	0.98450	0.98420	0.98389	0.98357	0.98325	0.98294	0.98261	0.98229	0.98196
11	0.98163	0.98129	0.98096	0.98061	0.98027	0.97992	0.97958	0.97922	0.97887	0.97851
12	0.97815	0.97778	0.97742	0.97705	0.97667	0.97630	0.97592	0.97553	0.97515	0.97476
13	0.97437	0.97398	0.97358	0.97318	0.97278	0.97237	0.97196	0.97155	0.97113	0.97072
14	0.97030	0.96987	0.96945	0.96902	0.96858	0.96815	0.96771	0.96727	0.96682	0.96638
15	0.96593	0.96547	0.96502	0.96456	0.96410	0.96363	0.96316	0.96269	0.96222	0.96174
16	0.96126	0.96078	0.96029	0.95981	0.95931	0.95882	0.95832	0.95782	0.95732	0.95681
17	0.95630	0.95579	0.95528	0.95476	0.95424	0.95372	0.95319	0.95266	0.95213	0.95159
18	0.95106	0.95052	0.94997	0.94943	0.94888	0.94832	0.94777	0.94721	0.94665	0.94609
19	0.94552	0.94495	0.94438	0.94380	0.94322	0.94264	0.94206	0.94147	0.94088	0.94029
20	0.93969	0.93909	0.93849	0.93789	0.93728	0.93667	0.93606	0.93544	0.93483	0.93420
21	0.93358	0.93295	0.93232	0.93169	0.93106	0.93042	0.92978	0.92913	0.92849	0.92784
22	0.92718	0.92653	0.92587	0.92521	0.92455	0.92388	0.92321	0.92254	0.92186	0.92119
23	0.92050	0.91982	0.91914	0.91845	0.91775	0.91706	0.91636	0.91566	0.91496	0.91425
24	0.91355	0.91283	0.91212	0.91140	0.91068	0.90996	0.90924	0.90851	0.90778	0.90704
25	0.90631	0.90557	0.90483	0.90408	0.90334	0.90259	0.90183	0.90108	0.90032	0.89956
26	0.89879	0.89803	0.89726	0.89649	0.89571	0.89493	0.89415	0.89337	0.89259	0.89180
27	0.89101	0.89021	0.88942	0.88862	0.88782	0.88701	0.88620	0.88539	0.88458	0.88377
28	0.88295	0.88213	0.88130	0.88048	0.87965	0.87882	0.87798	0.87715	0.87631	0.87546
29	0.87462	0.87377	0.87292	0.87207	0.87121	0.87036	0.86949	0.86863	0.86777	0.86690
30	0.86603	0.86515	0.86427	0.86340	0.86251	0.86163	0.86074	0.85985	0.85896	0.85806
31	0.85717	0.85627	0.85536	0.85446	0.85355	0.85264	0.85173	0.85081	0.84989	0.84897
32	0.84805	0.84712	0.84619	0.84526	0.84433	0.84339	0.84245	0.84151	0.84057	0.83962
33	0.83867	0.83772	0.83676	0.83581	0.83485	0.83389	0.83292	0.83195	0.83098	0.83001
34	0.82904	0.82806	0.82708	0.82610	0.82511	0.82413	0.82314	0.82214	0.82115	0.82015
35	0.81915	0.81815	0.81714	0.81614	0.81513	0.81412	0.81310	0.81208	0.81106	0.81004
36	0.80902	0.80799	0.80696	0.80593	0.80489	0.80386	0.80282	0.80178	0.80073	0.79968
37	0.79864	0.79758	0.79653	0.79547	0.79441	0.79335	0.79229	0.79122	0.79016	0.78908
38	0.78801	0.78694	0.78586	0.78478	0.78369	0.78261	0.78152	0.78043	0.77934	0.77824
39	0.77715	0.77605	0.77494	0.77384	0.77273	0.77162	0.77051	0.76940	0.76828	0.76717
40	0.76604	0.76492	0.76380	0.76267	0.76154	0.76041	0.75927	0.75813	0.75700	0.75585
41	0.75471	0.75356	0.75241	0.75126	0.75011	0.74896	0.74780	0.74664	0.74548	0.74431
42	0.74314	0.74198	0.74080	0.73963	0.73846	0.73728	0.73610	0.73491	0.73373	0.73254
43	0.73135	0.73016	0.72897	0.72777	0.72657	0.72537	0.72417	0.72297	0.72176	0.72055
44	0.71934	0.71813	0.71691	0.71569	0.71447	0.71325	0.71203	0.71080	0.70957	0.70834
45	0.70711	0.70587	0.70463	0.70339	0.70215	0.70091	0.69966	0.69842	0.69717	0.69591

Table 11 Natural Cosines (Continued)

	0.0	0.1	0.2	0.3	0.4	0.5	0.6	0.7	0.8	0.9
46	0.69466	0.69340	0.69214	0.69088	0.68962	0.68835	0.68709	0.68582	0.68455	0.68327
47	0.68200	0.68072	0.67944	0.67816	0.67688	0.67559	0.67430	0.67301	0.67172	0.67043
48	0.66913	0.66783	0.66653	0.66523	0.66393	0.66262	0.66131	0.66000	0.65869	0.65738
49	0.65606	0.65474	0.65342	0.65210	0.65077	0.64945	0.64812	0.64679	0.64546	0.64412
50	0.64279	0.64145	0.64011	0.63877	0.63742	0.63608	0.63473	0.63338	0.63203	0.63068
51	0.62932	0.62796	0.62660	0.62524	0.62388	0.62251	0.62115	0.61978	0.61841	0.61704
52	0.61566	0.61429	0.61291	0.61153	0.61015	0.60876	0.60738	0.60599	0.60460	0.60321
53	0.60182	0.60042	0.59902	0.59763	0.59622	0.59482	0.59342	0.59201	0.59061	0.58920
54	0.58779	0.58637	0.58496	0.58354	0.58212	0.58070	0.57928	0.57786	0.57643	0.57501
55	0.57358	0.57215	0.57071	0.56928	0.56784	0.56641	0.56497	0.56353	0.56208	0.56064
56	0.55919	0.55775	0.55630	0.55484	0.55339	0.55194	0.55048	0.54902	0.54756	0.54610
57	0.54464	0.54317	0.54171	0.54024	0.53877	0.53730	0.53583	0.53435	0.53288	0.53140
58	0.52992	0.52844	0.52696	0.52547	0.52399	0.52250	0.52101	0.51952	0.51803	0.51653
59	0.51504	0.51354	0.51204	0.51054	0.50904	0.50754	0.50603	0.50453	0.50302	0.50151
60	0.50000	0.49849	0.49697	0.49546	0.49394	0.49242	0.49090	0.48938	0.48786	0.48634
61	0.48481	0.48328	0.48175	0.48022	0.47869	0.47716	0.47562	0.47409	0.47255	0.47101
62	0.46947	0.46793	0.46639	0.46484	0.46330	0.46175	0.46020	0.45865	0.45710	0.45554
63	0.45399	0.45243	0.45088	0.44932	0.44776	0.44620	0.44464	0.44307	0.44151	0.43994
64	0.43837	0.43680	0.43523	0.43366	0.43209	0.43051	0.42894	0.42736	0.42578	0.42420
65	0.42262	0.42104	0.41945	0.41787	0.41628	0.41469	0.41310	0.41151	0.40992	0.40833
66	0.40674	0.40514	0.40355	0.40195	0.40035	0.39875	0.39715	0.39555	0.39394	0.39234
67	0.39073	0.38912	0.38752	0.38591	0.38430	0.38268	0.38107	0.37946	0.37784	0.37622
68	0.37461	0.37299	0.37137	0.36975	0.36812	0.36650	0.36488	0.36325	0.36162	0.36000
69	0.35837	0.35674	0.35511	0.35347	0.35184	0.35021	0.34857	0.34694	0.34530	0.34366
70	0.34202	0.34038	0.33874	0.33710	0.33545	0.33381	0.33216	0.33051	0.32887	0.32722
71	0.32557	0.32392	0.32227	0.32061	0.31896	0.31730	0.31565	0.31399	0.31233	0.31068
72	0.30902	0.30736	0.30570	0.30403	0.30237	0.30071	0.29904	0.29737	0.29571	0.29404
73	0.29237	0.29070	0.28903	0.28736	0.28569	0.28402	0.28234	0.28067	0.27899	0.27731
74	0.27564	0.27396	0.27228	0.27060	0.26892	0.26724	0.26556	0.26387	0.26219	0.26050
75	0.25882	0.25713	0.25545	0.25376	0.25207	0.25038	0.24869	0.24700	0.24531	0.24362
76	0.24192	0.24023	0.23853	0.23684	0.23514	0.23345	0.23175	0.23005	0.22835	0.22665
77	0.22495	0.22325	0.22155	0.21985	0.21814	0.21644	0.21474	0.21303	0.21132	0.20962
78	0.20791	0.20620	0.20450	0.20279	0.20108	0.19937	0.19766	0.19595	0.19423	0.19252
79	0.19081	0.18910	0.18738	0.18567	0.18395	0.18224	0.18052	0.17880	0.17708	0.17537
80	0.17365	0.17193	0.17021	0.16849	0.16677	0.16505	0.16333	0.16160	0.15988	0.15816
81	0.15643	0.15471	0.15299	0.15126	0.14954	0.14781	0.14608	0.14436	0.14263	0.14090
82	0.13917	0.13744	0.13572	0.13399	0.13226	0.13053	0.12880	0.12706	0.12533	0.12360
83	0.12187	0.12014	0.11840	0.11667	0.11494	0.11320	0.11147	0.10973	0.10800	0.10626
84	0.10453	0.10279	0.10106	0.09932	0.09758	0.09585	0.09411	0.09237	0.09063	0.08889
85	0.08716	0.08542	0.08368	0.08194	0.08020	0.07846	0.07672	0.07498	0.07324	0.07150
86	0.06976	0.06802	0.06627	0.06453	0.06279	0.06105	0.05931	0.05756	0.05582	0.05408
87	0.05234	0.05059	0.04885	0.04711	0.04536	0.04362	0.04188	0.04013	0.03839	0.03664
88	0.03490	0.03316	0.03141	0.02967	0.02792	0.02618	0.02443	0.02269	0.02094	0.01920
89	0.01745	0.01571	0.01396	0.01222	0.01047	0.00873	0.00698	0.00524	0.00349	0.00175

Table 12 Natural Tangents

	0.0	0.1	0.2	0.3	0.4	0.5	0.6	0.7	0.8	0.9
0	0.00000	0.00175	0.00349	0.00524	0.00698	0.00873	0.01047	0.01222	0.01396	0.01571
1	0.01746	0.01920	0.02095	0.02269	0.02444	0.02619	0.02793	0.02968	0.03143	0.03317
2	0.03492	0.03667	0.03842	0.04016	0.04191	0.04366	0.04541	0.04716	0.04891	0.05066
3	0.05241	0.05416	0.05591	0.05766	0.05941	0.06116	0.06291	0.06467	0.06642	0.06817
4	0.06993	0.07168	0.07344	0.07519	0.07695	0.07870	0.08046	0.08221	0.08397	0.08573
5	0.08749	0.08925	0.09101	0.09277	0.09453	0.09629	0.09805	0.09981	0.10158	0.10334
6	0.10510	0.10687	0.10863	0.11040	0.11217	0.11394	0.11570	0.11747	0.11924	0.12101
7	0.12278	0.12456	0.12633	0.12810	0.12988	0.13165	0.13343	0.13521	0.13698	0.13876
8	0.14054	0.14232	0.14410	0.14588	0.14767	0.14945	0.15124	0.15302	0.15481	0.15660
9	0.15838	0.16017	0.16196	0.16376	0.16555	0.16734	0.16914	0.17093	0.17273	0.17453
10	0.17633	0.17813	0.17993	0.18173	0.18353	0.18534	0.18714	0.18895	0.19076	0.19257
11	0.19438	0.19619	0.19801	0.19982	0.20164	0.20345	0.20527	0.20709	0.20891	0.21073
12	0.21256	0.21438	0.21621	0.21804	0.21986	0.22169	0.22353	0.22536	0.22719	0.22903
13	0.23087	0.23271	0.23455	0.23639	0.23823	0.24008	0.24193	0.24377	0.24562	0.24747
14	0.24933	0.25118	0.25304	0.25490	0.25676	0.25862	0.26048	0.26235	0.26421	0.26608
15	0.26795	0.26982	0.27169	0.27357	0.27545	0.27732	0.27921	0.28109	0.28297	0.28486
16	0.28675	0.28864	0.29053	0.29242	0.29432	0.29621	0.29811	0.30001	0.30192	0.30382
17	0.30573	0.30764	0.30955	0.31147	0.31338	0.31530	0.31722	0.31914	0.32106	0.32299
18	0.32492	0.32685	0.32878	0.33072	0.33266	0.33460	0.33654	0.33848	0.34043	0.34238
19	0.34433	0.34628	0.34824	0.35020	0.35216	0.35412	0.35608	0.35805	0.36002	0.36199
20	0.36397	0.36595	0.36793	0.36991	0.37190	0.37388	0.37588	0.37787	0.37986	0.38186
21	0.38386	0.38587	0.38787	0.38988	0.39190	0.39391	0.39593	0.39795	0.39997	0.40200
22	0.40403	0.40606	0.40809	0.41013	0.41217	0.41421	0.41626	0.41831	0.42036	0.42242
23	0.42447	0.42654	0.42860	0.43067	0.43274	0.43481	0.43689	0.43897	0.44105	0.44314
24	0.44523	0.44732	0.44942	0.45152	0.45362	0.45573	0.45784	0.45995	0.46206	0.46418
25	0.46631	0.46843	0.47056	0.47270	0.47483	0.47698	0.47912	0.48127	0.48342	0.48557
26	0.48773	0.48989	0.49206	0.49423	0.49640	0.49858	0.50076	0.50295	0.50514	0.50733
27	0.50953	0.51173	0.51393	0.51614	0.51835	0.52057	0.52279	0.52501	0.52724	0.52947
28	0.53171	0.53395	0.53620	0.53844	0.54070	0.54296	0.54522	0.54748	0.54975	0.55253
29	0.55431	0.55659	0.55888	0.56117	0.56347	0.56577	0.56808	0.57039	0.57271	0.57503
30	0.57735	0.57968	0.58201	0.58435	0.58670	0.58905	0.59140	0.59376	0.59612	0.59849
31	0.60086	0.60324	0.60562	0.60801	0.61040	0.61280	0.61520	0.61761	0.62003	0.62245
32	0.62487	0.62730	0.62973	0.63217	0.63462	0.63707	0.63953	0.64199	0.64446	0.64693
33	0.64941	0.65189	0.65438	0.65688	0.65938	0.66189	0.66440	0.66692	0.66944	0.67197
34	0.67451	0.67705	0.67960	0.68215	0.68471	0.68728	0.68985	0.69243	0.69502	0.69761
35	0.70021	0.70281	0.70542	0.70804	0.71066	0.71329	0.71593	0.71857	0.72122	0.72388
36	0.72654	0.72921	0.73189	0.73457	0.73726	0.73996	0.74267	0.74538	0.74810	0.75082
37	0.75355	0.75629	0.75904	0.76180	0.76456	0.76733	0.77010	0.77289	0.77568	0.77848
38	0.78129	0.78410	0.78692	0.78975	0.79259	0.79544	0.79829	0.80115	0.80402	0.80690
39	0.80978	0.81268	0.81558	0.81849	0.82141	0.82434	0.82727	0.83022	0.83317	0.83613
40	0.83910	0.84208	0.87507	0.84806	0.85107	0.85408	0.85710	0.86014	0.86318	0.86623
41	0.86929	0.87236	0.87543	0.87852	0.88162	0.88473	0.88784	0.89097	0.89410	0.89725
42	0.90040	0.90357	0.90674	0.90993	0.91313	0.91633	0.91955	0.92277	0.92601	0.92926
43	0.93252	0.93578	0.93906	0.94235	0.94565	0.94896	0.95229	0.95562	0.95897	0.96232
44	0.96569	0.96907	0.97246	0.97586	0.97927	0.98270	0.98613	0.98958	0.99304	0.99652
45	1.00000	1.00350	1.00701	1.01053	1.01406	1.01761	1.02117	1.02474	1.02832	1.03192

Table 12 Natural Tangents (Continued)

	0.0	0.1	0.2	0.3	0.4	0.5	0.6	0.7	0.8	0.9
46	1.03553	1.03915	1.04279	1.04644	1.05010	1.05378	1.05747	1.06117	1.06489	1.06862
47	1.07237	1.07613	1.07990	1.08369	1.08749	1.09131	1.09514	1.09899	1.10285	1.10672
48	1.11061	1.11452	1.11844	1.12238	1.12633	1.13029	1.13428	1.13828	1.14229	1.14632
49	1.15037	1.15443	1.15851	1.16261	1.16672	1.17085	1.17500	1.17916	1.18334	1.18754
50	1.19175	1.19599	1.20024	1.20451	1.20879	1.21310	1.21742	1.22176	1.22612	1.23050
51	1.23490	1.23931	1.24375	1.24820	1.25268	1.25717	1.26169	1.26622	1.27077	1.27535
52	1.27994	1.28456	1.28919	1.29385	1.29853	1.30323	1.30795	1.31269	1.31745	1.32224
53	1.32704	1.33187	1.33673	1.34160	1.34650	1.35142	1.35637	1.36134	1.36633	1.37134
54	1.37638	1.38145	1.38653	1.39165	1.39679	1.40195	1.40714	1.41235	1.41759	1.42286
55	1.42815	1.43347	1.43881	1.44418	1.44958	1.45501	1.46046	1.46595	1.47146	1.47699
56	1.48256	1.48816	1.49378	1.49944	1.50512	1.51084	1.51658	1.52235	1.52816	1.53400
57	1.53986	1.54576	1.55170	1.55766	1.56366	1.56969	1.57575	1.58184	1.58797	1.59414
58	1.60033	1.60657	1.61283	1.61914	1.62578	1.63185	1.63826	1.64471	1.65120	1.65772
59	1.66428	1.67088	1.67752	1.68419	1.69091	1.69766	1.70446	1.71129	1.71817	1.72509
60	1.73205	1.73905	1.74610	1.75319	1.76032	1.76749	1.77471	1.78198	1.78929	1.79665
61	1.80405	1.81150	1.81899	1.82654	1.83413	1.84177	1.84946	1.85720	1.86499	1.87283
62	1.88073	1.88867	1.89667	1.90472	1.91282	1.92098	1.92920	1.93746	1.94579	1.95417
63	1.96261	1.97111	1.97966	1.98828	1.99695	2.00569	2.01449	2.02335	2.03227	2.04125
64	2.05030	2.05942	2.06860	2.07785	2.08716	2.09654	2.10600	2.11552	2.12511	2.13477
65	2.14451	2.15432	2.16420	2.17416	2.18419	2.19430	2.20449	2.21475	2.22510	2.23553
66	2.24604	2.25663	2.26730	2.27806	2.28891	2.29984	2.31086	2.32197	2.33317	2.34447
67	2.35585	2.36733	2.37891	2.39058	2.40235	2.41421	2.42618	2.43825	2.45043	2.46270
68	2.47509	2.48758	2.50018	2.51289	2.52571	2.53865	2.55170	2.56487	2.57815	2.59156
69	2.60509	2.61874	2.63252	2.64642	2.66046	2.67462	2.68892	2.70335	2.71792	2.73263
70	2.74748	2.76247	2.77761	2.79289	2.80833	2.82391	2.83965	2.85555	2.87161	2.88783
71	2.90421	2.92076	2.93748	2.95437	2.97144	2.98868	3.00611	3.02372	3.04152	3.05950
72	3.07768	3.09606	3.11464	3.13341	3.15240	3.17159	3.19100	3.21063	3.23048	3.25055
73	3.27085	3.29139	3.31216	3.33317	3.35443	3.37594	3.39771	3.41973	3.44202	3.46458
74	3.48741	3.51053	3.53393	3.55761	3.58160	3.60588	3.63048	3.65538	3.68061	3.70616
75	3.73205	3.75828	3.78485	3.81177	3.83906	3.86671	3.89474	3.92316	3.95196	3.98117
76	4.01078	4.04081	4.07127	4.10216	4.13350	4.16530	4.19756	4.23030	4.26352	4.29724
77	4.33148	4.36623	4.40152	4.43735	4.47374	4.51071	4.54826	4.58641	4.62518	4.66458
78	4.70463	4.74534	4.78673	4.82882	4.87162	4.91516	4.95945	5.00451	5.05037	5.09704
79	5.14455	5.19293	5.24218	5.29235	5.34345	5.39552	5.44857	5.50264	5.55777	5.61397
80	5.67128	5.72974	5.78938	5.85024	5.91236	5.97576	6.04051	6.10664	6.17419	6.24321
81	6.31375	6.38587	6.45961	6.53503	6.61219	6.69116	6.77199	6.85475	6.93952	7.02637
82	7.11537	7.20661	7.30018	7.39616	7.49465	7.59575	7.69957	7.80622	7.91582	8.02848
83	8.14435	8.26355	8.38625	8.51259	8.64275	8.77689	8.91520	9.05789	9.20516	9.35724
84	9.51436	9.67680	9.84482	10.019	10.199	10.385	10.579	10.780	10.988	11.205
85	11.430	11.664	11.909	12.163	12.429	12.706	12.996	13.300	13.617	13.951
86	14.301	14.669	15.056	15.464	15.895	16.350	16.832	17.343	17.886	18.464
87	19.081	19.740	20.446	21.205	22.022	22.904	23.859	24.898	26.031	27.271
88	28.636	30.145	31.821	33.694	35.801	38.188	40.917	44.066	47.740	52.081
89	57.290	63.657	71.615	81.847	95.489	114.589	143.237	190.984	286.478	572.957

Table 13   Regular Polygons and Polyhedrons[*]

REGULAR POLYGONS

NUMBER OF SIDES	AREA, A	CIRCUMFERENCE, C	SIDE, s	RADIUS OF INSCRIBED CIRCLE, r	RADIUS OF CIRCUMSCRIBED CIRCLE, R
3	$\dfrac{\sqrt{3}}{4}s^2$	$3s$	s	$\dfrac{\sqrt{3}}{6}s$	$\dfrac{\sqrt{3}}{3}s$
	$3\sqrt{3}\,r^2$	$6\sqrt{3}\,r$	$2\sqrt{3}\,r$	r	$2r$
	$\dfrac{3\sqrt{3}}{4}R^2$	$3\sqrt{3}\,R$	$\sqrt{3}\,R$	$\dfrac{1}{2}R$	R
4	s^2	$4s$	s	$\dfrac{1}{2}s$	$\dfrac{\sqrt{2}}{2}s$
	$4r^2$	$8r$	$2r$	r	$\sqrt{2}\,r$
	$2R^2$	$4\sqrt{2}\,R$	$\sqrt{2}\,R$	$\dfrac{\sqrt{2}}{2}R$	R
5	$\dfrac{\sqrt{25+10\sqrt{5}}}{4}s^2$	$5s$	s	$\dfrac{\sqrt{25+10\sqrt{5}}}{10}s$	$\dfrac{\sqrt{50+10\sqrt{5}}}{10}s$
	$5\sqrt{5-2\sqrt{5}}\,r^2$	$10\sqrt{5-2\sqrt{5}}\,r$	$2\sqrt{5-2\sqrt{5}}\,r$	r	$(\sqrt{5}-1)r$
	$\dfrac{5}{8}\sqrt{10+2\sqrt{5}}\,R^2$	$\dfrac{5}{2}\sqrt{10-2\sqrt{5}}\,R$	$\dfrac{\sqrt{10-2\sqrt{5}}}{2}R$	$\dfrac{\sqrt{5}+1}{4}R$	R
6	$\dfrac{3\sqrt{3}}{2}s^2$	$6s$	s	$\dfrac{\sqrt{3}}{2}s$	s
	$2\sqrt{3}\,r^2$	$4\sqrt{3}\,r$	$\dfrac{2\sqrt{3}}{3}r$	r	$\dfrac{2\sqrt{3}}{3}r$
	$\dfrac{3\sqrt{3}}{2}R^2$	$6R$	R	$\dfrac{\sqrt{3}}{2}R$	R
8	$2(\sqrt{2}+1)s^2$	$8s$	s	$\dfrac{\sqrt{2}+1}{2}s$	$\dfrac{\sqrt{4+2\sqrt{2}}}{2}s$
	$8(\sqrt{2}-1)r^2$	$16(\sqrt{2}-1)r$	$2(\sqrt{2}-1)r$	r	$\sqrt{4-2\sqrt{2}}\,r$
	$2\sqrt{2}\,R^2$	$8\sqrt{2-\sqrt{2}}\,R$	$\sqrt{2-\sqrt{2}}\,R$	$\dfrac{\sqrt{2+\sqrt{2}}}{2}R$	R
10	$\dfrac{5\sqrt{5+2\sqrt{5}}}{2}s^2$	$10s$	s	$\dfrac{\sqrt{5+2\sqrt{5}}}{2}s$	$\dfrac{\sqrt{5}+1}{2}s$
	$2\sqrt{25-10\sqrt{5}}\,r^2$	$4\sqrt{25-10\sqrt{5}}\,r$	$\dfrac{2\sqrt{25-10\sqrt{5}}}{5}r$	r	$\dfrac{\sqrt{50-10\sqrt{5}}}{5}r$
	$\dfrac{5\sqrt{10-2\sqrt{5}}}{4}R^2$	$5(\sqrt{5}-1)R$	$\dfrac{\sqrt{5}-1}{2}R$	$\dfrac{\sqrt{10+2\sqrt{5}}}{4}R$	R
12	$3(2+\sqrt{3})s^2$	$12s$	s	$\dfrac{2+\sqrt{3}}{2}s$	$\dfrac{\sqrt{6}+\sqrt{2}}{2}s$
	$12(2-\sqrt{3})r^2$	$24(2-\sqrt{3})r$	$2(2-\sqrt{3})r$	r	$(\sqrt{6}-\sqrt{2})r$
	$3R^2$	$6(\sqrt{6}-\sqrt{2})R$	$\dfrac{\sqrt{6}-\sqrt{2}}{2}R$	$\dfrac{\sqrt{6}+\sqrt{2}}{4}R$	R

[*] Prepared by Dr. Edwin G. H. Comfort.

Table 13 Regular Polygons and Polyhedrons* (Continued)

REGULAR POLYHEDRONS

NUMBER OF FACES	VOLUME, V	SURFACE, S	EDGE, e	RADIUS OF INSCRIBED SPHERE, r	RADIUS OF CIRCUMSCRIBED SPHERE, R
4	$\dfrac{\sqrt{2}}{12}\,e^3$	$\sqrt{3}\,e^2$	e	$\dfrac{\sqrt{6}}{12}\,e$	$\dfrac{\sqrt{6}}{4}\,e$
	$8\sqrt{3}\,r^3$	$24\sqrt{3}\,r^2$	$2\sqrt{6}\,r$	r	$3r$
	$\dfrac{8\sqrt{3}}{27}\,R^3$	$\dfrac{8\sqrt{3}}{3}\,R^2$	$\dfrac{2\sqrt{6}}{3}\,R$	$\dfrac{1}{3}\,R$	R
6	e^3	$6e^2$	e	$\dfrac{1}{2}\,e$	$\dfrac{\sqrt{3}}{2}\,e$
	$8r^3$	$24r^2$	$2r$	r	$\sqrt{3}\,r$
	$\dfrac{8\sqrt{3}}{9}\,R^3$	$8R^2$	$\dfrac{2\sqrt{3}}{3}\,R$	$\dfrac{\sqrt{3}}{3}\,R$	R
8	$\dfrac{\sqrt{2}}{3}\,e^3$	$2\sqrt{3}\,e^2$	e	$\dfrac{\sqrt{6}}{6}\,e$	$\dfrac{\sqrt{2}}{2}\,e$
	$4\sqrt{3}\,r^3$	$12\sqrt{3}\,r^2$	$\sqrt{6}\,r$	r	$\sqrt{3}\,r$
	$\dfrac{4}{3}\,R^3$	$4\sqrt{3}\,R^2$	$\sqrt{2}\,R$	$\dfrac{\sqrt{3}}{3}\,R$	R
12	$\dfrac{15+7\sqrt{5}}{4}\,e^3$	$3\sqrt{25+10\sqrt{5}}\,e^2$	e	$\dfrac{\sqrt{250+110\sqrt{5}}}{20}\,e$	$\dfrac{\sqrt{3}+\sqrt{15}}{4}\,e$
	$10\sqrt{130-58\sqrt{5}}\,r^3$	$30\sqrt{130-58\sqrt{5}}\,r^2$	$\sqrt{50-22\sqrt{5}}\,r$	r	$\sqrt{15-6\sqrt{5}}\,r$
	$\dfrac{2(5\sqrt{3}+\sqrt{15})}{9}\,R^3$	$2\sqrt{50-10\sqrt{5}}\,R^2$	$\dfrac{\sqrt{15}-\sqrt{3}}{3}\,R$	$\dfrac{\sqrt{75+30\sqrt{5}}}{15}\,R$	R
20	$\dfrac{5(3+\sqrt{5})}{12}\,e^3$	$5\sqrt{3}\,e^2$	e	$\dfrac{3\sqrt{3}+\sqrt{15}}{12}\,e$	$\dfrac{\sqrt{10+2\sqrt{5}}}{4}\,e$
	$10(7\sqrt{3}-3\sqrt{15})r^3$	$30(7\sqrt{3}-3\sqrt{15})r^2$	$(3\sqrt{3}-\sqrt{15})r$	r	$\sqrt{15-6\sqrt{5}}\,r$
	$\dfrac{2\sqrt{10+2\sqrt{5}}}{3}\,R^3$	$2(5\sqrt{3}-\sqrt{15})R^2$	$\dfrac{\sqrt{50-10\sqrt{5}}}{5}\,R$	$\dfrac{\sqrt{75+30\sqrt{5}}}{15}\,R$	R

Table 14 Conversions

Power of 10	Prefix	Abbreviation
10^{-18}	atto	a
10^{-15}	femto	f
10^{-12}	pico	p
10^{-9}	nano	n
10^{-6}	micro	μ
10^{-3}	milli	m
10^{-2}	centi	c
10^{-1}	deci	d
10^{0}	base unit	—
10^{1}	deka	da
10^{2}	hecto	h
10^{3}	kilo	k
10^{6}	mega	M
10^{9}	giga	G
10^{12}	tera	T
10^{15}	peta	P
10^{18}	exa	E

LENGTH CONVERSIONS

Angstrom units (Å)	$\times\ 1 \times 10^{-10}$	= meters (m)
	$\times\ 1 \times 10^{-4}$	= micrometers (μm)
	$\times\ 1.650\ 763\ 73 \times 10^{-4}$	= wavelengths of orange-red line of krypton 86
Cables	$\times\ 120$	= fathoms
	$\times\ 720$	= feet (ft)
	$\times\ 219.456$	= meters (m)
Fathoms	$\times\ 6$	= feet (ft)
	$\times\ 1.828\ 8$	= meters (m)
Feet (ft)	$\times\ 12$	= inches (in.)
	$\times\ 0.3048$	= meters (m)
Furlongs	$\times\ 660$	= feet (ft)
	$\times\ 201.168$	= meters (m)
	$\times\ 220$	= yards (yd)
Inches (in.)	$\times\ 2.54 \times 10^{8}$	= Angstroms (Å)
	$\times\ 25.4$	= millimeters (mm)
	$\times\ 8.333\ 33 \times 10^{-2}$	= feet (ft)
Kilometers (km)	$\times\ 3.280\ 839 \times 10^{3}$	= feet (ft)
	$\times\ 0.539\ 956$	= nautical miles (nmi)
	$\times\ 0.621\ 371$	= statute miles (mi)
	$\times\ 1.093\ 613 \times 10^{3}$	= yards (yd)

Table 14   Conversions (Continued)

Light-years	$\times\ 9.460\ 55 \times 10^{12}$	$=$ kilometers (km)
	$\times\ 5.878\ 51 \times 10^{12}$	$=$ statute miles (mi)
Meters (m)	$\times\ 1 \times 10^{10}$	$=$ Angstroms (Å)
	$\times\ 3.280\ 839\ 9$	$=$ feet (ft)
	$\times\ 39.370\ 079$	$=$ inches (in.)
	$\times\ 1.093\ 61$	$=$ yards (yd)
Micrometers (μm)	$\times\ 10^4$	$=$ Angstroms (Å)
	$\times\ 10^{-4}$	$=$ centimeters (cm)
	$\times\ 10^{-6}$	$=$ meters (m)
Nautical miles (nmi)	$\times\ 8.439\ 049$	$=$ cables
(international)	$\times\ 6.076\ 115\ 49 \times 10^3$	$=$ feet (ft)
	$\times\ 1.852 \times 10^3$	$=$ meters (m)
	$\times\ 1.150\ 77$	$=$ statute miles (mi)
Statute miles (mi)	$\times\ 5.280 \times 10^3$	$=$ feet (ft)
	$\times\ 8$	$=$ furlongs
	$\times\ 6.336\ 0 \times 10^4$	$=$ inches (in.)
	$\times\ 1.609\ 34$	$=$ kilometers (km)
	$\times\ 8.689\ 7 \times 10^{-1}$	$=$ nautical miles (nmi)
Mils (mil)	$\times\ 10^{-3}$	$=$ inches (in.)
	$\times\ 2.54 \times 10^{-2}$	$=$ millimeters (mm)
	$\times\ 25.4$	$=$ micrometers (μm)
Yards (yd)	$\times\ 3$	$=$ feet (ft)
	$\times\ 9.144 \times 10^{-1}$	$=$ meters (m)
Feet per hour (ft/h)	$\times\ 3.048 \times 10^{-4}$	$=$ kilometers per hour (km/h)
	$\times\ 1.645\ 788 \times 10^{-4}$	$=$ knots (kn)
Feet per minute	$\times\ 0.3048$	$=$ meters per minute (m/min)
(ft/min)	$\times\ 5.08 \times 10^{-3}$	$=$ meters per second (m/s)
Feet per second (ft/s)	$\times\ 1.097\ 28$	$=$ kilometers per hour (km/h)
	$\times\ 18.288$	$=$ meters per minute (m/min)
Kilometers per hour	$\times\ 3.280\ 839 \times 10^3$	$=$ feet per hour (ft/h)
(km/h)	$\times\ 54.680\ 66$	$=$ feet per minute (ft/min)
	$\times\ 0.277\ 777$	$=$ meters per second (m/s)
	$\times\ 0.621\ 371$	$=$ statute miles per hour (mph)
Kilometers per minute	$\times\ 3.280\ 839 \times 10^3$	$=$ feet per minute (ft/min)
(km/min)	$\times\ 37.282\ 27$	$=$ miles per hour (mi/h)

Table 14   Conversions (Continued)

Knots (kn)	$\times$ 6.076 115 $\times$ 10³	= feet per hour (ft/h)
	$\times$ 101.268 5	= feet per minute (ft/min)
	$\times$ 1.687 809	= feet per second (ft/s)
	$\times$ 1.852	= kilometers per hour (km/h)
	$\times$ 30.866	= meters per minute (m/min)
	$\times$ 0.5144	= meters per second (m/s)
	$\times$ 1.150 77	= statute miles per hour (mph)
Meters per hour (m/h)	$\times$ 3.280 839	= feet per hour (ft/h)
	$\times$ 88	= feet per minute (ft/min)
	$\times$ 1.466	= feet per second (ft/s)
	$\times$ 1 $\times$ 10⁻³	= kilometers per hour (km/h)
	$\times$ 1.667 $\times$ 10⁻²	= meters per minute (m/min)
Feet per second² (ft/s²)	$\times$ 1.097 28	= kilometers per hour per second (km/h/s)
	$\times$ 0.304 8	= meters per second² (m/s²)

AREA CONVERSIONS

Acres	$\times$ 4.046 85 $\times$ 10⁻³	= square kilometers (km²)
	$\times$ 4.046 856 $\times$ 10³	= square meters (m²)
	$\times$ 4.356 0 $\times$ 10⁴	= square feet (ft²)
Ares	$\times$ 2.471 053 8 $\times$ 10⁻²	= acres
	$\times$ 1	= square dekameters (dam²)
	$\times$ 10²	= square meters (m²)
Barns (b)	$\times$ 1 $\times$ 10⁻²⁸	= square meters (m²)
Circular mils (cmil)	$\times$ 1 $\times$ 10⁻⁶	= circular inches
	$\times$ 5.067 074 8 $\times$ 10⁻⁴	= square millimeters (mm²)
	$\times$ 0.785 398 1	= square mils (mil²)
Hectares	$\times$ 2.471 05	= acres
	$\times$ 10²	= ares
	$\times$ 10⁴	= square meters (m²)
Square feet (ft²)	$\times$ 2.295 684 $\times$ 10⁻⁵	= acres
	$\times$ 9.290 3 $\times$ 10⁻⁴	= ares
	$\times$ 144	= square inches (in.²)
	$\times$ 9.290 304 $\times$ 10⁻²	= square meters (m²)
Square inches (in.²)	$\times$ 1.273 239 5 $\times$ 10⁶	= circular mils (cmil)
	$\times$ 6.944 4 $\times$ 10⁻³	= square feet (ft²)
	$\times$ 6.451 6 $\times$ 10⁻⁴	= square meters (m²)

Table 14 Conversions (Continued)

Square kilometers (km²)	× 247.105 38	= acres
	× 1.076 391 0 × 10⁷	= square feet (ft²)
Square meters (m²)	× 10.763 9	= square feet (ft²)
Square miles (mi²)	× 640	= acres
	× 2.787 828 8 × 10⁷	= square feet (ft²)
	× 2.589 988 1	= square kilometers (km²)
Square mils (mil²)	× 1.273 23	= circular mils (cmil)
	× 10⁻⁶	= square inches (in.²)

VOLUME CONVERSIONS

Acre-feet	× 1.233 481 4 × 10³	= cubic meters (m³)
Acre-inches	× 102.79	= cubic meters (m³)
Board-feet	× 2.359 737 × 10⁻³	= cubic meters (m³)
	× 144	= cubic inches (in.³)
Bushels (Imperial)	× 3.636 87 × 10⁻²	= cubic meters (m³)
	× 1.284 348	= cubic feet (ft³)
Bushels (U.S.) (bu)	× 3.523 907 × 10⁻²	= cubic meters (m³)
	× 1.244 456	= cubic feet (ft³)
	× 8	= U.S. dry gallons
	× 35.238 08	= liters (l)
Cords	× 128	= cubic feet (ft³)
	× 3.624 57	= cubic meters (m³)
Cubic feet (ft³)	× 2.831 6 × 10⁻²	= cubic meters (m³)
	× 28.316 85	= liters (l)
Cubic feet (ft³) of water	× 62.426 2	= pounds (lb) of water
Cubic feet per hour (ft³/h)	× 28.316 85	= liters per hour (l/h)
Cubic feet per minute (ft³/min)	× 0.471 934	= liters per second (l/s)
Cubic feet per second (ft³/s)	× 28.316 05	= liters per second (l/s)
Cubic inches (in.³)	× 1.638 706 4 × 10⁴	= cubic millimeters (mm³)
	× 1.638 706 4 × 10⁻⁵	= cubic meters (m³)
	× 1.638 7 × 10⁻²	= liters (l)
	× 16.386 71	= milliliters (ml)

Table 14   Conversions (Continued)

Cubic meters (m³)	× 8.107 13 × 10⁻⁴	= acre-feet
	× 35.314 667	= cubic feet (ft³)
Cubic millimeters (mm³)	× 10⁻³	= cubic centimeters (cm³)
	× 10⁻⁹	= cubic meters (m³)
	× 6.102 374 4 × 10⁻⁵	= cubic inches (in.³)
Cubic yards (yd³)	× 27	= cubic feet (ft³)
	× 0.764 555	= cubic meters (m³)
Gallons (Imperial)	× 277.4	= cubic inches (in.³)
	× 4.546 090	= liters (l)
	× 10	= pounds (lb) of water
Gallons (U.S. dry)	× 268.802 5	= cubic inches (in.³)
	× 4.404 884	= liters (l)
Gallons (U.S. liquid)	× 231	= cubic inches (in.³)
	× 3.785 412	= liters (l)
Kiloliters (kl)	× 35.314 67	= cubic feet (ft³)
	× 6.102 374 × 10⁴	= cubic inches (in.³)
	× 1.000	= cubic meters (m³)
	× 1.307 950 6	= cubic yards (yd³)
	× 219.969	= imperial gallons
	× 264.172	= U.S. gallons (liquid) gallons (gal)
Liters (l)	× 10³	= cubic centimeters (cm³)
	× 1.000 × 10⁶	= cubic millimeters (mm³)
	× 1.000 × 10⁻³	= cubic meters (m³)
	× 61.023 74	= cubic inches (in.³)
	× 3.531 5 × 10⁻²	= cubic feet (ft³)
	× 1.307 95 × 10⁻³	= cubic yards (yd³)
	× 0.219 969	= imperial gallons
	× 0.879 877	= imperial quarts
	× 0.264 172	= U.S. (liquid) gallons (gal)
	× 1.056 688	= U.S. quarts (qt)
Pints (Imperial)	× 0.125	= imperial gallons
	× 0.568 261	= liters (l)
	× 20	= imperial fluid ounces
	× 0.5	= imperial quarts
	× 568.260 9	= cubic centimeters (cm³)

Table 14   Conversions (Continued)

Pints (U.S.)	× 0.125	= U.S. (liquid) gallons (gal)
	× 0.473 176 5	= liters (l)
	× 16	= U.S. fluid ounces (oz)
	× 0.5	= U.S. quarts (qt)
	× 473.176 5	= cubic centimeters (cm³)
Quarts (Imperial)	× 1.136 52 × 10³	= cubic centimeters (cm³)
	× 69.354 8	= cubic inches (in.³)
	× 1.136 522 8	= liters (l)
Quarts (U.S.)	× 946.353	= cubic centimeters (cm³)
	× 57.75	= cubic inches (in.³)
	× 0.946 353	= liters (l)

TIME CONVERSIONS

(No attempt has been made in this brief treatment to correlate solar, mean solar, sidereal, and mean sidereal days.)

Mean solar days (d)	× 24	= mean solar hours (h)
Mean solar hours (h)	× 3.600 × 10³	= mean solar seconds (s)
	× 60	= mean solar minutes (min)

MASS CONVERSIONS

Grains (gr)	× 6.479 8 × 10⁻²	= grams (g)
	× 2.285 71 × 10⁻³	= avoirdupois ounces (oz)
Grams (g)	× 15.432 358	= grains (gr)
	× 3.527 396 × 10⁻²	= avoirdupois ounces (oz)
	× 2.204 62 × 10⁻³	= avoirdupois pounds (lb)
Kilograms (kg)	× 564.383 4	= avoirdupois drams
	× 2.204 622 6	= avoirdupois pounds (lb)
	× 9.842 065 × 10⁻⁴	= long tons
	× 10⁻³	= metric tons (t)
	× 1.102 31 × 10⁻³	= short tons (ton)
Avoirdupois ounces (oz)	× 28.349 5	= grams (g)
	× 6.25 × 10⁻²	= avoirdupois pounds (lb)
	× 0.911 458	= troy ounces
Avoirdupois pounds (lb)	× 256	= drams
	× 4.535 923 7 × 10²	= grams (g)
	× 0.453 592 4	= kilograms (kg)
	× 16	= ounces (oz)

Table 14   Conversions (Continued)

Long tons	$\times 2.24 \times 10^3$	= avoirdupois pounds (lb)
	$\times 1.106\ 046\ 9$	= metric tons (t)
	$\times 1.12$	= short tons (ton)
Metric tons (t)	$\times 10^3$	= kilograms (kg)
	$\times 2.204\ 622 \times 10^3$	= avoirdupois pounds (lb)
Short tons (ton)	$\times 2 \times 10^3$	= avoirdupois pounds (lb)
	$\times 907.184\ 74$	= kilograms (kg)

FORCE CONVERSIONS

Dynes (dyn)	$\times 10^{-5}$	= newtons (N)
Newtons (N)	$\times 10^5$	= dynes (dyn)
	$\times 0.224\ 808$	= pounds-force (lb)
Pounds (lb)	$\times 4.448\ 22$	= newtons (N)

ANGLE CONVERSIONS

Degrees	$\times 60$	= minutes
	$\times 1.745\ 329 \times 10^{-2}$	= radians (rad)
Degrees per foot	$\times 5.726\ 145 \times 10^{-4}$	= radians per centimeter (rad/cm)
Degrees per minute	$\times 2.908\ 8 \times 10^{-4}$	= radians per second (rad/s)
	$\times 4.629\ 629 \times 10^{-5}$	= revolutions per second (r/s)
Degrees per second	$\times 1.745\ 329\ 3 \times 10^{-2}$	= radians per second (rad/s)
	$\times 0.166$	= revolutions per minute (rpm)
	$\times 2.77 \times 10^{-3}$	= revolutions per second (r/s)
Minutes	$\times 1.667 \times 10^{-2}$	= degrees
	$\times 2.908\ 8 \times 10^{-4}$	= radians (rad)
	$\times 60$	= seconds
Radians	$\times 0.159\ 154$	= circumferences
	$\times 57.295\ 77$	= degrees
	$\times 3.437\ 746 \times 10^3$	= minutes
Seconds	$\times 2.777 \times 10^{-4}$	= degrees
	$\times 1.667 \times 10^{-2}$	= minutes
	$\times 4.848\ 136\ 8 \times 10^{-6}$	= radians (rad)
Steradians (sr)	$\times 0.159\ 154\ 9$	= hemispheres
	$\times 7.957\ 74 \times 10^{-2}$	= spheres
	$\times 0.636\ 619\ 7$	= spherical right angles

Table 14   Conversions (Continued)

ENERGY CONVERSIONS

British thermal units (Btu)	$\times$ 1.054 35 $\times 10^3$	= joules (J)
(thermochemical)	$\times$ 2.928 27 $\times 10^{-4}$	= kilowatthours (kWh)
	$\times$ 1.054 35 $\times 10^3$	= watt-seconds (W-s)
Foot-pounds-force (ft-lb)	$\times$ 1.355 818 0	= joules (J)
	$\times$ 0.138 255	= kilogramforce-meters (kg-m)
	$\times$ 3.766 16 $\times 10^{-7}$	= kilowatthours (kWh)
	$\times$ 1.355 818 0	= newton-meters (N-m)
Joules (J)	$\times$ 9.484 5 $\times 10^{-4}$	= British thermal units (Btu)
	$\times$ 0.737 562	= foot-pounds-force (ft-lb)
	$\times$ 0.101 971 6	= kilogramforce-meters (kg-m)
	$\times$ 2.777 7 $\times 10^{-7}$	= kilowatthours (kWh)
	$\times$ 1	= watt-seconds (W-s)
Kilogramforce-meters (kg-m)	$\times$ 9.287 7 $\times 10^{-3}$	= British thermal units (Btu)
	$\times$ 7.233 01	= foot-pounds-force (ft-lb)
	$\times$ 9.806 65	= joules (J)
	$\times$ 9.806 65	= newtonmeters (N-m)
	$\times$ 2.724 0 $\times 10^{-3}$	= watthours (Wh)
Kilowatthours (kWh)	$\times$ 3.409 52 $\times 10^3$	= British Thermal Units (Btu)
	$\times$ 2.655 22 $\times 10^6$	= foot-pounds-force (ft-lb)
	$\times$ 1.341 02	= horsepower-hours (hp-h)
	$\times$ 3.6 $\times 10^6$	= joules (J)
	$\times$ 3.670 98 $\times 10^5$	= kilogramforce-meters (kg-m)
Newton-meters (N-m)	$\times$ 0.101 971	= kilogramforce-meters (kg-m)
	$\times$ 0.737 562	= foot-pounds-force (ft-lb)
Watthours (Wh)	$\times$ 3.414 43	= British thermal units (Btu)
	$\times$ 2.655 22 $\times 10^3$	= foot-pounds-force (ft-lb)
	$\times$ 3.6 $\times 10^3$	= joules (J)
	$\times$ 3.670 98 $\times 10^2$	= kilogramforce-meters (kg-m)

POWER CONVERSIONS

British thermal units per hour (Btu/h)	$\times$ 2.928 7 $\times 10^{-4}$	= kilowatts (kW)
	$\times$ 0.292 875	= watts (W)
British thermal units per minute (Btu/min)	$\times$ 1.757 25 $\times 10^{-2}$	= kilowatts (kW)

Table 14   Conversions (Continued)

British thermal units per pound (Btu/lb)	$\times$ 2.324 4	= joules/gram (J/g)
British thermal units per second (Btu/s)	$\times$ 1.413 91	= horsepower (hp)
	$\times$ 107.514	= kilogram-meters/second (kg-m/s)
	$\times$ 1.054 35	= kilowatts (kW)
	$\times$ 1.054 35 $\times$ 10^3	= watts (W)
Foot-pounds-force per hour (ft-lb/h)	$\times$ 5.050 $\times$ 10^{-7}	= horsepower (hp)
	$\times$ 3.766 16 $\times$ 10^{-7}	= kilowatts (kW)
Foot-pounds-force per minute (ft-lb/min)	$\times$ 3.030 303 $\times$ 10^{-5}	= horsepower (hp)
	$\times$ 2.259 70 $\times$ 10^{-2}	= joules/second (J/s)
	$\times$ 2.259 70 $\times$ 10^{-5}	= kilowatts (kW)
Horsepower (hp)	$\times$ 42.435 6	= British thermal units per minute (Btu/min)
	$\times$ 550	= foot-pounds-force per second (ft-lb/s)
	$\times$ 0.746	= kilowatts (kW)
	$\times$ 746	= joules per second (J/s)
Kilogram-meters per second (kg-m/s)	$\times$ 9.806 65	= watts (W)
Kilowatts (kW)	$\times$ 3.414 43 $\times$ 10^3	= British thermal units per hour (Btu/h)
	$\times$ 2.655 22 $\times$ 10^6	= foot-pounds-force per hour (ft-lb/h)
	$\times$ 4.425 37 $\times$ 10^4	= foot-pounds-force per minute (ft-lb/min)
	$\times$ 737.562	= foot-pounds-force per second (ft-lb/s)
	$\times$ 1.019 726 $\times$ 10^7	= gram-centimeters/second (g-cm/s)
	$\times$ 1.341 02	= horsepower (hp)
	$\times$ 3.6 $\times$ 10^6	= joules per hour (J/h)
	$\times$ 10^3	= joules per second (J/s)
	$\times$ 3.671 01 $\times$ 10^5	= kilogram-meters per hour (kg-m/h)
	$\times$ 999.835	= international watt
Watts (W)	$\times$ 44.253 7	= foot-pounds-force per minute (ft-lb/min)
	$\times$ 1.341 02 $\times$ 10^{-3}	= horsepower (hp)
	$\times$ 1	= joules per second (J/s)

Table 14   Conversions (Continued)

PRESSURE CONVERSIONS

Atmospheres (standard) (atm)	× 1.013 25	= bars (bar)
	× 1.033 23 × 10³	= grams per square centimeter (g/cm²)
	× 1.033 23 × 10⁷	= grams per square meter (g/m²)
	× 14.696 0	= pounds per square inch (lb/in.²)
	× 760	= torrs
	× 101	= kilopascals (kPa)
Bars (bar)	× 0.986 923	= standard atmospheres (atm)
	× 10⁶	= baryes
	× 1.019 716 × 10⁷	= grams per square meter (g/m²)
	× 1.019 716 × 10⁴	= kilograms-force per square meter (kg/m²)
	× 14.503 8	= pounds-force per square inch (lb/in.²)
Baryes	× 10⁻⁶	= bars (bar)
Inches of mercury (inHg)	× 3.386 4 × 10⁻²	= bars (bar)
	× 345.316	= kilograms-force per square meter (kg/m²)
	× 70.726 2	= pounds-force per square foot (lb/ft²)
Pascal (Pa)	× 1	= newton per square meter (N/m²)

ELECTRICAL CONVERSIONS

Amperes (A)	× 1	= coulombs per second (C/s)
Ampere-hours (Ah)	× 3.6 × 10³	= coulombs (C)
Bels (B)	× 10	= decibels (dB)
Coulombs (C)	× 6.241 96 × 10¹⁸	= electron charges
Megamhos per centimeter (Mmho/cm)	× 2.54	= megasiemens per inch (MS/in.)
Megamhos per inch (Mmho/in.)	× 0.393 700 79	= megasiemens per centimeter (MS/cm)
Nepers (Np)	× 8.686	= decibels (dB)
Mhos (mho)	× 1	= siemens (S)

Table 14   Conversions (Continued)

LIGHTING CONVERSIONS

Candelas (cd)	× 1	= lumens per steradian (lm/sr)
Footcandles (fc)	× 1	= lumens per square foot (lm/ft²)
	× 10.763 9	= lumens per square meter (lm/m²) [lux (lx)]
Footlamberts (fL)	× 1	= lumens per square foot (lm/ft²)
Lamberts (L)	× 0.318 30	= candelas per square centimeter (cd/cm²)
	× 295.719	= candelas per square foot (cd/ft²)
	× 2.053 60	= candelas per square inch (cd/in.²)
	× 929.03	= footlamberts (fL)
	× 1	= lumens per square centimeter (lm/cm²)
Lumens (lm)	× 7.957 74 × 10⁻²	= spherical candlepower
Lumens per square centimeter (lm/cm²)	× 1	= lamberts (L)
Lumens per square centimeter steradian [lm/(cm²-sr)]	× 3.141 59	= lamberts (L)
Lumens per square foot (lm/ft²)	× 1	= footcandles (fc)
	× 1	= footlamberts (fL)
	× 10.763 910	= lumens per square meter (lm/m²)
Lumens per square meter (lm/m²)	× 9.290 30 × 10⁻²	= footcandles (fc)
Lux (lx)	× 9.290 30 × 10⁻²	= footcandles (fc)
	× 1	= lumens per square meter (lm/m²)
Phots (ph)	× 10⁴	= lux (lx)

Table 14 Conversions (Continued)

MAGNETIC CONVERSIONS

Gausses (G)	$\times 1$	= lines per square centimeter
	$\times 1$	= maxwells per square centimeter (Mx/cm²)
Gilberts (Gb)	$\times 0.795\ 774\ 72$	= ampere-turns (A)
Kilolines	$\times 10^3$	= maxwells (Mx)
	$\times 10^{-5}$	= webers (Wb)
Lines	$\times 1$	= maxwells (Mx)
Lines per square centimeter	$\times 1$	= gausses (G)
Lines per square inch	$\times 0.155$	= gausses (G)
	$\times 10^{-8}$	= webers per square inch (Wb/in.²)
Maxwells (Mx)	$\times 1$	= lines
Maxwells per square centimeter (Mx/cm²)	$\times 6.451\ 6$	= maxwells per square inch (Mx/in.²)
Oersteds (Oe)	$\times 2.021\ 267$	= ampere-turns per inch (A/in.)
	$\times 125.7$	= ampere-turns per meter (A/m)
	$\times 1$	= gilberts per centimeter (Gb/cm)
Teslas (T)	$\times 1$	= webers per square meter (Wb/m²)
	$\times 10^4$	= gausses (G)
Webers (Wb)	$\times 10^8$	= lines
Webers per square centimeter (Wb/cm²)	$\times 10^8$	= gausses (G)
	$\times 6.451\ 6 \times 10^8$	= lines per square inch

ANSWERS TO ODD-NUMBERED EXERCISES

Chapter 1

1-3 24: 2, 2, 2, 3
120: 2, 2, 2, 3, 5
720: 2, 2, 2, 2, 3, 3, 5
5040: 2, 2, 2, 2, 3, 3, 5, 7
10: 2, 5
70: 2, 5, 7
770: 2, 5, 7, 11

1-7 Common factors: 2, 2, 2
gcd = 8

1-9 Common factors: 2, 5, 11
gcd = 110

1-11 Common factor: 5
gcd: 5

1-13 gcd = 30

1-17 −11

1-19 162

1-21 363

1-23 64

1-27 11,037,600 accidents per year

1-29 1700 rpm

1-31 24 teeth

1-33 60 rpm

1-35 54 rpm

1-37 2.5 A

1-39 $R = E/I$

1-41 rpm of $D = d \times$ rpm of d/D

Chapter 2

2-1 $\frac{1}{3}; \frac{1}{3}; \frac{3}{5}; \frac{1}{3}$

2-3 $\frac{3}{4}$

2-5 $2\frac{1}{8}$

2-7 $5\frac{1}{7}$

2-9 $9\frac{7}{9}$

2-11 $3\frac{1}{6}$

2-13
(a) $3\frac{5}{7}$
(b) $5\frac{4}{9}$
(c) $3\frac{5}{7}$
(d) $2\frac{17}{240}$
(e) $15\frac{11}{12}$
(f) $24\frac{13}{18}$
(g) $49\frac{1}{4}$
(h) $28\frac{11}{12}$
(i) $3\frac{2}{5}$
(j) $4\frac{23}{28}$
(k) $210\frac{11}{12}$
(l) $15\frac{15}{28}$

2-15 $1\frac{9}{10}$ ohms

2-17 $522\frac{1}{3}$ lb

2-19
(a) $30\frac{4}{7}$
(b) $2\frac{71}{140}$
(c) $9\frac{4}{5}$
(d) $23\frac{23}{26}$
(e) $8\frac{3}{5}$
(f) $6\frac{3}{8}$

2-21 Sum: $4\frac{61}{72}$; difference: $2\frac{19}{72}$

2-23 $3\frac{11}{24}$ in.

2-25 $20\frac{7}{8}$ ohms

2-27 $3\frac{3}{4}$

2-29 $2\frac{5}{8}$

2-31 $1\frac{2}{3}$

2-33 $1\frac{7}{15}$

2-35 16

2-37 32

2-39 $13\frac{11}{18}$

2-41 $115\frac{9}{32}$

2-43 $2\frac{1}{3}$

2-45 32

2-47 $4\frac{1}{2}$

2-49 $2\frac{1}{3}$

2-51 9 kg, 18 kg, $6\frac{3}{4}$ kg, $22\frac{1}{2}$ kg, 900 kg

2-53 44 cm, 88 cm, $\frac{1}{7}$ cm

2-55 144 kg copper; 24 kg tin; 6 kg zinc

2-57 94,500 revolutions

2-59 150 gal

2-61 $43,104

2-63 $\frac{5}{48}; \frac{35}{48}$

2-65 One-fourth as large; one-eighth as large; one-seventh as large

2-67 10 ft by 10 ft; $16\frac{1}{2}$ ft by $18\frac{1}{2}$ ft; $21\frac{1}{2}$ ft by 24 ft; $12\frac{1}{4}$ ft by $17\frac{1}{4}$ ft

2-69 1   **2-71** $\frac{2}{3}$ **2-73** $\frac{28}{213}$

2-75 $\frac{14}{33}$

2-77 $1\frac{39}{40}$

2-79 $\frac{1}{34}$

2-93 $\frac{11}{16}$ in.; $2\frac{3}{4}$ in.

2-95 Choose one 16-ft board and one 10-ft board.

2-97 **(a)** $0°$ **(d)** $-17\frac{7}{9}°$ **(g)** $95°$

       (b) $37°$ **(e)** $40\frac{5}{9}°$ **(h)** $149°$

       (c) $21\frac{1}{9}°$ **(f)** $14°$

Chapter 3

3-1 **(a)** 0.023; 0.03; 0.000 000 97

       (b) 600.000 004 1

       (c) 10.0019

       (d) 50.000 000 085

       (e) 800,000,000.000 009 6

3-3 **(a)** Eight-tenths

       (b) Ninety-hundredths

       (c) Four hundred seven thousandths

       (d) Five thousand nine millionths

       (e) Forty-five and thirteen ten-thousandths

       (f) Twenty-one and two hundred two thousand two millionths

       (g) Five and twenty one thousand three hundred fifty-seven hundred-thousandths

       (h) Twelve thousand and twelve hundred-thousandths

3-5 **(a)** $\frac{5}{8}$ **(c)** $\frac{47}{60}$ **(e)** $\frac{29}{64}$

       (b) $\frac{7}{8}$ **(d)** $\frac{1}{800}$

3-7 No. $00 = \frac{11}{32}$ in.; No. $2 = \frac{17}{64}$ in.; No. $4 = \frac{15}{64}$ in.; No. $7 = \frac{3}{16}$ in.; No. $13 = \frac{3}{32}$ in.; No. $28 = \frac{1}{64}$ in.

3-9 **(a)** 126.7814

       (b) 535.121 12

       (c) 260.631 960 5

3-11 **(b)** 0.301 03

       (c) 0.698 97

       (d) 0.015 588

       (e) 1.7274

       (f) 0.289 80

       (g) 20.6375

3-13 99.090

3-15 **(a)** 0.013 394

       (b) 38.353 036

       (c) 0.016 065 5

       (d) 1786.564

       (e) 0.000 001 140

3-15 **(f)** 8.803 089

3-17 **(a)** 11.22

       (b) 785.022 420

       (c) 0.531 30

       (d) 0.088 685 1

       (e) 0.988 637 65

       (f). 38.834 817 187 5

3-19 **(a)** 25.

       (b) 199.692 668

3-21 **(a)** 1.9695

       (b) 27.3871

       (c) 7.9257

       (d) 13.8305

       (e) 0.4710

       (f) 0.0812

3-23   **(a)** 3000.   **(d)** 1795.20

         (b) 2764.608   **(e)** 55.2508

         (c) 0.0216   **3-25** 6.465 in.

3-27   23.240 625 lb copper; 1.758 750 lb tin; 0.125 625 lb phosphorus

3-29 $0.30   **3-33** 1.609 km; 0.402 km; 0.0914 km

3-31   153 turns   **3-35** 8.555 lb

3-37   96.52 lb tin; 29.21 lb copper; 1057.91 lb zinc; 48.26 lb antimony; 38.10 lb lead

3-39 $192 paid; $64 remains   **3-47** 2.1248 cm   **3-57** 0.763

3-41   0.02 ft³; 0.86 ft³   **3-49** $3360.00   **3-59** 0.366

3-43   **(a)** 6.168 in.   **3-51** 0.0040 cm/rev **3-61** 0.418

         (b) 93 sheets   **3-53** 0.200   **3-63** 14.5

         (c) 3034.125 lb   **3-55** 0.664   **3-65** 19.5

3-45 $1117.51

Chapter 4

4-1   5; 12; 22; 35   **4-9**   6.15%; $18\frac{1}{3}$%

4-3   0.64; 1.60; 7.04; 64   **4-11** 18; 48; 75; 288

4-5   50%; 25%; 20%   **4-13** 3; 30; 120; 9000

4-7 $66\frac{2}{3}$%; $8\frac{1}{3}$%; 5%   **4-15** 10

4-17 30   **4-27** $4445   **4-37** 40,000

4-19 $33\frac{1}{3}$ **4-29** 480,000 cars   **4-39** 4.4%

4-21 $14\frac{2}{7}$ **4-31** 960   **4-41** $108.28

4-23 9.5%   **4-33** $15.25   **4-43** 1.04%

4-25 $32,500   **4-35** 17.86%   **4-45** $11,450.00

4-47   3,000.8825 lb iron; 240.0575 lb nickel; 34.06 lb other materials

4-49 $5809.51

4-51   0.0599% error in largest measurement; 0.0634% error in smallest measurement

4-53 $5.83

4-55   copper 3.64%, 0.0864 kg; antimony 8.18%, 0.1944 kg; tin 88.18%, 2.0952 kg

4-57 $2.62 per mat

4-59   57.44% *A*; 42.56% *B*

4-61   0.0474% error

4-63   0.918 lb carbon; 0.086 lb silicon; 0.014 lb phosphorus; 0.275 lb manganese; 0.015 lb sulfur; 75.192 lb iron

4-65   2,491 ft

4-67   **(a)** $225.09   **(e)** $175.24   **(i)** $650.94

         (b) $447.16   **(f)** $320.80   **(j)** $103.96

         (c) $160.32   **(g)** $222.59

         (d) $869.36   **(h)** $1693.00

4-69 $245.25

4-71

| | PAYMENT | | OUTSTANDING |
MONTH	INTEREST	PRINCIPAL	PRINCIPAL
May 1	$133.33	$121.67	$8378.33
June 1	$111.71	$123.29	$8255.04
July 1	$110.07	$124.93	$8130.11
August 1	$108.40	$126.60	$8003.51
September 1	$106.71	$128.29	$7875.22

Chapter 5

5-1	3:2; 1:3; 2:3					
5-3	227.27 mi					
5-5	80 gallons alcohol; 60 gallons water					
5-7	9:2					
5-9	5:7; 5:16; 5:3					
5-11	23:192					
5-13	$2027.03, $2837.84, $4054.05, $6081.08					
5-15	89.1 hp	5-33	410.7 m³	5-51	(a)	11.1480 m²
5-17	13	5-35	0.317 in.		(b)	33.8156 m²
5-19	20	5-37	45%		(c)	222.96 m²
5-21	11.76 years	5-39	$72.00		(d)	1393.5 m²
5-23	6 days	5-41	2.15 lb		(e)	1.6722 m²
5-25	40 revolutions	5-43	4.42 sp gr	5-53	(b)	1.35 kg
5-27	31.3̄ gallons	5-45	750 liters, 1380 kg		(c)	33.75 kg
5-29	$24000.00	5-47	30.03 lb		(d)	1350 kg
5-31	$13.02	5-49	6750 gallons		(e)	3.60 kg

Chapter 6

6-1	39.82 cm	6-15	74 cm
6-3	66.793 cm	6-17	30 yds 1 ft
6-5	118.23 cm	6-19	48 m
6-7	348 6 mm	6-21	60 sq ft
6-9	119.41 m	6-23	6.69 m²
6-11	16.47 mm	6-25	8.58 m²
6-13	41.28 m	6-27	22.67 cm
		6-29	156,000 m

6-31	(a)	1.5 ft	(f)	3.280 ft
	(b)	21.98 ft	(g)	13.5 ft
	(c)	10.8 ft	(h)	1 ft
	(d)	0.583 ft		
	(e)	42.650 ft		

6-33 0.023 in.

6-35 2.207 lb

6-37 68.18 min

6-39 26.82 m/s

6-41 121.92 cm

6-43 8.284 ohms

6-45 29.3 sec

6-47 36,500 kg/in²; 5657 kg/cm²; 0.00146 mm/in; 0.00051 mm/cm

6-49 24.7 min

6-51 (f) is the most accurate; (c) is the least accurate, because it has only two significant figures

	(a)	**(b)**	**(c)**
absolute error	1000 mi	10 ft	0.00001 cm
relative error	0.353%	0.362%	2.778%
	(d)	**(e)**	**(f)**
absolute error	0.1 cm	0.01 ft	0.001 km
relative error	0.394%	0.305%	0.062%

6-53 47,000 sq ft   **6-55** 10.6 in, 10.65 in

Chapter 7

7-1 (a) 9^2; 81
(b) 7^3; 343
(c) 11^4; 14,641

7-3 1296

7-5 55

7-7 36.36

7-9 74

7-11 38.26

7-13 55.28

7-15 0.0350

7-17 30.01

7-19 1.189

7-21 6.737

7-23 6.993

7-25 3.482

7-27 33.36

7-29 0.1792

7-31 0.1421

7-33 53.85

7-35 244.9

7-37 (a) 55
(b) 98
(c) 115
(d) 145
(e) 74
(f) 109
(g) 121
(h) 153
(i) 416

7-37 (j) 0.717
(k) 0.035
(l) 9.89
(m) 43.6

7-39 1.187

7-41 11.89

7-43 8.738

7-45 67.58

7-47 31.64

7-49 0.3513

7-51 0.088 88

7-53 0.4369

7-55 115.1

7-57 244.9

Chapter 8

8-1 (a) 27
(b) 174
(c) 7518
(d) 986
(e) 6

8-3 11, 12; 1, $\frac{2}{3}$, 0.9

8-5 7, 6

8-7 $v^3 \left(\dfrac{5}{x} + \dfrac{6}{y} \right)$

8-9 (a) $9a^3b^3c$
(b) $3^2y^2z^4$
(c) $4(2^2)a^2c$
(d) $5a^2b^2c^2 + 10ab^2c - 15ab^2c$

8-9 **(e)** $2^3(a+b)^2$
  (f) $11^3(ax)^2$
  (g) a^5b^3
  (h) $(ab-c)^3$

8-11 **(a)** 5×10^5
  (b) 48×10^7
  (c) 123×10^3
  (d) 101×10^5
  (e) 3.4681×10^9
  (f) $9.463\ 78 \times 10^{15}$

8-13 **(a)** 108
  (b) 9000
  (c) 264,600

8-15 42; 14.4; 80; 100

8-17 26; $n+m$

8-19 100 yd; $10c$ yd; 600 yd;
  $600d$ yd

8-21 $20c$; cp

8-23 83

8-25 5791

8-27 **(a)** 72
  (b) 17
  (c) 20,000
  (d) 1266
  (e) 20,000
  (f) 289

8-29 40,000

8-31 2500

8-33 159

8-35 640

8-37 51

8-39 343

8-41 **(a)** 40π
  (b) 40π
  (c) 44π

8-43 32

8-45 9

8-47 $54\frac{5}{9}$

8-49 72

8-51 6

8-53 Area equals the radius squared times π

8-55 $A = \frac{1}{2}bh$; 14 units²; several

8-57 **(a)** Interest equals the principal times the rate times the time.
  (b) Amount equals the principal plus the principal times the rate times the time.
  (c) Amount equals the principal plus the interest.
  (d) Principal equals the interest divided by the rate and the time.
  (e) Time equals the interest divided by the principal and the rate.
  (f) Rate equals the interest divided by the principal and the time.
  (g) Rate equals the amount minus the principal divided by the principal and the time.
  (h) $I = prt$; \$120
  (i) $P = I/rt$; \$344.83
  (j) $r = \dfrac{I}{pt}$; 12%
  (k) $t = \dfrac{I}{pr}$; 2.5 years

8-59 $4(L + \$5000) = H + L$

8-61 $V = lbh$; 14,400 ft³

8-63 $x + a$; $x + a + c$

8-65 8; $2n + 2$; $2n - 2$

8-67 $x - 2, x, x + 2$

8-69 $100d$

Chapter 9

9-1 14; -10; 0; 33; -5; -5; 11; 36; -6; 0; -1; -2; 0; $+3$; 0

9-3 -5; -11; -20; -1

9-5 75; -7000; 7000; -2000; \$0.30; 80%

9-7   9

9-9   0

9-11 $9\frac{1}{6}$

9-13   \$10.75

9-15 -50

9-17 $53°$

9-19 $-2mn - 12pq$

9-21 $2a + 17$

9-23 $2x - 4t + 21$

9-25 $11s + 3t - 3st$

9-27 $2x^3 - 8x^2y + 3xy^2$

9-29 $-8ab - 4b^2$

9-31 $-2x^2y - xy^2$

9-33 $25xy^4 + 3y^5 - 2x^4y - 5x^3y^2$

9-35 $570x^3 - 640y^3 + 490xyz + 64x^2y + 275xy^2$

9-37 $4m + 8n + p$

9-39 $8ab - 6abc - 7bc^2 + 18$

9-41 $3abx - 6abxy - 10aby^2 + 2abx^2$

9-43   **(a)** $a + b$ **(h)** $2a - b + 4$

      (b) $-9x + 3$ **(i)**   16

      (c) $13y - 14$ **(j)**   21

      (d) $x^2 + 3z - 18$ **(k)**   0

      (e) $-3m + n$ **(l)** $4x + 9y$

      (f) $2x + y - 2z$ **(m)** $-a - 3b - 10$

      (g) $-4n + 8v - 7u - 16w$

9-45 $-5a - 10b - 1$

9-47 $16a - 16b + 15c$

9-49 $a - (2b - 3c)$

9-51 $8a + b - (64a^2 + 16a + 1)$

9-53 $16z^2 + 10w - (14z^3w - 15w^2 + 5z^2)$

9-55   **(a)** $4 + (-4x^2 + 4x - 1); \; 4 - (4x^2 - 4x + 1)$

      (b) $10 + (-3x + y - 3z); \; 10 - (3x - y + 3z)$

      (c) $9u^2 + (9u - 18v + 24); \; 9u^2 - (-9u + 18v - 24)$

      (d) $16a^2 + 4b^2 + (-16ab - 9abc - c^2); \; 16a^2 + 4b^2 - (16ab + 9abc + c^2)$

      (e) $81x + (-9xy + 4y - 1); \; 81x - (9xy - 4y + 1)$

      (f) $41ab + (-4ac + bc - 4abc); \; 41ab - (4ac - bc + 4abc)$

Chapter 10

10-1 $x = 1$ 10-7 $y = 3$ 10-13 $x = 12$

10-3 $x = 0.5$ 10-9 $y = 2$ 10-15 $y = 13$

10-5 $x = 2$ 10-11 $x = 2$

10-17 $s - 5; \; s - T; \; s - (5 + T); \; s + (5 + T)$

10-19 $\dfrac{d}{8}$ mi; $\dfrac{dt}{8}$ mi; $\dfrac{d}{8}\left(t + \dfrac{m}{60}\right)$ mi   10-37 $\dfrac{\$140}{x}$

10-21 $2000; \; 100a; \; 100; \; 10d; \; 10d + 5n$ 10-39   Dress \$37.00; hat \$26.00

10-23 $a - 4, \; a - 2, \; a, \; a + 2, \; a + 4$ 10-41   4 ft

10-25 $x = 6$ 10-43   250, 50

10-27 $x = 49.7$ liters   10-45   70, 72, 74

10-29 $x = 1$ 10-47   21, 22, 23

10-31 $x = 3$ cm   10-49   250 ohms

10-33 $x = 6$ in.   10-51   \$630, \$1260, \$1710

10-35 $x = 70$ lb

Chapter 11

11-1	$+60$	**11-13**	576	**11-25**	x^4y^8
11-3	$-64;\ 125;\ -1331$	**11-15**	-780	**11-27**	x^5y^5
11-5	$+1$	**11-17**	0	**11-29**	s^8t^{12}
11-7	-108	**11-19**	0	**11-31**	$-2P^3Q^4$
11-9	8000	**11-21**	$20a^3b^5$	**11-33**	$-48a^7b^6$
11-11	108	**11-23**	$-40ab^7c^{13}$	**11-35**	$-30x^3 - 36x^2$

11-37 $-20ay + 16az$ **11-69** $32x^5 - 16x^4 - 16x^3 + 8x^2 + 2x - 1$

11-39 $20x^3y^3z^3 - 50x^2y^4z^2 + 40x^2y^3z^2$ **11-71** $x^6 - y^6$

11-41 $x^2 - 2x + 1$ **11-73** $a^6 - 3a^4x^2 + 3a^2x^4 - x^6$

11-43 $x^2 + 7x + 12$ **11-75** $9t^2 - 16u^2$

11-45 $x^2 - 10x - 11$ **11-77** 0

11-47 $3x^2 - 15x + 12$ **11-79** $2xy - 4y^2$

11-49 $10x^2 - 101x + 10$ **11-81** $14x^3 + 31x^2 - 94x + 24$

11-51 $25x^2 - 16$ **11-83** $3s^2 + 28s + 32$

11-53 $x^3 - y^3$ **11-85** $abc - 2ac^2 - 2bc^2 + 4c^3$

11-55 $b^4 - 25b^2$ **11-87** $x = 7$

11-57 $9x^4 - 16b^4$ **11-89** $x = 10$

11-59 $-49x^2 + 98x^3 - 49x^2$ **11-91** $x = -10$

11-61 $a^2x^4 - b^2y^4$ **11-93** $x = 4/15$

11-63 $4a^2 + 12ax + 9x^2 - y^2$ **11-95** $x = 12$

11-65 $x^3 - x + y^3 - y$ **11-97** $x = 14/3$

11-67 $4x^4 + 4x^2y^2 + y^4 - 4z^4$ **11-99** $x = 5$

11-101 Square: 22 ft $\times$ 22 ft; rectangle: 23 ft $\times$ 25 ft

11-103 $s = \$12{,}000 + p(n - 1)$

11-105 \$120

Chapter 12

12-1	-3	**12-15**	$-21b^3d$
12-3	$-3b$	**12-17**	$6x^2 - 8xy + 10x^3;\ -3x + 4y - 5x^2$
12-5	$+3x^3t$	**12-19**	$5(41) - 4 + 7(41)^2$
12-7	$-2a^4b^5$	**12-21**	$5yz - xy + 3y$
12-9	$-2x^3y^2$	**12-23**	$-3xy + xy^2 + 3y$
12-11	$+3w^2$	**12-25**	$2y^2 - 3y + 1$
12-13	$+7n^3x^2$		

12-27 $\dfrac{a^4b^3}{6} - \dfrac{4a^3b^2}{3} + \dfrac{2a^2b}{3}$

12-29 $(1 + a)^4 - b(1 + a)^3 + 7(1 + a)^2;\ (1 + a)^3 - b(1 + a)^2 + 7(1 + a)^1;$
$a(1 + a)^3 - ab(1 + a)^2 + 7a(1 + a)$

12-31 $-1 + 2(a + b)^5;\ 3(m - n)(a + b)^3 - 6(m - n)(a + b)^8$

12-33	$x - 4$	**12-41**	$x + 1$	**12-49**	2.25 mi
12-35	$2 - x$	**12-43**	$x^2 + 1$	**12-51**	$7(x - 3)$
12-37	$x - 10$	**12-45**	$9x^4 + 6x^2y^2 + 4y^2$	**12-53**	$6x(2y - 5z)$
12-39	$x - 7$	**12-47**	$x + 2$	**12-55**	$4uv^2(uv - 3)$

12-57	$2abc(6c - 2a + 3b)$	**12-111**	$9 + 6y + y^2 - x^2$
12-59	$13(1 - 2hk - 3uv)$	**12-113**	$a^4 - b^4 - 2ab^3 - a^2b^2$
12-61	$100m(m - 2n + 3n^2)$	**12-115**	$a^2 + 14a + 49 - b^2$
12-63	$17(A^2 - 3B^2)$	**12-117**	9991
12-65	$(x - 2)(a + b)$	**12-119**	$489{,}996$
12-67	$(2x - 3y)(xy + 1)$	**12-121**	$(3x + y)(3x - y)$
12-69	$10x(s + t)[x - 2y(s + t)^2]$	**12-123**	$(5a + 8c)(5a - 8c)$
12-71	$x^2 + 12x + 36$	**12-125**	$(xy + 2yz)(xy - 2yz)$
12-73	$4x^2 - 24xy + 36y^2$	**12-127**	$(2ab + 3c)(2ab - 3c)$
12-75	$4b^4 + 4b^2 + 1$	**12-129**	$(Abc - 80)(Abc + 80)$
12-77	$4a^2x^2 - 12axby + 9b^2y^2$	**12-131**	$(18ax^2 - 7)(18ax^2 + 7)$
12-79	25	**12-133**	$(a + b + c)(a + b - c)$
12-81	2401	**12-135**	$(2 + x + 2y)(2 - x - 2y)$
12-83	$(x + 1)^2$	**12-137**	$(2b - 3c + 4x)(2b - 3c - 4x)$
12-85	$(2y - 1)^2$	**12-139**	$(2x + 3)(x + 4)$
12-87	$9(v - 1)^2$	**12-141**	$2(3x + 5)(x + 2)$
12-89	$(3 + x^4)^2$	**12-143**	$(6x + 5)(x + 2)$
12-91	$2(2x^2 + 3xy + 4y^2)$	**12-145**	$(4x + 5)(3x - 4)$
12-93	$(0.4t + 1)^2$	**12-147**	$3(5x + 3y)(3x - 7y)$
12-95	$(2x^3 + 3y^2)^2$	**12-149**	$4(x + y)^2$
12-97	$(4S - 5T)^2$	**12-151**	$(5x + 4)(2x + 3)$
12-99	$[x + (a + b)]^2$	**12-153**	$(4 + x^2)(2 + x)(2 - x)$
12-101	$x^2 - 25$	**12-155**	$(3a - 2)(2a + 1)$
12-103	$25x^2y^2 - 36$	**12-157**	$(2x + 5)(7x - 3)$
12-105	$9x^2y^2v^2 - 16a^2b^2$	**12-159**	$m(n - 3)^2$
12-107	$121a^2x^2t^4v^6 - w^8$	**12-161**	$(a + b + 2c)(a - b - 2c)$
12-109	$a^2 + 8a + 16 - b^2$	**12-163**	$(x - y + 5)(x - y - 5)$

12-165	$(3 + a - 2b)[9 - 3(a - 2b) + (a - 2b)^2]$
12-167	$2x^2y(y - 2)(y^2 + 2y + 4)$
12-169	$(ax + b)(ax - b)$
12-171	$Ax^2(x - 1)(x^2 + x + 1)$
12-173	$(a + x - y)(a - x + y)$
12-175	$[(a - 2b) + (a + 2b)][(a - 2b)^2 - (a - 2b)(a + 2b) + (a + 2b)^2]$
12-177	24 ft **12-179** 1 ft

Chapter 13

13-1	$\dfrac{28x}{21}$	**13-11**	$\dfrac{8a^2c^4}{b^3}$
13-3	$\dfrac{15t}{3(6u - 7v)}$	**13-13**	$\dfrac{(a - b)(a + b)}{a}$
13-5	$\dfrac{2x - 3xy - 2y + 3y^2}{x^2 - y^2}$	**13-15**	$\dfrac{m - 4}{m - 8}$
13-7	$\frac{7}{11}$	**13-17**	$\dfrac{a - b}{5ab}$
13-9	$\dfrac{2xz^2}{3}$	**13-19**	$\dfrac{(x - 2)}{(x - 5)}$

13-21 $\frac{1}{4}$

13-23 $\dfrac{(x + y)}{(x^2 - xy + y^2)}$

13-25 $\dfrac{24}{(h - 2k)}$

13-27 $1\frac{4}{21}$

13-29 $\dfrac{a^2 + 3a - 6b + 2ab}{(a + 2b)(a - 2b)}$

13-31 $\dfrac{10 + ax + 2a}{(x + 2)^2}$

13-33 $\dfrac{71x}{12y}$

13-35 $\dfrac{16x - 7}{18}$

13-37 $\dfrac{13x + 3}{20}$

13-39 $\dfrac{ab - 3c}{4ab}$

13-41 $\dfrac{-22ab + 19}{8}$

13-43 $\dfrac{2x - 5}{(x - 6)(x - 2)(x - 3)}$

13-45 $\dfrac{x^2 + 9x + 10}{(x + 1)(x - 2)(x + 3)(x + 2)}$

13-47 $\dfrac{f'f''F + ff''F + ff'F + ff'f''}{ff'f''F}$

13-49 $P = \dfrac{2A + 2b^2}{b}$

13-51 $\dfrac{2am}{(a + c)(a - c)}$

13-53 $\frac{1}{3}$

13-55 $\dfrac{x}{2}$

13-57 $\dfrac{21xyz^5}{8a^3b^5}$

13-59 $y + 1$

13-61 $\dfrac{q^3}{6xy}$

13-63 $2(1 - b)$

13-65 $\dfrac{2y^2}{5}$

13-67 $(x + 1)^2$

13-69 $\dfrac{(x + 4)^4}{(x - 4)^4}$

13-71 $\dfrac{a^4 - b^4}{a^2b}$

13-73 $\dfrac{1}{b}$

13-75 $\dfrac{c(3 + a)(x + y)}{z(2 + a)(b - a)}$

13-77 $\dfrac{x - 2}{x - 3}$

13-79 $\dfrac{x + 5}{x}$

13-81 $\dfrac{x}{2x + 1}$

13-83 $\dfrac{x + y}{-y}$

13-85 $\dfrac{(h - 1)(-h + 5)}{h^3 - 3h + 8}$

13-87 -1

13-89 $\dfrac{-8ay}{a^2 - y^2}$

13-91 a^2

13-93 a

13-95 $3\frac{3}{7}$ hr

13-97 $\dfrac{10m}{m + 10}$

13-99   800 mph

Chapter 14

14-1 $a = 5$

14-3 $u = 5$

14-5 $t = -6$

14-7 $x = -18$

14-9 $s = -1$

14-11 $x = \dfrac{-1}{19}$

14-13 $x = 33\frac{1}{3}$

14-15 $x = 7\frac{5}{6}$

14-17 $x = 0$

14-19 $a = -3$

14-21 $a = \dfrac{-2}{3}$

14-23 $a = 1\frac{11}{26}$

14-25 $y = -7\frac{3}{4}$

14-27 $y = \frac{1}{3}$

14-29 16, 26

14-31 7, 8

14-33 $x = -2,\ x = 4$

14-35 $x = \frac{3}{2},\ x = -\frac{2}{3},\ x = 1$

14-37 $x = 0,\ x = -4,\ x = \frac{1}{2}$

14-39 $x = \pm 3,\ x = \pm 2$

14-41 $s = \pm 5$

14-43 $-9, 7$

14-45 $11, -4$

14-47 3 ft

14-49 8 rods by 12 rods

14-51 $p = \dfrac{S}{h},\ h = \dfrac{S}{p}$

14-53 $h = \dfrac{V}{\pi r^2}$

14-55 $R = \dfrac{A}{4\pi^2 r}$

14-57 $a^2 = \dfrac{T}{6},\ a = \sqrt{\dfrac{T}{6}}$

14-59 $r = \sqrt{\dfrac{V}{\pi h}}$

14-61 $h = \dfrac{v^2}{2g}$

14-63 $P = \dfrac{A}{rt + 1},\ r = \dfrac{A - P}{Pt},\ t = \dfrac{A - P}{Pr}$

14-65 **(a)** 3 cm × 5 cm

(b) 4 cm × 6 cm

(c) 5 cm × 7 cm

Chapter 15

15-1 $x = 24$

15-3 $x = 24$

15-5 $y = 17\frac{10}{19}$

15-7 $u = 48$

15-9 $t = 37\frac{5}{19}$

15-11 $h = 23$

15-13 $x = -2\frac{12}{25}$

15-15 $x = 3$

15-17 $x = \dfrac{-12}{7}$

15-19 $x = 12$

15-21 $x = 5$

15-23 **(a)** $b = \dfrac{4}{x + 3}$

(b) $x = \dfrac{4 - 3b}{b}$

15-25 $x = a$

15-27 $w = \dfrac{3A}{2h}$

15-29 $r^2 = R^2 - \dfrac{V}{\pi h}$, $r = \sqrt{R^2 - \dfrac{V}{\pi h}}$

15-31 $P = \dfrac{2(T - A)}{S}$

15-35 $P = \dfrac{WD_1}{D}$

15-37 $E = \dfrac{I(R + nr)}{n}$; $R = \dfrac{En}{I} - nr$; $r = \dfrac{E}{I} - \dfrac{R}{n}$; $n = \dfrac{IR}{(E - Ir)}$

15-39 $k = \dfrac{abl}{bl + al + ab}$; 2644

15-43 $R_0 = \dfrac{R_1}{1 + at}$; $a = \dfrac{R_1 - R_0}{R_0 t}$; $t = \dfrac{R_1 - R_0}{R_0 a}$

15-45 $x = \dfrac{FT^2}{16m}$

15-47 $r_1 = \dfrac{Cr}{(Kr - C)}$

15-49 $t = \dfrac{273(p_t v_t - p_0 v_0)}{p_0 v_0}$

15-51 Father \$80, daughter \$40

15-53 \$576

15-55 60

15-57 \$5000

15-59 \$7000 at 5%, \$4000 at 6%

15-61 3.5 years

15-63 2880 ft³ oxygen, 11,520 ft³ nitrogen

15-65 \$684, \$456, \$342

15-67 $x = 100$

15-69 $1016.\overline{6}$ mi

15-71 (a) $r_4 = 6.53$
(b) $r_3 = 208.\overline{3}$
(c) $r_2 = 12.358$
(d) $r_1 = 28.905$

15-73 (a) 17.2°C
(b) -33.9°C

15-75 Wrought iron, 1550°C; steel, 1350°C; cast iron, 1210°C; silver, 1000°C; lead, 327°C; tin, 246°C

15-77 $-40°$ **15-81** 36 kW   **15-85** 10.5 hp

15-79 5600 W   **15-83** 8 kW   **15-87** 38.8 hp

15-89 $P = \dfrac{33,000H}{LAN}$; $L = \dfrac{33,000H}{PAN}$; $A = \dfrac{33,000H}{PLN}$; $N = \dfrac{33,000H}{PLA}$

15-91 137 hp   **15-101** 165,000 ft · lb; $3\frac{1}{3}$ hp

15-93 35.368 lb/in.²   **15-103** 18.2 kg

15-95 $H = 2.58D^2N$ **15-105** 1450 lb

15-97 191 hp   **15-107** 8100 lb, 3100 lb

15-99 1100 ft/min

Chapter 16

16-9 The lines pass through the point $(4, 3)$

16-11 The lines pass through the point $(0, 2)$

16-13 The lines pass through the point $(1, \frac{3}{2})$

16-15 The lines pass through the point $(2, 2)$

16-17 A curve which passes through these points:

x	0	1	2	3	4	-1	-2	-3	-4
y	-3	-1	5	15	29	-1	5	15	29

16-19 A curve which passes through these points:

x	0	1	-1	2	-2	3	-3	4	-4
y	-1	0	-2	7	-9	26	-28	63	-65

16-21 Let $A = \$100$ and $P = \$75$; then $A = P(1 + rt)$ means that $rt = \frac{1}{3}$; the curve has these coordinates:

r	6	3	1	$\frac{1}{3}$	$\frac{1}{9}$	$\frac{1}{18}$
t	$\frac{1}{18}$	$\frac{1}{9}$	$\frac{1}{3}$	1	3	6

16-23 A curve which passes through these points:

r	V
0	0
1	4.19
2	33.51
3	113.10
4	268.08
5	523.60
6	904.78
7	1436.76
8	2144.66

16-25 **(a)** A straight line which passes through these points:

x	1	2	4	-4	8
y	.25	.5	1	-1	2

(b) A straight line which passes through these points:

x	1	2	4	-4	8
y	.5	1	2	-2	4

(c) A straight line which passes through these points:

x	0	1	2	-2
y	0	4	8	-8

(d) A straight line which passes through these points:

x	0	1	2	-1	-2
y	0	-2	-4	2	4

16-27　**(a)**　A straight line which passes through these points:

$$\frac{x\,|\,0\ 2\ -6\ -12}{y\,|\,6\ 7\ \ 3\ \ \ \ 0}$$

(b)　A straight line which passes through these points:

$$\frac{x\,|\,0\ 1\ -4\ -8}{y\,|\,4\ 5\ \ 2\ \ \ 0}$$

(c)　A straight line which passes through these points:

$$\frac{x\,|\ \ 0\ \ \ 4\ 8\ 12}{y\,|\,-4\ -2\ 0\ \ 2}$$

(d)　A straight line which passes through these points:

$$\frac{x\,|\,0\ 2\ -2\ 4}{y\,|\,0\ 1\ -1\ 2}$$

16-29　**(a)**　A straight line parallel to the y axis and passing through $x = 0$
　　　　(b)　A straight line parallel to the y axis and passing through $x = 5$
　　　　(c)　A straight line parallel to the y axis and passing through $x = -6$
　　　　(d)　A straight line parallel to the x axis and passing through $y = 0$
　　　　(e)　A straight line parallel to the x axis and passing through $y = -4$
　　　　(f)　A straight line parallel to the x axis and passing through $y = 2\frac{1}{2}$

16-31　$x = 1$; product $= -1$

16-33　**(a)**　minimum: $(1,-1)$; x intercepts: 0, 2
　　　　(b)　minimum: $(-1,-1)$; x intercepts: 0, -2
　　　　(c)　minimum: $(3,-1)$; x intercepts: 4, 2
　　　　(d)　minimum: $(-3,-1)$; x intercepts -4, -2

16-35　36 cm²

16-37　Maximum value $= \dfrac{S^2}{4}$, each side $= \dfrac{S}{2}$

Chapter 17

17-1　**(a)**　$x = 3, y = 1$　　　　**(g)**　$x = 7, y = 11$
　　　　(b)　$x = 3, y = 2$　　　　**(h)**　$x = \frac{2}{5}, y = \frac{1}{11}$
　　　　(c)　$x = 7, y = -2$　　　**(i)**　$x = \frac{5}{4}, y = \frac{1}{4}$
　　　　(d)　$x = \frac{1}{2}, y = \frac{1}{3}$　　　　**(j)**　$x = 1, y = 2$
　　　　(e)　$x = -\frac{2}{5}, y = \frac{3}{5}$　　　**(k)**　$x = 1.5, y = 2$
　　　　(f)　$x = \frac{5}{3}, y = \frac{6}{5}$　　　　**(l)**　$x = 1, y = 2$

17-3　$\frac{4}{15}$

17-5　12 ft

17-7　suit \$116, coat \$165

17-9　96 rugs

17-11　6, 4, 2

17-13 $T = \dfrac{\pi LFD}{12S}$

17-15 $x = 0.3975,\ y = 0.4525,\ z = 0.3125$

17-17 48, 70, 60

17-19 $fw_2/p,\ fw_1/p$

17-21 6 oz

17-23 3.53 oz of 70% silver and 8.47 oz of 87% silver

17-25 3, 4, 5

Chapter 18

18-1 $16x^8y^{12}$

18-3 $15{,}625x^{24}y^{30}$

18-5 $256h^4k^{16}m^4$

18-7 $\dfrac{4096x^6}{15{,}625y^6}$

18-9 $2x^2y^3$

18-11 $\dfrac{\sqrt{2}xy^2}{3hk^2}$

18-13 $2xy^2z^3$

18-15 $\dfrac{4m^2}{n^3}$

18-17 $8x^2y^3$

18-19 0.432 97

18-21 2.635 79

18-23 3.772 62

18-25 7.000 00 − 10

18-27 3.967 45

18-29 2.432 97

18-31 3.002 17

18-33 1.722 63

18-35 3.288 03

18-37 7.436 16 − 10

18-39 (a) 1,000,000

(b) 4800

(c) 625

(d) 20

(e) 1

(f) 0.03

(g) 50.16

(h) 56.011

(i) 0.0009

(j) 32,385

(k) 7855.9

(l) 24.853

18-41 9.870

18-43 0.785

18-45 9.878

18-47 0.4733

18-49 1.007

18-51 3.038

18-53 3.947

18-55 1.188×10^{15}

18-57 0.0720

18-59 5027.22

18-61 $1669.87, $1854.58

18-63 159 months

18-65 1739.65

Chapter 19

19-1 $x = 3,\ x = -3$

19-3 $x = +6,\ x = -6$

19-5 $x = \frac{1}{2},\ x = -\frac{1}{2}$

19-7 $x = +6,\ x = -6$

19-9 $x = +a,\ x = -a$

19-11 $u = 4,\ u = -4$

19-13 $t = \pm 2.5$

19-15 $d = \pm\sqrt{\dfrac{mna}{F}}$

19-17 $r = 6.1$ cm

19-19 $\pm 5, \pm 18$

19-21 $x = 1,\ x = 2$

19-23 $x = 2,\ x = 9$

19-25 $x = 4,\ x = -5$

19-27 $x = 7,\ x = 11$

19-29 $x = -\frac{1}{2},\ x = -\frac{2}{3}$

19-31 $x = \frac{1}{3},\ x = \frac{5}{4}$

19-33 $x = \dfrac{-6}{5},\ x = \dfrac{5}{2}$

19-35 $4x$

19-37 16

19-39 $36x$

19-41 x

19-43 $x = 2, x = -6$

19-45 $x = 32, x = -2$

19-47 $x = 3, x = \dfrac{-13}{3}$

19-49 $x = 2, x = -5$

19-51 $x = -1 + \sqrt{2}, x = -1 - \sqrt{2}$

19-53 $x = \dfrac{-1 + \sqrt{101}}{5}, x = \dfrac{-1 - \sqrt{101}}{5}$

19-55 (a) $x = 1, x = 4$

(b) $x = -\frac{1}{2}, x = -2$

(c) $x = \frac{1}{3}, x = 3$

(d) $x = 1, x = \frac{3}{2}$

(e) $x = \frac{1}{6}, x = \frac{2}{5}$

(f) $x = \frac{2}{3}, x = \frac{3}{2}$

(g) $u = -1 \pm \sqrt{3}$

(h) $h = \dfrac{+3 \pm \sqrt{13}}{2}$

(i) $t = \dfrac{-3 \pm \sqrt{41}}{4}$

19-55 (j) $s = \dfrac{4 \pm \sqrt{76}}{6}$

(k) $y = \frac{2}{7}, y = 3$

(l) $p = \dfrac{1 \pm \sqrt{10}}{9}$

(m) $k = \dfrac{+.2 \pm \sqrt{1.04}}{.5}$

(n) $z = \dfrac{-2.1 \pm \sqrt{5.85}}{2.4}$

(o) $U = \frac{1}{3}, U = -1$

(p) $N = \dfrac{-1}{a}, N = \dfrac{-1}{b}$

19-57 $R = \dfrac{-H}{6} \pm \sqrt{\dfrac{126T + 11H^2}{396}}$

19-59 $3, 4; -3, -4$

19-61 39 articles

19-63 12 days

19-65 2 ft

19-67 12 in.

19-69 $10

19-71 $96

19-73 2 in.

19-75 6.3 h, 8.8 h

Chapter 20

20-1 $A = s^2$

20-3 $V = lwh$

20-5 No

20-7 (a) $V = hr^2$

(b) $W = \dfrac{1}{d^2}$

(c) $P = Av^2$

(d) $F = \dfrac{m_1 m_2}{d^2}$

20-9 Inversely

20-11 500 in.^{2}

20-13 2,144.77 ft^3

20-15 40 ft; 12 s

20-17 $H = kI^2Rt$

20-19 0.079 ft^2

20-21 2.11 oz

20-23 4.369 ohms

20-25 3,904.7 ft

20-27 0.0123 coulomb

20-29 32.73 microfarads

20-31 $L = \dfrac{Kbd^2}{W}; b = \dfrac{WL}{Kd^2}; d = \sqrt{\dfrac{WL}{Kb}}; K = \dfrac{WL}{bd^2}$

20-33	15.4 ft	**(c)**	11.7 in.
20-35	$6\frac{1}{4}$ in.	**(d)**	8.13 ft
20-37	11,875 lb	20-41	9:12.5:43.2
20-39	**(a)** 888.89 lb	20-43	9.72 in.
	(b) 333.33 lb	20-45	$42,250

Chapter 21

21-5 Yes
21-7 Acute angles: 30°, 45°, 60°, 1°
Obtuse angles: 120°, 125°
Right angle sum: 30°, 60°
Complementary angles: 30°, 60°
Supplementary angles: 60°, 120°
21-9 45°
21-11 30°, 60°
21-13 30°, 60°
21-15 $x = 30°$
21-17 35°, 55°
21-23 The triangles are congruent. The angles are equal respectively.
21-27 60°, 90°, 120°, 135° each, respectively.
Perimeters will vary according to drawings.
21-29 No
21-31 14, 8
21-33 180°; 360°; 720°; 1080°

Chapter 22

22-1 **(a)** 126 in.² **(e)** 312 cm²
(b) 112 cm² **(f)** 432 square rods
(c) 84 m² **(g)** 364.5 in.²
(d) 1920 in.²
22-5 The two triangles have equal bases and altitudes.
22-7 No, the dimensions vary.
22-9 15 ft
22-11 **(a)** $A = 7.5$ ft². Yes; regardless of the shape, the dimensions are the same.
(b) $A = 50$ m²
(c) $h = 160$ cm
(d) $b = 24$ ft

22-13	29.25 ft²	22-27	17.5 ft²
22-15	10,800 cm²	22-29	**(a)** 39 cm by 33 cm
22-17	56 ft²		**(b)** 42.9 cm by 36.3 cm
22-19	27,000 blocks	22-31	0.105 m²
22-21	$196.50	22-33	27.5 m²
22-23	$2628.90	22-35	2200 cm²
22-25	280 ft²		

Chapter 23

23-1   15 ft   23-5   2.449 m   23-9   140.584 mi
23-3   1.732 m   23-7   8.276 mph   23-11 13 m
23-13 13.416 ft
23-15 (a) 28.284 m

      (b) $\sqrt{a^2 + b^2 + c^2}$

      (c) $2\sqrt{a^2 + b^2 + c^2}$

23-17 BC (fitter) = 25.4999; BC (actual) = 25.4558; 0.173% error
23-19 12.162 cm
23-21 85.446 ft
23-23 $b = 20.493$ m; $a = 12.493$ m
23-25 0.0156 in./ft; 0.312 in./ft; 0.5676 in./ft; 0.9096 in./ft
23-27 6.667 cm/m   23-39   3.897 km²
23-29 $\frac{1}{16}$, in./in.   23-41   7 ft
23-31 1 in.   23-43   76.287 in.
23-33 150 m; no   23-45   6.928 in.; 8 in.
23-35 7 ft 10$\frac{7}{8}$ in.   23-47   1.732 in.
23-37 (a) 7.194 ft   23-49   2.75 in.; 3.175 in.
      (b) 28.775 in.   23-51 $\frac{1}{9}$ in.; 18 turns
      (c) 41.568 km   23-53 $\frac{1}{9}$ in.; $\frac{1}{18}$ in.
      (d) 13.857 m   23-55   37.818 turns
                                     23-57 $\frac{8}{23}$ in., 1$\frac{1}{23}$ in.

Chapter 24

24-1   Answer by construction.
24-3   Yes
24-5   (a) $C = 6.2832$ ft; $A = 3.1416$ ft²
      (b) $C = 9.4248$ ft; $A = 7.0686$ ft²
      (c) $C = 14.1372$ m; $A = 15.9044$ m²
      (d) $C = 62.832$ ft; $A = 314.16$ ft²
      (e) $C = 37.6992$ km; $A = 133.0976$ km²
      (f) $C = 6.817$ cm; $A = 36.4975$ cm²
      (g) $C = 36$ ft; $A = 1017.878$ ft²
      (h) $C = 16$ ft; $A = 50.2656$ ft
      (i) $C = 2$ ft; $A = 6.2832$ ft
24-7   238.761 cm²
24-9   2.618 in.
24-11 4 ft
24-13 3.1416 in.; 18.8496 in.
24-15 1515.33 kg
24-17 56.25 times
24-19 36
24-23 3.55 mi

24-25 **(a)** $C = 157.08$ cm; $A = 1963.5$ cm²
(b) $C = 222.145$ cm; $A = 3927$ cm²
(c) $A = 1963.5$ cm²

24-27	7.5398 m		**24-63**	76.449 lb/in.²
24-29	16.971 in.		**24-65**	519.413 cm
24-31	49.365 mi; 34.907 mi		**24-67**	29
24-33	720.288 rpm		**24-69**	22.108 m
24-35	41.888 cm		**24-71**	38.197 rpm
24-37	0.292 ft		**24-73**	The areas are equal
24-39	1.2 m		**24-75**	13.11 in.
24-41	50.125 ft		**24-77**	4.060 in.; 0.065 in.

24-43 **(a)** 6.545 in.²
(b) 104.72 km²
(c) 785.4 in.²
(d) 1074.717 ft²
(e) 5000 m²

24-79 657.38 in.²
24-81 208.84 hp
24-83 **(a)** 11.546 in.
(b) 21.213 in.
(c) 40 cm

24-45	74,237.24 gallons		**24-85**	9:4; 9:4
24-47	12.649 in.		**24-87**	1.2374 in.
24-49	38.064 in.		**24-89**	13.6 in.
24-51	$14.422; x = \sqrt{R^2 + r^2}$		**24-91**	1120 rpm
24-53	57.2958°		**24-93**	10.667 in.
24-55	190.985 rpm		**24-95**	12,544 cmil
24-57	299.211 rpm		**24-97**	25,464.73 cmil
24-59	54.641 cm		**24-99**	211,600 cmil
24-61	4.555 cm			

Chapter 25

25-1 **(a)** 6.48 cm; 272.098 cm³
(b) 8.956 cm; 376.133 cm³
(c) 225.898 cm³
(d) 265.972 cm³

25-3	10 cm; 7957.27 cm³	**25-11**	39,000 gallons	**25-19**	6 ft	
25-5	14.574 cm	**25-13**	70.5 in.³	**25-21**	1:17.3	
25-7	9180 bricks	**25-15**	3072	**25-23**	24 oz	
25-9	1318 bushels	**25-17**	1.6 cm			

25-25 $(2\pi + 8)$ in.³; $(4\pi + 24)$ in.²
25-27 31.38 min
25-29 916,730.32 cm
25-31 1257.975 kg
25-33 10,204 washers
25-35 1113.36 yd³
25-37 $17\frac{7}{8}$ in.
25-39 39.2 ft/s
25-41 5.71 in.

25-43 **(a)** 11,875 lb
(b) 6,720 lb
25-45 $667.98
25-47 $V = abc - 4br^2(1 - \frac{\pi}{4})$
25-49 5.555 ft; 99.19 gal
25-51 6.3 in.
25-53 **(a)** 39.62 ft²
(b) 183.26 gal

25-55   0.959 in.³

25-57   1,060.29 lb; 3512 lb

25-59   933 ft²; yes

25-61   1016.111 kg

25-63   12,784 gallons;
          100,405.5 lb

25-65   283 gallons; 912 gallons;
          1841 gallons

25-67   169.6464 in.³

25-69   41.88 tons

25-71 $100\,\pi\sqrt{3}$

25-73   502.2 lb

25-77   35.357 in.²

25-79   188.496 gallons

25-81   yes; 26.808 kg

25-83   1.707 lb

25-85   approx. 32,170,000,000 acres

25-89   69.32 gallons

25-91 $\dfrac{5\,\pi\,d^2}{4}$

25-93   15.35 loads

25-95   0.32 oz

25-97   13,159 ft²

25-99   2705 lb

25-101   24.16 in.

25-103   (a) $V = \frac{1}{6}x(x^2 + 4x^2 + x^2)$

          (b) $V = \frac{1}{6}h(r^2\pi + 4r^2\pi + \pi r)$

          (c) $V = \frac{1}{3}\pi h r^2$

          (d) $V = \frac{2}{3}\pi r^3$

25-105   3360 in.³

Chapter 26

26-3   65° first quadrant;
         240°, third quadrant;
         374°, first quadrant;
         180°, second and third quadrants;
         375°, first quadrant;
         210°, third quadrant;
         15°, first quadrant;
         790°, first quadrant

26-5   (a) 0.25°
         (b) 0.5°
         (c) 0.75°
         (d) 0.004 167°
         (e) 0.013 89°
         (f) 0.2908°
         (g) 0.4875°
         (h) 7.3067°

26-7   (a) second and third quadrants
         (b) first and fourth quadrants
         (c) first and second quadrants
         (d) third and fourth quadrants
         (e) first quadrant
         (f) second and third quadrants
         (g) first quadrant
         (h) first quadrant

26-9   (a) 18°
         (b) 20°
         (c) 30°

26-9   (d) 36°
         (e) 45°
         (f) 60°
         (g) 72°
         (h) 90°
         (i) 120°
         (j) 135°
         (k) 180°
         (l) 210°
         (m) 225°
         (n) 270°
         (o) 300°
         (p) 315°
         (q) 330°
         (r) 360°
         (s) 540°
         (t) 1080°
         (u) 1080°

26-11   (a) 60°; $\pi/3$ rad
         (b) 90°; $\pi/2$ rad
         (c) 108°; $3\pi/5$ rad
         (d) 120°; $2\pi/3$ rad
         (e) 135°; $3\pi/4$ rad
         (f) 150°; $5\pi/6$ rad

26-13 6π; $\pi/2$; 360π

26-15   30°, 75°, 75°

Chapter 27

27-1 (a) first and second

(b) third and fourth

(c) between first and fourth; second and third

(d) first and fourth

(e) second and third

(f) between first and second; third and fourth

(g) (1) first and third, (2) second and fourth, (3) between first and fourth, between second and third

(h) first

(i) none

27-3 sine, tangent, cotangent, and cosecant

27-5 fourth

27-7 second

27-9 sine and tangent are zero, cosine and secant are unity, cosecant and cotangent are not defined

27-11 45° line: sine = cosine, tangent = cotangent, secant = cosecant; 225° line: sine = cosine, tangent = cotangent, secant = cosecant

27-13 No; no; yes, any angle in the fourth quadrant

27-15 (a) $\sin A = a/c$ $\qquad\sin B = b/c$

$\cos A = b/c$ $\qquad\cos B = a/c$

$\tan A = a/b$ $\qquad\tan B = b/a$

$\cot A = b/a$ $\qquad\cot B = a/b$

$\sec A = c/b$ $\qquad\sec B = c/a$

$\csc A = c/a$ $\qquad\csc B = c/b$

(b) $\sin A = n/p$ $\qquad\sin B = m/p$

$\cos A = m/p$ $\qquad\cos B = n/p$

$\tan A = n/m$ $\qquad\tan B = m/n$

$\cot A = m/n$ $\qquad\cot B = n/m$

$\sec A = p/m$ $\qquad\sec B = p/n$

$\csc A = p/n$ $\qquad\csc B = p/m$

(c) $\sin A = CB/AB$ $\qquad\sin B = AC/AB$

$\cos A = AC/AB$ $\qquad\cos B = CB/AB$

$\tan A = CB/AC$ $\qquad\tan B = AC/CB$

$\cot A = AC/CB$ $\qquad\cot B = CB/AC$

$\sec A = AB/AC$ $\qquad\sec B = AB/CB$

$\csc A = AB/CB$ $\qquad\csc B = AB/AC$

(d) $\sin A = 6/7$ $\qquad\sin B = \sqrt{13}/7$

$\cos A = \sqrt{13}/7$ $\qquad\cos B = 6/7$

$\tan A = 6/\sqrt{13}$ $\qquad\tan B = \sqrt{13}/6$

$\cot A = \sqrt{13}/6$ $\qquad\cot B = 6/\sqrt{13}$

$\sec A = 7/\sqrt{13}$ $\qquad\sec B = 7/6$

$\csc A = 7/6$ $\qquad\csc B = 7/\sqrt{13}$

(e) $\sin A = 3/5$ $\qquad\sin B = 4/5$

$\cos A = 4/5$ $\qquad\cos B = 3/5$

$\tan A = 3/4$ $\qquad\tan B = 4/3$

$$\cot A = 4/3 \qquad \cot B = 3/4$$
$$\sec A = 5/4 \qquad \sec B = 5/3$$
$$\csc A = 5/3 \qquad \csc B = 5/4$$

27-17 $a = \frac{3}{4}$

27-19 $a = 8$

27-21 $b = 6\frac{6}{7}$

27-23 $a = 24.5;\ b = 96.95$

27-25   30°

27-27   22.5°

Chapter 28

28-1   (a) 0.061 05

(b) 0.135 72

(c) 0.439 94

(d) 0.549 02

(e) 0.979 92

(f) 0.707 11

(g) 1.0000

(h) 0.649 45

(i) 0.845 26

(j) 0.866 03

(k) 0.500 00

28-3   (a) 0.061 16

(b) 0.136 98

(c) 0.468 43

28-3   (d) 0.656 88

(e) 4.915 16

(f) 1.000 00

(g) infinity

(h) 0.854 08

(i) 1.581 84

(j) 1.732 05

(k) 0.577 35

28-5   56.4°

28-7   45°

28-9   26.74°

28-11   38.316°

28-13   30°

28-15   69.7°

28-17   22.622°

28-19   36.767°

28-21   45°

28-23   30.2°

28-25   22.549°

28-27   38.35°

28-29 $R = 32.989\ 55$

28-31 $R = 39.693\ 22$

28-33   30,056.7

28-35   669.7

28-37   5.011275 km

28-39   2.11 m/s

28-41   474.8 ft; 10°: 1885 ft; 20°: 7312 ft; 30°: 15,630 ft; 45°: 31,250 ft; 60°: 46,860 ft; 90°: 62,500 ft

28-43   28.6 mi

28-47 $E = \dfrac{I}{d^2};\ d = \sqrt{\dfrac{I}{E}};\ I = Ed^2$

28-49   (a) 0.3276 lm/ft²

(b) 1.2 ft

(c) 106.8 cd

28-51

	Angle	Sine	Cosine	Tangent
(a)	140°	+0.642 79	−0.766 04	−0.839 10
(b)	170°	+0.173 65	−0.984 81	−0.176 33
(c)	190°	−0.173 65	−0.984 81	+0.176 33
(d)	250°	−0.939 69	−0.342 02	+2.747 48
(e)	280°	−0.984 81	+0.173 65	−5.671 28
(f)	340°	−0.342 02	+0.939 69	−0.363 97
(g)	460°	+0.984 81	−0.173 65	−5.671 28
(h)	1220°	+0.642 79	−0.766 04	−0.839 10
(i)	3890°	−0.939 69	+0.342 02	−2.747 48
(j)	−1190°	−0.939 69	−0.342 02	+2.747 48
(k)	−915°	+0.258 82	−0.965 93	−0.267 95
(l)	−1420.66	+0.331 05	+0.943 61	+0.350 84

Chapter 29

29-1 $a = 185.606$ cm
$b = 74.502$ cm
$B = 21.87°$

29-3 $b = 4160.899$ m
$c = 4277.337$ m
$B = 76.59°$

29-5 $a = 24.606\ 62$ ft
$b = 18.341\ 26$ ft
$B = 36.70°$

29-7 $b = 1298.476$ ft
$c = 2185.534$ ft
$A = 53.55°$

29-9 $b = 0.000\ 055\ 9$ km
$c = 0.000\ 801$ km

29-11 $B = 4°$
$c = 2.001\ 08$ cm
$A = 88.12°$
$B = 1.88°$

29-13 Not a right triangle

29-15 $b = 2.0737$ cm
$A = 46.568°$
$B = 43.432°$

29-17 $a = 1.189$ ft
$A = 43.333°$
$B = 46.667°$

29-19 $39.09°$

29-21 2091.7 m

29-23 21.98 ft

29-25 Equal sides each 61.588 99 cm; base angles each 60.85°

29-27 9.465°   **29-33** 49.7052 cm

29-29 15.3846%; 8.746°   **29-35** 16.657 m

29-31 338.6925 ft   **29-37** 35.2648 rods

29-39 $OF = 0.309$ in.
$BF = 0.951$ in.
$OG = 0.809$ in.
$GC = 0.588$ in.

29-41 B: 0.882 in., 1.21 4 in.
C: 1.427 in., 0.464 in.

29-43 $AD = 2.873$ cm
$CD = 2.410$ cm

29-43 $DB = 1.502$ cm

29-45 32.664 36 cm

29-47 94.455 ft

29-49 858.795 ft

29-51 35.264°

29-53 Height = 356.6119 ft
Distance = 389.174 ft

29-57 81.565 84 m

29-59 9.2718; projections are determined by the most convenient trigonometric function

29-61 Force driven forward = $0.9659p$ lb
Force driven upward = $0.2588p$ lb

29-63 452.5908 kg

29-65 Force moving upward = 47.8828 kg
Force a long plane = 131.5566 kg

29-67 134.482 lb

29-69 −56.498°; 227.37 lb

29-71 27.42°

29-73 Horizontal: 259.9212 lb
Vertical: 317.4054 lb

Chapter 30

30-3 Yes; shift either curve 90°   **30-9** Yes; shift either curve 90°

Chapter 31

31-1   **(a)** $b = 17.51$
              $c = 17.693$
              $C = 74.5°$
    (b) $b = 205.153$
              $c = 236.226$
              $C = 123.2°$
    (c) $b = 227.7$
              $c = 157.4$
              $C = 25.2°$
              $K = 3{,}973.6$ square units
    (d) $a = 63.12$
              $c = 92.9$
              $A = 38°$
              $K = 2856.9$ square units
    (e) $b = 57.13$
              $c = 47.59$
              $B = 44.41°$
              $K = 1339.107$ square units

31-3   **(a)** $b = 5.2915$
              $A = 40.9°$
              $C = 79.1°$
    (b) $c = 14.95$
              $A = 77.35°$
              $B = 43.55°$
    (c) 10.392

31-5 102.3 ft^2
31-7 668.8 rods
31-9 859.005 ft
31-11 104.52 ft
31-15 90.84 m
31-19 203.448 m
31-21 1338.1 m
31-23 4.609

31-25 **(a)** 46 kg in the same direction as F_1 and F_2
      (b) 10 kg in the direction of F_2
      (c) 33.28 kg in a direction such that the angle between F_2 and F is 32.73°
      (d) 44.51 kg in a direction such that the angle between F_2 and F is 11.66°

31-27 298.5 kg at an angle of 29.23° with the 145-kg force

31-29 $\dfrac{\sqrt{2F^2(1 + \cos A)}}{\frac{1}{2}A}$

31-31 $F = 188.9$ in a direction such that the angle between F and the force of 70 is 19.32°

31-33 $F = 155$; the two forces coincide at 60° acting in opposite directions

31-35 4140 W
31-37 27.266°
31-39 2191.6 W
31-41 5246 W
31-43 **(a)** 1.335 88

      (b) 1.519 18

      (c) 2.417 62
31-47 48.59°
31-49 0.9112 in.
31-51 2.8893 in.
31-53 **(a)** 0.4358
      (b) 0.8683

31-53 **(c)** 1.2941
      (d) 1.7101
      (e) 2.1131
      (f) 2.5000
      (g) 2.8679

      (h) 3.2140

      (i) 3.5356
      (j) 3.8302
      (k) 4.0958
      (l) 4.3302
      (m) 4.5316
      (n) 4.6985

31-53 **(o)** 4.8297

   (p) 4.9241

   (q) 4.9810

31-55 $\sin \theta = \dfrac{x}{10}$

31-57 approximately 104,429 mm²

31-59 2436.57 cm²

31-63 220 rad/s

31-65 18.75 rad/s

31-67 6°; 0.1047 rad

31-69 5940 ft/min

31-71 about 531×10^{11} mi

31-73 18.18 s

31-75 399.9998 ft

31-77 19,103.33 ft

31-79 54.8 mi; 0.795°

31-81 44.3677 mi; 156.86 mi

31-83 approximately 23 mi

31-85 24.18°

31-87 20.705°

31-89 69.135°

INDEX